The Self Organization of Disordered Systems

" Examples of Self Organziation In Nature "

Edited by Paul F. Kisak

Contents

1 Self-organization   1
     1.1 Overview   1
         1.1.1 Principles of self-organization   1
     1.2 History of the idea   2
         1.2.1 Developing views   2
     1.3 Examples   3
         1.3.1 Physics   3
         1.3.2 Chemistry   4
         1.3.3 Biology   4
         1.3.4 Computer Science   5
         1.3.5 Cybernetics   5
         1.3.6 Human society   6
         1.3.7 Psychology and education   7
         1.3.8 Traffic flow   9
         1.3.9 Methodology   9
     1.4 Criticism   10
     1.5 See also   10
     1.6 References   10
     1.7 Further reading   13
     1.8 External links   14
         1.8.1 Dissertations and theses on self-organization   15

2 Statistical fluctuations   16
     2.1 Description   16
     2.2 Examples   16
     2.3 See also   16

3 Positive feedback   17
     3.1 Overview   18
         3.1.1 Basic   18

	3.1.2 Hysteresis	18
3.2	Terminology	19
3.3	Examples and applications	19
	3.3.1 In electronics	19
	3.3.2 Switches	21
	3.3.3 In biology	22
	3.3.4 In psychology	23
	3.3.5 In economics	23
	3.3.6 In meteorology	24
	3.3.7 In climatology	24
	3.3.8 In sociology	25
	3.3.9 In chemistry	25
3.4	See also	25
	3.4.1 Similar terminology	25
	3.4.2 Analogous concepts	25
	3.4.3 Examples	25
3.5	References	25
3.6	Further reading	27
3.7	External links	27

4 Chaos theory — 28

4.1	Introduction	28
4.2	Chaotic dynamics	29
	4.2.1 Sensitivity to initial conditions	29
	4.2.2 Topological mixing	30
	4.2.3 Density of periodic orbits	30
	4.2.4 Strange attractors	30
	4.2.5 Minimum complexity of a chaotic system	31
	4.2.6 Jerk systems	31
4.3	Spontaneous order	32
4.4	History	32
4.5	Distinguishing random from chaotic data	34
4.6	Applications	35
	4.6.1 Computer science	35
	4.6.2 Biology	35
	4.6.3 Other areas	36
4.7	See also	36
4.8	References	36
4.9	Scientific literature	40

	4.9.1 Articles	40
	4.9.2 Textbooks	40
	4.9.3 Semitechnical and popular works	41
4.10	External links	42

5 Extremal principles in non-equilibrium thermodynamics — 43

- 5.1 Fluctuations, entropy, 'thermodynamics forces', and reproducible dynamical structure — 43
- 5.2 Local thermodynamic equilibrium — 44
- 5.3 Linear and non-linear processes — 44
- 5.4 Continuous and discontinuous motions of fluids — 44
- 5.5 Historical development — 44
  - 5.5.1 W. Thomson, Baron Kelvin — 44
  - 5.5.2 Helmholtz — 45
  - 5.5.3 J.W. Strutt, Baron Rayleigh — 45
  - 5.5.4 Korteweg — 45
  - 5.5.5 Onsager — 45
  - 5.5.6 Prigogine — 45
  - 5.5.7 Casimir — 45
  - 5.5.8 Ziman — 46
  - 5.5.9 Ziegler — 46
  - 5.5.10 Gyarmati — 46
  - 5.5.11 Paltridge — 46
- 5.6 Speculated thermodynamic extremum principles for energy dissipation and entropy production — 46
  - 5.6.1 Prigogine's proposed theorem of minimum entropy production for very slow purely diffusive transfer — 47
  - 5.6.2 Faster transfer with convective circulation: second entropy — 47
  - 5.6.3 Speculated principles of maximum entropy production and minimum energy dissipation — 47
  - 5.6.4 Prospects — 48
- 5.7 See also — 48
- 5.8 References — 48

6 Self-assembly — 52

- 6.1 Self-assembly in chemistry and materials science — 52
  - 6.1.1 Distinctive features — 53
  - 6.1.2 Examples — 53
  - 6.1.3 Properties — 53
- 6.2 Self-assembly at the macroscopic scale — 54
- 6.3 Consistent concepts of self-organization and self-assembly — 54
- 6.4 See also — 55

	6.5	References	55
	6.6	External links and further reading	56

7 Cellular automaton — 57

- 7.1 Overview . . . 58
- 7.2 History . . . 59
- 7.3 Classification . . . 60
  - 7.3.1 Reversible . . . 61
  - 7.3.2 Totalistic . . . 61
  - 7.3.3 Related automata . . . 61
- 7.4 Elementary cellular automata . . . 62
- 7.5 Rule space . . . 63
- 7.6 Biology . . . 63
- 7.7 Chemical types . . . 63
- 7.8 Applications . . . 64
  - 7.8.1 Computer processors . . . 64
  - 7.8.2 Cryptography . . . 64
  - 7.8.3 Error correction coding . . . 64
- 7.9 Modeling physical reality . . . 64
- 7.10 Specific rules . . . 65
- 7.11 Problems solved . . . 65
- 7.12 See also . . . 65
- 7.13 Reference notes . . . 65
- 7.14 References . . . 67
- 7.15 External links . . . 68

8 Emergence — 69

- 8.1 In philosophy . . . 70
  - 8.1.1 Definitions . . . 70
  - 8.1.2 Strong and weak emergence . . . 70
  - 8.1.3 Objective or subjective quality . . . 72
- 8.2 In religion, art and humanities . . . 72
- 8.3 Emergent properties and processes . . . 72
- 8.4 Emergent structures in nature . . . 73
  - 8.4.1 Non-living, physical systems . . . 74
  - 8.4.2 Living, biological systems . . . 75
- 8.5 In humanity . . . 76
  - 8.5.1 Spontaneous order . . . 76
  - 8.5.2 Computer AI . . . 78

		8.5.3 Language 78

- 8.5.3 Language 78
- 8.5.4 Emergent change processes 78
- 8.6 See also 78
- 8.7 References 79
- 8.8 Bibliography 80
- 8.9 Further reading 80
- 8.10 External links 82

9 Gauss's principle of least constraint — 83
- 9.1 Hertz's principle of least curvature 83
- 9.2 See also 83
- 9.3 References 83
- 9.4 External links 84

10 Self-organized criticality — 85
- 10.1 Overview 85
- 10.2 Examples of self-organized critical dynamics 86
- 10.3 See also 86
- 10.4 References 86
- 10.5 Further reading 87

11 Spontaneous order — 89
- 11.1 History 89
- 11.2 Examples 89
  - 11.2.1 Markets 89
  - 11.2.2 Game studies 90
  - 11.2.3 Anarchism 90
  - 11.2.4 Sobornost 90
  - 11.2.5 Recent developments 90
- 11.3 See also 90
- 11.4 References 91
- 11.5 External links 91

12 Metastability — 92
- 12.1 Statistical physics and thermodynamics 92
  - 12.1.1 States of matter 92
  - 12.1.2 Condensed matter and macromolecules 93
- 12.2 Quantum mechanics 93
  - 12.2.1 Nuclear physics 93
  - 12.2.2 Atomic and molecular physics 93

          12.2.3 Chemistry 94
          12.2.4 Electron systems in biochemistry 94
   12.3 Electronic circuits 94
   12.4 Computational neuroscience 94
   12.5 See also 94
   12.6 References 94

13 Butterfly effect 96
   13.1 History 96
   13.2 Illustration 97
   13.3 Theory and mathematical definition 97
   13.4 In physical systems 98
          13.4.1 In weather 98
          13.4.2 In quantum mechanics 98
   13.5 In popular culture 98
   13.6 See also 98
   13.7 References 99
   13.8 Further reading 100
   13.9 External links 100

14 Self-organized criticality control 101
   14.1 Self-organized criticality control schemes 101
   14.2 Applications 101
   14.3 See also 101
   14.4 References 102

15 Spontaneous magnetization 103
   15.1 Overview 103
   15.2 Temperature dependence 103
   15.3 Notes and references 103
   15.4 Further reading 103

16 Crystallization 104
   16.1 Process 104
   16.2 Crystallization in nature 105
   16.3 Methods 105
          16.3.1 Typical equipment 105
   16.4 Thermodynamic view 106
   16.5 Crystallization dynamics 106
          16.5.1 Nucleation 106

16.5.2 Crystal growth 107
16.5.3 Crystal size distribution 107
16.6 Main crystallization processes 107
16.6.1 Cooling crystallization 108
16.6.2 Evaporative crystallization 109
16.6.3 DTB crystallizer 109
16.7 See also 109
16.8 References 109
16.9 Further reading 110
16.10 External links 110

17 Phase transition — 113
17.1 Types of phase transition 113
17.2 Classifications 115
17.2.1 Ehrenfest classification 115
17.2.2 Modern classifications 115
17.3 Characteristic properties 115
17.3.1 Phase coexistence 115
17.3.2 Critical points 116
17.3.3 Symmetry 116
17.3.4 Order parameters 116
17.3.5 Relevance in cosmology 116
17.3.6 Critical exponents and universality classes 117
17.3.7 Critical slowing down and other phenomena 118
17.3.8 Percolation theory 118
17.3.9 Phase transitions in biological systems 118
17.4 See also 118
17.5 References 118
17.5.1 Further reading 120
17.6 External links 120

18 Critical opalescence — 121
18.1 External links 121

19 Percolation — 122
19.1 Background 122
19.2 Examples 122
19.3 See also 122
19.4 References 123

- 19.5 Further reading . . . 123
- 19.6 External links . . . 123

20 Structure formation — 124
- 20.1 Overview . . . 124
  - 20.1.1 Very early universe . . . 124
  - 20.1.2 Growth of structure . . . 124
  - 20.1.3 Recombination . . . 124
- 20.2 Very early universe . . . 125
  - 20.2.1 The horizon problem . . . 125
- 20.3 Primordial plasma . . . 126
  - 20.3.1 Acoustic oscillations . . . 126
- 20.4 Linear structure . . . 126
- 20.5 Nonlinear structure . . . 127
- 20.6 Gas evolution . . . 127
- 20.7 Modelling structure formation . . . 128
  - 20.7.1 Cosmological perturbations . . . 128
  - 20.7.2 Inflation and initial conditions . . . 128
- 20.8 See also . . . 128
- 20.9 References . . . 128

21 De Sitter universe — 130
- 21.1 Mathematical expression . . . 130
- 21.2 Potential for the Universe . . . 130
- 21.3 Relative expansion . . . 130
- 21.4 Modelling cosmic inflation . . . 131
- 21.5 See also . . . 131
- 21.6 References . . . 131

22 Diffusion-limited aggregation — 132
- 22.1 Artwork based on diffusion-limited aggregation . . . 133
- 22.2 See also . . . 133
- 22.3 References . . . 133
- 22.4 External links . . . 133

23 Reaction–diffusion system — 135
- 23.1 One-component reaction–diffusion equations . . . 135
- 23.2 Two-component reaction–diffusion equations . . . 136
- 23.3 Three- and more-component reaction–diffusion equations . . . 137
- 23.4 Applications and universality . . . 137

23.5	Experiments	137
23.6	Numerical treatments	137
23.7	See also	138
23.8	Some examples of reaction-diffusion equations	138
23.9	References	138
23.10	External links	139

24 Molecular self-assembly — 140

24.1	Supramolecular systems	140
24.2	Biological systems	141
24.3	Nanotechnology	141
	24.3.1 DNA nanotechnology	141
24.4	Two-dimensional monolayers	141
24.5	See also	141
24.6	References	141
24.7	External and further reading	142

25 Autocatalysis — 144

25.1	Chemical reactions	144
	25.1.1 Chemical equilibrium	144
	25.1.2 Far from equilibrium	144
	25.1.3 Autocatalytic reactions	145
25.2	Creation of order	145
	25.2.1 Background	145
	25.2.2 Temporal order	146
	25.2.3 Spatial order	147
25.3	Biological example	147
25.4	Phase transitions	148
25.5	Asymmetric autocatalysis	148
25.6	Role in origin of life	148
25.7	Examples of autocatalytic reactions	148
25.8	See also	149
25.9	References	149
25.10	External links	149

26 Liquid crystal — 150

26.1	History	150
26.2	Design of liquid crystalline materials	152
26.3	Liquid-crystal phases	152

26.3.1 Thermotropic liquid crystals 153
26.3.2 Lyotropic liquid crystals 155
26.3.3 Metallotropic liquid crystals 156
26.3.4 Laboratory analysis of mesophases 156
26.4 Biological liquid crystals 156
26.5 Mineral liquid crystals 157
26.6 Pattern formation in liquid crystals 157
26.7 Theoretical treatment of liquid crystals 157
26.7.1 Director 157
26.7.2 Order parameter 157
26.7.3 Onsager hard-rod model 158
26.7.4 Maier–Saupe mean field theory 158
26.7.5 McMillan's model 158
26.7.6 Elastic continuum theory 159
26.8 External influences on liquid crystals 159
26.8.1 Electric and magnetic field effects 159
26.8.2 Surface preparations 159
26.8.3 Fredericks transition 159
26.9 Effect of chirality 160
26.10 Applications of liquid crystals 160
26.11 See also 161
26.12 References 161
26.13 External links 164

27 Grid complex 165
27.1 Nomencluture 165
27.2 Application 165
27.3 References 165

28 Colloidal crystal 166
28.1 Introduction 166
28.2 Origins 167
28.3 Trends 167
28.4 Bulk crystals 167
28.4.1 Aggregation 168
28.4.2 Viscoelasticity 168
28.4.3 Phase transitions 168
28.4.4 Phonon dispersion 168
28.4.5 Kossel lines 168

        28.4.6 Growth rates 168

        28.4.7 Microgravity 168

    28.5 Thin films 169

        28.5.1 Long-range order 169

        28.5.2 Mobile lattice defects 169

    28.6 Non-spherical colloid based crystals 169

    28.7 Applications 169

        28.7.1 Photonics 169

        28.7.2 Self-assembly 169

    28.8 See also 170

    28.9 References 170

    28.10 Further reading 172

    28.11 External links 172

29 **Self-assembled monolayer** **173**

    29.1 Types 173

    29.2 Preparation 174

    29.3 Characterization 174

        29.3.1 Defects 174

        29.3.2 Nanoparticle properties 175

    29.4 Kinetics 175

    29.5 Patterning 176

        29.5.1 1. Locally attract 176

        29.5.2 2. Locally remove 176

        29.5.3 3. Modify tail groups 177

    29.6 Applications 177

        29.6.1 Thin-film SAMs 177

        29.6.2 Patterned SAMs 177

        29.6.3 Metal organic superlattices 178

    29.7 References 178

    29.8 Further reading 180

    29.9 External links 180

30 **Micelle** **181**

    30.1 History 181

    30.2 Solvation 182

    30.3 Energy of formation 182

    30.4 Micelle packing parameter 182

    30.5 Block copolymer micelles 183

- 30.5.1 Dynamic micelles 183
- 30.5.2 Kinetically frozen micelles 183
- 30.6 Inverse/reverse micelles 183
- 30.7 Supermicelles 183
- 30.8 Uses 184
- 30.9 See also 184
- 30.10 References 184

31 Copolymer — 186
- 31.1 Types of copolymers 187
    - 31.1.1 Graft copolymers 187
    - 31.1.2 Block copolymers 187
- 31.2 Copolymer equation 188
- 31.3 Copolymer engineering 188
- 31.4 See also 189
- 31.5 References 189
- 31.6 External links 189

32 Langmuir–Blodgett film — 190
- 32.1 Historical background 190
- 32.2 Physical insight 191
- 32.3 Pressure–area characteristics 191
- 32.4 Applications 192
- 32.5 See also 193
- 32.6 References 193
- 32.7 Bibliography 193
- 32.8 External links 194

33 Biological organisation — 195
- 33.1 Levels 195
- 33.2 Fundamentals 196
- 33.3 See also 196
- 33.4 Notes 196
    - 33.4.1 References 196
- 33.5 External links 197

34 Protein folding — 198
- 34.1 Known facts 198
    - 34.1.1 Relationship between folding and amino acid sequence 198
    - 34.1.2 Disruption of the native state 199

- 34.1.3 Incorrect protein folding and neurodegenerative disease 200
- 34.1.4 Effect of external factors on the folding of proteins 200
- 34.1.5 The Levinthal paradox and kinetics 200
- 34.2 Experimental techniques for studying protein folding 200
  - 34.2.1 Protein nuclear magnetic resonance spectroscopy 200
  - 34.2.2 Circular dichroism 201
  - 34.2.3 Dual polarisation interferometry 201
  - 34.2.4 Vibrational circular dichroism of proteins 201
  - 34.2.5 Studies of folding with high time resolution 201
  - 34.2.6 Proteolysis 201
  - 34.2.7 Optical tweezers 201
- 34.3 Computational methods for studying protein folding 202
  - 34.3.1 Energy landscape of protein folding 202
  - 34.3.2 Modeling of protein folding 202
- 34.4 See also 202
- 34.5 References 203
- 34.6 External links 205

35 Lipid bilayer 206

- 35.1 Structure and organization 207
  - 35.1.1 Cross section analysis 207
  - 35.1.2 Asymmetry 208
  - 35.1.3 Phases and phase transitions 208
  - 35.1.4 Surface chemistry 209
- 35.2 Biological roles 209
  - 35.2.1 Containment and separation 209
  - 35.2.2 Signaling 209
- 35.3 Characterization methods 210
- 35.4 Transport across the bilayer 211
  - 35.4.1 Passive diffusion 211
  - 35.4.2 Ion pumps and channels 212
  - 35.4.3 Endocytosis and exocytosis 212
  - 35.4.4 Electroporation 213
- 35.5 Mechanics 213
- 35.6 Fusion 214
- 35.7 Model systems 215
- 35.8 Commercial applications 215
- 35.9 History 216
- 35.10 See also 216

- 35.11 References . . . 217
- 35.12 External links . . . 221

36 Homeostasis 222
- 36.1 Biological . . . 222
- 36.2 Examples of some of the better understood physiological homeostats . . . 223
  - 36.2.1 The core body temperature homeostat . . . 223
  - 36.2.2 The blood glucose homeostat . . . 223
  - 36.2.3 The plasma ionized calcium homeostat . . . 224
  - 36.2.4 The blood partial pressure of oxygen and carbon dioxide homeostats . . . 224
  - 36.2.5 The blood oxygen content homeostat . . . 224
  - 36.2.6 The arterial blood pressure homeostat . . . 225
  - 36.2.7 The extracellular sodium concentration homeostat . . . 225
  - 36.2.8 The extracellular potassium concentration homeostat . . . 226
  - 36.2.9 The volume of body water homeostat . . . 226
  - 36.2.10 The extracellular fluid pH homeostat . . . 227
- 36.3 Homeostatic breakdown . . . 227
- 36.4 Examples from technology . . . 228
- 36.5 Biosphere . . . 228
- 36.6 Predictive . . . 228
- 36.7 Other fields . . . 228
  - 36.7.1 Risk . . . 228
  - 36.7.2 Stress . . . 229
- 36.8 History of discovery . . . 229
- 36.9 See also . . . 229
- 36.10 Foot note . . . 229
- 36.11 References . . . 230
- 36.12 Further reading . . . 231
- 36.13 External links . . . 231

37 Pattern formation 232
- 37.1 Examples . . . 232
  - 37.1.1 Biology . . . 232
  - 37.1.2 Chemistry . . . 233
  - 37.1.3 Physics . . . 233
  - 37.1.4 Mathematics . . . 233
  - 37.1.5 Computer graphics . . . 233
- 37.2 References . . . 234
- 37.3 Bibliography . . . 234

37.4 External links 234

38 Morphogenesis 235

38.1 History 235

38.2 Molecular basis 235

38.3 Cellular basis 236

    38.3.1 Cell-cell adhesion 236

    38.3.2 Extracellular matrix 236

    38.3.3 Cell contractility 236

38.4 See also 237

38.5 Notes 237

38.6 References 237

38.7 Sources 237

38.8 External links 237

39 Abiogenesis 238

39.1 Early geophysical conditions 239

    39.1.1 The earliest biological evidence for life on Earth 240

39.2 Conceptual history 240

    39.2.1 Spontaneous generation 241

    39.2.2 The origin of the terms *biogenesis* and *abiogenesis* 241

    39.2.3 Louis Pasteur and Charles Darwin 242

    39.2.4 "Primordial soup" hypothesis 242

    39.2.5 Proteinoid microspheres 243

39.3 Current models 244

39.4 Chemical origin of organic molecules 245

    39.4.1 Chemical synthesis 245

    39.4.2 Autocatalysis 246

    39.4.3 Information theory 247

    39.4.4 Homochirality 247

39.5 Self-enclosement, reproduction, duplication and the RNA world 247

    39.5.1 Protocells 247

    39.5.2 RNA world 248

    39.5.3 RNA synthesis and replication 249

    39.5.4 Pre-RNA world 250

39.6 Origin of biological metabolism 250

    39.6.1 Iron–sulfur world 250

    39.6.2 Zn-world hypothesis 251

    39.6.3 Deep sea vent hypothesis 251

- 39.6.4 Thermosynthesis 253
- 39.7 Other models of abiogenesis 253
  - 39.7.1 Clay hypothesis 254
  - 39.7.2 Gold's "deep-hot biosphere" model 254
  - 39.7.3 Panspermia 254
  - 39.7.4 Extraterrestrial organic molecules 255
  - 39.7.5 Lipid world 256
  - 39.7.6 Polyphosphates 257
  - 39.7.7 PAH world hypothesis 257
  - 39.7.8 Radioactive beach hypothesis 257
  - 39.7.9 Thermodynamic dissipation 257
  - 39.7.10 Multiple genesis 258
  - 39.7.11 Fluctuating hydrothermal pools on volcanic islands 259
- 39.8 See also 259
- 39.9 Notes 259
- 39.10 References 259
- 39.11 Bibliography 272
- 39.12 Further reading 275
- 39.13 External links 276
  - 39.13.1 Video resources 277

40 Hypercycle (chemistry)   278

- 40.1 Model formulation 279
  - 40.1.1 Model evolution 279
  - 40.1.2 Error threshold problem 279
  - 40.1.3 Hypercycle models 279
  - 40.1.4 Alternative concepts 280
- 40.2 Mathematical model 280
  - 40.2.1 Elementary hypercycle 280
  - 40.2.2 Hypercycle with translation 281
- 40.3 Evolution of hypercycles 281
  - 40.3.1 Formation of the first hypercycles 281
  - 40.3.2 Evolutionary dynamics 282
  - 40.3.3 Evolutionary dynamics: a mathematical model 282
- 40.4 Compartmentalization and genome integration 282
- 40.5 Hypercycles and ribozymes 283
- 40.6 Related problems and reformulations 284
- 40.7 See also 285
- 40.8 References 285

40.9 External links 287

41 Autocatalytic set 288
- 41.1 Formal definition 288
  - 41.1.1 Definition 288
  - 41.1.2 Example 289
- 41.2 Probability that a random set is autocatalytic 289
- 41.3 Formal limitations 289
- 41.4 Linguistic aspects 289
- 41.5 Non-autonomous autocatalytic sets 289
- 41.6 References 290
- 41.7 See also 290

42 Multi-agent system 291
- 42.1 Concept 291
  - 42.1.1 Characteristics 292
  - 42.1.2 Self-organization and self-steering 292
  - 42.1.3 Systems paradigms 292
  - 42.1.4 Properties 292
- 42.2 Study of multi-agent systems 292
- 42.3 Frameworks 293
- 42.4 Applications in the real world 293
- 42.5 See also 293
- 42.6 References 293
- 42.7 Further reading 294
- 42.8 External links 295

43 Self-organizing network 296
- 43.1 SON architectural types 296
  - 43.1.1 Distributed SON 296
  - 43.1.2 Centralized SON 296
  - 43.1.3 Hybrid SON 296
- 43.2 SON sub-functions 296
  - 43.2.1 Self-configuration functions 296
  - 43.2.2 Self-optimization functions 297
  - 43.2.3 Self-healing functions 297
- 43.3 Introduction of SON 297
- 43.4 References 297
- 43.5 Literature 297

44 Dual-phase evolution — 298

- 44.1 Introduction — 298
- 44.2 The DPE mechanism — 298
  - 44.2.1 Underlying network — 298
  - 44.2.2 Phase shifts — 298
  - 44.2.3 Selection and variation — 299
  - 44.2.4 System memory — 299
- 44.3 Examples — 299
  - 44.3.1 Social networks — 299
  - 44.3.2 Socio-economics — 299
  - 44.3.3 Forest ecology — 300
  - 44.3.4 Search algorithms — 300
- 44.4 Related processes — 300
- 44.5 See also — 300
- 44.6 References — 301

45 Molecular assembler — 302

- 45.1 Nanofactories — 302
- 45.2 Self-replication — 303
- 45.3 Drexler and Smalley debate — 303
- 45.4 Regulation — 304
- 45.5 Formal scientific review — 304
- 45.6 Grey goo — 304
- 45.7 In fiction — 304
- 45.8 See also — 305
- 45.9 References — 305
- 45.10 External links — 305

46 Critical mass (sociodynamics) — 306

- 46.1 History — 306
  - 46.1.1 Predecessors — 306
- 46.2 Logic of collective action and common good — 306
- 46.3 Gender politics — 307
- 46.4 Interactive media — 307
  - 46.4.1 Markus essay — 307
  - 46.4.2 Fax machine example — 307
- 46.5 See also — 308

43.6 External links — 297

CONTENTS   xix

    46.6 References 308
    46.7 Further reading 308

47 Herd behavior   310
    47.1 In animals 310
    47.2 Symmetry-breaking 310
    47.3 In human societies 310
        47.3.1 Stock market bubbles 311
        47.3.2 In crowds 311
        47.3.3 Everyday decision-making 311
    47.4 In Marketing 311
        47.4.1 Herd Behavior in Brand and Product success 312
        47.4.2 Herd Behavior in Social Marketing 312
    47.5 See also 313
    47.6 Notes 313
    47.7 References 313
    47.8 Further reading 314

48 Groupthink   316
    48.1 History 316
    48.2 Symptoms 317
    48.3 Causes 318
    48.4 Prevention 318
    48.5 Empirical findings and meta-analysis 319
    48.6 Case studies 320
        48.6.1 Politics and military 320
        48.6.2 Corporate world 321
        48.6.3 Sports 321
    48.7 Recent developments 321
        48.7.1 Ubiquity model 321
        48.7.2 Reexamination 322
        48.7.3 Reformulation 322
        48.7.4 Sociocognitive theory 322
    48.8 See also 322
    48.9 References 323
    48.10 Further reading 324

49 Joint attention   326
    49.1 Humans 326

- 49.1.1 Levels of joint attention ... 326
- 49.1.2 Gaze ... 327
- 49.1.3 Intention ... 327
- 49.1.4 Language comprehension ... 327
- 49.1.5 Language production ... 328
- 49.1.6 Relationship to socio-emotional development ... 328
- 49.1.7 Developmental markers in infancy ... 328
- 49.1.8 Individuals with disabilities ... 329
- 49.2 Other animals ... 329
  - 49.2.1 Definitions in non-human animals ... 329
  - 49.2.2 Dyadic joint attention ... 329
  - 49.2.3 Shared gaze ... 329
- 49.3 See also ... 329
- 49.4 References ... 329
- 49.5 Text and image sources, contributors, and licenses ... 332
  - 49.5.1 Text ... 332
  - 49.5.2 Images ... 343
  - 49.5.3 Content license ... 351

Chapter 1

Self-organization

Self-organization in micron-sized $Nb_3O_7(OH)$ cubes during a hydrothermal treatment at 200 °C. Initially amorphous cubes gradually transform into ordered 3D meshes of crystalline nanowires as summarized in the model below.[1]

Self-organization is a process where some form of overall order or coordination arises out of the local interactions between smaller component parts of an initially disordered system. The process of self-organization can be spontaneous, and it is not necessarily controlled by any auxiliary agent outside of the system. It is often triggered by random fluctuations that are amplified by positive feedback. The resulting organization is wholly decentralized or distributed over all the components of the system. As such, the organization is typically robust and able to survive and, even, self-repair substantial damage or perturbations. Chaos theory discusses self-organization in terms of islands of predictability in a sea of chaotic unpredictability. Self-organization occurs in a variety of physical, chemical, biological, robotic, social, and cognitive systems. Examples of its realization can be found in crystallization, thermal convection of fluids, chemical oscillation, animal swarming, and neural networks.

1.1 Overview

Self-organization is realized[2] in the physics of non-equilibrium processes, and in chemical reactions, where it is often described as self-assembly. The concept of self-organization has proven useful in the description of biological systems,[3] from the subcellular to the ecosystem level.[4] Cited examples of self-organizing behaviour also appear in the literature of many other disciplines, both in the natural sciences and in the social sciences such as economics or anthropology. Self-organization has also been observed in mathematical systems such as cellular automata.[5] Sometimes the notion of self-organization becomes conflated with that of the related concept of emergence.[6] Properly defined, however, there may be instances of self-organization without emergence and emergence without self-organization.

Self-organization usually relies on three basic ingredients:[7]

1. strong dynamical non-linearity, often though not necessarily involving positive and negative feedback
2. balance of exploitation and exploration
3. multiple interactions

1.1.1 Principles of self-organization

The cybernetician William Ross Ashby formulated the original principle of self-organization in 1947.[8][9] It states that any deterministic dynamic system will automatically evolve towards a state of equilibrium that can be described in terms of an attractor in a basin of surrounding states. Once there, the further evolution of the system is constrained to remain in the attractor. This constraint on the system as a whole implies a form of mutual dependency or coordination between its constituent components or "subsystems". In Ashby's terms, each subsystem has adapted to the environment formed by all other subsystems.

The cybernetician Heinz von Foerster formulated the principle of "order from noise" in 1960.[10] It notes that self-organization is facilitated by random perturbations ("noise") that let the system explore a variety of states in its state space. This increases the chance that the system would arrive into the basin of a "strong" or "deep" attractor, from which it would then quickly enter the attractor itself. The thermodynamicist Ilya Prigogine formulated a similar principle as "order through fluctuations"[11] or "order out of chaos".[12] It is applied in the method of simulated annealing that is used in problem solving and in machine learning.

1.2 History of the idea

The idea that the dynamics of a system can lead to an increase of the system's organization has a long history. One of the earliest statements of this idea was by the philosopher Descartes, in the fifth part of his *Discourse on Method*, where he presents it hypothetically. Descartes further elaborated on the idea at great length in his unpublished work *The World*.

The ancient atomists believed that a designing intelligence is unnecessary to effect natural order, arguing that given enough time and space and matter, organization is ultimately inevitable, although there is no preferred tendency for this to happen. What Descartes introduced was the idea that the ordinary laws of nature *tend* to produce organization (For related history, see Aram Vartanian, *Diderot and Descartes*).

The economic concept of the "invisible hand" due to Adam Smith can be understood as an attempt to describe the influence of the market as a spontaneous order on people's actions.

Beginning with the 18th century, natural scientists sought to understand the "universal laws of form" in order to explain the observed forms of living organisms. Because of its association with Lamarckism, their ideas fell into disrepute until the early 20th century, when pioneers such as D'Arcy Wentworth Thompson revived them. The modern understanding is that there are indeed universal laws, arising from fundamental physics and chemistry, that govern growth and form in biological systems.

Sadi Carnot and Rudolf Clausius discovered the Second Law of Thermodynamics in the 19th century. It states that total entropy, sometimes understood as disorder, will always increase over time in an isolated system. This means that a system cannot spontaneously increase its order, without an external relationship that decreases order elsewhere in the system (e.g. through consuming the low-entropy energy of a battery and diffusing high-entropy heat).

Originally, the term "self-organizing" was used by Immanuel Kant in his *Critique of Judgment*, where he argued that teleology is a meaningful concept only if there exists such an entity whose parts or "organs" are simultaneously ends and means. Such a system of organs must be able to behave as if it has a mind of its own, that is, it is capable of governing itself.

The term "self-organizing" was introduced to contemporary science in 1947 by the psychiatrist and engineer W. Ross Ashby.[8] It was taken up by the cyberneticians Heinz von Foerster, Gordon Pask, Stafford Beer, and von Foerster organized a conference on "The Principles of Self-Organization" at the University of Illinois' Allerton Park in June, 1960 which led to a series of conferences on Self-Organizing Systems.[13] Norbert Wiener also took up the idea in the second edition of his *Cybernetics: or Control and Communication in the Animal and the Machine* (1961).

Self-organization as a word and concept was used by those associated with general systems theory in the 1960s, but did not become commonplace in the scientific literature until its adoption by physicists and researchers in the field of complex systems in the 1970s and 1980s.[14] After Ilya Prigogine's 1977 Nobel Prize, the *thermodynamic concept of self-organization* received some attention of the public, and scientific researchers started to migrate from the *cybernetic view* to the *thermodynamic view*.[15]

1.2.1 Developing views

Other views of self-organization in physical systems interpret it as a strictly accumulative construction process, commonly displaying an "S" curve history of development. As discussed somewhat differently by different researchers, local complex systems for exploiting energy gradients evolve from seeds of organization, through a succession of natural starting and ending phases for inverting their directions of development. The accumulation of working processes which their exploratory parts construct as they exploit their gradient becomes the "learning", "organization" or "design" of the system as a physical artifact, such for an ecology or economy. For example, A. Bejan's books and papers describe his approach as "Constructal Theory".[16][17][18] P. F. Henshaw's work on decoding net-energy system construction processes termed "Natural Systems Theory", uses various analytical methods to quantify and map them such as System Energy Assessment[19] for taking true quantitative measures of whole complex energy using systems, and for anticipating their successions, such as Models Learning Change[20] to permit adapting models to their emerging inverted designs. G. Y. Georgiev's work is utilizing the principle of least (stationary) action in Physics, to define organization of a complex system as the state of the constraints determining the total action of the elements in a system. Organization is then defined numerically as the reciprocal of the average action per one element and one edge crossing, if the system is described as a network. The elementary quantum of action, Planck's constant, is used to make the measure dimensionless and to define it as inversely proportional to the number of quanta of action expended by the elements for one edge crossing. The mechanism of self-organization is the interaction between the elements and the constraints, which leads to constraint minimization. This is consistent with the Gauss principle of least constraint. More elements minimize the constraints faster, another aspect of the mechanism, which is through quantity accumulation. As a result, the paths of the elements are straightened, which is consistent with Hertz's principle of least curvature. The state of a system with least average sum of actions of its elements is defined as its attractor. In open systems, where there is constant inflow and outflow of energy and elements, this final state is never reached, but the system always tends toward it.[15] This method can help describe, quantify, manage, design and predict future behavior of complex systems, to achieve the highest rates of self-organization to improve their quality, which is the numerical value of their organization. It can be applied to complex systems in physics, chemistry, biology, ecology, economics, cities, network theory and others, where they are present.[15][21][22]

1.3 Examples

The following list summarizes and classifies the instances of self-organization found in different disciplines. As the list grows, it becomes increasingly difficult to determine whether these phenomena are all fundamentally the same process, or the same label applied to several different processes. Self-organization, despite its intuitive simplicity as a concept, has proven notoriously difficult to define and pin down formally or mathematically, and it is entirely possible that any precise definition might not include all the phenomena to which the label has been applied.

The farther a phenomenon is removed from physics, the more controversial the idea of self-organization *as understood by physicists* becomes. Also, even when self-organization is clearly present, attempts at explaining it through physics or statistics are usually criticized as reductionistic.

Similarly, when ideas about self-organization originate in, say, biology or social science, the farther one tries to take the concept into chemistry, physics or mathematics, the more resistance is encountered, usually on the grounds that it implies direction in fundamental physical processes. However the tendency of hot bodies to get cold (see Thermodynamics) and by Le Chatelier's Principle—the statistical mechanics extension of Newton's Third Law—to oppose this tendency should be noted.

1.3.1 Physics

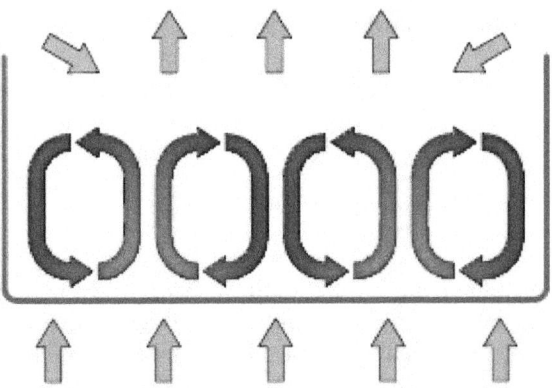

Convection cells in a gravity field

There are several broad classes of physical processes that can be described as self-organization. Such examples from physics include:

- structural (order-disorder, first-order) phase transitions, and spontaneous symmetry breaking such as
  - spontaneous magnetization, crystallization (see crystal growth, and liquid crystal) in the classical domain and
  - the laser, superconductivity and Bose–Einstein condensation, in the quantum domain (but with macroscopic manifestations)
- second-order phase transition, associated with "critical points" at which the system exhibits scale-invariant structures. Examples of these include:
  - critical opalescence of fluids at the critical point
  - percolation in random media
- structure formation in thermodynamic systems away from equilibrium. The theory of dissipative structures of Prigogine and Hermann Haken's Synergetics were developed to unify the understanding of these phenomena, which include lasers, turbulence and convective instabilities (e.g., Bénard cells) in fluid dynamics,
  - structure formation in astrophysics and cosmology (including star formation, planetary systems formation, galaxy formation)
  - self-similar expansion
  - Diffusion-limited aggregation
  - percolation
  - reaction-diffusion systems, such as Belousov–Zhabotinsky reaction

- self-organizing dynamical systems: complex systems made up of small, simple units connected to each other usually exhibit self-organization

  - Self-organized criticality (SOC)

- In tribology, friction coupled with other simultaneous effects, such as heat transfer, wear, and material diffusion. can lead to self-organized patterns at the frictional interface, ranging from stick-slip patterns to in-situ formed tribofilms and surface roughness adjustment of two materials in contact.

- In spin foam system and loop quantum gravity that was proposed by Lee Smolin. The main idea is that the evolution of space in time should be robust in general. Any fine-tuning of cosmological parameters weaken the independency of the fundamental theory. Philosophically, it can be assumed that in the early time, there has not been any agent to tune the cosmological parameters. Smolin and his colleagues in a series of works show that, based on the loop quantization of spacetime, in the very early time, a simple evolutionary model (similar to the sand pile model) behaves as a power law distribution on both the size and area of avalanche.

  - Although, this model, which is restricted only on the frozen spin networks, exhibits a non-stationary expansion of the universe. However, it is the first serious attempt toward the final ambitious goal of determining the cosmic expansion and inflation based on a self-organized criticality theory in which the parameters are not tuned, but instead are determined from within the complex system.[23]

- A laser can also be characterized as a self organized system to the extent that normal states of thermal equilibrium characterized by electromagnetic energy absorption are stimulated out of equilibrium in a reverse of the absorption process. "If the matter can be forced out of thermal equilibrium to a sufficient degree, so that the upper state has a higher population than the lower state (population inversion), then more stimulated emission than absorption occurs, leading to coherent growth (amplification or gain) of the electromagnetic wave at the transition frequency."[24]

1.3.2 Chemistry

Self-organization in chemistry includes:

1. molecular self-assembly

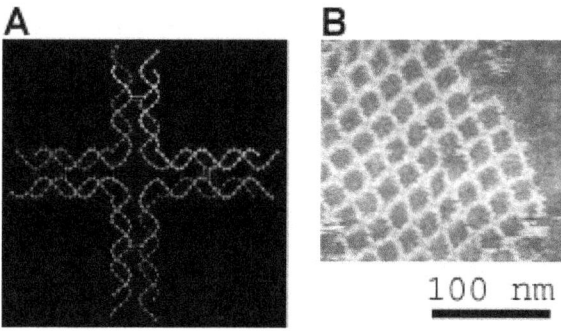

The DNA structure at left (schematic shown) will self-assemble into the structure visualized by atomic force microscopy at right. Image from Strong.[25]

2. reaction-diffusion systems and oscillating chemical reactions
3. autocatalytic networks (see: autocatalytic set)
4. liquid crystals
5. grid complexes
6. colloidal crystals
7. self-assembled monolayers
8. micelles
9. microphase separation of block copolymers
10. Langmuir-Blodgett films

1.3.3 Biology

Birds flocking, an example of self-organization in biology

Main article: Biological organisation

According to *Scott Camazine.. [et al.]*:

The following is an incomplete list of the diverse phenomena which have been described as self-organizing in biology.

1. spontaneous folding of proteins and other biomacromolecules
2. formation of lipid bilayer membranes
3. homeostasis (the self-maintaining nature of systems from the cell to the whole organism)
4. pattern formation and morphogenesis, or how the living organism develops and grows. See also embryology.
5. the coordination of human movement, e.g. seminal studies of bimanual coordination by Kelso
6. the creation of structures by social animals, such as social insects (bees, ants, termites), and many mammals
7. flocking behaviour (such as the formation of flocks by birds, schools of fish, etc.)
8. the origin of life itself from self-organizing chemical systems, in the theories of hypercycles and autocatalytic networks
9. the organization of Earth's biosphere in a way that is broadly conducive to life (according to the controversial Gaia hypothesis)

1.3.4 Computer Science

Gosper's Glider Gun creating "gliders" in the cellular automaton Conway's Game of Life.[27]

As mentioned above, phenomena from mathematics and computer science such as cellular automata, random graphs, and some instances of evolutionary computation and artificial life exhibit features of self-organization. In swarm robotics, self-organization is used to produce emergent behavior. In particular the theory of random graphs has been used as a justification for self-organization as a general principle of complex systems. In the field of multi-agent systems, understanding how to engineer systems that are capable of presenting self-organized behavior is a very active research area.

Algorithms

Many optimization algorithms can be considered as a self-organization system because the aim of the optimization is to find the optimal solution to a problem. If the solution is considered as a state of the iterative system, the optimal solution is essentially the selected, converged state or structure of the system, driven by the algorithm based on the system landscape.[28][29] In fact, one can view all optimization algorithms as a self-organization system.

Networks

Self-organization is an important component for a successful ability to establish networking whenever needed. Such mechanisms are also referred to as Self-organizing networks. Intensified work in the latter half of the first decade of the 21st century was mainly due to interest from the wireless communications industry. It is driven by the plug and play paradigm, and that wireless networks need to be relatively simpler to manage than they used to be.

Only certain kinds of networks are self-organizing. The best known examples are small-world networks and scale-free networks. These emerge from bottom-up interactions, and appear to be limitless in size. In contrast, there are top-down hierarchical networks, which are not self-organizing. These are typical of organizations, and have severe size limits.

In many natural systems, self-organization results from repeated phase shifts in their underlying network of connections. Such phase shifts alter the balance between internal processes (e.g. selection and variation). They give rise to the phenomenon of dual-phase evolution.

1.3.5 Cybernetics

Wiener regarded the automatic serial identification of a black box and its subsequent reproduction as sufficient to meet the condition of self-organization.[30] The importance of phase locking or the "attraction of frequencies", as he called it, is discussed in the 2nd edition of his

"Cybernetics".[31] Drexler sees self-replication as a key step in nano and universal assembly.

By contrast, the four concurrently connected galvanometers of W. Ross Ashby's Homeostat hunt, when perturbed, to converge on one of many possible stable states.[32] Ashby used his state counting measure of variety[33] to describe stable states and produced the "Good Regulator"[34] theorem which requires internal models for self-organized endurance and stability (e.g. Nyquist stability criterion).

Warren McCulloch proposed "Redundancy of Potential Command"[35] as characteristic of the organization of the brain and human nervous system and the necessary condition for self-organization.

Heinz von Foerster proposed Redundancy, $R = 1 - H/H_{max}$, where H is entropy.[36][37] In essence this states that unused potential communication bandwidth is a measure of self-organization.

In the 1970s Stafford Beer considered this condition as necessary for autonomy which identifies self-organization in persisting and living systems. Using Variety analyses he applied his neurophysiologically derived recursive Viable System Model to management. It consists of five parts: the monitoring of performance of the survival processes (1), their management by recursive application of regulation (2), homeostatic operational control (3) and development (4) which produce maintenance of identity (5) under environmental perturbation. Focus is prioritized by an alerting "algedonic loop" feedback: a sensitivity to both pain and pleasure produced from under-performance or over-performance relative to a standard capability.[38]

In the 1990s Gordon Pask pointed out von Foerster's H and Hmax were not independent and interacted via countably infinite recursive concurrent spin processes[39] (he favoured the Bohm interpretation) which he called concepts (liberally defined in *any* medium, "productive and, incidentally reproductive"). His strict definition of concept "a procedure to bring about a relation"[40] permitted his theorem "Like concepts repel, unlike concepts attract"[41] to state a general spin based **Principle of Self-organization**. His edict, an exclusion principle, "There are No Doppelgangers"[42][39] means no two concepts can be the same (all interactions occur with different perspectives making time incommensurable for actors). This means, after sufficient duration as differences assert, all concepts will attract and coalesce as pink noise and entropy increases (and see Big Crunch, self-organized criticality). The theory is applicable to all organizationally closed or homeostatic processes that produce enduring and coherent products (where spins have a fixed average phase relationship and also in the sense of Rescher Coherence Theory of Truth with the proviso that the sets and their members exert repulsive forces at their boundaries) through interactions: evolving, learning and adapting.

Pask's Interactions of Actors "hard carapace" model is reflected in some of the ideas of emergence and coherence. It requires a knot emergence topology that produces radiation during interaction with a unit cell that has a prismatic tensegrity structure. Laughlin's contribution to emergence reflects some of these constraints.

1.3.6 Human society

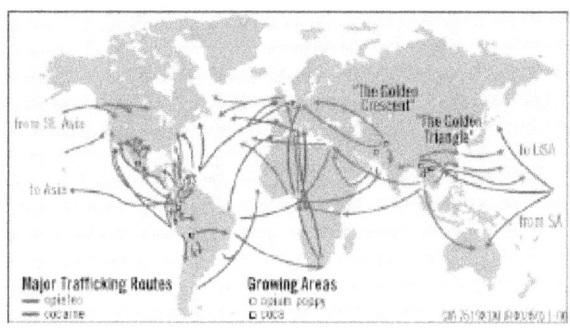

Social self-organization in international drug routes

The self-organizing behaviour of social animals and the self-organization of simple mathematical structures both suggest that self-organization should be expected in human society. Tell-tale signs of self-organization are usually statistical properties shared with self-organizing physical systems (see Zipf's law, power law, Pareto principle). Examples such as critical mass, herd behaviour, groupthink and others, abound in sociology, economics, behavioral finance and anthropology.[43] The theory of human social self-organization is also known as spontaneous order theory.

In social theory the concept of self-referentiality has been introduced as a sociological application of self-organization theory by Niklas Luhmann (1984). For Luhmann the elements of a social system are self-producing communications, i.e. a communication produces further communications and hence a social system can reproduce itself as long as there is dynamic communication. For Luhmann human beings are sensors in the environment of the system. Luhmann developed an evolutionary theory of Society and its subsystems, using functional *analyses* and systems *theory*.[44]

Self-organization in human and computer networks can give rise to a decentralized, distributed, self-healing system, protecting the security of the actors in the network by limiting the scope of knowledge of the entire system held by each individual actor. The Underground Railroad is a good example of this sort of network. The networks that arise from drug trafficking exhibit similar self-organizing properties. The Sphere College Project seeks to apply self-organization to adult education. Parallel examples exist in the world of privacy-preserving computer networks such as Tor. In each case, the network as a whole exhibits distinctive synergistic behavior through the combination of the behaviors of individual actors in the network. Usually the growth of such networks is fueled by an ideology or sociological force that is adhered to or shared by all participants in the network.[15]

Economics

In economics, a market economy is sometimes said to be self-organizing. Paul Krugman has written on the role that market self-organization plays in the business cycle in his book "The Self Organizing Economy".[45] Friedrich Hayek coined the term *catallaxy*[46] to describe a "self-organizing system of voluntary co-operation", in regards to the spontaneous order of the free market economy. Neo-classical economists hold that imposing central planning usually makes the self-organized economic system less efficient. On the other end of the spectrum, economists consider that market failures are so significant that self-organization produces bad results and that the state should direct production and pricing. Most economists adopt an intermediate position and recommend a mixture of market economy and command economy characteristics (sometimes called a mixed economy). When applied to economics, the concept of self-organization can quickly become ideologically imbued.[15][47]

Collective intelligence

Visualization of links between pages on a wiki. This is an example of collective intelligence through collaborative editing.

Non-thermodynamic concepts of entropy and self-organization have been explored by many theorists. Cliff Joslyn and colleagues and their so-called "global brain" projects. Marvin Minsky's "Society of Mind" and the no-central editor in charge policy of the open sourced internet encyclopedia, called Wikipedia, are examples of applications of these principles — see collective intelligence.

Donella Meadows, who codified twelve leverage points that a self-organizing system could exploit to organize itself, was one of a school of theorists who saw human creativity as part of a general process of adapting human lifeways to the planet and taking humans out of conflict with natural processes. See Gaia philosophy, deep ecology, ecology movement and Green movement for similar self-organizing ideals. (The connections between self-organisation and Gaia theory and the environmental movement are explored in the book *The Unity of Nature* by Alan Marshall).

1.3.7 Psychology and education

Self-organised learning

Enabling others to "learn how to learn"[48] is usually misconstrued as instructing them[49] how to successfully submit to being taught. Whilst fully accepting that we can always learn from others, particularly those with more and/or different experience than ourselves; self-organised learning (SOL) repudiates any idea[50] that this reduces to accepting that "the expert knows best" or that there is ever "the one best method." It offers an alternative definition of learning as "the construction of personally significant, relevant and viable meaning."

This more democratic 'bottom up' approach to learning is to be frequently tested experientially[51] by the learner(s) as being more "meaningful, constructive and creatively effective for me or us."

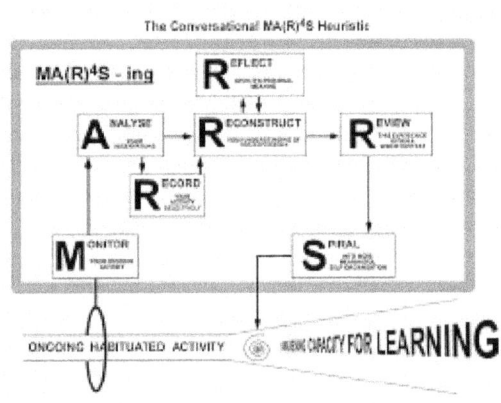

Cybernetic algorithm

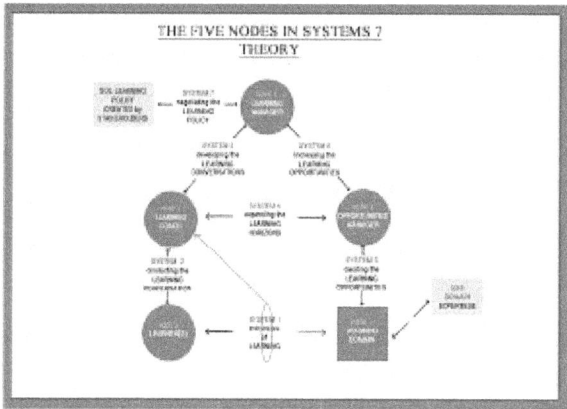

Systems algorithm

Since human learning may be achieved by one person,[52] or groups of learners working together;[53] SOL is not only a more rewarding and effective way of living one's personal life; it is also applicable in any group of people living, playing and/or working together.

As many young children, pupils, students and lifelong learners eventually become ruefully aware, this 'testing out of what I have learned' needs to be carried out in each learner(s) whole process of living, and so it extends well beyond the confines of specific learning environments (home, school, university, etc.), and eventually beyond the reaches of the controllers of these environments (parents, teachers, employers, etc.)[54]

SOL needs to be tested, and intermittently revised, through the ongoing personal experience[55] of the learner(s) themselves in their ever-expanding outer and inner lives.

Whilst internal life may cease to expand, the external environment does not. If a learner allows themselves to become progressively more other-organised, they become less able to recognise and respond to varying needs for change. Unfortunately this is often the current reported experience of many during, and hence after their parenting, schooling and/or higher education.

But, this SOL way of understanding the learning process need not be restricted by either consciousness or language.[56] Nor is it restricted to humans, since analogous directional self-organizing (learning?) processes are reported variously within the life sciences and even within the less-living sciences, for example, of physics and chemistry: (as is clearly articulated in other sections of this 'Self-organization' Section).

Since SOL is as yet only very superficially recognised within psychology and education, it is useful to place it more firmly within the human public mind-pool[57] of achievement, knowledge, experience and understanding. SOL can also be placed within a hierarchy of scientific explanatory concepts, for example:

1. Cause and Effect (requires "other things being equal")

2. Cybernetics[40] (incorporates item 1 in this list) with greater complexity, providing internal feedback and feed-forward controls: but still implying a sealed boundary. (i.e. other things being equal)

3. Systems Theory[58] (incorporates item 2 in this list, and opens the boundaries)

4. Self-organized System (incorporates item 3 in this list) and attributes this property to the interaction, patterning and coordination among the sub-systems of the system in question; in response to flow across its boundaries

5. Self-Organised Learning (SOL)[59] (incorporates item 4 in this list) but also requires that the parts each systematically respond, change and develop in the light of their experience, whilst self-organizing in the developing experiential interest of the whole).

SOL not only involves self-organization of the first order, i.e. what is mostly experienced as learning from experience without much conscious awareness of the process. At a second level of SOL consciousness enables us, (possibly uniquely among living beings) to reflect upon and thus self-organise the very process of self-organisation itself, (See 'Cybernetic algorithm' figure). It also enables organisations small and large to self-organise themselves, (see 'System algorithm' figure).

Once this approach to human learning is acknowledged, then we can re-set science into its place within the total human mind-pool. A mind-pool of human know-how and feel-how as an ever expanding and hopefully self-organizing resource.

6. Learning Conversation (incorporates item 5 in this list) and yet is at the same time its major tool. The Learning Conversation is a two-way process between SOLers, even within one person (conversing with oneself). Whilst not necessarily requiring language i.e. dialogue; it does require that the each participant really attempts to represent their meaning to the other(s), and that they all attempt to create personally significant, relevant and viable meaning in themselves in response to the others representations. So art, drama, music, computer programs, maths problems, ???, etc., can all create different, if limited, forms of Learning Conversation which really only become fully functional when at least two humans really attempt to fully communicate, and effectively share their understanding. That is achieve shared meaning in an event that approximates to what Maslow called a creative encounter[60]

7. Conversational Science[61] (will require item 6 in this list, the main method of SOL) among all seekers after significant, relevant and viable shared meaning. Science and many other human activities still need major paradigm shifts if we are to achieve Self-Organised Living. It also requires equal stakeholder-ship for each converser. Thus SOL can be seen as necessary but not sufficient for science to contribute positively to the benefit of the society, within which it may have only spasmodically been conversing successfully (SOL wise). Until, perhaps, both science and society as a whole will become Self-Organised Learners (SOLers) continually learning from their own shared experience and using what they learn in the shared interest of all concerned.

1.3.8 Traffic flow

The self-organizing behaviour of drivers in traffic flow determines almost all traffic spatiotemporal phenomena observed in real traffic data like traffic breakdown at a highway bottleneck, highway capacity, the emergence of moving traffic jams, etc. Self-organization in traffic flow is extremely complex spatiotemporal dynamic process. For this reason, only in 1996–2002 spatiotemporal self-organization effects in traffic have been understood in real measured traffic data and explained by Boris Kerner's three-phase traffic theory.

1.3.9 Methodology

In many complex systems in nature, there are global phenomena that are the irreducible result of local interactions between components whose individual study would not allow us to see the global properties of the whole combined system. Thus, a growing number of researchers think that many properties of language are not directly encoded by any of the components involved, but are the self-organized outcomes of the interactions of the components.

Building mathematical models in the context of research into language origins and the evolution of languages is enjoying growing popularity in the scientific community, because it is a crucial tool for studying the phenomena of language in relation to the complex interactions of its components. These systems are put to two main types of use: 1) they serve to evaluate the internal coherence of verbally expressed theories already proposed by clarifying all their hypotheses and verifying that they do indeed lead to the proposed conclusions ; 2) they serve to explore and generate new theories, which themselves often appear when one simply tries to build an artificial system reproducing the verbal behavior of humans.

As it were, the construction of operational models to test proposed hypotheses in linguistics is gaining much contemporary attention. An operational model is one which defines the set of its assumptions explicitly and above all shows how to calculate their consequences, that is, to prove that they lead to a certain set of conclusions.

In the emergence of language

The emergence of language in the human species has been described in a game-theoretic framework based on a model of senders and receivers of information. The evolution of certain properties of language such as inference follow from this sort of framework (with the parameters stating that information transmitted can be partial or redundant, and the underlying assumption that the sender and receiver each want to take the action in his/her best interest). Likewise, models have shown that compositionality, a central component of human language, emerges dynamically during linguistic evolution, and need not be introduced by biological evolution. Tomasello (1999) argues that through one evolutionary step, the ability to sustain culture, the groundwork for the evolution of human language was laid. The ability to ratchet cultural advances cumulatively allowed for the complex development of human cognition unseen in other animals.

In language acquisition

Within a species' ontogeny, the acquisition of language has also been shown to self-organize. Through the ability to see others as intentional agents (theory of mind), and actions such as 'joint attention,' human children have the scaffolding they need to learn the language of those around them.

In articulatory phonology

Articulatory phonology takes the approach that speech production consists of a coordinated series of gestures, called 'constellations,' which are themselves dynamical systems. In this theory, linguistic contrast comes from the distinction between such gestural units, which can be described on a low-dimensional level in the abstract. However, these structures are necessarily context-dependent in real-time production. Thus the context-dependence emerges naturally from the dynamical systems themselves. This statement is controversial, however, as it suggests a universal phonetics which is not evident across languages.[62] Cross-linguistic patterns show that what can be treated as the same gestural units produce different contextualised patterns in different languages.[63] Articulatory Phonology fails to attend to the acoustic output of the gestures themselves (meaning that many typological patterns remain unexplained).[64] Freedom among listeners in the weighting of perceptual cues in the acoustic signal has a more fundamental role to play in the emergence of structure.[65] The realization of the perceptual contrasts by means of articulatory movements means that articulatory considerations do play a role,[66] but these are purely secondary.

In diachrony and synchrony

Several mathematical models of language change rely on self-organizing or dynamical systems. Abrams and Strogatz (2003) produced a model of language change that focused on "language death" – the process by which a speech community merges into the surrounding speech communities. Nakamura et al. (2008) proposed a variant of this model that incorporates spatial dynamics into language contact transactions in order to describe the emergence of creoles. Both of these models proceed from the assumption that language change, like any self-organizing system, is a large-scale act or entity (in this case the creation or death of a language, or changes in its boundaries) that emerges from many actions on a micro-level. The microlevel in this example is the everyday production and comprehension of language by speakers in areas of language contact.

1.4 Criticism

Heinz Pagels, in a balanced, but ultimately negative 1985 book review of Ilya Prigogine and Isabelle Stengers' *Order Out of Chaos* in *Physics Today*, appeals to authority:[67]

In theology, Thomas Aquinas (1225–1274) in his *Summa Theologica* assumes a teleological created universe in rejecting the idea that something can be a self-sufficient cause of its own organization:[68]

("The body of the Article" consists of the *quinque viae*.)

1.5 See also

- Ant mill
- Autowave
- Biology concepts: Bow tie (biology) – evolution – morphogenesis – homeostasis – Gaia Hypothesis
- Causality
- Chemistry concepts: reaction-diffusion – autocatalysis
- Complex systems concepts: emergence – evolutionary computation – artificial life – self-organized criticality – "edge of chaos" – spontaneous order – metastability – Chaos theory – Butterfly effect
- Computer science concepts: swarm intelligence
- Constructal law
- Dual-phase evolution
- Self-organized criticality control
- Free energy principle
- Free will
- Information theory
- Language – Operator grammar
- Mathematics concepts: fractal – random graph – power law – small world phenomenon – cellular automata
- Organization of the artist
- Philosophical concepts: tectology – Religious naturalism
- Physics concepts: thermodynamics – non-equilibrium thermodynamics – constructal theory – statistical mechanics – phase transition – dissipative structures – turbulence
- Social concepts: participatory organization
- Spontaneous order
- Stigmergy
- Systems theory concepts: cybernetics – autopoiesis – polytely
- Santiago theory of cognition
- Thermodynamics concepts: Second Law of Thermodynamics – Heat death of the Universe

1.6 References

[1] Betzler, S. B.; Wisnet, A.; Breitbach, B.; Mitterbauer, C.; Weickert, J.; Schmidt-Mende, L.; Scheu, C. (2014). "Template-free synthesis of novel, highly-ordered 3D hierarchical Nb$_3$O$_7$(OH) superstructures with semiconductive and photoactive properties". *Journal of Materials Chemistry A* 2 (30): 12005. doi:10.1039/C4TA02202E.

[2] Glansdorff, P., Prigogine, I. (1971). *Thermodynamic Theory of Structure, Stability and Fluctuations*, Wiley-Interscience, London. ISBN 0-471-30280-5

1.6. REFERENCES

[3] Witzany G (2014). Biological Self-Organization. International Journal of Signs and Semiotic Systems 3: 1-11.

[4] Compare: Camazine, Scott (2003). *Self-organization in Biological Systems*. Princeton studies in complexity (reprint ed.). Princeton University Press. ISBN 9780691116242. Retrieved 2016-04-05.

[5] Ilachinski, Andrew (2001). *Cellular Automata: A Discrete Universe*. World Scientific. p. 247. ISBN 9789812381835. Retrieved 2016-04-05. We have already seen ample evidence for what is arguably the single most impressive general property of CA, namely their capacity for self-organization.

[6] Bernard Feltz et al (2006). *Self-organization and Emergence in Life Sciences*. ISBN 9781402039164. p. 1.

[7] Bonabeau, Eric; Dorigo, Marco and Theraulaz, Guy (1999). *Swarm intelligence: from natural to artificial systems*. ISBN 0195131592. pp. 9–11.

[8] Ashby, W. R. (1947). "Principles of the Self-Organizing Dynamic System". *The Journal of General Psychology* 37 (2): 125–8. doi:10.1080/00221309.1947.9918144. PMID 20270223.

[9] Ashby, W. R. (1962). "Principles of the self-organizing system", pp. 255–278 in *Principles of Self-Organization*. Heinz von Foerster and George W. Zopf, Jr. (eds.) U.S. Office of Naval Research.

[10] Von Foerster, H. (1960). [Retrieved from http://e1020.pbworks.com/f/fulltext.pdf "On self-organizing systems and their environments"], pp. 31–50 in *Self-organizing systems*. M.C. Yovits and S. Cameron (eds.), Pergamon Press, London

[11] Nicolis, G. and Prigogine, I. (1977). *Self-organization in nonequilibrium systems: From dissipative structures to order through fluctuations*. Wiley, New York.

[12] Prigogine, I. and Stengers, I. (1984). *Order out of chaos: Man's new dialogue with nature*. Bantam Books.

[13] Asaro, P. (2007). "Heinz von Foerster and the Bio-Computing Movements of the 1960s" in Albert Müller and Karl H. Müller (eds.) *An Unfinished Revolution? Heinz von Foerster and the Biological Computer Laboratory BCL 1958–1976*. Vienna, Austria: Edition Echoraum.

[14] As an indication of the increasing importance of this concept, when queried with the keyword self-organ*, *Dissertation Abstracts* finds nothing before 1954, and only four entries before 1970. There were 17 in the years 1971–1980; 126 in 1981–1990; and 593 in 1991–2000.

[15] Biel, R.; Mu-Jeong Kho (November 2009). "The Issue of Energy within a Dialectical Approach to the Regulationist Problematique" (PDF). *Recherches & Régulation Working Papers, RR Série ID 2009-1*, Association Recherche & Régulation (http://theorie-regulation.org): 1–21. Retrieved 2013-11-09. External link in |publisher= (help)

[16] Bejan, A.; Lorente, S. (2006). "Constructal theory of generation of configuration in nature and engineering". *Journal of Applied Physics* 100 (4): 041301. Bibcode:2006JAP...100d1301B. doi:10.1063/1.2221896.

[17] Bejan, Adrian; Zane, Peder (January 24, 2012). *Design in Nature: How the Constructal Law Governs Evolution in Biology, Physics, Technology, and Social Organization*. New York: Doubleday. p. 11. ISBN 978-0-307-744340.

[18] Bejan, Adrian (May 24, 2016). *The Physics of Life: The Evolution of Everything*. St. Martin's Press. p. 272. ISBN 1250078822.

[19] Henshaw, King; Zarnikau (2011). "System Energy Assessment (SEA), Defining a Standard Measure of EROI for Energy Businesses as Whole Systems". *Sustainability* 3 (10): 1908–1943. doi:10.3390/su3101908.

[20] Henshaw, P. F. (2010). "Models Learning Change". *Cosmos and History* 6 (1).

[21] Georgiev, Georgi Yordanov (2012) "A quantitative measure, mechanism and attractor for self-organization in networked complex systems", pp. 90–95 in *Lecture Notes in Computer Science* (LNCS 7166), F. A. Kuipers and P. E. Heegaard (Eds.): IFIP International Federation for Information Processing, Proceedings of the Sixth International Workshop on Self-Organizing Systems (IWSOS 2012), Springer-Verlag (2012).

[22] Georgiev, Georgi Yordanov; Georgiev, Iskren Yordanov (2002). "The least action and the metric of an organized system". *Open Systems and Information Dynamics* 9 (4): 371–380. arXiv:1004.3518. Bibcode:2010arXiv1004.3518G. doi:10.1023/a:1021858318296.

[23] Ansari M. H. (2004) Self-organized theory in quantum gravity. arxiv.org

[24] Zeiger, H. J. and Kelley, P. L. (1991) "Lasers", pp. 614–619 in *The Encyclopedia of Physics*, Second Edition, edited by Lerner, R. and Trigg, G., VCH Publishers.

[25] Strong, M. (2004). "Protein Nanomachines". *PLoS Biol.* 2 (3): e73–e74. doi:10.1371/journal.pbio.0020073. PMC 368168. PMID 15024422.

[26] Camazine, Deneubourg, Franks, Sneyd, Theraulaz, Bonabeau, *Self-Organization in Biological Systems*, Princeton University Press, 2003. ISBN 0-691-11624-5 --ISBN 0-691-01211-3 (pbk.) p. 8

[27] Dennett, Daniel (1995), *Darwin's Dangerous Idea*, Penguin Books, London, ISBN 978-0-14-016734-4

[28] Yang, X. S.; Deb, S.; Loomes, M.; Karamanoglu, M. (2013). "A framework for self-tuning optimization algorithm". *Neural Computing and Applications* 23 (7–8): 2051. doi:10.1007/s00521-013-1498-4.

[29] X. S. Yang (2014) *Nature-Inspired Optimization Algorithms*, Elsevier.

[30] Wiener, Norbert (1962) "The mathematics of self-organising systems". *Recent developments in information and decision processes*, Macmillan, N. Y. and Chapter X in *Cybernetics, or control and communication in the animal and the machine*, The MIT Press.

[31] *Cybernetics, or control and communication in the animal and the machine*, The MIT Press, Cambridge, Massachusetts and Wiley, NY, 1948. 2nd Edition 1962 "Chapter X "Brain Waves and Self-Organizing Systems"pp 201–202.

[32] Ashby, William Ross (1952) *Design for a Brain*, Chapter 5 Chapman & Hall

[33] Ashby, William Ross (1956) *An Introduction to Cybernetics*, Part Two Chapman & Hall

[34] Conant, R. C.; Ashby, W. R. (1970). "Every good regulator of a system must be a model of that system" (PDF). *Int. J. Systems Sci.* **1** (2): 89–97. doi:10.1080/00207727008920220.

[35] *Embodiments of Mind* MIT Press (1965)"

[36] von Foerster, Heinz; Pask, Gordon (1961). "A Predictive Model for Self-Organizing Systems, Part I". *Cybernetica* **3**: 258–300.

[37] von Foerster, Heinz; Pask, Gordon (1961). "A Predictive Model for Self-Organizing Systems, Part II". *Cybernetica* **4**: 20–55.

[38] "Brain of the Firm" Alan Lane (1972) see also Viable System Model also in "Beyond Dispute " Wiley Stafford Beer 1994 "Redundancy of Potential Command" pp. 157–158.

[39] Pask, Gordon (1996). "Heinz von Foerster's Self-Organisation, the Progenitor of Conversation and Interaction Theories" (PDF). *Systems Research* **13** (3): 349–362. doi:10.1002/(sici)1099-1735(199609)13:3<349::aid-sres103>3.3.co;2-7.

[40] Pask, G. (1973). *Conversation, Cognition and Learning. A Cybernetic Theory and Methodology*. Elsevier

[41] Green, N. (2001). "On Gordon Pask". *Kybernetes* **30** (5/6): 673. doi:10.1108/03684920110391913.

[42] Pask, Gordon (1993) *Interactions of Actors (IA), Theory and Some Applications*.

[43] *Interactive models for self organization and biological systems* Center for Models of Life, Niels Bohr Institute, Denmark

[44] Luhmann, Niklas (1995) *Social Systems*. Stanford, California: Stanford University Press. ISBN 0804726256. p. 410.

[45] Krugman, P. (1995) *The Self Organizing Economy*. Blackwell Publishers. ISBN 1557866996

[46] Hayek, F. (1976) *Law, Legislation and Liberty, Volume 2: The Mirage of Social Justice*. University of Chicago Press.

[47] Marshall, A. (2002) *The Unity of Nature*, Chapter 5. Imperial College Press. ISBN 1860943306.

[48] Rogers.C. (1969). *Freedom to Learn*. Merrill

[49] Feynman, R. P. (1987) *Elementary Particles and the Laws of Physics*. The Dyrac 1997 Memorial Lecture. Cambridge University Press. ISBN 9780521658621.

[50] Illich. I. (1971) *A Celebration of Awareness*. Penguin Books.

[51] Harri-Augstein E. S. (2000) *The University of Learning in transformation*

[52] Schumacher, E. F. (1997) *This I Believe and Other Essays (Resurgence Book)*. ISBN 1870098668.

[53] Revans R. W. (1982) *The Origins and Growth of Action Learning* Chartwell-Bratt, Bromley

[54] Thomas L.F. and Harri-Augstein S. (1993) "On Becoming a Learning Organisation" in *Report of a 7 year Action Research Project with the Royal Mail Business*. CSHL Monograph

[55] Rogers C.R. (1971) *On Becoming a Person*. Constable, London

[56] Prigogyne I. & Sengers I. (1985) *Order out of Chaos* Flamingo Paperbacks. London

[57] Capra F (1989) *Uncommon Wisdom* Flamingo Paperbacks. London

[58] Bohm D. (1994) *Thought as a System*. Routledge.

[59] Harri-Augstein E. S. and Thomas L. F. (1991)*Learning Conversations: The SOL way to personal and organizational growth*. Routledge

[60] Maslow, A. H. (1964). *Religions, values, and peak-experiences*, Columbus: Ohio State University Press.

[61] *Conversational Science* Thomas L.F. and Harri-Augstein E.S. (1985)

[62] Sole, M-J. (1992). "Phonetic and phonological processes: nasalization". *Language & Speech* **35**: 29–43.

[63] Ladefoged, Peter (2003) "Commentary: some thoughts on syllables – an old-fashioned interlude", pp. 269–276 in *Papers in laboratory Phonology VI*. Local, John, Richard Ogden & Ros Temple (eds.). Cambridge University Press.

[64] see papers in *Phonetica* **49**, 1992, special issue on Articulatory Phonology

[65] Ohala, John J. (1996). "Speech perception is hearing sounds, not tongues". *Journal of the Acoustical Society of America* **99** (3): 1718–1725. Bibcode:1996ASAJ...99.1718O. doi:10.1121/1.414696. PMID 8819861.

[66] Lindblom, B. (1999). *Emergent phonology* (PDF). Proceedings of the Twenty-fifth Annual Meeting of the Berkeley Linguistics Society, University of California, Berkeley.

[67] Pagels, H. R. (January 1, 1985). "Is the irreversibility we see a fundamental property of nature?" (PDF). *Physics Today*: 97–99.

[68] Article 3. Whether God exists? newadvent.org

1.7 Further reading

- W. Ross Ashby (1966), *Design for a Brain*, Chapman & Hall, 2nd edition.

- Amoroso, Richard (2005) *The Fundamental Limit and Origin of Complexity in Biological Systems*.

- Per Bak (1996), *How Nature Works: The Science of Self-Organized Criticality*, Copernicus Books.

- Philip Ball (1999), *The Self-Made Tapestry: Pattern Formation in Nature*, Oxford University Press.

- Stafford Beer, Self-organization as autonomy: *Brain of the Firm* 2nd edition Wiley 1981 and *Beyond Dispute* Wiley 1994.

- A. Bejan (2000), *Shape and Structure, from Engineering to Nature*, Cambridge University Press, Cambridge, UK, 324 pp.

- Mark Buchanan (2002), *Nexus: Small Worlds and the Groundbreaking Theory of Networks* W. W. Norton & Company.

- Scott Camazine, Jean-Louis Deneubourg, Nigel R. Franks, James Sneyd, Guy Theraulaz, & Eric Bonabeau (2001) *Self-Organization in Biological Systems*, Princeton Univ Press.

- Falko Dressler (2007), *Self-Organization in Sensor and Actor Networks*, Wiley & Sons.

- Manfred Eigen and Peter Schuster (1979), *The Hypercycle: A principle of natural self-organization*, Springer.

- Myrna Estep (2003), *A Theory of Immediate Awareness: Self-Organization and Adaptation in Natural Intelligence*, Kluwer Academic Publishers.

- Myrna L. Estep (2006), *Self-Organizing Natural Intelligence: Issues of Knowing, Meaning, and Complexity*, Springer-Verlag.

- J. Doyne Farmer et al. (editors) (1986), "Evolution, Games, and Learning: Models for Adaptation in Machines and Nature", in: *Physica D*, Vol 22.

- Carlos Gershenson and Francis Heylighen (2003). "When Can we Call a System Self-organizing?" In Banzhaf, W, T. Christaller, P. Dittrich, J. T. Kim, and J. Ziegler, Advances in Artificial Life, 7th European Conference, ECAL 2003, Dortmund, Germany, pp. 606–614. LNAI 2801. Springer.

- Hermann Haken (1983) *Synergetics: An Introduction. Nonequilibrium Phase Transition and Self-Organization in Physics, Chemistry, and Biology*, Third Revised and Enlarged Edition, Springer-Verlag.

- F.A. Hayek *Law, Legislation and Liberty*, RKP, UK.

- Francis Heylighen (2001): "The Science of Self-organization and Adaptivity".

- Henrik Jeldtoft Jensen (1998), *Self-Organized Criticality: Emergent Complex Behaviour in Physical and Biological Systems*, Cambridge Lecture Notes in Physics 10, Cambridge University Press.

- Steven Berlin Johnson (2001), *Emergence: The Connected Lives of Ants, Brains, Cities, and Software*.

- Stuart Kauffman (1995), *At Home in the Universe*, Oxford University Press.

- Stuart Kauffman (1993), *Origins of Order: Self-Organization and Selection in Evolution* Oxford University Press.

- J. A. Scott Kelso (1995), *Dynamic Patterns: The self-organization of brain and behavior*, The MIT Press, Cambridge, Massachusetts.

- J. A. Scott Kelso & David A Engstrom (2006), "*The Complementary Nature*", The MIT Press, Cambridge, Massachusetts.

- Alex Kentsis (2004), *Self-organization of biological systems: Protein folding and supramolecular assembly*, Ph.D. Thesis, New York University.

- E.V.Krishnamurthy(2009)", Multiset of Agents in a Network for Simulation of Complex Systems", in "Recent advances in Nonlinear Dynamics and synchronization, ,(NDS-1) -Theory and applications, Springer Verlag, New York,2009. Eds. K.Kyamakya et al.

- Paul Krugman (1996), *The Self-Organizing Economy*, Cambridge, Massachusetts, and Oxford: Blackwell Publishers.

- Elizabeth McMillan (2004) "Complexity, Organizations and Change".

- Marshall, A (2002) The Unity of Nature, Imperial College Press: London (esp. chapter 5)

- Müller, J.-A., Lemke, F. (2000), *Self-Organizing Data Mining*.

- Gregoire Nicolis and Ilya Prigogine (1977) *Self-Organization in Non-Equilibrium Systems*, Wiley.

- Heinz Pagels (1988), *The Dreams of Reason: The Computer and the Rise of the Sciences of Complexity*, Simon & Schuster.

- Gordon Pask (1961), *The cybernetics of evolutionary processes and of self organizing systems*, 3rd. International Congress on Cybernetics, Namur, Association Internationale de Cybernetique.

- Christian Prehofer ea. (2005), "Self-Organization in Communication Networks: Principles and Design Paradigms", in: *IEEE Communications Magazine*, July 2005.

- Mitchell Resnick (1994), *Turtles, Termites and Traffic Jams: Explorations in Massively Parallel Microworlds*, Complex Adaptive Systems series, MIT Press.

- Lee Smolin (1997), *The Life of the Cosmos* Oxford University Press.

- Ricard V. Solé and Brian C. Goodwin (2001), *Signs of Life: How Complexity Pervades Biology*, Basic Books.

- Ricard V. Solé and Jordi Bascompte (2006), *Selforganization in Complex Ecosystems*, Princeton U. Press

- Steven Strogatz (2004), *Sync: The Emerging Science of Spontaneous Order*, Theia.

- D'Arcy Thompson (1917), *On Growth and Form*, Cambridge University Press, 1992 Dover Publications edition.

- Tom De Wolf, Tom Holvoet (2005), *Emergence Versus Self-Organisation: Different Concepts but Promising When Combined*, In Engineering Self Organising Systems: Methodologies and Applications, Lecture Notes in Computer Science, volume 3464, pp 1–15.

- K. Yee (2003), "Ownership and Trade from Evolutionary Games", International Review of Law and Economics, 23.2, 183–197.

- Louise B. Young (2002), *The Unfinished Universe*

- Mikhail Prokopenko (ed.) (2008), *Advances in Applied Self-organizing Systems*, Springer.

- Alfred Hübler (2009), "Digital wires," Complexity, 14.5,7–9,

- Rüdiger H. Jung (2010), *Self-organization* In: Helmut K. Anheier, Stefan Toepler, Regina List (editors): *International Encyclopedia of Civil Society*. Springer Science + Business Media LLC, New York 2010, ISBN 978-0-387-93996-4, p. 1364–1370.

1.8 External links

- Self-organization at Scholarpedia, curated by Hermann Haken.

- Max Planck Institute for Dynamics and Self-Organization, Göttingen

- PDF file on self-organized common law with references

- An entry on self-organization at the *Principia Cybernetica* site

- The Science of Self-organization and Adaptivity, a review paper by Francis Heylighen

- The *Self-Organizing Systems (SOS) FAQ* by Chris Lucas, from the USENET newsgroup comp.theory.self-org.sys

- David Griffeath, *Primordial Soup Kitchen* (graphics, papers)

- nlin.AO, nonlinear preprint archive, (electronic preprints in adaptation and self-organizing systems)

- Structure and Dynamics of Organic Nanostructures

- Metal organic coordination networks of oligopyridines and Cu on graphite

- *Selforganization in complex networks* The Complex Systems Lab, Barcelona

- Computational Mechanics Group at the Santa Fe Institute

- "Organisation must grow" (1939) W. Ross Ashby journal page 759, from The W. Ross Ashby Digital Archive

- Cosma Shalizi's notebook on self-organization from 2003-06-20, used under the GFDL with permission from author.

- Connectivism:SelfOrganization

- UCLA Human Complex Systems Program

- "Interactions of Actors (IA), Theory and Some Applications" 1993 Gordon Pask's theory of learning, evolution and self-organization (in draft).

1.8. EXTERNAL LINKS

- The Cybernetics Society
- Scott Camazine's webpage on self-organization in biological systems
- Mikhail Prokopenko's page on Information-driven Self-organisation (IDSO)
- Lakeside Labs Self-Organizing Networked Systems A platform for science and technology, Klagenfurt, Austria.
- Watch 32 discordant metronomes synch up all by themselves theatlantic.com

1.8.1 Dissertations and theses on self-organization

- Gershenson, Carlos. (2007). "Design and control of Self-organizing Systems" (PhD thesis).
- de Boer, Bart. (1999). Self-Organisation in Vowel Systems Vrije Universiteit Brussel AI-lab (PhD thesis).

Chapter 2

Statistical fluctuations

Statistical fluctuations are fluctuations in quantities derived from many identical random processes. They are fundamental and unavoidable. It can be proved that the relative fluctuations reduce as the square root of the number of identical processes.

Statistical fluctuations are responsible for many results of statistical mechanics and thermodynamics, including phenomena such as shot noise in electronics.

- Quantum fluctuation
- Thermal fluctuations
- Universal conductance fluctuations

2.1 Description

When a number of random processes occur, it can be shown that the outcomes fluctuate (vary in time) and that the fluctuations are inversely proportional to the square root of the number of processes.

2.2 Examples

As an example that will be familiar to all, if a fair coin is tossed many times and the number of heads and tails counted, the ratio of heads to tails will be very close to 1 (about as many heads as tails); but after only a few throws, outcomes with a significant excess of heads over tails or vice versa are common; if an experiment with a few throws is repeated over and over, the outcomes will fluctuate a lot.

An electric current so small that not many electrons are involved flowing through a p-n junction is susceptible to statistical fluctuations as the actual number of electrons per unit time (the current) will fluctuate; this produces detectable and unavoidable electrical noise known as shot noise.

2.3 See also

- Primordial fluctuations

Chapter 3

Positive feedback

Alarm or panic can spread by positive feedback among a herd of animals to cause a stampede.

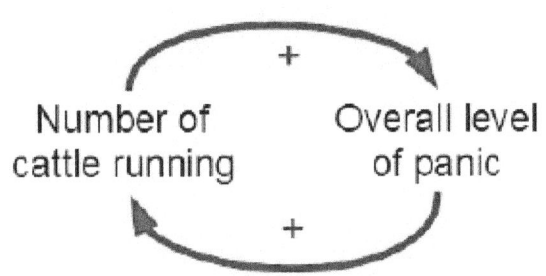

Causal loop diagram that depicts the causes of a stampede as a positive feedback loop.

In sociology a network effect can quickly create the positive feedback of a bank run. The above photo is of the UK Northern Rock 2007 bank run. See also viral video.

Positive feedback is a process that occurs in a feedback loop in which the effects of a small disturbance on a system include an increase in the magnitude of the perturbation.[1] That is, *A produces more of B which in turn produces more of A*.[2] In contrast, a system in which the results of a change act to reduce or counteract it has negative feedback.[1][3] Both concepts play an important role in science and engineering, including biology, chemistry, and cybernetics.

Mathematically, positive feedback is defined as a positive loop gain around a closed loop of cause and effect.[1][3] That is, positive feedback is in phase with the input, in the sense that it adds to make the input larger.[4][5] Positive feedback tends to cause system instability. When the loop gain is positive and above 1, there will typically be exponential growth, increasing oscillations, chaotic behavior or other divergences from equilibrium.[3] System parameters will typically accelerate towards extreme values, which may damage or destroy the system, or may end with the system latched into a new stable state. Positive feedback may be controlled by signals in the system being filtered, damped, or limited, or it can be cancelled or reduced by adding negative feedback.

Positive feedback is used in digital electronics to force voltages away from intermediate voltages into '0' and '1' states. On the other hand, thermal runaway is a positive feedback that can destroy semiconductor junctions. Positive feedback in chemical reactions can increase the rate of reactions, and in some cases can lead to explosions. Positive feedback in mechanical design causes tipping-point, or 'over-centre', mechanisms to snap into position, for example in switches and locking pliers. Out of control, it can cause bridges to collapse. Positive feedback in economic systems can cause boom-then-bust cycles. A familiar example of positive feedback is the loud squealing or howling sound produced by audio feedback in public address systems: the microphone picks up sound from its own loudspeakers, amplifies it, and sends it through the speakers again.

3.1 Overview

Positive feedback enhances or amplifies an effect by it having an influence on the process which gave rise to it. For example, when part of an electronic output signal returns to the input, and is in phase with it, the system gain is increased.[6] The feedback from the outcome to the originating process can be direct, or it can be via other state variables.[3] Such systems can give rich qualitative behaviors, but whether the feedback is instantaneously positive or negative in sign has an extremely important influence on the results.[3] Positive feedback reinforces and negative feedback moderates the original process. *Positive* and *negative* in this sense refer to loop gains greater than or less than zero, and do not imply any value judgements as to the desirability of the outcomes or effects.[7] A key feature of positive feedback is thus that small disturbances get bigger. When a change occurs in a system, positive feedback causes further change, in the same direction.

3.1.1 Basic

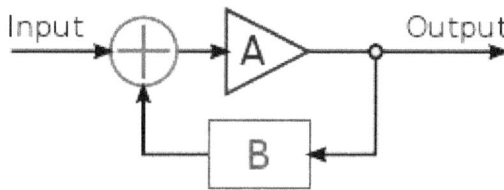

A basic feedback system can be represented by this block diagram. In the diagram the + symbol is an adder and A and B are arbitrary causal functions.

A simple feedback loop is shown in the diagram. If the loop gain AB is positive, then a condition of *positive* or *regenerative* feedback exists.

If the functions A and B are linear and AB is smaller than unity, then the overall system gain from the input to output is finite, but can be very large as AB approaches unity.[8] In that case, it can be shown that the overall or "closed loop" gain from input to output is:

$G_c = A/(1 - AB)$

When AB > 1, the system is unstable, so does not have a well-defined gain; the gain may be called infinite.

Thus depending on the feedback, state changes can be convergent, or divergent. The result of positive feedback is to augment changes, so that small perturbations may result in big changes.

A system in equilibrium in which there is positive feedback to any change from its current state may be unstable, in which case the equilibrium is said to be in an unstable equilibrium. The magnitude of the forces that act to move such a system away from its equilibrium are an increasing function of the "distance" of the state from the equilibrium.

Positive feedback does not necessarily imply instability of an equilibrium, for example stable *on* and *off* states may exist in positive-feedback architectures.[9]

3.1.2 Hysteresis

Main article: Hysteresis

In the real world, positive feedback loops typically do not

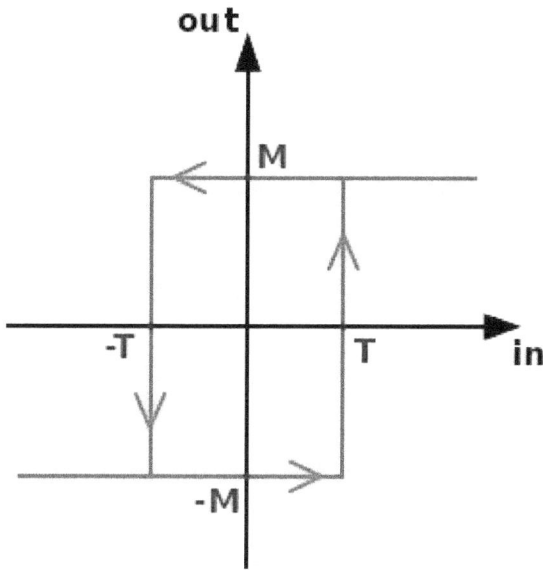

Hysteresis causes the output value to depend on the history of the input

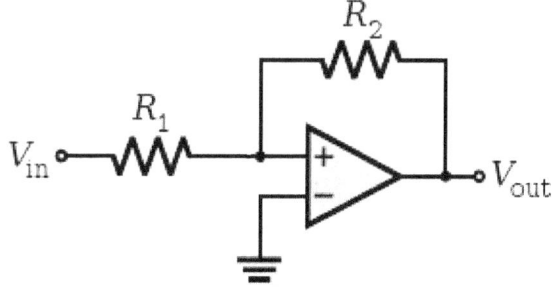

In a Schmitt trigger circuit, feedback to the non-inverting input of an amplifier pushes the output directly away from the applied voltage towards the maximum or minimum voltage the amplifier can generate.

cause ever-increasing growth, but are modified by limiting

effects of some sort. According to Donella Meadows:

> "Positive feedback loops are sources of growth, explosion, erosion, and collapse in systems. A system with an unchecked positive loop ultimately will destroy itself. That's why there are so few of them. Usually a negative loop will kick in sooner or later."[10]

Hysteresis, in which the starting point affects where the system ends up, can be generated by positive feedback. When the gain of the feedback loop is above 1, then the output moves away from the input: if it is above the input, then it moves towards the nearest positive limit, while if it is below the input then it moves towards the nearest negative limit.

Once it reaches the limit, it will be stable. However, if the input goes past the limit, then the feedback will change sign and the output will move in the opposite direction until it hits the opposite limit. The system therefore shows bistable behaviour.

3.2 Terminology

The terms *positive* and *negative* were first applied to feedback before World War II. The idea of positive feedback was already current in the 1920s with the introduction of the regenerative circuit.[11]

Friis & Jensen (1924) described regeneration in a set of electronic amplifiers as a case where *the "feed-back" action is positive* in contrast to negative feed-back action, which they mention only in passing.[12] Harold Stephen Black's classic 1934 paper first details the use of negative feedback in electronic amplifiers. According to Black:

> "Positive feed-back increases the gain of the amplifier, negative feed-back reduces it."[13]

According to Mindell (2002) confusion in the terms arose shortly after this:

> "...Friis and Jensen had made the same distinction Black used between 'positive feed-back' and 'negative feed-back', based not on the sign of the feedback itself but rather on its effect on the amplifier's gain. In contrast, Nyquist and Bode, when they built on Black's work, referred to negative feedback as that with the sign reversed. Black had trouble convincing others of the utility of his invention in part because confusion existed over basic matters of definition."[11](p121)

3.3 Examples and applications

3.3.1 In electronics

A vintage style regenerative radio receiver. Due to the controlled use of positive feedback, sufficient amplification can be derived from a single vacuum tube or valve (centre).

Regenerative circuits were invented and patented in 1914[14] for the amplification and reception of very weak radio signals. Carefully controlled positive feedback around a single transistor amplifier can multiply its gain by 1,000 or more.[15] Therefore, a signal can be amplified 20,000 or even 100,000 times in one stage, that would normally have a gain of only 20 to 50. The problem with regenerative amplifiers working at these very high gains is that they easily become unstable and start to oscillate. The radio operator has to be prepared to tweak the amount of feedback fairly continuously for good reception. Modern radio receivers use the superheterodyne design, with many more amplification stages, but much more stable operation and no positive feedback.

The oscillation that can break out in a regenerative radio circuit is used in electronic oscillators. By the use of tuned circuits or a piezoelectric crystal (commonly quartz), the signal that is amplified by the positive feedback remains linear and sinusoidal. There are several designs for such harmonic oscillators, including the Armstrong oscillator, Hartley oscillator, Colpitts oscillator, and the Wien bridge oscillator. They all use positive feedback to create oscillations.[16]

Many electronic circuits, especially amplifiers, incorporate negative feedback. This reduces their gain, but improves their linearity, input impedance, output impedance, and

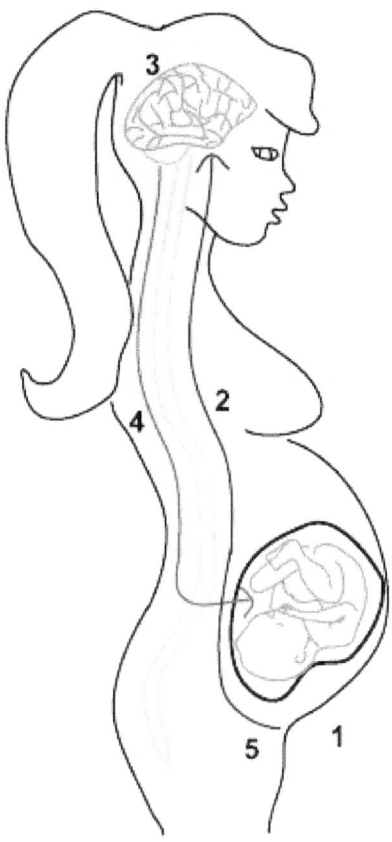

Positive feedback is the amplification of a body's response to a stimulus. For example, in childbirth, when the head of the fetus pushes up against the cervix (1) it stimulates a nerve impulse from the cervix to the brain (2). When the brain is notified, it signals the pituitary gland to release a hormone called Oxytocin (3). Oxytocin is then carried via the bloodstream to the uterus (4) causing contractions, pushing the fetus towards the cervix eventually inducing childbirth.

bandwidth, and stabilises all of these parameters, including the closed-loop gain. These parameters also become less dependent on the details of the amplifying device itself, and more dependent on the feedback components, which are less likely to vary with manufacturing tolerance, age and temperature. The difference between positive and negative feedback for AC signals is one of phase: if the signal is fed back out of phase, the feedback is negative and if it is in phase the feedback is positive. One problem for amplifier designers who use negative feedback is that some of the components of the circuit will introduce phase shift in the feedback path. If there is a frequency (usually a high frequency) where the phase shift reaches 180°, then the designer must ensure that the amplifier gain at that frequency is very low (usually by low-pass filtering). If the loop gain (the product of the amplifier gain and the extent of the pos-

itive feedback) at any frequency is greater than one, then the amplifier will oscillate at that frequency (Barkhausen stability criterion). Such oscillations are sometimes called parasitic oscillations. An amplifier that is stable in one set of conditions can break into parasitic oscillation in another. This may be due to changes in temperature, supply voltage, adjustment of front-panel controls, or even the proximity of a person or other conductive item. Amplifiers may oscillate gently in ways that are hard to detect without an oscilloscope, or the oscillations may be so extensive that only a very distorted or no required signal at all gets through, or that damage occurs. Low frequency parasitic oscillations have been called 'motorboating' due to the similarity to the sound of a low-revving exhaust note.[17]

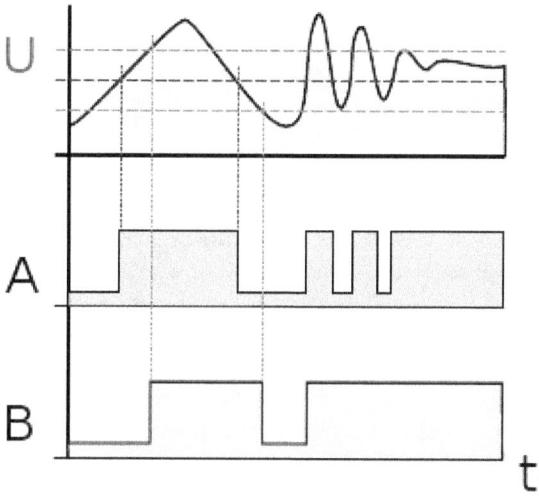

The effect of using a Schmitt trigger (B) instead of a comparator (A)

Digital electronic circuits are sometimes designed to benefit from positive feedback. Normal logic gates usually rely simply on gain to push digital signal voltages away from intermediate values to the values that are meant to represent boolean '0' and '1'. When an input voltage is expected to vary in an analogue way, but sharp thresholds are required for later digital processing, the Schmitt trigger circuit uses positive feedback to ensure that if the input voltage creeps gently above the threshold, the output is forced smartly and rapidly from one logic state to the other. One of the corollaries of the Schmitt trigger's use of positive feedback is that, should the input voltage move gently down again past the same threshold, the positive feedback will hold the output in the same state with no change. This effect is called hysteresis: the input voltage has to drop past a different, lower threshold to 'un-latch' the output and reset it to its original digital value. By reducing the extent of the positive feedback, the hysteresis-width can be reduced, but it can not entirely be eradicated. The Schmitt trigger is, to

3.3. EXAMPLES AND APPLICATIONS

some extent, a latching circuit.[18]

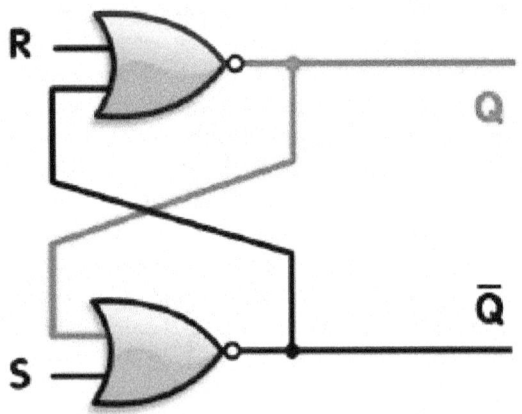

Illustration of an R-S ('reset-set') flip-flop made from two digital nor gates with positive feedback. Red and black mean logical '1' and '0', respectively.

An electronic flip-flop, or "latch", or "bistable multivibrator", is a circuit that due to high positive feedback is not stable in a balanced or intermediate state. Such a bistable circuit is the basis of one bit of electronic memory. The flip-flop uses a pair of amplifiers, transistors, or logic gates connected to each other so that positive feedback maintains the state of the circuit in one of two unbalanced stable states after the input signal has been removed, until a suitable alternative signal is applied to change the state.[19] Computer random access memory (RAM) can be made in this way, with one latching circuit for each bit of memory.[20]

Thermal runaway occurs in electronic systems because some aspect of a circuit is allowed to pass more current when it gets hotter, then the hotter it gets, the more current it passes, which heats it some more and so it passes yet more current. The effects are usually catastrophic for the device in question. If devices have to be used near to their maximum power-handling capacity, and thermal runaway is possible or likely under certain conditions, improvements can usually be achieved by careful design.[21]

Audio and video systems can demonstrate positive feedback. If a microphone picks up the amplified sound output of loudspeakers in the same circuit, then howling and screeching sounds of audio feedback (at up to the maximum power capacity of the amplifier) will be heard, as random noise is re-amplified by positive feedback and filtered by the characteristics of the audio system and the room. Microphones are not the only transducers subject to this effect. Record deck pickup cartridges can do the same, usually in the low frequency range below about 100 Hz, manifesting as a low rumble. Jimi Hendrix helped to develop the controlled and musical use of audio feedback in electric guitar

A phonograph turntable is prone to acoustic feedback

playing,[22] and later Brian May was a famous proponent of the technique.[23]

Video feedback.

Similarly, if a video camera is pointed at a monitor screen that is displaying the camera's own signal, then repeating patterns can be formed on the screen by positive feedback. This video feedback effect was used in the opening sequences to the first ten series of the television program *Doctor Who*.

3.3.2 Switches

In electrical switches, including bimetallic strip based thermostats, the switch usually has hysteresis in the switching action. In these cases hysteresis is mechanically achieved via positive feedback within a tipping point mechanism. The positive feedback action minimises the length of time arcing occurs for during the switching and also holds the contacts in an open or closed state.[24]

3.3.3 In biology

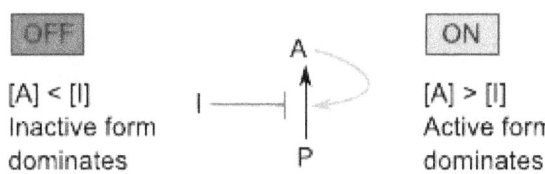

Positive feedback is a mechanism by which an output is enhanced, such as protein levels. However, in order to avoid any fluctuation in the protein level, the mechanism is inhibited stochastically (I), therefore when the concentration of the activated protein (A) is past the threshold ([I]), the loop mechanism is activated and the concentration of A increases exponentially if d[A]=k [A]

In physiology

A number of examples of positive feedback systems may be found in physiology.

- One example is the onset of contractions in childbirth, known as the Ferguson reflex. When a contraction occurs, the hormone oxytocin causes a nerve stimulus, which stimulates the hypothalamus to produce more oxytocin, which increases uterine contractions. This results in contractions increasing in amplitude and frequency.[25](pp924–925)

- Another example is the process of blood clotting. The loop is initiated when injured tissue releases signal chemicals that activate platelets in the blood. An activated platelet releases chemicals to activate more platelets, causing a rapid cascade and the formation of a blood clot.[25](pp392–394)

- Lactation also involves positive feedback in that as the baby suckles on the nipple there is a nerve response into the spinal cord and up into the hypothalamus of the brain, which then stimulates the pituitary gland to produce more prolactin to produce more milk.[25](p926)

- A spike in estrogen during the follicular phase of the menstrual cycle causes ovulation.[25](p907)

- The generation of nerve signals is another example, in which the membrane of a nerve fibre causes slight leakage of sodium ions through sodium channels, resulting in a change in the membrane potential, which in turn causes more opening of channels, and so on. So a slight initial leakage results in an explosion of sodium leakage which creates the nerve action potential.[25](p59)

- In excitation–contraction coupling of the heart, an increase in intracellular calcium ions to the cardiac myocyte is detected by ryanodine receptors in the membrane of the sarcoplasmic reticulum which transport calcium out into the cytosol in a positive feedback physiological response.

In most cases, such feedback loops culminate in countersignals being released that suppress or breaks the loop. Childbirth contractions stop when the baby is out of the mother's body. Chemicals break down the blood clot. Lactation stops when the baby no longer nurses.[25]

In gene regulation

Positive feedback is a well studied phenomenon in gene regulation, where it is most often associated with bistability. Positive feedback occurs when a gene activates itself directly or indirectly via a double negative feedback loop. Genetic engineers have constructed and tested simple positive feedback networks in bacteria to demonstrate the concept of bistability.[26] A classic example of positive feedback is the lac operon in *E. coli*. Positive feedback plays an integral role in cellular differentiation, development, and cancer progression, and therefore, positive feedback in gene regulation can have significant physiological consequences. Random motions in molecular dynamics coupled with positive feedback can trigger interesting effects, such as create population of phenotypically different cells from the same parent cell.[27] This happens because noise can become amplified by positive feedback. Positive feedback can also occur in other forms of cell signaling, such as enzyme kinetics or metabolic pathways.[28]

In evolutionary biology

Positive feedback loops have been used to describe aspects of the dynamics of change in biological evolution. For example, beginning at the macro level, Alfred J. Lotka (1945) argued that the evolution of the species was most essentially a matter of selection that fed back energy flows to capture more and more energy for use by living systems.[29] At the human level, Richard Alexander (1989) proposed that social competition between and within human groups fed back to the selection of intelligence thus constantly producing more and more refined human intelligence.[30] Crespi (2004) discussed several other examples of positive feedback loops in evolution.[31] The analogy of Evolutionary arms races provide further examples of positive feedback in biological systems.[32]

It has been shown that changes in biodiversity through the Phanerozoic correlate much better with hyperbolic

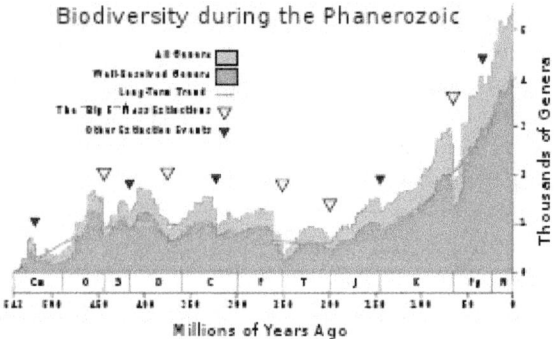

During the Phanerozoic the biodiversity shows a steady but not monotonic increase from near zero to several thousands of genera.

model (widely used in demography and macrosociology) than with exponential and logistic models (traditionally used in population biology and extensively applied to fossil biodiversity as well). The latter models imply that changes in diversity are guided by a first-order positive feedback (more ancestors, more descendants) and/or a negative feedback arising from resource limitation. Hyperbolic model implies a second-order positive feedback. The hyperbolic pattern of the world population growth has been demonstrated (see below) to arise from a second-order positive feedback between the population size and the rate of technological growth. The hyperbolic character of biodiversity growth can be similarly accounted for by a positive feedback between the diversity and community structure complexity. It has been suggested that the similarity between the curves of biodiversity and human population probably comes from the fact that both are derived from the interference of the hyperbolic trend (produced by the positive feedback) with cyclical and stochastic dynamics.[33][34]

Immune system

A cytokine storm, or **hypercytokinemia** is a potentially fatal immune reaction consisting of a positive feedback loop between cytokines and immune cells, with highly elevated levels of various cytokines.[35] In normal immune function, positive feedback loops can be utilized to enhance the action of B lymphocytes. When a B cell binds its antibodies to an antigen and becomes activated, it begins releasing antibodies and secreting a complement protein called C3. Both C3 and a B cell's antibodies can bind to a pathogen, and when a B cell has its antibodies bind to a pathogen with C3, it speeds up that B cell's secretion of more antibodies and more C3, thus creating a positive feedback loop.[36]

Cell death

Apoptosis is a caspase-mediated process of cellular death, whose aim is the removal of long-lived or damaged cells. A failure of this process has been implicated in prominent conditions such as cancer or Parkinson's disease. The very core of the apoptotic process is the auto-activation of caspases, which may be modeled via a positive-feedback loop. This positive feedback exerts an auto-activation of the effector caspase by means of intermediate caspases. When isolated from the rest of apoptotic pathway, this positive-feedback presents only one stable steady state, regardless of the number of intermediate activation steps of the effector caspase.[9] When this core process is complemented with inhibitors and enhancers of caspases effects, this process presents bistability, thereby modeling the alive and dying states of a cell.[37]

3.3.4 In psychology

Winner (1996) described gifted children as driven by positive feedback loops involving setting their own learning course, this feeding back satisfaction, thus further setting their learning goals to higher levels and so on.[38] Winner termed this positive feedback loop as a "rage to master." Vandervert (2009a, 2009b) proposed that the child prodigy can be explained in terms of a positive feedback loop between the output of thinking/performing in working memory, which then is fed to the cerebellum where it is streamlined, and then fed back to working memory thus steadily increasing the quantitative and qualitative output of working memory.[39][40] Vandervert also argued that this working memory/cerebellar positive feedback loop was responsible for language evolution in working memory.

3.3.5 In economics

Market dynamics

According to the theory of reflexivity advanced by George Soros, price changes are driven by a positive feedback process whereby investors' expectations are influenced by price movements so their behaviour acts to reinforce movement in that direction until it becomes unsustainable, whereupon the feedback drives prices in the opposite direction.[41]

Systemic risk

Systemic risk is the risk that an amplification or leverage or positive feedback process presents to a system. This is usually unknown, and under certain conditions this process can amplify exponentially and rapidly lead to destructive or chaotic behavior. A Ponzi scheme is a good example of a positive-feedback system: funds from new investors are used to pay out unusually high returns, which in turn attract more new investors, causing rapid growth toward collapse. W. Brian Arthur has also studied and written on positive feedback in the economy (e.g. W. Brian Arthur, 1990).[42] Hyman Minsky proposed a theory that certain credit expansion practices could make a market economy into "a deviation amplifying system" that could suddenly collapse,[43] sometimes called a "Minsky moment".

Simple systems that clearly separate the inputs from the outputs are not prone to systemic risk. This risk is more likely as the complexity of the system increases, because it becomes more difficult to see or analyze all the possible combinations of variables in the system even under careful stress testing conditions. The more efficient a complex system is, the more likely it is to be prone to systemic risks, because it takes only a small amount of deviation to disrupt the system. Therefore, well-designed complex systems generally have built-in features to avoid this condition, such as a small amount of friction, or resistance, or inertia, or time delay to decouple the outputs from the inputs within the system. These factors amount to an inefficiency, but they are necessary to avoid instabilities.

The 2010 Flash Crash incident was blamed on the practice of high-frequency trading (HFT),[44] although whether HFT really increases systemic risk remains controversial.

Human population growth

Agriculture and human population can be considered to be in a positive feedback mode,[45] which means that one drives the other with increasing intensity. It is suggested that this positive feedback system will end sometime with a catastrophe, as modern agriculture is using up all of the easily available phosphate and is resorting to highly efficient monocultures which are more susceptible to systemic risk.

Technological innovation and human population can be similarly considered, and this has been offered as an explanation for the apparent hyperbolic growth of the human population in the past, instead of a simpler exponential growth.[46] It is proposed that the growth rate is accelerating because of second-order positive feedback between population and technology.[47](p133–160) Technological growth increases the carrying capacity of land for people, which leads to more population, and so more potential inventors in further technological growth.[47](p146)

Prejudice, social institutions and poverty

Gunnar Myrdal described a vicious circle of increasing inequalities, and poverty, which is known as "circular cumulative causation".[48]

3.3.6 In meteorology

Drought intensifies through positive feedback. A lack of rain decreases soil moisture, which kills plants and/or causes them to release less water through transpiration. Both factors limit evapotranspiration, the process by which water vapor is added to the atmosphere from the surface, and add dry dust to the atmosphere, which absorbs water. Less water vapor means both low dew point temperatures and more efficient daytime heating, decreasing the chances of humidity in the atmosphere leading to cloud formation. Lastly, without clouds, there cannot be rain, and the loop is complete.

3.3.7 In climatology

See also: Climate change feedback

Climate "forcings" may push a climate system in the direction of warming or cooling,[49] for example, increased atmospheric concentrations of greenhouse gases cause warming at the surface. Forcings are external to the climate system and feedbacks are internal processes of the system. Some feedback mechanisms act in relative isolation to the rest of the climate system while others are tightly coupled.[50] Forcings, feedbacks and the dynamics of the climate system determine how much and how fast the climate changes. The main positive feedback in global warming is the tendency of warming to increase the amount of water vapor in the atmosphere, which in turn leads to further warming.[51] The main negative feedback comes from the Stefan–Boltzmann law, the amount of heat radiated from the Earth into space is proportional to the fourth power of the temperature of Earth's surface and atmosphere.

Other examples of positive feedback subsystems in climatology include:

- A warmer atmosphere will melt ice and this changes the albedo which further warms the atmosphere.

- Methane hydrates can be unstable so that a warming ocean could release more methane, which is also a greenhouse gas.

The Intergovernmental Panel on Climate Change (IPCC) Fourth Assessment Report states that "Anthropogenic warming could lead to some effects that are abrupt or irreversible, depending upon the rate and magnitude of the climate change."[52]

3.3.8 In sociology

A self-fulfilling prophecy is a social positive feedback loop between beliefs and behavior: if enough people believe that something is true, their behavior can make it true, and observations of their behavior may in turn increase belief. A classic example is a bank run.

Another sociological example of positive feedback is the network effect. When more people are encouraged to join a network this increases the reach of the network therefore the network expands ever more quickly. A viral video is an example of the network effect in which links to a popular video are shared and redistributed, ensuring that more people see the video and then re-publish the links. This is the basis for many social phenomena, including Ponzi schemes and chain letters. In many cases population size is the limiting factor to the feedback effect.

3.3.9 In chemistry

If a chemical reaction causes the release of heat, and the reaction itself happens faster at higher temperatures, then there is a high likelihood of positive feedback. If the heat produced is not removed from the reactants fast enough, thermal runaway can occur and very quickly lead to a chemical explosion.

3.4 See also

- Chain reaction
- Donella Meadows' twelve leverage points to intervene in a system
- Hyperbolic growth
- Reflexivity (social theory)
- Stability criterion
- Strategic complementarity
- System dynamics
- Technological singularity
- Thermal runaway

3.4.1 Similar terminology

- Vicious/virtuous circle: in social and financial systems, a complex of events that reinforces itself through a feedback loop.

- Positive reinforcement: a situation in operant conditioning where a consequence increases the frequency of a behaviour.
- Praise of performance: a term often applied in the context of performance appraisal,[53] although this usage is disputed.[54]
- Self-reinforcing feedback: a term used in systems dynamics to avoid confusion with the "praise" usage.[55]

3.4.2 Analogous concepts

- Matthew effect
- Self-fulfilling prophecy
- Virtuous circle and vicious circle

3.4.3 Examples

- Autocatalysis
- Meander

3.5 References

[1] Ben Zuckerman & David Jefferson (1996). *Human Population and the Environmental Crisis*. Jones & Bartlett Learning. p. 42. ISBN 9780867209662.

[2] Keesing, R.M. (1981). Cultural anthropology: A contemporary perspective (2nd ed.) p.149. Sydney: Holt, Rinehard & Winston, Inc.

[3] Bernard P. Zeigler; Herbert Praehofer; Tag Gon Kim Section (2000). "3.3.2 Feedback in continuous systems". *Theory of Modeling and Simulation: Integrating Discrete Event and Continuous Complex Dynamic Systems*. Academic Press. p. 55. ISBN 9780127784557. A positive feedback loop is one with an even number of negative influences [around the loop].

[4] S W Amos; R W Amos (2002). *Newnes Dictionary of Electronics* (4th ed.). Newnes. p. 247. ISBN 9780750656429.

[5] Rudolf F. Graf (1999). *Modern Dictionary of Electronics* (7th ed.). Newnes. p. 276. ISBN 9780750698665.

[6] "Positive feedback". *Oxford English Dictionary*. Oxford University Press. Retrieved 15 April 2014.

[7] "Feedback". *Glossary*. Metadesigners Network. Retrieved 15 April 2014.

[8] Electronics circuits and devices second edition. Ralph J. Smith

[9] Lopez-Caamal, Fernando; Middleton, Richard H.; Huber, Heinrich (February 2014). "Equilibria and stability of a class of positive feedback loops". *Journal of Mathematical Biology* **68**: 609–645. doi:10.1007/s00285-013-0644-z.

[10] Donella Meadows, *Leverage Points: Places to Intervene in a System*, 1999

[11] Mindell, David A. (2002). *Between Human and Machine : Feedback, Control, and Computing before Cybernetics*. Baltimore, MD: Johns Hopkins University Press.

[12] Friis, H. T.; Jensen, A. G. (April 1924), "High Frequency Amplifiers", *Bell System Technical Journal* **3**: 181–205, doi:10.1002/j.1538-7305.1924.tb01354.x

[13] Black, H. S. (January 1934), "Stabilized feed-back amplifiers", *Electrical Engineering* **53**: 114–120, doi:10.1109/ee.1934.6540374

[14] US 1113149, Armstrong, E. H., "Wireless receiving system"

[15] Kitchin, Charles. "A Short Wave Regenerative Receiver Project". Retrieved 23 September 2010.

[16] "Sinewave oscillators". *EDUCYPEDIA - electronics*. Retrieved 23 September 2010.

[17] Self, Douglas (2009). *Audio Power Amplifier Design Handbook*. Focal Press. pp. 254–255. ISBN 978-0-240-52162-6.

[18] "CMOS Schmitt Trigger—A Uniquely Versatile Design Component" (PDF). *Fairchild Semiconductor Application Note 140*. Fairchild Semiconductors. 1975. Retrieved 29 September 2010.

[19] Strandh, Robert. "Latches and flip-flops". Laboratoire Bordelais de Recherche en Informatique. Retrieved 4 November 2010.

[20] Wayne, Storr. "Sequential Logic Basics: SR Flip-Flop". Electronics-Tutorials.ws. Retrieved 29 September 2010.

[21] Sharma, Bijay Kumar (2009). "Analog Electronics Lecture 4 Part C RC coupled Amplifier Design Procedure". Retrieved 29 September 2010.

[22] Shadwick, Keith (2003). *Jimi Hendrix, Musician*. Backbeat Books. p. 92. ISBN 0-87930-764-1.

[23] May, Brian. "Burns Brian May Tri-Sonic Pickups". House Music & Duck Productions. Retrieved 2 February 2011.

[24] "Positive Feedback and Bistable Systems" (PDF). University of Washington. * Non-Hysteretic Switches, Memoryless Switches: These systems have no memory, that is, once the input signal is removed, the system returns to its original state. * Hysteretic Switches, Bistability: Bistable systems, in contrast, have memory. That is, when switched to one state or another, these systems remain in that state unless forced to change back. The light switch is a common example of a bistable system from everyday life. All bistable systems are based around some form of positive feedback loop.

[25] Guyton, Arthur C. (1991) *Textbook of Medical Physiology*. (8th ed). Philadelphia: W.B. Saunders. ISBN 0-7216-3994-1

[26] Hasty, J.; McMillen, D.; Collins, J. J. (2002). "Engineered gene circuits". *Nature* **420** (6912): 224–230. doi:10.1038/nature01257.

[27] Veening, J.; Smits, W. K.; Kuipers, O. P. "Bistability, Epigenetics, and Bet-Hedging in Bacteria". *Annual Review of Microbiology* **62** (1): 193–210. doi:10.1146/annurev.micro.62.081307.163002.

[28] Bagowski, C. P.; Ferrell, J. E. (2001). "Bistability in the JNK cascade". *Current Biology* **11** (15): 1176–1182. doi:10.1016/S0960-9822(01)00330-X.

[29] Lotka, A (1945). "The law of evolution as a maximal principle". *Human Biology* **17**: 168–194.

[30] Alexander, R. (1989). Evolution of the human psyche. In P. Millar & C. Stringer (Eds.), The human revolution: Behavioral and biological perspectives on the origins of modern humans (pp. 455-513). Princeton: Princeton University Press.

[31] Crespi, B. J. (2004). "Vicious circles: positive feedback in major evolutionary and ecological transitions". *Trends in Ecology and Evolution* **19**: 627–633. doi:10.1016/j.tree.2004.10.001.

[32] Dawkins, R. 1991. *The Blind Watchmaker* London: Penguin. Note: W.W. Norton also published this book, and some citations may refer to that publication. However, the text is identical, so it depends on which book is at hand

[33] Markov A., Korotayev A.Phanerozoic marine biodiversity follows a hyperbolic trend // Palaeoworld. Volume 16, Issue 4, December 2007, Pages 311-318

[34] Markov, A.; Korotayev, A. (2008). "Hyperbolic growth of marine and continental biodiversity through the Phanerozoic and community evolution". *Journal of General Biology* **69** (3): 175–194.

[35] Osterholm, Michael T. (2005-05-05). "Preparing for the Next Pandemic". *The New England Journal of Medicine* **352** (18): 1839–1842. doi:10.1056/NEJMp058068. PMID 15872196.

[36] Paul, William E. (September 1993). "Infectious Diseases and the Immune System". *Scientific American*: 93.

[37] Eissing, Thomas (2014). "Bistability analyses of a caspase activation model for receptor-induced apoptosis". *Journal of Biological Chemistry*: 36892–36897.

[38] Winner, E. (1996). *Gifted children: Myths and Realities*. New York: Basic Books. ISBN 0465017606.

[39] Vandervert, L. (2009a). Working memory, the cognitive functions of the cerebellum and the child prodigy. In L.V. Shavinina (Ed.), International handbook on giftedness (pp. 295-316). The Netherlands: Springer Science.

[40] Vandervert, L. (2009b). "The emergence of the child prodigy 10,000 years ago: An evolutionary and developmental explanation". *Journal of Mind and Behavior* 30 (1–2): 15–32.

[41] Azzopardi, Paul V. (2010), *Behavioural Technical Analysis*, Harriman House Limited, p. 116, ISBN 9780857190680

[42] Arthur, W. Brian (1990). "Positive Feedbacks in the Economy". *Scientific American* **262** (2): 80.

[43] The Financial Instability Hypothesis by Hyman P. Minsky, Working Paper No. 74, May 1992, pp. 6–8

[44] "Findings Regarding the Market Events of May 6, 2010" (PDF). 2010-09-30.

[45] Brown, A. Duncan (2003), *Feed or Feedback: Agriculture, Population Dynamics and the State of the Planet*, Utrecht: International Books, ISBN 978-90-5727-048-2

[46] B.M. Dolgonosov. "On the reasons of hyperbolic growth in the biological and human world systems" Institute of Water Problems, Russian Academy of Sciences, Gubkina 3, Moscow 119991, Russia, March 2010. doi:10.1016/j.ecolmodel.2010.03.028

[47] Korotayev A. Compact Mathematical Models of World System Development, and How they can Help us to Clarify our Understanding of Globalization Processes. *Globalization as Evolutionary Process: Modeling Global Change*. Edited by George Modelski, Tessaleno Devezas, and William R. Thompson. London: Routledge, 2007. P. 133-160.

[48] Berger, Sebastian. "Circular Cumulative Causation (CCC) à la Myrdal and Kapp — Political Institutionalism for Minimizing Social Costs" (PDF). Retrieved 26 November 2011.

[49] US NRC (2012), *Climate Change: Evidence, Impacts, and Choices*, US National Research Council (US NRC), p.9. Also available as PDF

[50] *Understanding Climate Change Feedbacks*, U.S. National Academy of Sciences

[51] http://www.ipcc.ch/publications_and_data/ar4/wg1/en/ch8s8-6-3-1.html

[52] IPCC. "Climate Change 2007: Synthesis Report. Contribution of Working Groups I, II and III to the Fourth Assessment Report of the Intergovernmental Panel on Climate Change. Pg 53" (PDF).

[53] *Positive feedback occurs when one is told he has done something well or correctly.* Tom Coens and Mary Jenkins, "Abolishing Performance Appraisals", p116.

[54] *..."positive feedback" does not mean "praise" and "negative feedback" does not mean "criticism". Positive feedback denotes a self-reinforcing process ... Telling someone your opinion does not constitute feedback unless they act on your suggestions and thus lead you to revise your view.* John D.Sterman, Business Dynamics: Systems Thinking and Modeling for a Complex World McGraw Hill/Irwin, 2000. p14. ISBN 978-0-07-238915-9

[55] Peter M. Senge (1990). *The Fifth Discipline: The Art and Practice of the Learning Organization*. New York: Doubleday. p. 424. ISBN 0-385-26094-6.

3.6 Further reading

- Norbert Wiener (1948), *Cybernetics or Control and Communication in the Animal and the Machine*, Paris, Hermann et Cie - MIT Press, Cambridge, MA.
- Katie Salen and Eric Zimmerman. *Rules of Play*. MIT Press. 2004. ISBN 0-262-24045-9. Chapter 18: Games as Cybernetic Systems.

3.7 External links

- Quotations related to Positive feedback at Wikiquote

Chapter 4

Chaos theory

For other uses, see Chaos Theory (disambiguation).

Chaos theory is the field of study in mathematics that stud-

A plot of the Lorenz attractor for values r = 28, σ = 10, b = 8/3

ies the behavior of dynamical systems that are highly sensitive to initial conditions—a response popularly referred to as the butterfly effect.[1] Small differences in initial conditions (such as those due to rounding errors in numerical computation) yield widely diverging outcomes for such dynamical systems, rendering long-term prediction impossible in general.[2] This happens even though these systems are deterministic, meaning that their future behavior is fully determined by their initial conditions, with no random elements involved.[3] In other words, the deterministic nature of these systems does not make them predictable.[4][5] This behavior is known as **deterministic chaos**, or simply **chaos**. The theory was summarized by Edward Lorenz as:[6]

> Chaos: When the present determines the future, but the approximate present does not approximately determine the future.

Chaotic behavior exists in many natural systems, such as weather and climate.[7][8] It also occurs spontaneously in some systems with artificial components, such as road traffic.[9] This behavior can be studied through analysis of a chaotic mathematical model, or through analytical techniques such as recurrence plots and Poincaré maps. Chaos theory has applications in several disciplines, including meteorology, sociology, physics, computer science, engineering, economics, biology, ecology, and philosophy.

4.1 Introduction

Chaos theory concerns deterministic systems whose behavior can in principle be predicted. Chaotic systems are predictable for a while and then 'appear' to become random. The amount of time for which the behavior of a chaotic

4.2. CHAOTIC DYNAMICS

system can be effectively predicted depends on three things: How much uncertainty we are willing to tolerate in the forecast, how accurately we are able to measure its current state, and a time scale depending on the dynamics of the system, called the Lyapunov time. Some examples of Lyapunov times are: chaotic electrical circuits, about 1 millisecond; weather systems, a few days (unproven); the solar system, 50 million years. In chaotic systems, the uncertainty in a forecast increases exponentially with elapsed time. Hence, mathematically, doubling the forecast time more than squares the proportional uncertainty in the forecast. This means, in practice, a meaningful prediction cannot be made over an interval of more than two or three times the Lyapunov time. When meaningful predictions cannot be made, the system appears to be random.[10]

4.2 Chaotic dynamics

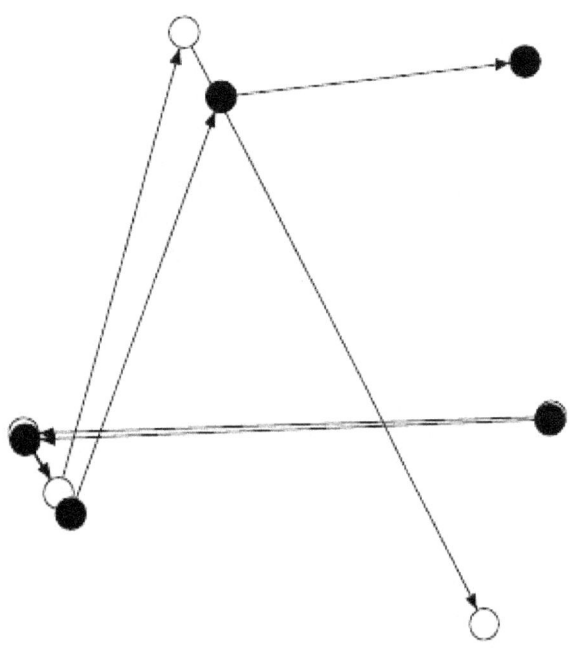

The map defined by $x \to 4x(1-x)$ *and* $y \to (x+y) \bmod 1$ *displays sensitivity to initial x positions. Here, two series of x and y values diverge markedly over time from a tiny initial difference. Note, however, that the y coordinate is defined modulo one at each step, so the square region is actually depicting a cylinder, and the two points are closer than they look*

In common usage, "chaos" means "a state of disorder".[11] However, in chaos theory, the term is defined more precisely. Although no universally accepted mathematical definition of chaos exists, a commonly used definition originally formulated by Robert L. Devaney says that, for a dynamical system to be classified as chaotic, it must have these properties:[12]

1. it must be sensitive to initial conditions
2. it must be topologically mixing
3. it must have dense periodic orbits

In some cases, the last two properties in the above have been shown to actually imply sensitivity to initial conditions.[13][14] In these cases, while it is often the most practically significant property, "sensitivity to initial conditions" need not be stated in the definition.

If attention is restricted to intervals, the second property implies the other two.[15] An alternative, and in general weaker, definition of chaos uses only the first two properties in the above list.[16]

4.2.1 Sensitivity to initial conditions

Main article: Butterfly effect

Sensitivity to initial conditions means that each point in a chaotic system is arbitrarily closely approximated by other points with significantly different future paths, or trajectories. Thus, an arbitrarily small change, or perturbation, of the current trajectory may lead to significantly different future behavior.

Sensitivity to initial conditions is popularly known as the "butterfly effect", so-called because of the title of a paper given by Edward Lorenz in 1972 to the American Association for the Advancement of Science in Washington, D.C., entitled *Predictability: Does the Flap of a Butterfly's Wings in Brazil set off a Tornado in Texas?*. The flapping wing represents a small change in the initial condition of the system, which causes a chain of events leading to large-scale phenomena. Had the butterfly not flapped its wings, the trajectory of the system might have been vastly different.

A consequence of sensitivity to initial conditions is that if we start with only a finite amount of information about the system (as is usually the case in practice), then beyond a certain time the system will no longer be predictable. This is most familiar in the case of weather, which is generally predictable only about a week ahead.[17] Of course, this does not mean that we cannot say anything about events far in the future; some restrictions on the system are present. With weather, we know that the temperature will never reach 100 °C or fall to −130 °C on earth, but we are not able to say exactly what day we will have the hottest temperature of the year.

In more mathematical terms, the Lyapunov exponent measures the sensitivity to initial conditions. Given two starting trajectories in the phase space that are infinitesimally close, with initial separation $\delta \mathbf{Z}_0$ end up diverging at a rate given by

$$|\delta \mathbf{Z}(t)| \approx e^{\lambda t}|\delta \mathbf{Z}_0|$$

where t is the time and λ is the Lyapunov exponent. The rate of separation depends on the orientation of the initial separation vector, so a whole spectrum of Lyapunov exponents exist. The number of Lyapunov exponents is equal to the number of dimensions of the phase space, though it is common to just refer to the largest one. For example, the maximal Lyapunov exponent (MLE) is most often used because it determines the overall predictability of the system. A positive MLE is usually taken as an indication that the system is chaotic.

Also, other properties relate to sensitivity of initial conditions, such as measure-theoretical mixing (as discussed in ergodic theory) and properties of a K-system.[5]

4.2.2 Topological mixing

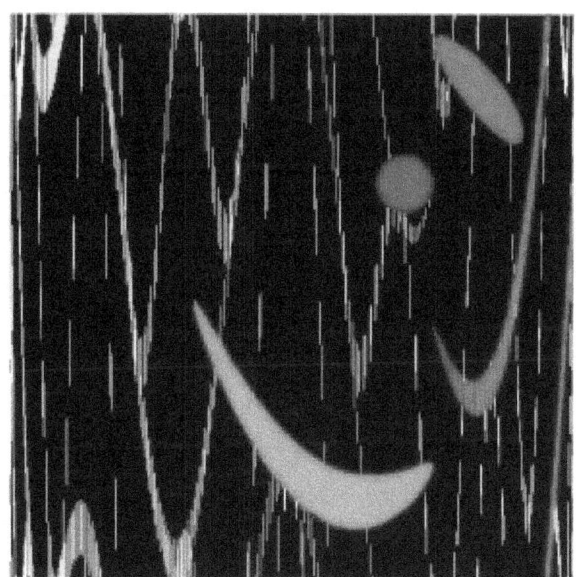

The map defined by x → 4 x (1 − x) *and* y → (x + y) mod 1 *also displays topological mixing. Here, the blue region is transformed by the dynamics first to the purple region, then to the pink and red regions, and eventually to a cloud of vertical lines scattered across the space.*

Topological mixing (or **topological transitivity**) means that the system will evolve over time so that any given region or open set of its phase space will eventually overlap with any other given region. This mathematical concept of "mixing" corresponds to the standard intuition, and the mixing of colored dyes or fluids is an example of a chaotic system.

Topological mixing is often omitted from popular accounts of chaos, which equate chaos with only sensitivity to initial conditions. However, sensitive dependence on initial conditions alone does not give chaos. For example, consider the simple dynamical system produced by repeatedly doubling an initial value. This system has sensitive dependence on initial conditions everywhere, since any pair of nearby points will eventually become widely separated. However, this example has no topological mixing, and therefore has no chaos. Indeed, it has extremely simple behavior: all points except 0 will tend to positive or negative infinity.

4.2.3 Density of periodic orbits

For a chaotic system to have dense periodic orbits means that every point in the space is approached arbitrarily closely by periodic orbits.[18] The one-dimensional logistic map defined by $x \to 4 x (1 - x)$ is one of the simplest systems with density of periodic orbits. For example, $\frac{5-\sqrt{5}}{8} \to \frac{5+\sqrt{5}}{8} \to \frac{5-\sqrt{5}}{8}$ (or approximately $0.3454915 \to 0.9045085 \to 0.3454915$) is an (unstable) orbit of period 2, and similar orbits exist for periods 4, 8, 16, etc. (indeed, for all the periods specified by Sharkovskii's theorem).[19]

Sharkovskii's theorem is the basis of the Li and Yorke[20] (1975) proof that any one-dimensional system that exhibits a regular cycle of period three will also display regular cycles of every other length, as well as completely chaotic orbits.

4.2.4 Strange attractors

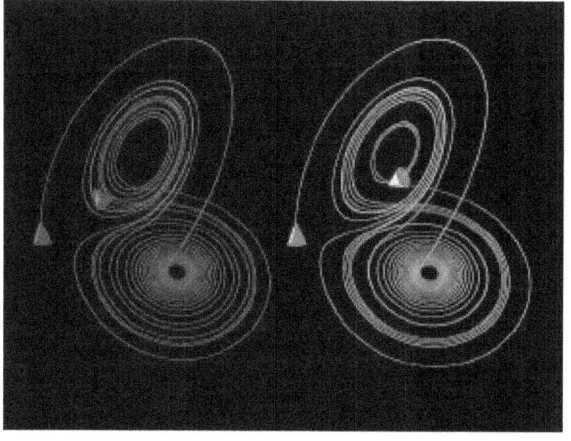

The Lorenz attractor displays chaotic behavior. These two plots demonstrate sensitive dependence on initial conditions within the region of phase space occupied by the attractor.

Some dynamical systems, like the one-dimensional logistic map defined by $x \to 4 x (1 - x)$, are chaotic everywhere, but in many cases chaotic behavior is found only in a subset of phase space. The cases of most interest arise when the chaotic behavior takes place on an attractor, since then a large set of initial conditions will lead to orbits that converge to this chaotic region.

An easy way to visualize a chaotic attractor is to start with a point in the basin of attraction of the attractor, and then simply plot its subsequent orbit. Because of the topological transitivity condition, this is likely to produce a picture of the entire final attractor, and indeed both orbits shown in the figure on the right give a picture of the general shape of the Lorenz attractor. This attractor results from a simple three-dimensional model of the Lorenz weather system. The Lorenz attractor is perhaps one of the best-known chaotic system diagrams, probably because it was not only one of the first, but it is also one of the most complex and as such gives rise to a very interesting pattern, that with a little imagination, looks like the wings of a butterfly.

Unlike fixed-point attractors and limit cycles, the attractors that arise from chaotic systems, known as strange attractors, have great detail and complexity. Strange attractors occur in both continuous dynamical systems (such as the Lorenz system) and in some discrete systems (such as the Hénon map). Other discrete dynamical systems have a repelling structure called a Julia set which forms at the boundary between basins of attraction of fixed points – Julia sets can be thought of as strange repellers. Both strange attractors and Julia sets typically have a fractal structure, and the fractal dimension can be calculated for them.

4.2.5 Minimum complexity of a chaotic system

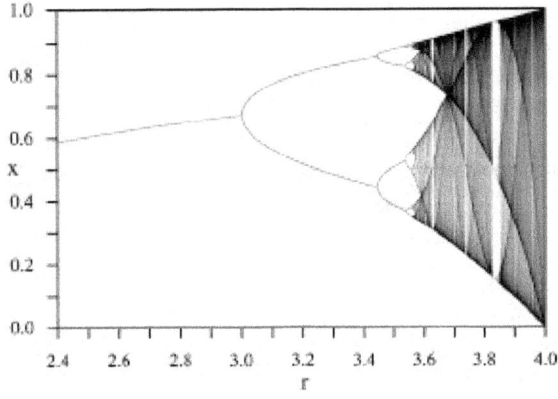

Bifurcation diagram of the logistic map $x \to r\,x\,(1-x)$. *Each vertical slice shows the attractor for a specific value of* r. *The diagram displays period-doubling as* r *increases, eventually producing chaos.*

Discrete chaotic systems, such as the logistic map, can exhibit strange attractors whatever their dimensionality. In contrast, for continuous dynamical systems, the Poincaré–Bendixson theorem shows that a strange attractor can only arise in three or more dimensions. Finite-dimensional linear systems are never chaotic; for a dynamical system to display chaotic behavior, it has to be either nonlinear or infinite-dimensional.

The Poincaré–Bendixson theorem states that a two-dimensional differential equation has very regular behavior. The Lorenz attractor discussed above is generated by a system of three differential equations such as:

$$\frac{dx}{dt} = \sigma y - \sigma x,$$
$$\frac{dy}{dt} = \rho x - xz - y,$$
$$\frac{dz}{dt} = xy - \beta z.$$

where x, y, and z make up the system state, t is time, and σ, ρ, β are the system parameters. Five of the terms on the right hand side are linear, while two are quadratic; a total of seven terms. Another well-known chaotic attractor is generated by the Rossler equations which have only one nonlinear term out of seven. Sprott [21] found a three-dimensional system with just five terms, that had only one nonlinear term, which exhibits chaos for certain parameter values. Zhang and Heidel [22][23] showed that, at least for dissipative and conservative quadratic systems, three-dimensional quadratic systems with only three or four terms on the right-hand side cannot exhibit chaotic behavior. The reason is, simply put, that solutions to such systems are asymptotic to a two-dimensional surface and therefore solutions are well behaved.

While the Poincaré–Bendixson theorem shows that a continuous dynamical system on the Euclidean plane cannot be chaotic, two-dimensional continuous systems with non-Euclidean geometry can exhibit chaotic behavior.[24] Perhaps surprisingly, chaos may occur also in linear systems, provided they are infinite dimensional.[25] A theory of linear chaos is being developed in a branch of mathematical analysis known as functional analysis.

4.2.6 Jerk systems

In physics, jerk is the third derivative of position, with respect to time. As such, differential equations of the form

$$J\left(\dddot{x}, \ddot{x}, \dot{x}, x\right) = 0$$

are sometimes called *Jerk equations*. It has been shown, that a jerk equation, which is equivalent to a system of three first order, ordinary, non-linear differential equations is in a certain sense the minimal setting for solutions showing chaotic behaviour. This motivates mathematical interest in jerk systems. Systems involving a fourth or higher derivative are called accordingly hyperjerk systems.[26]

A jerk system's behavior is described by a jerk equation, and for certain jerk equations, simple electronic circuits may be designed which model the solutions to this equation. These circuits are known as jerk circuits.

One of the most interesting properties of jerk circuits is the possibility of chaotic behavior. In fact, certain well-known chaotic systems, such as the Lorenz attractor and the Rössler map, are conventionally described as a system of three first-order differential equations, but which may be combined into a single (although rather complicated) jerk equation. Nonlinear jerk systems are in a sense minimally complex systems to show chaotic behaviour; there is no chaotic system involving only two first-order, ordinary differential equations (the system resulting in an equation of second order only).

An example of a jerk equation with nonlinearity in the magnitude of x is:

$$\frac{d^3 x}{dt^3} + A\frac{d^2 x}{dt^2} + \frac{dx}{dt} - |x| + 1 = 0.$$

Here, A is an adjustable parameter. This equation has a chaotic solution for $A=3/5$ and can be implemented with the following jerk circuit; the required nonlinearity is brought about by the two diodes:

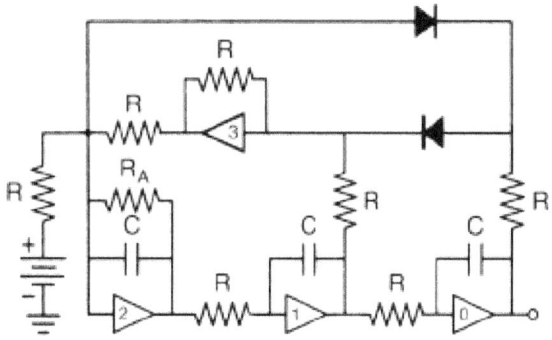

In the above circuit, all resistors are of equal value, except $R_A = R/A = 5R/3$, and all capacitors are of equal size. The dominant frequency will be $1/2\pi RC$. The output of op amp 0 will correspond to the x variable, the output of 1 will correspond to the first derivative of x and the output of 2 will correspond to the second derivative.

4.3 Spontaneous order

Under the right conditions, chaos will spontaneously evolve into a lockstep pattern. In the Kuramoto model, four conditions suffice to produce synchronization in a chaotic system. Examples include the coupled oscillation of Christiaan Huygens' pendulums, fireflies, neurons, the London Millennium Bridge resonance, and large arrays of Josephson junctions.[27]

4.4 History

Barnsley fern created using the chaos game. Natural forms (ferns, clouds, mountains, etc.) may be recreated through an Iterated function system (IFS).

An early proponent of chaos theory was Henri Poincaré. In the 1880s, while studying the three-body problem, he found that there can be orbits that are nonperiodic, and yet not forever increasing nor approaching a fixed point.[28][29] In 1898 Jacques Hadamard published an influential study of the chaotic motion of a free particle gliding frictionlessly on a surface of constant negative curvature, called "Hadamard's billiards".[30] Hadamard was able to show that all trajectories are unstable, in that all particle trajectories diverge exponentially from one another, with a positive Lyapunov exponent.

Chaos theory got its start in the field of ergodic theory. Later studies, also on the topic of nonlinear differential equations, were carried out by George David Birkhoff,[31] Andrey Nikolaevich Kolmogorov,[32][33][34] Mary Lucy Cartwright and John Edensor Littlewood,[35] and Stephen Smale.[36] Except for Smale, these studies were all directly inspired by physics: the three-body problem in the case of Birkhoff, turbulence and astronomical problems in the case of Kolmogorov, and radio engineering in the case of Cartwright and Littlewood. Although chaotic planetary motion had not been observed, experimentalists had encountered turbulence in fluid motion and nonperiodic oscillation in radio circuits without the benefit of a theory to explain what they were seeing.

Despite initial insights in the first half of the twentieth century, chaos theory became formalized as such only after mid-century, when it first became evident to some scientists that linear theory, the prevailing system theory at that time, simply could not explain the observed behavior of certain experiments like that of the logistic map. What had been attributed to measure imprecision and simple "noise" was considered by chaos theorists as a full component of the studied systems.

The main catalyst for the development of chaos theory was the electronic computer. Much of the mathematics of chaos theory involves the repeated iteration of simple mathematical formulas, which would be impractical to do by hand. Electronic computers made these repeated calculations practical, while figures and images made it possible to visualize these systems. As a graduate student in Chihiro Hayashi's laboratory at Kyoto University, Yoshisuke Ueda was experimenting with analog computers and noticed, on Nov. 27, 1961, what he called "randomly transitional phenomena". Yet his advisor did not agree with his conclusions at the time, and did not allow him to report his findings until 1970.[37][38]

Turbulence in the tip vortex from an airplane wing. Studies of the critical point beyond which a system creates turbulence were important for chaos theory, analyzed for example by the Soviet physicist Lev Landau, who developed the Landau-Hopf theory of turbulence. David Ruelle and Floris Takens later predicted, against Landau, that fluid turbulence could develop through a strange attractor, a main concept of chaos theory.

An early pioneer of the theory was Edward Lorenz whose interest in chaos came about accidentally through his work on weather prediction in 1961.[7] Lorenz was using a simple digital computer, a Royal McBee LGP-30, to run his weather simulation. He wanted to see a sequence of data again and to save time he started the simulation in the middle of its course. He was able to do this by entering a printout of the data corresponding to conditions in the middle of his simulation which he had calculated last time. To his surprise the weather that the machine began to predict was completely different from the weather calculated before. Lorenz tracked this down to the computer printout. The computer worked with 6-digit precision, but the printout rounded variables off to a 3-digit number, so a value like 0.506127 was printed as 0.506. This difference is tiny and the consensus at the time would have been that it should have had practically no effect. However, Lorenz had discovered that small changes in initial conditions produced large changes in the long-term outcome.[39] Lorenz's discovery, which gave its name to Lorenz attractors, showed that even detailed atmospheric modelling cannot, in general, make precise long-term weather predictions.

In 1963, Benoit Mandelbrot found recurring patterns at every scale in data on cotton prices.[40] Beforehand he had studied information theory and concluded noise was patterned like a Cantor set: on any scale the proportion of noise-containing periods to error-free periods was a constant – thus errors were inevitable and must be planned for by incorporating redundancy.[41] Mandelbrot described both the "Noah effect" (in which sudden discontinuous changes can occur) and the "Joseph effect" (in which persistence of a value can occur for a while, yet suddenly change afterwards).[42][43] This challenged the idea that changes in price were normally distributed. In 1967, he published "How long is the coast of Britain? Statistical self-similarity and fractional dimension", showing that a coastline's length varies with the scale of the measuring instrument, resembles itself at all scales, and is infinite in length for an infinitesimally small measuring device.[44] Arguing that a ball of twine appears to be a point when viewed from far away (0-dimensional), a ball when viewed from fairly near (3-dimensional), or a curved strand (1-dimensional), he argued that the dimensions of an object are relative to the observer and may be fractional. An object whose irregularity is constant over different scales ("self-similarity") is a fractal (examples include the Menger sponge, the Sierpiński gasket, and the Koch curve or "snowflake", which is infinitely long yet encloses a finite space and has a fractal dimension of circa 1.2619). In 1982 Mandelbrot published *The Fractal Geometry of Nature*, which became a classic of chaos theory. Biological systems such as the branching of the circulatory and bronchial systems proved to fit a fractal model.[45]

In December 1977, the New York Academy of Sciences organized the first symposium on Chaos, attended by David Ruelle, Robert May, James A. Yorke (coiner of the term "chaos" as used in mathematics), Robert Shaw, and the meteorologist Edward Lorenz. The following year, independently Pierre Coullet and Charles Tresser with the article "Iterations d'endomorphismes et groupe de renormalisation" and Mitchell Feigenbaum with the article "Quantitative Universality for a Class of Nonlinear Transformations" described logistic maps.[46][47] They notably discovered the universality in chaos, permitting the application of chaos theory to many different phenomena.

In 1979, Albert J. Libchaber, during a symposium organized in Aspen by Pierre Hohenberg, presented his experimental observation of the bifurcation cascade that leads to chaos and turbulence in Rayleigh–Bénard convection systems. He was awarded the Wolf Prize in Physics in 1986 along with Mitchell J. Feigenbaum for their inspiring achievements.[48]

In 1986, the New York Academy of Sciences co-organized with the National Institute of Mental Health and the Office of Naval Research the first important conference on chaos in biology and medicine. There, Bernardo Huberman presented a mathematical model of the eye tracking disorder among schizophrenics.[49] This led to a renewal of physiology in the 1980s through the application of chaos theory, for example, in the study of pathological cardiac cycles.

In 1987, Per Bak, Chao Tang and Kurt Wiesenfeld published a paper in *Physical Review Letters*[50] describing for the first time self-organized criticality (SOC), considered to be one of the mechanisms by which complexity arises in nature.

Alongside largely lab-based approaches such as the Bak–Tang–Wiesenfeld sandpile, many other investigations have focused on large-scale natural or social systems that are known (or suspected) to display scale-invariant behavior. Although these approaches were not always welcomed (at least initially) by specialists in the subjects examined, SOC has nevertheless become established as a strong candidate for explaining a number of natural phenomena, including earthquakes (which, long before SOC was discovered, were known as a source of scale-invariant behavior such as the Gutenberg–Richter law describing the statistical distribution of earthquake sizes, and the Omori law[51] describing the frequency of aftershocks), solar flares, fluctuations in economic systems such as financial markets (references to SOC are common in econophysics), landscape formation, forest fires, landslides, epidemics, and biological evolution (where SOC has been invoked, for example, as the dynamical mechanism behind the theory of "punctuated equilibria" put forward by Niles Eldredge and Stephen Jay Gould).

Given the implications of a scale-free distribution of event sizes, some researchers have suggested that another phenomenon that should be considered an example of SOC is the occurrence of wars. These investigations of SOC have included both attempts at modelling (either developing new models or adapting existing ones to the specifics of a given natural system), and extensive data analysis to determine the existence and/or characteristics of natural scaling laws.

In the same year, James Gleick published *Chaos: Making a New Science*, which became a best-seller and introduced the general principles of chaos theory as well as its history to the broad public, though his history under-emphasized important Soviet contributions.[52] Initially the domain of a few, isolated individuals, chaos theory progressively emerged as a transdisciplinary and institutional discipline, mainly under the name of nonlinear systems analysis. Alluding to Thomas Kuhn's concept of a paradigm shift exposed in *The Structure of Scientific Revolutions* (1962), many "chaologists" (as some described themselves) claimed that this new theory was an example of such a shift, a thesis upheld by Gleick.

The availability of cheaper, more powerful computers broadens the applicability of chaos theory. Currently, chaos theory continues to be a very active area of research,[53] involving many different disciplines (mathematics, topology, physics, social systems, population modeling, biology, meteorology, astrophysics, information theory, computational neuroscience, etc.).

4.5 Distinguishing random from chaotic data

It can be difficult to tell from data whether a physical or other observed process is random or chaotic, because in practice no time series consists of a pure "signal." There will always be some form of corrupting noise, even if it is present as round-off or truncation error. Thus any real time series, even if mostly deterministic, will contain some (pseudo-)randomness.[54][55]

All methods for distinguishing deterministic and stochastic processes rely on the fact that a deterministic system always evolves in the same way from a given starting point.[54][56] Thus, given a time series to test for determinism, one can

1. pick a test state;
2. search the time series for a similar or nearby state; and
3. compare their respective time evolutions.

Define the error as the difference between the time evolution of the test state and the time evolution of the nearby state. A deterministic system will have an error that either remains small (stable, regular solution) or increases exponentially with time (chaos). A stochastic system will have a randomly distributed error.[57]

Essentially, all measures of determinism taken from time series rely upon finding the closest states to a given test state (e.g., correlation dimension, Lyapunov exponents, etc.). To define the state of a system, one typically relies on phase space embedding methods such as Poincaré plots.[58] Typically one chooses an embedding dimension and investigates the propagation of the error between two nearby states. If the error looks random, one increases the dimension. If the dimension can be increased to obtain a deterministically looking error, then analysis is done. Though it may sound simple, one complication is that as the dimension increases, the search for a nearby state requires a lot more computation time and a lot of data (the amount of data required increases exponentially with embedding dimension) to find a suitably close candidate. If the embedding dimension (number of measures per state) is chosen too small (less than the "true" value), deterministic data can appear to be random, but in theory there is no problem choosing the dimension too large – the method will work.

When a nonlinear deterministic system is attended by external fluctuations, its trajectories present serious and permanent distortions. Furthermore, the noise is amplified due to the inherent nonlinearity and reveals totally new dynamical properties. Statistical tests attempting to separate noise from the deterministic skeleton or inversely isolate the deterministic part risk failure. Things become worse when the deterministic component is a nonlinear feedback system.[59] In presence of interactions between nonlinear deterministic components and noise, the resulting nonlinear series can display dynamics that traditional tests for nonlinearity are sometimes not able to capture.[60]

The question of how to distinguish deterministic chaotic systems from stochastic systems has also been discussed in philosophy. It has been shown that they might be observationally equivalent.[61]

4.6 Applications

Chaos theory was born from observing weather patterns, but it has become applicable to a variety of other situations. Some areas benefiting from chaos theory today are geology, mathematics, microbiology, biology, computer science, economics,[63][64][65] engineering,[66] finance,[67][68] algorithmic trading,[69][70][71] meteorology, philosophy, physics, politics, population dynamics,[72] psychology,[9] and robotics. A few categories are listed below with examples, but this is by no means a comprehensive

A conus textile shell, similar in appearance to Rule 30, a cellular automaton with chaotic behaviour.[62]

list as new applications are appearing.

4.6.1 Computer science

Chaos theory is not new to computer science and has been used for many years in cryptography. One type of encryption, secret key or symmetric key, relies on diffusion and confusion, which is modeled well by chaos theory.[73] Another type of computing, DNA computing, when paired with chaos theory, offers a more efficient way to encrypt images and other information.[74] Robotics is another area that has recently benefited from chaos theory. Instead of robots acting in a trial-and-error type of refinement to interact with their environment, chaos theory has been used to build a predictive model.[75] Chaotic dynamics have been exhibited by passive walking biped robots.[76]

4.6.2 Biology

For over a hundred years, biologists have been keeping track of populations of different species with population models. Most models are continuous, but recently scientists have been able to implement chaotic models in certain populations.[77] For example, a study on models of Canadian lynx showed there was chaotic behavior in the population growth.[78] Chaos can also be found in ecological systems, such as hydrology. While a chaotic model for hydrology has its shortcomings, there is still much to be learned from looking at the data through the lens of chaos theory.[79] Another biological application is found in cardiotocography. Fetal surveillance is a delicate balance of obtaining accurate information while being as noninvasive as possible. Better models of warning signs of fetal hypoxia can be obtained through chaotic modeling.[80]

4.6.3 Other areas

In chemistry, predicting gas solubility is essential to manufacturing polymers, but models using particle swarm optimization (PSO) tend to converge to the wrong points. An improved version of PSO has been created by introducing chaos, which keeps the simulations from getting stuck.[81] In celestial mechanics, especially when observing asteroids, applying chaos theory leads to better predictions about when these objects will come in range of Earth and other planets.[82] In quantum physics and electrical engineering, the study of large arrays of Josephson junctions benefitted greatly from chaos theory.[83] Closer to home, coal mines have always been dangerous places where frequent natural gas leaks cause many deaths. Until recently, there was no reliable way to predict when they would occur. But these gas leaks have chaotic tendencies that, when properly modeled, can be predicted fairly accurately.[84]

Chaos theory can be applied outside of the natural sciences. By adapting a model of career counseling to include a chaotic interpretation of the relationship between employees and the job market, better suggestions can be made to people struggling with career decisions.[85] Modern organizations are increasingly seen as open complex adaptive systems, with fundamental natural nonlinear structures, subject to internal and external forces which may be sources of chaos. The chaos metaphor—used in verbal theories—grounded on mathematical models and psychological aspects of human behavior provides helpful insights to describing the complexity of small work groups, that go beyond the metaphor itself.[86]

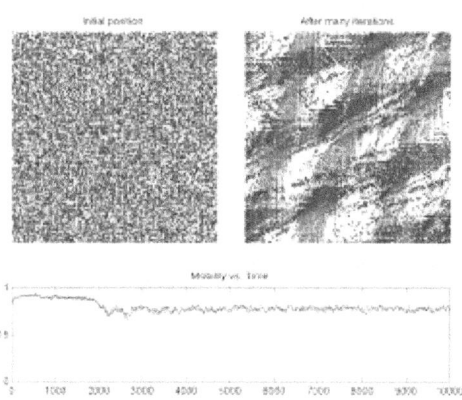

The red cars and blue cars take turns to move; the red ones only move upwards, and the blue ones move rightwards. Every time, all the cars of the same colour try to move one step if there is no car in front of it. Here, the model has self-organized in a somewhat geometric pattern where there are some traffic jams and some areas where cars can move at top speed.

It is possible that economic models can also be improved through an application of chaos theory, but predicting the health of an economic system and what factors influence it most is an extremely complex task.[87] Economic and financial systems are fundamentally different from those in the classical natural sciences since the former are inherently stochastic in nature, as they result from the interactions of people, and thus pure deterministic models are unlikely to provide accurate representations of the data. The empirical literature that tests for chaos in economics and finance presents very mixed results, in part due to confusion between specific tests for chaos and more general tests for non-linear relationships.[88]

Traffic forecasting is another area that greatly benefits from applications of chaos theory. Better predictions of when traffic will occur would allow measures to be taken for it to be dispersed before the traffic starts, rather than after. Combining chaos theory principles with a few other methods has led to a more accurate short-term prediction model (see the plot of the BML traffic model at right).[89]

Chaos theory also finds applications in psychology. For example, in modeling group behavior in which heterogeneous members may behave as if sharing to different degrees what in Wilfred Bion's theory is a basic assumption, the group dynamics is the result of the individual dynamics of the members: each individual reproduces the group dynamics in a different scale, and the chaotic behavior of the group is reflected in each member.[90]

4.7 See also

4.8 References

[1] Boeing (2015). "Chaos Theory and the Logistic Map". Retrieved 2015-07-16.

[2] Kellert, Stephen H. (1993). *In the Wake of Chaos: Unpredictable Order in Dynamical Systems*. University of Chicago Press. p. 32. ISBN 0-226-42976-8.

[3] Kellert 1993, p. 56

[4] Kellert 1993, p. 62

[5] Werndl, Charlotte (2009). "What are the New Implications of Chaos for Unpredictability?". *The British Journal for the Philosophy of Science* 60 (1): 195–220. doi:10.1093/bjps/axn053.

[6] Danforth, Christopher M. (April 2013). "Chaos in an Atmosphere Hanging on a Wall". *Mathematics of Planet Earth 2013*. Retrieved 4 April 2013.

[7] Lorenz, Edward N. (1963). "Deterministic non-periodic flow". *Journal of the Atmospheric Sciences* 20 (2): 130–

141. Bibcode:1963JAtS...20..130L. doi:10.1175/1520-0469(1963)020<0130:DNF>2.0.CO;2.

[8] Ivancevic, Vladimir G.; Tijana T. Ivancevic (2008). *Complex nonlinearity: chaos, phase transitions, topology change, and path integrals*. Springer. ISBN 978-3-540-79356-4.

[9] Safonov, Leonid A.; Tomer, Elad; Strygin, Vadim V.; Ashkenazy, Yosef; Havlin, Shlomo (2002). "Multifractal chaotic attractors in a system of delay-differential equations modeling road traffic". *Chaos: An Interdisciplinary Journal of Nonlinear Science* 12 (4): 1006. Bibcode:2002Chaos..12.1006S. doi:10.1063/1.1507903. ISSN 1054-1500.

[10] *Sync: The Emerging Science of Spontaneous Order*, Steven Strogatz, Hyperion, New York, 2003, pages 189-190.

[11] Definition of chaos at Wiktionary;

[12] Hasselblatt, Boris; Anatole Katok (2003). *A First Course in Dynamics: With a Panorama of Recent Developments*. Cambridge University Press. ISBN 0-521-58750-6.

[13] Elaydi, Saber N. (1999). *Discrete Chaos*. Chapman & Hall/CRC. p. 117. ISBN 1-58488-002-3.

[14] Basener, William F. (2006). *Topology and its applications*. Wiley. p. 42. ISBN 0-471-68755-3.

[15] Vellekoop, Michel; Berglund, Raoul (April 1994). "On Intervals, Transitivity = Chaos". *The American Mathematical Monthly* 101 (4): 353–5. doi:10.2307/2975629. JSTOR 2975629.

[16] Medio, Alfredo; Lines, Marji (2001). *Nonlinear Dynamics: A Primer*. Cambridge University Press. p. 165. ISBN 0-521-55874-3.

[17] Watts, Robert G. (2007). *Global Warming and the Future of the Earth*. Morgan & Claypool. p. 17.

[18] Devaney 2003

[19] Alligood, Sauer & Yorke 1997

[20] Li, T.Y.; Yorke, J.A. (1975). "Period Three Implies Chaos" (PDF). *American Mathematical Monthly* 82 (10): 985–92. Bibcode:1975AmMM...82..985L. doi:10.2307/2318254.

[21] Sprott, J.C. (1997). "Simplest dissipative chaotic flow". *Physics Letters A* 228 (4–5): 271–274. Bibcode:1997PhLA..228..271S. doi:10.1016/S0375-9601(97)00088-1.

[22] Fu, Z.; Heidel, J. (1997). "Non-chaotic behaviour in three-dimensional quadratic systems". *Nonlinearity* 10 (5): 1289–1303. Bibcode:1997Nonli..10.1289F. doi:10.1088/0951-7715/10/5/014.

[23] Heidel, J.; Fu, Z. (1999). "Nonchaotic behaviour in three-dimensional quadratic systems II. The conservative case". *Nonlinearity* 12 (3): 617–633. Bibcode:1999Nonli..12..617H. doi:10.1088/0951-7715/12/3/012.

[24] Rosario, Pedro (2006). *Underdetermination of Science: Part I*. Lulu.com. ISBN 1411693914.

[25] Bonet, J.; Martínez-Giménez, F.; Peris, A. (2001). "A Banach space which admits no chaotic operator". *Bulletin of the London Mathematical Society* 33 (2): 196–8. doi:10.1112/blms/33.2.196.

[26] K. E. Chlouverakis and J. C. Sprott, Chaos Solitons & Fractals 28, 739-746 (2005), Chaotic Hyperjerk Systems, http://sprott.physics.wisc.edu/pubs/paper297.htm

[27] Steven Strogatz, *Sync: The Emerging Science of Spontaneous Order*, Hyperion, 2003.

[28] Poincaré, Jules Henri (1890). "Sur le problème des trois corps et les équations de la dynamique. Divergence des séries de M. Lindstedt". *Acta Mathematica* 13: 1–270. doi:10.1007/BF02392506.

[29] Diacu, Florin; Holmes, Philip (1996). *Celestial Encounters: The Origins of Chaos and Stability*. Princeton University Press.

[30] Hadamard, Jacques (1898). "Les surfaces à courbures opposées et leurs lignes géodesiques". *Journal de Mathématiques Pures et Appliquées* 4: 27–73.

[31] George D. Birkhoff, *Dynamical Systems*, vol. 9 of the American Mathematical Society Colloquium Publications (Providence, Rhode Island: American Mathematical Society, 1927)

[32] Kolmogorov, Andrey Nikolaevich (1941). "Local structure of turbulence in an incompressible fluid for very large Reynolds numbers". *Doklady Akademii Nauk SSSR* 30 (4): 301–5. Bibcode:1941DoSSR..30..301K. Reprinted in: Kolmogorov, A. N. (1991). "The Local Structure of Turbulence in Incompressible Viscous Fluid for Very Large Reynolds Numbers". *Proceedings of the Royal Society A* 434 (1890): 9–13. Bibcode:1991RSPSA.434....9K. doi:10.1098/rspa.1991.0075.

[33] Kolmogorov, A. N. (1941). "On degeneration of isotropic turbulence in an incompressible viscous liquid". *Doklady Akademii Nauk SSSR* 31 (6): 538–540. Reprinted in: Kolmogorov, A. N. (1991). "Dissipation of Energy in the Locally Isotropic Turbulence". *Proceedings of the Royal Society A* 434 (1890): 15–17. Bibcode:1991RSPSA.434...15K. doi:10.1098/rspa.1991.0076.

[34] Kolmogorov, A. N. (1954). "Preservation of conditionally periodic movements with small change in the Hamiltonian function". *Doklady Akademii Nauk SSSR*. Lecture Notes in Physics 98: 527–530. Bibcode:1979LNP....93...51K. doi:10.1007/BFb0021737. ISBN 3-540-09120-3. See also Kolmogorov–Arnold–Moser theorem

[35] Cartwright, Mary L.; Littlewood, John E. (1945). "On non-linear differential equations of the second order, I: The equation $y'' + k(1-y^2)y' + y = b\lambda k\cos(\lambda t + a)$, k large". *Journal of the London Mathematical Society* 20 (3): 180–9.

doi:10.1112/jlms/s1-20.3.180. See also: Van der Pol oscillator

[36] Smale, Stephen (January 1960). "Morse inequalities for a dynamical system". *Bulletin of the American Mathematical Society* **66**: 43–49. doi:10.1090/S0002-9904-1960-10386-2.

[37] Abraham & Ueda 2001, See Chapters 3 and 4

[38] Sprott 2003, p. 89

[39] Gleick, James (1987). *Chaos: Making a New Science*. London: Cardinal. p. 17. ISBN 0-434-29554-X.

[40] Mandelbrot, Benoît (1963). "The variation of certain speculative prices". *Journal of Business* **36** (4): 394–419. doi:10.1086/294632. JSTOR 2350970.

[41] Berger J.M.; Mandelbrot B. (1963). "A new model for error clustering in telephone circuits". *IBM Journal of Research and Development* **7**: 224–236. doi:10.1147/rd.73.0224.

[42] Mandelbrot, B. (1977). *The Fractal Geometry of Nature*. New York: Freeman. p. 248.

[43] See also: Mandelbrot, Benoît B.; Hudson, Richard L. (2004). *The (Mis)behavior of Markets: A Fractal View of Risk, Ruin, and Reward*. New York: Basic Books. p. 201.

[44] Mandelbrot, Benoît (5 May 1967). "How Long Is the Coast of Britain? Statistical Self-Similarity and Fractional Dimension". *Science* **156** (3775): 636–8. Bibcode:1967Sci...156..636M. doi:10.1126/science.156.3775.636. PMID 17837158.

[45] Buldyrev, S.V.; Goldberger, A.L.; Havlin, S.; Peng, C.K.; Stanley, H.E. (1994). "Fractals in Biology and Medicine: From DNA to the Heartbeat". In Bunde, Armin; Havlin, Shlomo. *Fractals in Science*. Springer. pp. 49–89. ISBN 3-540-56220-6.

[46] Feigenbaum, Mitchell (July 1978). "Quantitative universality for a class of nonlinear transformations". *Journal of Statistical Physics* **19** (1): 25–52. Bibcode:1978JSP....19...25F. doi:10.1007/BF01020332.

[47] Coullet, Pierre, and Charles Tresser. "Iterations d'endomorphismes et groupe de renormalisation." Le Journal de Physique Colloques 39.C5 (1978): C5-25

[48] "The Wolf Prize in Physics in 1986.".

[49] Huberman, B.A. (July 1987). "A Model for Dysfunctions in Smooth Pursuit Eye Movement". *Annals of the New York Academy of Sciences*. **504** Perspectives in Biological Dynamics and Theoretical Medicine: 260–273. Bibcode:1987NYASA.504..260H. doi:10.1111/j.1749-6632.1987.tb48737.x.

[50] Bak, Per; Tang, Chao; Wiesenfeld, Kurt; Tang; Wiesenfeld (27 July 1987). "Self-organized criticality: An explanation of the 1/f noise". *Physical Review Letters* **59** (4): 381–4. Bibcode:1987PhRvL..59..381B. doi:10.1103/PhysRevLett.59.381. However, the conclusions of this article have been subject to dispute. "?".. See especially: Laurson, Lasse; Alava, Mikko J.; Zapperi, Stefano (15 September 2005). "Letter: Power spectra of self-organized critical sand piles". *Journal of Statistical Mechanics: Theory and Experiment* **0511**. L001.

[51] Omori, F. (1894). "On the aftershocks of earthquakes". *Journal of the College of Science, Imperial University of Tokyo* **7**: 111–200.

[52] Gleick, James (August 26, 2008). *Chaos: Making a New Science*. Penguin Books. ISBN 0143113453.

[53] Motter, A. E.; Campbell, D. K. (2013). "Chaos at fifty". *Phys. Today* **66** (5): 27–33. doi:10.1063/pt.3.1977.

[54] Provenzale, A.; Smith; Vio; Murante; et al. (1992). "Distinguishing between low-dimensional dynamics and randomness in measured time-series". *Physica D* **58**: 31–49. Bibcode:1992PhyD...58...31P. doi:10.1016/0167-2789(92)90100-2.

[55] Brock, W.A. (October 1986). "Distinguishing random and deterministic systems: Abridged version". *Journal of Economic Theory* **40**: 168–195. doi:10.1016/0022-0531(86)90014-1.

[56] Sugihara G., May R.; May (1990). "Nonlinear forecasting as a way of distinguishing chaos from measurement error in time series" (PDF). *Nature* **344** (6268): 734–741. Bibcode:1990Natur.344..734S. doi:10.1038/344734a0. PMID 2330029.

[57] Casdagli, Martin (1991). "Chaos and Deterministic versus Stochastic Non-linear Modelling". *Journal of the Royal Statistical Society, Series B* **54** (2): 303–328. JSTOR 2346130.

[58] Broomhead, D.S.; King, G.P.; King (June–July 1986). "Extracting qualitative dynamics from experimental data". *Physica D* **20** (2-3): 217–236. Bibcode:1986PhyD...20..217B. doi:10.1016/0167-2789(86)90031-X.

[59] Kyrtsou C (2008). "Re-examining the sources of heteroskedasticity: the paradigm of noisy chaotic models". *Physica A* **387** (27): 6785–9. Bibcode:2008PhyA..387.6785K. doi:10.1016/j.physa.2008.09.008.

[60] Kyrtsou, C. (2005). "Evidence for neglected linearity in noisy chaotic models". *International Journal of Bifurcation and Chaos* **15** (10): 3391–4. Bibcode:2005IJBC...15.3391K. doi:10.1142/S0218127405013964.

4.8. REFERENCES

[61] Werndl, Charlotte (2009). "Are Deterministic Descriptions and Indeterministic Descriptions Observationally Equivalent?". *Studies in History and Philosophy of Modern Physics* **40** (3): 232–242. doi:10.1016/j.shpsb.2009.06.004.

[62] Stephen Coombes (February 2009). "The Geometry and Pigmentation of Seashells" (PDF). *www.maths.nottingham.ac.uk*. University of Nottingham. Retrieved 2013-04-10.

[63] Kyrtsou C.; Labys W. (2006). "Evidence for chaotic dependence between US inflation and commodity prices". *Journal of Macroeconomics* **28** (1): 256–266. doi:10.1016/j.jmacro.2005.10.019.

[64] Kyrtsou C., Labys W.; Labys (2007). "Detecting positive feedback in multivariate time series: the case of metal prices and US inflation". *Physica A* **377** (1): 227–229. Bibcode:2007PhyA..377..227K. doi:10.1016/j.physa.2006.11.002.

[65] Kyrtsou, C.; Vorlow, C. (2005). "Complex dynamics in macroeconomics: A novel approach". In Diebolt, C.; Kyrtsou, C. *New Trends in Macroeconomics*. Springer Verlag.

[66] Applying Chaos Theory to Embedded Applications

[67] Hristu-Varsakelis, D.; Kyrtsou, C. (2008). "Evidence for nonlinear asymmetric causality in US inflation, metal and stock returns". *Discrete Dynamics in Nature and Society* **2008**: 1–7. doi:10.1155/2008/138547. 138547.

[68] Kyrtsou, C.; M. Terraza, (2003). "Is it possible to study chaotic and ARCH behaviour jointly? Application of a noisy Mackey-Glass equation with heteroskedastic errors to the Paris Stock Exchange returns series". *Computational Economics* **21** (3): 257–276. doi:10.1023/A:1023939610962.

[69] Williams, Bill Williams, Justine (2004). *Trading chaos : maximize profits with proven technical techniques* (2nd ed.). New York: Wiley. ISBN 9780471463085.

[70] Peters, Edgar E. (1994). *Fractal market analysis : applying chaos theory to investment and economics* (2. print. ed.). New York u.a.: Wiley. ISBN 978-0471585244.

[71] Peters, / Edgar E. (1996). *Chaos and order in the capital markets : a new view of cycles, prices, and market volatility* (2nd ed.). New York: John Wiley & Sons. ISBN 978-0471139386.

[72] Dilão, R.; Domingos, T. (2001). "Periodic and Quasi-Periodic Behavior in Resource Dependent Age Structured Population Models". *Bulletin of Mathematical Biology* **63** (2): 207–230. doi:10.1006/bulm.2000.0213. PMID 11276524.

[73] Wang, Xingyuan; Zhao, Jianfeng (2012). "An improved key agreement protocol based on chaos". *Commun. Nonlinear Sci. Numer. Simul.* **15** (12): 4052–4057. Bibcode:2010CNSNS..15.4052W. doi:10.1016/j.cnsns.2010.02.014.

[74] Babaei, Majid (2013). "A novel text and image encryption method based on chaos theory and DNA computing". *Natural Computing. an International Journal* **12** (1): 101–107. doi:10.1007/s11047-012-9334-9.

[75] Nehmzow, Ulrich; Keith Walker (Dec 2005). "Quantitative description of robot–environment interaction using chaos theory". *Robotics and Autonomous Systems* **53** (3–4): 177–193. doi:10.1016/j.robot.2005.09.009.

[76] Goswami, Ambarish; Thuilot, Benoit; Espiau, Bernard (1998). "A Study of the Passive Gait of a Compass-Like Biped Robot: Symmetry and Chaos". *The International Journal of Robotics Research* **17** (12): 1282–1301. doi:10.1177/027836499801701202.

[77] Eduardo, Liz; Ruiz-Herrera, Alfonso (2012). "Chaos in discrete structured population models". *SIAM Journal on Applied Dynamical Systems* **11** (4): 1200–1214. doi:10.1137/120868980.

[78] Lai, Dejian (1996). "Comparison study of AR models on the Canadian lynx data: a close look at BDS statistic". *Computational Statistics \& Data Analysis* **22** (4): 409–423. doi:10.1016/0167-9473(95)00056-9.

[79] Sivakumar, B (31 January 2000). "Chaos theory in hydrology: important issues and interpretations". *Journal of Hydrology* **227** (1–4): 1–20. Bibcode:2000JHyd..227....1S. doi:10.1016/S0022-1694(99)00186-9.

[80] Bozóki, Zsolt (February 1997). "Chaos theory and power spectrum analysis in computerized cardiotocography". *European Journal of Obstetrics & Gynecology and Reproductive Biology* **71** (2): 163–168. doi:10.1016/s0301-2115(96)02628-0.

[81] Li, Mengshan; Xingyuan Huanga; Hesheng Liua; Bingxiang Liub; Yan Wub; Aihua Xiongc; Tianwen Dong (25 October 2013). "Prediction of gas solubility in polymers by back propagation artificial neural network based on self-adaptive particle swarm optimization algorithm and chaos theory". *Fluid Phase Equilibria* **356**: 11–17. doi:10.1016/j.fluid.2013.07.017.

[82] Morbidelli, A. (2001). "Chaotic diffusion in celestial mechanics". *Regular & Chaotic Dynamics* **6** (4): 339–353. doi:10.1070/rd2001v006n04abeh000182.

[83] Steven Strogatz, *Sync: The Emerging Science of Spontaneous Order*, Hyperion, 2003

[84] Dingqi, Li; Yuanping Chenga; Lei Wanga; Haifeng Wanga; Liang Wanga; Hongxing Zhou (May 2011). "Prediction method for risks of coal and gas outbursts based on spatial chaos theory using gas desorption index of drill cuttings". *Mining Science and Technology* **21** (3): 439–443.

[85] Pryor, Robert G. L.; Norman E. Aniundson; Jim E. H. Bright (June 2008). "Probabilities and Possibilities: The Strategic Counseling Implications of the Chaos Theory of Careers". *The Career Development Quarterly* **56**: 309–318. doi:10.1002/j.2161-0045.2008.tb00096.x.

[86] Dal Forno, Arianna; Merlone, Ugo (2013). "Chaotic Dynamics in Organization Theory". In Bischi, Gian Italo; Chiarella, Carl; Shusko, Irina. *Global Analysis of Dynamic Models in Economics and Finance*. Springer-Verlag. pp. 185–204. ISBN 978-3-642-29503-4.

[87] Juárez, Fernando (2011). "Applying the theory of chaos and a complex model of health to establish relations among financial indicators". *Procedia Computer Science 3*: 982–986. doi:10.1016/j.procs.2010.12.161.

[88] Brooks, Chris (1998). "Chaos in foreign exchange markets: a sceptical view". *Computational Economics* 11: 265–281. doi:10.1023/A:1008650024944. ISSN 1572-9974.

[89] Wang, Jin; Qixin Shi (February 2013). "Short-term traffic speed forecasting hybrid model based on Chaos–Wavelet Analysis-Support Vector Machine theory". *Transportation Research Part C: Emerging Technologies* 27: 219–232. doi:10.1016/j.trc.2012.08.004.

[90] Dal Forno, Arianna; Merlone, Ugo (2013). "Nonlinear dynamics in work groups with Bion's basic assumptions". *Nonlinear Dynamics, Psychology, and Life Sciences* 17 (2): 295–315. ISSN 1090-0578.

4.9 Scientific literature

4.9.1 Articles

- Sharkovskii, A.N. (1964). "Co-existence of cycles of a continuous mapping of the line into itself". *Ukrainian Math. J.* 16: 61–71.

- Li, T.Y.; Yorke. J.A. (1975). "Period Three Implies Chaos". *American Mathematical Monthly* 82 (10): 985–92. Bibcode:1975AmMM...82..985L. doi:10.2307/2318254.

- Crutchfield; Tucker; Morrison; J.D.; Packard; N.H.; Shaw; R.S (December 1986). "Chaos". *Scientific American* 255 (6): 38–49 (bibliography p.136). Bibcode:1986SciAm.255...38T. Online version (Note: the volume and page citation cited for the online text differ from that cited here. The citation here is from a photocopy, which is consistent with other citations found online, but which don't provide article views. The online content is identical to the hardcopy text. Citation variations will be related to country of publication).

- Kolyada, S.F. (2004). "Li-Yorke sensitivity and other concepts of chaos". *Ukrainian Math. J.* 56 (8): 1242–57. doi:10.1007/s11253 005-0055-4.

- Strelioff, C.; Hübler, A. (2006). "Medium-Term Prediction of Chaos" (PDF). *Phys. Rev. Lett.* 96 (4): 044101. Bibcode:2006PhRvL..96d4101S. doi:10.1103/PhysRevLett.96.044101. PMID 16486826. 044101.

- Hübler, A.; Foster, G.; Phelps, K. (2007). "Managing Chaos: Thinking out of the Box" (PDF). *Complexity* 12 (3): 10–13. doi:10.1002/cplx.20159.

- Motter, Adilson E.; Campbell, David K. (2013). "Chaos at 50". *Physics Today* 66: 27. doi:10.1063/PT.3.1977.

4.9.2 Textbooks

- Alligood, K.T.; Sauer, T.; Yorke, J.A. (1997). *Chaos: an introduction to dynamical systems*. Springer-Verlag. ISBN 0-387-94677-2.

- Baker, G. L. (1996). *Chaos, Scattering and Statistical Mechanics*. Cambridge University Press. ISBN 0-521-39511-9.

- Badii, R.; Politi A. (1997). *Complexity: hierarchical structures and scaling in physics*. Cambridge University Press. ISBN 0-521-66385-7.

- Bunde; Havlin, Shlomo, eds. (1996). *Fractals and Disordered Systems*. Springer. ISBN 3642848702. and Bunde; Havlin, Shlomo, eds. (1994). *Fractals in Science*. Springer. ISBN 3-540-56220-6.

- Collet, Pierre, and Eckmann, Jean-Pierre (1980). *Iterated Maps on the Interval as Dynamical Systems*. Birkhauser. ISBN 0-8176-4926-3.

- Devaney, Robert L. (2003). *An Introduction to Chaotic Dynamical Systems* (2nd ed.). Westview Press. ISBN 0-8133-4085-3.

- Gollub, J. P.; Baker, G. L. (1996). *Chaotic dynamics*. Cambridge University Press. ISBN 0-521-47685-2.

- Guckenheimer, John; Holmes, Philip (1983). *Nonlinear Oscillations, Dynamical Systems, and Bifurcations of Vector Fields*. Springer-Verlag. ISBN 0-387-90819-6.

- Gulick, Denny (1992). *Encounters with Chaos*. McGraw-Hill. ISBN 0-07-025203-3.

- Gutzwiller, Martin (1990). *Chaos in Classical and Quantum Mechanics*. Springer-Verlag. ISBN 0-387-97173-4.

- Hoover, William Graham (2001) [1999]. *Time Reversibility, Computer Simulation, and Chaos*. World Scientific. ISBN 981-02-4073-2.

4.9. SCIENTIFIC LITERATURE

- Kautz, Richard (2011). *Chaos: The Science of Predictable Random Motion*. Oxford University Press. ISBN 978-0-19-959458-0.

- Kiel, L. Douglas; Elliott, Euel W. (1997). *Chaos Theory in the Social Sciences*. Perseus Publishing. ISBN 0-472-08472-0.

- Moon, Francis (1990). *Chaotic and Fractal Dynamics*. Springer-Verlag. ISBN 0-471-54571-6.

- Ott, Edward (2002). *Chaos in Dynamical Systems*. Cambridge University Press. ISBN 0-521-01084-5.

- Strogatz, Steven (2000). *Nonlinear Dynamics and Chaos*. Perseus Publishing. ISBN 0-7382-0453-6.

- Sprott, Julien Clinton (2003). *Chaos and Time-Series Analysis*. Oxford University Press. ISBN 0-19-850840-9.

- Tél, Tamás; Gruiz, Márton (2006). *Chaotic dynamics: An introduction based on classical mechanics*. Cambridge University Press. ISBN 0-521-83912-2.

- Teschl, Gerald (2012). *Ordinary Differential Equations and Dynamical Systems*. Providence: American Mathematical Society. ISBN 978-0-8218-8328-0.

- Thompson JM, Stewart HB (2001). *Nonlinear Dynamics And Chaos*. John Wiley and Sons Ltd. ISBN 0-471-87645-3.

- Tufillaro; Reilly (1992). *An experimental approach to nonlinear dynamics and chaos*. Addison-Wesley. ISBN 0-201-55441-0.

- Wiggins, Stephen (2003). *Introduction to Applied Dynamical Systems and Chaos*. Springer. ISBN 0-387-00177-8.

- Zaslavsky, George M. (2005). *Hamiltonian Chaos and Fractional Dynamics*. Oxford University Press. ISBN 0-19-852604-0.

4.9.3 Semitechnical and popular works

- Christophe Letellier, *Chaos in Nature*, World Scientific Publishing Company, 2012, ISBN 978-981-4374-42-2.

- Abraham, Ralph H.; Ueda, Yoshisuke, eds. (2000). *The Chaos Avant-Garde: Memoirs of the Early Days of Chaos Theory*. World Scientific. ISBN 978-981-238-647-2.

- Barnsley, Michael F. (2000). *Fractals Everywhere*. Morgan Kaufmann. ISBN 978-0-12-079069-2.

- Bird, Richard J. (2003). *Chaos and Life: Complexit and Order in Evolution and Thought*. Columbia University Press. ISBN 978-0-231-12662-5.

- John Briggs and David Peat, *Turbulent Mirror: : An Illustrated Guide to Chaos Theory and the Science of Wholeness*, Harper Perennial 1990, 224 pp.

- John Briggs and David Peat, *Seven Life Lessons of Chaos: Spiritual Wisdom from the Science of Change*, Harper Perennial 2000, 224 pp.

- Cunningham, Lawrence A. (1994). "From Random Walks to Chaotic Crashes: The Linear Genealogy of the Efficient Capital Market Hypothesis". *George Washington Law Review* **62**: 546.

- Predrag Cvitanović, *Universality in Chaos*, Adam Hilger 1989, 648 pp.

- Leon Glass and Michael C. Mackey, *From Clocks to Chaos: The Rhythms of Life*, Princeton University Press 1988, 272 pp.

- James Gleick, *Chaos: Making a New Science*, New York: Penguin, 1988. 368 pp.

- John Gribbin. *Deep Simplicity*. Penguin Press Science. Penguin Books.

- L Douglas Kiel, Euel W Elliott (ed.), *Chaos Theory in the Social Sciences: Foundations and Applications*, University of Michigan Press, 1997, 360 pp.

- Arvind Kumar, *Chaos, Fractals and Self-Organisation; New Perspectives on Complexity in Nature*, National Book Trust, 2003.

- Hans Lauwerier, *Fractals*, Princeton University Press, 1991.

- Edward Lorenz, *The Essence of Chaos*, University of Washington Press, 1996.

- Alan Marshall (2002) The Unity of Nature: Wholeness and Disintegration in Ecology and Science, Imperial College Press: London

- Heinz-Otto Peitgen and Dietmar Saupe (Eds.), *The Science of Fractal Images*, Springer 1988, 312 pp.

- Clifford A. Pickover, *Computers, Pattern, Chaos, and Beauty: Graphics from an Unseen World*, St Martins Pr 1991.

- Ilya Prigogine and Isabelle Stengers, *Order Out of Chaos*, Bantam 1984.

- Heinz-Otto Peitgen and P. H. Richter, *The Beauty of Fractals : Images of Complex Dynamical Systems*, Springer 1986, 211 pp.

- David Ruelle, *Chance and Chaos*, Princeton University Press 1993.

- Ivars Peterson, *Newton's Clock: Chaos in the Solar System*, Freeman, 1993.

- Ian Roulstone; John Norbury (2013). *Invisible in the Storm: the role of mathematics in understanding weather*. Princeton University Press. ISBN 0691152721.

- David Ruelle, *Chaotic Evolution and Strange Attractors*, Cambridge University Press, 1989.

- Peter Smith, *Explaining Chaos*, Cambridge University Press, 1998.

- Ian Stewart, *Does God Play Dice?: The Mathematics of Chaos*, Blackwell Publishers, 1990.

- Steven Strogatz, *Sync: The emerging science of spontaneous order*, Hyperion, 2003.

- Yoshisuke Ueda, *The Road To Chaos*, Aerial Pr, 1993.

- M. Mitchell Waldrop, *Complexity : The Emerging Science at the Edge of Order and Chaos*, Simon & Schuster, 1992.

- Sawaya, Antonio (2010). *Financial time series analysis : Chaos and neurodynamics approach*.

4.10 External links

- Hazewinkel, Michiel, ed. (2001), "Chaos", *Encyclopedia of Mathematics*, Springer, ISBN 978-1-55608-010-4

- Nonlinear Dynamics Research Group with Animations in Flash

- The Chaos group at the University of Maryland

- The Chaos Hypertextbook. An introductory primer on chaos and fractals

- ChaosBook.org An advanced graduate textbook on chaos (no fractals)

- Society for Chaos Theory in Psychology & Life Sciences

- Nonlinear Dynamics Research Group at CSDC, Florence Italy

- Interactive live chaotic pendulum experiment, allows users to interact and sample data from a real working damped driven chaotic pendulum

- Nonlinear dynamics: how science comprehends chaos, talk presented by Sunny Auyang, 1998.

- Nonlinear Dynamics. Models of bifurcation and chaos by Elmer G. Wiens

- Gleick's *Chaos* (excerpt)

- Systems Analysis, Modelling and Prediction Group at the University of Oxford

- A page about the Mackey-Glass equation

- High Anxieties — The Mathematics of Chaos (2008) BBC documentary directed by David Malone

- The chaos theory of evolution - article published in Newscientist featuring similarities of evolution and non-linear systems including fractal nature of life and chaos.

- Jos Leys, Étienne Ghys et Aurélien Alvarez, *Chaos, A Mathematical Adventure*. Nine films about dynamical systems, the butterfly effect and chaos theory, intended for a wide audience.

Chapter 5

Extremal principles in non-equilibrium thermodynamics

Energy dissipation and entropy production extremal principles are ideas developed within non-equilibrium thermodynamics that attempt to predict the likely steady states and dynamical structures that a physical system might show. The search for extremum principles for non-equilibrium thermodynamics follows their successful use in other branches of physics.[1][2][3][4][5][6] According to Kondepudi (2008),[7] and to Grandy (2008),[8] there is no general rule that provides an extremum principle that governs the evolution of a far-from-equilibrium system to a steady state. According to Glansdorff and Prigogine (1971, page 16),[9] irreversible processes usually are not governed by global extremal principles because description of their evolution requires differential equations which are not self-adjoint, but local extremal principles can be used for local solutions. Lebon Jou and Casas-Vásquez (2008)[10] state that "In non-equilibrium ... it is generally not possible to construct thermodynamic potentials depending on the whole set of variables". Šilhavý (1997)[11] offers the opinion that "... the extremum principles of thermodynamics ... do not have any counterpart for [non-equilibrium] steady states (despite many claims in the literature)." It follows that any general extremal principle for a non-equilibrium problem will need to refer in some detail to the constraints that are specific for the structure of the system considered in the problem.

5.1 Fluctuations, entropy, 'thermodynamics forces', and reproducible dynamical structure

Apparent 'fluctuations', which appear to arise when initial conditions are inexactly specified, are the drivers of the formation of non-equilibrium dynamical structures. There is no special force of nature involved in the generation of such fluctuations. Exact specification of initial conditions would require statements of the positions and velocities of all particles in the system, obviously not a remotely practical possibility for a macroscopic system. This is the nature of thermodynamic fluctuations. They cannot be predicted in particular by the scientist, but they are determined by the laws of nature and they are the singular causes of the natural development of dynamical structure.[9]

It is pointed out[12][13][14][15] by W.T. Grandy Jr that entropy, though it may be defined for a non-equilibrium system, is when strictly considered, only a macroscopic quantity that refers to the whole system, and is not a dynamical variable and in general does not act as a local potential that describes local physical forces. Under special circumstances, however, one can metaphorically think as if the thermal variables behaved like local physical forces. The approximation that constitutes classical irreversible thermodynamics is built on this metaphoric thinking.

As indicated by the " " marks of Onsager (1931),[1] such a metaphorical but not categorically mechanical force, the thermal "force", X_{th}, 'drives' the conduction of heat. For this so-called "thermodynamic force", we can write

$$X_{th} = -\frac{1}{T}\nabla T$$

Actually this thermal "thermodynamic force" is a manifestation of the degree of inexact specification of the microscopic initial conditions for the system, expressed in the thermodynamic variable known as temperature, T. Temperature is only one example, and all the thermodynamic macroscopic variables constitute inexact specifications of the initial conditions, and have their respective "thermodynamic forces". These inexactitudes of specification are the source of the apparent fluctuations that drive the generation of dynamical structure, of the very precise but still less than perfect reproducibility of non-equilibrium experiments, and of the place of entropy in thermodynamics. If one did not know of such inexactitude of specification, one might find the origin of the fluctuations mysterious. What is meant here by "inexactitude of specification" is not that the mean values of the macroscopic variables are inexactly specified, but that the use of macroscopic variables to describe processes that actually occur by the motions and interactions of microscopic objects such as molecules is necessarily lacking in the molecular detail of the processes, and is thus inexact. There are many microscopic states compatible with a single macroscopic state, but only the latter is specified, and that is specified exactly for the purposes of the theory.

It is reproducibility in repeated observations that identifies dynamical structure in a system. E.T. Jaynes[16][17][18][19] explains how this reproducibility is why entropy is so important in this topic: entropy is a measure of experimental reproducibility. The entropy tells how many times one would have to repeat the experiment in order to expect to see a departure from the usual reproducible result. When the process goes on in a system with less than a 'practically infinite' number (much much less than Avogadro's or Loschmidt's numbers) of molecules, the thermodynamic reproducibility fades, and fluctuations become easier to see.[20][21]

According to this view of Jaynes, it is a common and mystificatory abuse of language, that one often sees reproducibility of dynamical structure called "order".[8][22] Dewar[22] writes "Jaynes considered reproducibility - rather than disorder - to be the key idea behind the second law of thermodynamics (Jaynes 1963,[23] 1965,[19] 1988,[24] 1989[25])." Grandy (2008)[8] in section 4.3 on page 55 clarifies the distinction between the idea that entropy is related to order (which he considers to be an "unfortunate" "mischaracterization" that needs "debunking"), and the aforementioned idea of Jaynes that entropy is a measure of experimental reproducibility of process (which Grandy regards as correct). According to this view, even the admirable book of Glansdorff and Prigogine (1971)[9] is guilty of this unfortunate abuse of language.

5.2 Local thermodynamic equilibrium

Various principles have been proposed by diverse authors for over a century. According to Glansdorff and Prigogine (1971, page 15),[9] in general, these principles apply only to systems that can be described by thermodynamical variables, in which dissipative processes dominate by excluding large deviations from statistical equilibrium. The thermodynamical variables are defined subject to the kinematical requirement of local thermodynamic equilibrium. This means that collisions between molecules are so frequent that chemical and radiative processes do not disrupt the local Maxwell-Boltzmann distribution of molecular velocities.

5.3 Linear and non-linear processes

Dissipative structures can depend on the presence of non-linearity in their dynamical régimes. Autocatalytic reactions provide examples of non-linear dynamics, and may lead to the natural evolution of self-organized dissipative structures.

5.4 Continuous and discontinuous motions of fluids

Much of the theory of classical non-equilibrium thermodynamics is concerned with the spatially continuous motion of fluids, but fluids can also move with spatial discontinuities. Helmholtz (1868)[26] wrote about how in a flowing fluid, there can arise a zero fluid pressure, which sees the fluid broken asunder. This arises from the momentum of the fluid flow, showing a different kind of dynamical structure from that of the conduction of heat or electricity. Thus for example: water from a nozzle can form a shower of droplets (Rayleigh 1878,[27] and in section 357 et seq. of Rayleigh (1896/1926)[28]); waves on the surface of the sea break discontinuously when they reach the shore (Thom 1975[29]). Helmholtz pointed out that the sounds of organ pipes must arise from such discontinuity of flow, occasioned by the passage of air past a sharp-edged obstacle; otherwise the oscillatory character of the sound wave would be damped away to nothing. The definition of the rate of entropy production of such a flow is not covered by the usual theory of classical non-equilibrium thermodynamics. There are many other commonly observed discontinuities of fluid flow that also lie beyond the scope of the classical theory of non-equilibrium thermodynamics, such as: bubbles in boiling liquids and in effervescent drinks; also protected towers of deep tropical convection (Riehl, Malkus 1958[30]), also called penetrative convection (Lindzen 1977[31]).

5.5 Historical development

5.5.1 W. Thomson, Baron Kelvin

William Thomson, later Baron Kelvin, (1852 a,[32] 1852 b[33]) wrote

"II. When heat is created by any unreversible process (such as friction), there is a *dissipation* of mechanical energy, and a full *restoration* of it to its primitive condition is impossible.

III. When heat is diffused by *conduction*, there is a *dissipation* of mechanical energy, and perfect *restoration* is impossible.

IV. When radiant heat or light is absorbed, otherwise than in vegetation, or in a chemical reaction, there is a *dissipation* of mechanical energy, and perfect *restoration* is impossible."

In 1854, Thomson wrote about the relation between two previously known non-equilibrium effects. In the Peltier effect, an electric current driven by an external electric field across a bimetallic junction will cause heat to be carried across the junction when the temperature gradient is constrained to zero. In the Seebeck effect, a flow of heat driven by a temperature gradient across such a junction will cause an electromotive force across the junction when the electric current is constrained to zero. Thus thermal and electric effects are said to be coupled. Thomson (1854)[34] proposed a theoretical argument, partly based on the work of Carnot and Clausius, and in those days partly simply speculative, that the coupling constants of these two effects would be found experimentally to be equal. Experiment later confirmed this proposal. It was later one of the ideas that led Onsager to his results as noted below.

5.5.2 Helmholtz

In 1869, Hermann von Helmholtz stated,[35] subject to a certain kind of boundary condition, a principle of least viscous dissipation of kinetic energy: "For a steady flow in a viscous liquid, with the speeds of flow on the boundaries of the fluid being given steady, in the limit of small speeds, the currents in the liquid so distribute themselves that the dissipation of kinetic energy by friction is minimum."[36]

In 1878, Helmholtz,[37] like Thomson also citing Carnot and Clausius, wrote about electric current in an electrolyte solution with a concentration gradient. This shows a non-equilibrium coupling, between electric effects and concentration-driven diffusion. Like Thomson (Kelvin) as noted above, Helmholtz also found a reciprocal relation, and this was another of the ideas noted by Onsager.

5.5.3 J.W. Strutt, Baron Rayleigh

Rayleigh (1873)[38] (and in Sections 81 and 345 of Rayleigh (1896/1926)[28]) introduced the dissipation function for the description of dissipative processes involving viscosity. More general versions of this function have been used by many subsequent investigators of the nature of dissipative processes and dynamical structures. Rayleigh's dissipation function was conceived of from a mechanical viewpoint, and it did not refer in its definition to temperature, and it needed to be 'generalized' to make a dissipation function suitable for use in non-equilibrium thermodynamics.

Studying jets of water from a nozzle, Rayleigh (1878,[27] 1896/1926[28]) noted that when a jet is in a state of conditionally stable dynamical structure, the mode of fluctuation most likely to grow to its full extent and lead to another state of conditionally stable dynamical structure is the one with the fastest growth rate. In other words, a jet can settle into a conditionally stable state, but it is likely to suffer fluctuation so as to pass to another, less unstable, conditionally stable state. He used like reasoning in a study of Bénard convection.[39] These physically lucid considerations of Rayleigh seem to contain the heart of the distinction between the principles of minimum and maximum rates of dissipation of energy and entropy production, which have been developed in the course of physical investigations by later authors.

5.5.4 Korteweg

Korteweg (1883)[40] gave a proof "that in any simply connected region, when the velocities along the boundaries are given, there exists, as far as the squares and products of the velocities may be neglected, only one solution of the equations for the steady motion of an incompressible viscous fluid, and that this solution is always stable." He attributed the first part of this theorem to Helmholtz, who had shown that it is a simple consequence of a theorem that "if the motion be steady, the currents in a viscous [incompressible] fluid are so distributed that the loss of [kinetic] energy due to viscosity is a minimum, on the supposition that the velocities along boundaries of the fluid are given." Because of the restriction to cases in which the squares and products of the velocities can be neglected, these motions are below the threshold for turbulence.

5.5.5 Onsager

Great theoretical progress was made by Onsager in 1931[1][41] and in 1953.[42][43]

5.5.6 Prigogine

Further progress was made by Prigogine in 1945[44] and later.[9][45] Prigogine (1947)[44] cites Onsager (1931).[1][41]

5.5.7 Casimir

Casimir (1945)[46] extended the theory of Onsager.

5.5.8 Ziman

Ziman (1956)[47] gave very readable account. He proposed the following as a general principle of the thermodynamics of irreversible processes: "*Consider all distributions of currents such that the intrinsic entropy production equals the extrinsic entropy production for the given set of forces. Then, of all current distributions satisfying this condition, the steady state distribution makes the entropy production a maximum.*" He commented that this was a known general principle, discovered by Onsager, but was "not quoted in any of the books on the subject". He notes the difference between this principle and "Prigogine's theorem, which states, crudely speaking, that if not all the forces acting on a system are fixed the free forces will take such values as to make the entropy production a minimum." Prigogine was present when this paper was read and he is reported by the journal editor to have given "notice that he doubted the validity of part of Ziman's thermodynamic interpretation".

5.5.9 Ziegler

Hans Ziegler extended the Melan-Prager non-equilibrium theory of materials to the non-isothermal case.[48]

5.5.10 Gyarmati

Gyarmati (1967/1970)[2] gives a systematic presentation, and extends Onsager's principle of least dissipation of energy, to give a more symmetric form known as Gyarmati's principle. Gyarmati (1967/1970)[2] cites 11 papers or books authored or co-authored by Prigogine.

Gyarmati (1967/1970)[2] also gives in Section III 5 a very helpful precis of the subtleties of Casimir (1945)).[46] He explains that the Onsager reciprocal relations concern variables which are even functions of the velocities of the molecules, and notes that Casimir went on to derive antisymmetric relations concerning variables which are odd functions of the velocities of the molecules.

5.5.11 Paltridge

The physics of the earth's atmosphere includes dramatic events like lightning and the effects of volcanic eruptions, with discontinuities of motion such as noted by Helmholtz (1868).[26] Turbulence is prominent in atmospheric convection. Other discontinuities include the formation of raindrops, hailstones, and snowflakes. The usual theory of classical non-equilibrium thermodynamics will need some extension to cover atmospheric physics. According to Tuck (2008),[49] "On the macroscopic level, the way has been pioneered by a meteorologist (Paltridge 1975,[50] 2001[51]). Initially Paltridge (1975)[50] used the terminology "minimum entropy exchange", but after that, for example in Paltridge (1978),[52] and in Paltridge (1979)[53]), he used the now current terminology "maximum entropy production" to describe the same thing. This point is clarified in the review by Ozawa, Ohmura, Lorenz, Pujol (2003).[54] Paltridge (1978)[52] cited Busse's (1967)[55] fluid mechanical work concerning an extremum principle. Nicolis and Nicolis (1980)[56] discuss Paltridge's work, and they comment that the behaviour of the entropy production is far from simple and universal. This seems natural in the context of the requirement of some classical theory of non-equilibrium thermodynamics that the threshold of turbulence not be crossed. Paltridge himself nowadays tends to prefer to think in terms of the dissipation function rather than in terms of rate of entropy production.

5.6 Speculated thermodynamic extremum principles for energy dissipation and entropy production

Jou, Casas-Vazquez, Lebon (1993)[57] note that classical non-equilibrium thermodynamics "has seen an extraordinary expansion since the second world war", and they refer to the Nobel prizes for work in the field awarded to Lars Onsager and Ilya Prigogine. Martyushev and Seleznev (2006)[4] note the importance of entropy in the evolution of natural dynamical structures: "Great contribution has been done in this respect by two scientists, namely Clausius, ... , and Prigogine." Prigogine in his 1977 Nobel Lecture[58] said: "... non-equilibrium may be a source of order. Irreversible processes may lead to a new type of dynamic states of matter which I have called "dissipative structures"." Glansdorff and Prigogine (1971)[9] wrote on page xx: "Such 'symmetry breaking instabilities' are of special interest as they lead to a spontaneous 'self-organization' of the system both from the point of view of its *space order* and its *function*."

Analyzing the Rayleigh-Bénard convection cell phenomenon, Chandrasekhar (1961)[59] wrote "Instability occurs at the minimum temperature gradient at which a balance can be maintained between the kinetic energy dissipated by viscosity and the internal energy released by the buoyancy force." With a temperature gradient greater than the minimum, viscosity can dissipate kinetic energy as fast as it is released by convection due to buoyancy, and a steady state with convection is stable. The steady state with convection is often a pattern of macroscopically visible hexagonal cells with convection up or down in the middle or at the 'walls' of each cell, depending on the temperature dependence of the quantities; in the atmosphere under various conditions it seems that either is possible. (Some details are discussed by Lebon, Jou, and Casas-Vásquez (2008)[10] on pages 143-158.) With a temperature gradient less than the minimum, viscosity and heat conduction are so effective that convection cannot keep going.

Glansdorff and Prigogine (1971)[9] on page xv wrote "Dissipative structures have a quite different [from equilibrium structures] status: they are formed and maintained through the effect of exchange of energy and matter in non-equilibrium conditions." They were referring to the dissipation function of Rayleigh (1873)[38] that was used also by Onsager (1931, I,[1] 1931, II[41]). On pages 78–80 of their book[9] Glansdorff and Prigogine (1971) consider the stability of laminar flow that was pioneered by Helmholtz; they concluded that at a stable steady state of sufficiently slow laminar flow, the dissipation function was minimum.

These advances have led to proposals for various extremal principles for the "self-organized" régimes that are possible for systems governed by classical linear and non-linear non-equilibrium thermodynamical laws, with stable stationary régimes being particularly investigated. Convection introduces effects of momentum which appear as non-linearity in the dynamical equations. In the more restricted case of no convective motion, Prigogine wrote of "dissipative structures". Šilhavý (1997)[11] offers the opinion that "... the extremum principles of [equilibrium] thermodynamics ... do not have any counterpart for [non-equilibrium] steady states (despite many claims in the literature)."

5.6.1 Prigogine's proposed theorem of minimum entropy production for very slow purely diffusive transfer

In 1945 Prigogine [44] (see also Prigogine (1947)[60]) proposed a "Theorem of Minimum Entropy Production" which applies only to the purely diffusive linear regime, with negligible inertial terms, near a stationary thermodynamically non-equilibrium state. Prigogine's proposal is that the rate of entropy production is locally minimum at every point. The proof offered by Prigogine is open to serious criticism.[61] A critical and unsupportive discussion of Prigogine's proposal is offered by Grandy (2008).[8] It has been shown by Barbera that the total whole body entropy production cannot be minimum, but this paper did not consider the pointwise minimum proposal of Prigogine.[62] A proposal closely related to Prigogine's is that the pointwise rate of entropy production should have its maximum value minimized at the steady state. This is compatible, but not identical, with the Prigogine proposal.[63] Moreover, N. W. Tschoegl proposes a proof, perhaps more physically motivated than Prigogine's, that would if valid support the conclusion of Helmholtz and of Prigogine, that under these restricted conditions, the entropy production is at a pointwise minimum.[64]

5.6.2 Faster transfer with convective circulation: second entropy

In contrast to the case of sufficiently slow transfer with linearity between flux and generalized force with negligible inertial terms, there can be heat transfer that is not very slow. Then there is consequent non-linearity, and heat flow can develop into phases of convective circulation. In these cases, the time rate of entropy production has been shown to be a non-monotonic function of time during the approach to steady state heat convection. This makes these cases different from the near-thermodynamic-equilibrium regime of very-slow-transfer with linearity. Accordingly, the local time rate of entropy production, defined according to the local thermodynamic equilibrium hypothesis, is not an adequate variable for prediction of the time course of far-from-thermodynamic equilibrium processes. The principle of minimum entropy production is not applicable to these cases.

To cover these cases, there is needed at least one further state variable, a non-equilibrium quantity, the so-called second entropy. This appears to be a step towards generalization beyond the classical second law of thermodynamics, to cover non-equilibrium states or processes. The classical law refers only to states of thermodynamic equilibrium, and local thermodynamic equilibrium theory is an approximation that relies upon it. Still it is invoked to deal with phenomena near but not at thermodynamic equilibrium, and has some uses then. But the classical law is inadequate for description of the time course of processes far from thermodynamic equilibrium. For such processes, a more powerful theory is needed, and the second entropy is part of such a theory.[65][66]

5.6.3 Speculated principles of maximum entropy production and minimum energy dissipation

Onsager (1931, I)[1] wrote: "Thus the vector field J of the heat flow is described by the condition that the rate of increase of entropy, less the dissipation function, be a maximum." Careful note needs to be taken of the opposite signs of the rate of entropy production and of the dissipation function, appearing in the left-hand side of Onsager's equation (5.13) on Onsager's page 423.[1]

Although largely unnoticed at the time, Ziegler proposed

an idea early with his work in the mechanics of plastics in 1961,[67] and later in his book on thermomechanics revised in 1983,[3] and in various papers (e.g., Ziegler (1987),[68]). Ziegler never stated his principle as a universal law but he may have intuited this. He demonstrated his principle using vector space geometry based on an "orthogonality condition" which only worked in systems where the velocities were defined as a single vector or tensor, and thus, as he wrote[3] at p. 347, was "impossible to test by means of macroscopic mechanical models", and was, as he pointed out, invalid in "compound systems where several elementary processes take place simultaneously".

In relation to the earth's atmospheric energy transport process, according to Tuck (2008),[49] "On the macroscopic level, the way has been pioneered by a meteorologist (Paltridge 1975,[50] 2001[69])." Initially Paltridge (1975)[50] used the terminology "minimum entropy exchange", but after that, for example in Paltridge (1978),[52] and in Paltridge (1979),[70] he used the now current terminology "maximum entropy production" to describe the same thing. The logic of Paltridge's earlier work is open to serious criticism.[8] Nicolis and Nicolis (1980)[71] discuss Paltridge's work, and they comment that the behaviour of the entropy production is far from simple and universal. Later work by Paltridge focuses more on the idea of a dissipation function than on the idea of rate of production of entropy.[69]

Sawada (1981),[72] also in relation to the Earth's atmospheric energy transport process, postulating a principle of largest amount of entropy increment per unit time, cites work in fluid mechanics by Malkus and Veronis (1958)[73] as having "proven a principle of maximum heat current, which in turn is a maximum entropy production for a given boundary condition", but this inference is not logically valid. Again investigating planetary atmospheric dynamics, Shutts (1981)[74] used an approach to the definition of entropy production, different from Paltridge's, to investigate a more abstract way to check the principle of maximum entropy production, and reported a good fit.

5.6.4 Prospects

Until recently, prospects for useful extremal principles in this area have seemed clouded. C. Nicolis (1999)[75] concludes that one model of atmospheric dynamics has an attractor which is not a regime of maximum or minimum dissipation; she says this seems to rule out the existence of a global organizing principle, and comments that this is to some extent disappointing; she also points to the difficulty of finding a thermodynamically consistent form of entropy production. Another top expert offers an extensive discussion of the possibilities for principles of extrema of entropy production and of dissipation of energy: Chapter 12 of Grandy (2008)[8] is very cautious, and finds difficulty in defining the 'rate of internal entropy production' in many cases, and finds that sometimes for the prediction of the course of a process, an extremum of the quantity called the rate of dissipation of energy may be more useful than that of the rate of entropy production; this quantity appeared in Onsager's 1931[1] origination of this subject. Other writers have also felt that prospects for general global extremal principles are clouded. Such writers include Glansdorff and Prigogine (1971), Lebon, Jou and Casas-Vásquez (2008), and Šilhavý (1997). A recent proposal may perhaps by-pass those clouded prospects.[65]

5.7 See also

- Non-equilibrium thermodynamics
- Dissipative system
- Self-organization
- Autocatalytic reactions and order creation
- Fluctuation theorem
- Fluctuation dissipation theorem

5.8 References

[1] Onsager, L. (1931). Reciprocal relations in irreversible processes, I, *Physical Review* 37:405-426

[2] Gyarmati, I. (1970). *Non-equilibrium Thermodynamics: Field Theory and Variational Principles*, Springer, Berlin; translated, by E. Gyarmati and W.F. Heinz, from the original 1967 Hungarian *Nemegyensulyi Termodinamika*, Muszaki Konyvkiado, Budapest.

[3] Ziegler, H., (1983). *An Introduction to Thermomechanics*, North-Holland, Amsterdam, ISBN 0-444-86503-9

[4] Martyushev, L.M., Seleznev, V.D. (2006). Maximum entropy production principle in physics, chemistry and biology, *Physics Reports* 426: 1-45

[5] Martyushev, I.M., Nazarova, A.S., Seleznev, V.D. (2007). On the problem of the minimum entropy production in the nonequilibrium stationary state, *Journal of Physics A: Mathematical and Theoretical* 40: 371-380.

[6] Hillert, M., Agren, J. (2006). Extremum principles for irreversible processes, *Acta Materialia* 54: 2063-2066

[7] Kondepudi, D. (2008)., *Introduction to Modern Thermodynamics*, Wiley, Chichester UK, ISBN 978-0-470-01598-8, page 172.

5.8. REFERENCES

[8] Grandy, W.T., Jr (2008). *Entropy and the Time Evolution of Macroscopic Systems*, Oxford University Press, Oxford, ISBN 978-0-19-954617-6.

[9] Glansdorff, P., Prigogine, I. (1971). *Thermodynamic Theory of Structure, Stability and Fluctuations*, Wiley-Interscience, London. ISBN 0-471-30280-5

[10] Lebon, G., Jou, J., Casas-Vásquez (2008). *Understanding Non-equilibrium Thermodynamics. Foundations, Applications, Frontiers*, Springer, Berlin, ISBN 978-3-540-74251-7.

[11] Šilhavý, M. (1997). *The Mechanics and Thermodynamics of Continuous Media*, Springer, Berlin, ISBN 3-540-58378-5, page 209.

[12] Grandy, W.T., Jr (2004). Time evolution in macroscopic systems. I: Equations of motion. *Found. Phys.* 34: 1-20. See .

[13] Grandy, W.T., Jr (2004). Time evolution in macroscopic systems. II: The entropy. *Found. Phys.* 34: 21-57. See .

[14] Grandy, W.T., Jr (2004). Time evolution in macroscopic systems. III: Selected applications. *Found. Phys.* 34: 771-813. See .

[15] Grandy 2004 see also .

[16] Jaynes, E.T. (1957). Information theory and statistical mechanics, *Physical Review* 106: 620-630.

[17] Jaynes, E.T. (1957). Information theory and statistical mechanics. II, *Physical Review* 108: 171-190.

[18] Jaynes, E.T. (1985). Macroscopic prediction, in *Complex Systems - Operational Approaches in Neurobiology*, edited by H. Haken, Springer-Verlag, Berlin, pp. 254-269, ISBN 3-540-15923-1.

[19] Jaynes, E.T. (1965). Gibbs vs Boltzmann Entropies, *American Journal of Physics* 33: 391-398.

[20] Evans, D.J., Searles, D.J. (2002). The fluctuation theorem, *Advances in Physics* 51: 1529-1585

[21] Wang, G.M., Sevick, E.M., Mittag, E., Searles, D.J., Evans, D.J. (2002) Experimental demonstration of violations of the Second Law of Thermodynamics for small systems and short time scales, *Physical Review Letters* 89: 050601-1 - 050601-4.

[22] Dewar, R.C. (2005). Maximum entropy production and non-equilibrium statistical mechanics, pp. 41-55 in *Non-equilibrium Thermodynamics and the Production of Entropy*, edited by A. Kleidon, R.D. Lorenz, Springer, Berlin. ISBN 3-540-22495-5.

[23] Jaynes, E.T. (1963). pp. 181-218 in *Brandeis Summer Institute 1962, Statistical Physics*, edited by K.W. Ford, Benjamin, New York.

[24] Jaynes, E.T. (1988). The evolution of Carnot's Principle, pp. 267-282 in *Maximum-entropy and Bayesian methods in science and engineering*, edited by G.J. Erickson, C.R. Smith, Kluwer, Dordrecht, volume 1, ISBN 90-277-2793-7.

[25] Jaynes, E.T. (1989). Clearing up mysteries, the original goal, pp. 1-27 in *Maximum entropy and Bayesian methods*, Kluwer, Dordrecht.

[26] Helmholtz, H. (1868). On discontinuous movements of fluids, *Philosophical Magazine* series 4, vol. 36: 337-346, translated by F. Guthrie from *Monatsbericht der koeniglich preussischen Akademie der Wissenschaften zu Berlin* April 1868, page 215 et seq.

[27] Strutt, J.W. (Baron Rayleigh) (1878). On the instability of jets, *Proceedings of the London Mathematical Society* 10: 4-13.

[28] Strutt, J.W. (Baron Rayleigh) (1896/1926). Section 357 et seq. *The Theory of Sound*, Macmillan, London, reprinted by Dover, New York, 1945.

[29] Thom, R. (1975). *Structural Stability and Morphogenesis: An outline of a general theory of models*, translated from the French by D.H. Fowler, W.A. Benjamin, Reading Ma, ISBN 0-8053-9279-3

[30] Riehl, H., Malkus, J.S. (1958). On the heat balance in the equatorial trough zone, *Geophysica* 6: 503-538.

[31] Lindzen, R.S. (1977). Some aspects of convection in meteorology, pp. 128-141 in *Problems of Stellar Convection*, volume 71 of *Lecture Notes in Physics*, Springer, Berlin, ISBN 978-3-540-08532-4.

[32] Thomson, William (1852 a). "On a Universal Tendency in Nature to the Dissipation of Mechanical Energy" Proceedings of the Royal Society of Edinburgh for April 19, 1852 [This version from Mathematical and Physical Papers, vol. i, art. 59, pp. 511.]

[33] Thomson, W. (1852 b). On a universal tendency in nature to the dissipation of mechanical energy, *Philosophical Magazine* 4: 304-306.

[34] Thomson, W. (1854). On a mechanical theory of thermo-electric currents, *Proceedings of the Royal Society of Edinburgh* pp. 91-98.

[35] Helmholtz, H. (1869/1871). Zur Theorie der stationären Ströme in reibenden Flüssigkeiten, *Verhandlungen des naturhistorisch-medizinischen Vereins zu Heidelberg*, Band V: 1-7. Reprinted in Helmholtz, H. (1882), *Wissenschaftliche Abhandlungen*, volume 1, Johann Ambrosius Barth, Leipzig, pages 223-230

[36] from page 2 of Helmholtz 1869/1871, translated by Wikipedia editor.

[37] Helmholtz, H. (1878). Ueber galvanische Ströme, verursacht durch Concentrationsunterschiede; Folgeren aus der mechanischen Wärmetheorie, Wiedermann's *Annalen der Physik und Chemie* 3: 201-216.

[38] Strutt, J.W. (Baron Rayleigh) (1873). Some theorems relating to vibrations, *Proceedings of the London Mathematical Society* 4:357-368.

[39] Strutt, J.W. (Baron Rayleigh) (1916). On convection currents in a horizontal layer of fluids, when the higher temperature is on the under side, *The London, Edinburgh, and Dublin Philosophical Magazine* series 6, volume 32: 529-546.

[40] Korteweg, D.J., (1883). On a general theorem of the stability of the motion of a viscous fluid, *The London, Edinburgh and Dublin Philosophical Journal of Science* 16: 112-118.

[41] Onsager, L. (1931). Reciprocal relations in irreversible processes. II, *Physical Review* 38: 2265-2279

[42] Onsager, L., Machlup, S. (1953). Fluctuations and Irreversible Processes, *Physical Review* 91: 1505-1512.

[43] Machlup, S., Onsager, L., (1953). Fluctuations and Irreversible Processes. II. Systems with kinetic energy, *Physical Review* 91: 1512-1515.

[44] Prigogine, I. (1945). Modération et transformations irréversibles des systèmes ouverts, *Bulletin de la Classe des Sciences., Académie Royale de Belgique* 31: 600-606.

[45] Prigogine, I. (1947). *Étude thermodynamique des Phenomènes Irréversibles*, Desoer, Liège.

[46] Casimir, H.B.G. (1945). On Onsager's principle of microscopic reversibility, *Reviews of Modern Physics* 17:343-350

[47] Ziman, J.M. (1956). The general variational principle of transport theory, *Canadian Journal of Physics* 34: 1256-1273.

[48] T. Inoue (2002). Metallo-Thermo-Mechanics–Application to Quenching. *In* G. Totten, M. Howes, and T. Inoue (eds.), Handbook of Residual Stress. pp. 296-311, ASM International, Ohio.

[49] Tuck, Adrian F. (2008) *Atmospheric Turbulence: a molecular dynamics perspective*, Oxford University Press. ISBN 978-0-19-923653-4. See page 33.

[50] Paltridge, G.W. (1975). Global dynamics and climate - a system of minimum entropy exchange, *Quarterly Journal of the Royal Meteorological Society 101*:475-484.

[51] Paltridge G.W.(2001). A physical basis for a maximum of thermodynamic dissipation of the climate system, *Quarterly Journal of the Royal Meteorological Society* 127:305-313.

[52] Paltridge, G.W. (1978). The steady-state format of global climate, *Quarterly Journal of the Royal Meteorological Society* 104: 927-945.

[53] Paltridge, G.W. (1979). Climate and thermodynamic systems of maximum dissipation, *Nature* 279:630-631.

[54] Ozawa, H., Ohmura, A., Lorenz, R.D., Pujol, T. (2003). The Second Law of Thermodynamics and the Global Climate System: A Review of the Maximum Entropy Production Principle, *Reviews of Geophysics*, 41, 4: 1-24.

[55] Busse, F.H.(1967). The stability of finite amplitude cellular convection and its relation to an extremum principle, *Journal of Fluid Mechanics* 30(4): 625-649.

[56] Nicolis, G., Nicolis, C. (1980). On the entropy balance of the earth-atmosphere system, *Quarterly Journal of the Royal Meteorological Society* 125:1859-1878.

[57] Jou, D., Casas-Vázquez, J., Lebon, G. (1993). *Extended Irreversible Thermodynamics*, Springer, Berlin, ISBN 3-540-55874-8, ISBN 0-387-55874-8.

[58] Prigogine, I. (1977). Time, Structure and Fluctuations, Nobel Lecture.

[59] Chandrasekhar, S. (1961). *Hydrodynamic and Hydromagnetic Stability*, Clarendon Press, Oxford.

[60] Prigogine, I. (1947). *Étude thermodynamique des Phenomènes Irreversibles*, Desoer, Liege.

[61] Lavenda, B.H. (1978). *Thermodynamics of Irreversible Processes*, Macmillan, London, ISBN 0-333-21616-4.

[62] Barbera, E. (1999). On the principle of minimum entropy production for Navier-Stokes-Fourier fluids, *Continuum Mech. Thermodyn.*, 11: 327–330.

[63] Struchtrup, H., Weiss, W. (1998). Maximum of the local entropy production becomes minimal in stationary processes, *Phys. Rev. Lett.*, 80: 5048–5051.

[64] Tschoegl, N.W. (2000). *Fundamentals of Equilibrium and Steady-State Thermodynamics*, Elsevier, Amsterdam, ISBN 0-444-50426-5, Chapter 30, pp. 213–215.

[65] Attard, P. (2012). "Optimising Principle for Non-Equilibrium Phase Transitions and Pattern Formation with Results for Heat Convection". arXiv:1208.5105.

[66] Attard, P. (2012). *Non-Equilibrium Thermodynamics and Statistical Mechanics: Foundations and Applications*, Oxford University Press, Oxford UK, ISBN 978-0-19-966276-0.

[67] Ziegler, H. (1961). "Zwei Extremalprinzipien der irreversiblen Thermodynamik". *Ingenieur-Archiv* 30 (6): 410–416. doi:10.1007/BF00531783.

[68] Ziegler, H.; Wehrli, C. (1987). "On a principle of maximal rate of entropy production". *J. Non-Equilib. Thermodyn* 12 (3): 229–243. Bibcode:1987JNET...12..229Z. doi:10.1515/jnet.1987.12.3.229.

[69] Paltridge, Garth W. (2001). "A physical basis for a maximum of thermodynamic dissipation of the climate system". *Quarterly Journal of the Royal Meteorological Society* 127 (572): 305. Bibcode:2001QJRMS.127..305P. doi:10.1002/qj.49712757203.

5.8. REFERENCES

[70] Paltridge, Garth W. (1979). "Climate and thermodynamic systems of maximum dissipation". *Nature* **279** (5714): 630. Bibcode:1979Natur.279..630P. doi:10.1038/279630a0.

[71] Nicolis, G.; Nicolis, C. (1980). "On the entropy balance of the earth-atmosphere system". *Quarterly Journal of the Royal Meteorological Society* **125** (557): 1859–1878. Bibcode:1999QJRMS.125.1859N. doi:10.1002/qj.49712555718.

[72] Sawada, Y. (1981). A thermodynamic variational principle in nonlinear non-equilibrium phenomena, *Progress of Theoretical Physics* 66: 68-76.

[73] Malkus, W.V.R.; Veronis, G. (1958). "Finite amplitude cellular convection". *Journal of Fluid Mechanics* **4** (3): 225–260. Bibcode:1958JFM.....4..225M. doi:10.1017/S0022112058000410.

[74] Shutts, G.J. (1981). "Maximum entropy production states in quasi-geostrophic dynamical models". *Quarterly Journal of the Royal Meteorological Society* **107** (453): 503–520. doi:10.1256/smsqj.45302.

[75] Nicolis, C. (1999). "Entropy production and dynamical complexity in a low-order atmospheric model". *Quarterly Journal of the Royal Meteorological Society* **125** (557): 1859–1878. Bibcode:1999QJRMS.125.1859N. doi:10.1002/qj.49712555718.

Chapter 6

Self-assembly

For other uses, see Self-construction (disambiguation).

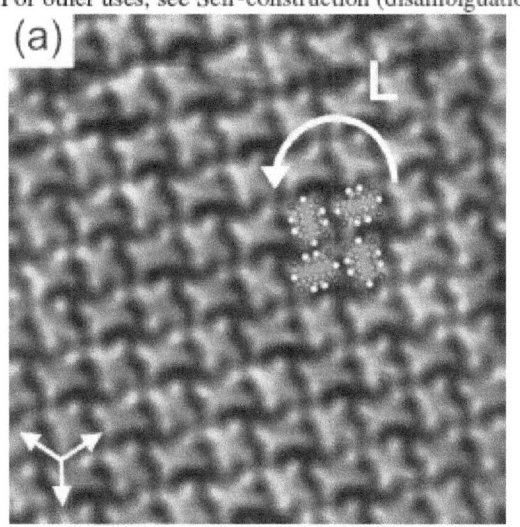

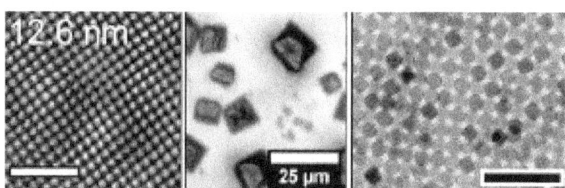

Iron oxide nanoparticles can be dispersed in an organic solvent (toluene). Upon its evaporation, they may self-assemble (left and right panels) into micron-sized mesocrystals (center) or multilayers (right). Each dot in the left image is a traditional "atomic" crystal shown in the image above. Scale bars: 100 nm (left), 25 μm (center), 50 nm (right).[1]

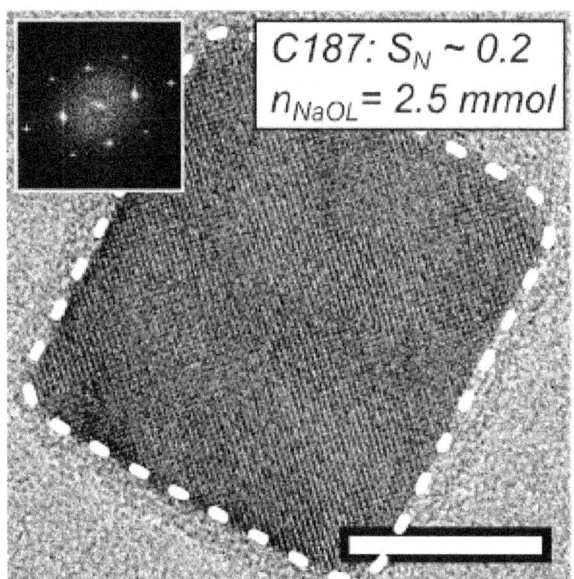

Electron microscopy image of an iron oxide nanoparticle. Regularly arranged dots within the dashed border are Fe atoms. Left inset is the corresponding electron diffraction pattern. Scale bar: 10 nm.[1]

Self-assembly is a process in which a disordered system of pre-existing components forms an organized structure or pattern as a consequence of specific, local interactions among the components themselves, without external direction. When the constitutive components are molecules, the process is termed molecular self-assembly.

Self-assembly can be classified as either static or dynamic. In *static* self-assembly, the ordered state forms as a system approaches equilibrium, reducing its free energy. However, in *dynamic* self-assembly, patterns of pre-existing components organized by specific local interactions are not commonly described as "self-assembled" by scientists in the associated disciplines. These structures are better described as "self-organized".

6.1 Self-assembly in chemistry and materials science

Self-assembly (SA) in the classic sense can be defined as *the spontaneous and reversible organization of molecular units*

6.1. SELF-ASSEMBLY IN CHEMISTRY AND MATERIALS SCIENCE

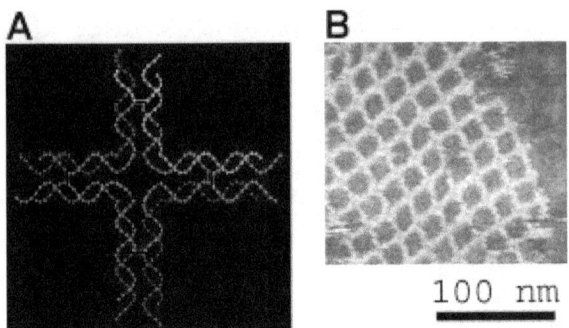

The DNA structure at left (schematic shown) will self-assemble into the structure visualized by atomic force microscopy at right. Image from Strong.[3]

into ordered structures by non-covalent interactions. The first property of a self-assembled system that this definition suggests is the spontaneity of the self-assembly process: the interactions responsible for the formation of the self-assembled system act on a strictly local level—in other words, *the nanostructure builds itself*.

6.1.1 Distinctive features

At this point, one may argue that any chemical reaction driving atoms and molecules to assemble into larger structures, such as precipitation, could fall into the category of SA. However, there are at least three distinctive features that make SA a distinct concept.

Order

First, the self-assembled structure must have a higher order than the isolated components, be it a shape or a particular task that the self-assembled entity may perform. This is generally not true in chemical reactions, where an ordered state may proceed towards a disordered state depending on thermodynamic parameters.

Interactions

The second important aspect of SA is the key role of slack interactions (e.g. Van der Waals, capillary, $\pi-\pi$, hydrogen bonds) with respect to more "traditional" covalent, ionic, or metallic bonds. Although typically less energetic by a factor of 10, these weak interactions play an important role in materials synthesis. It can be instructive to note how slack interactions hold a prominent place in materials, especially in biological systems, although they are often considered marginal with respect to "strong" (i.e. covalent, etc.) interactions. For instance, they determine the physical proper-

ties of liquids, the solubility of solids, and the organization of molecules in biological membranes.

Building blocks

The third distinctive feature of SA is that the building blocks are not only atoms and molecules, but span a wide range of nano- and mesoscopic structures, with different chemical compositions, shapes and functionalities.[4] Research into possible three-dimensional shapes of self-assembling micrites examines Platonic solids (regular polyhedral). The term 'micrite' was created by DARPA to refer to sub-millimeter sized microrobots, whose self-organizing abilities may be compared with those of slime mold.[5][6] Recent examples of novel building blocks include polyhedra and patchy particles. These nanoscale building blocks (NBBs) can in turn be synthesised through conventional chemical routes or by other SA strategies such as Directional Entropic Forces.

6.1.2 Examples

Important examples of SA in materials science include the formation of molecular crystals, colloids, lipid bilayers, phase-separated polymers, and self-assembled monolayers.[7][8] The folding of polypeptide chains into proteins and the folding of nucleic acids into their functional forms are examples of self-assembled biological structures. Recently, the three-dimensional macroporous structure was prepared via self-assembly of diphenylalanine derivative under cryoconditions, the obtained material can find the application in the field of regenerative medicine or drug delivery system.[9] P. Chen et al. demonstrated a microscale self-assembly method using the air-liquid interface established by Faraday wave as a template. This self-assembly method can be used for generation of diverse sets of symmetrical and periodic patterns from microscale materials such as hydrogels, cells, and cell spheroids.[10]

6.1.3 Properties

SA extends the scope of chemistry aiming at synthesising products with order and functionality properties, extending chemical bonds to weak interactions and encompassing the self-assembly of NBBs on all length scales.[11] In covalent synthesis and polymerisation, the scientist links atoms together in any desired conformation, which does not necessarily have to be the energetically most favoured position; self-assembling molecules, on the other hand, adopt a structure at the thermodynamic minimum, finding the best combination of interactions between subunits but not forming covalent bonds between them. In self-assembling structures, the scientist must predict this minimum, not merely place the atoms in the location desired.

Another characteristic common to nearly all self-assembled systems is their thermodynamic stability. For SA to take place without intervention of external forces, the process must lead to a lower Gibbs free energy, thus self-assembled structures are thermodynamically more stable than the single, unassembled components. A direct consequence is the general tendency of self-assembled structures to be relatively free of defects. An example is the formation of two-dimensional superlattices composed of an orderly arrangement of micrometre-sized polymethylmethacrylate (PMMA) spheres, starting from a solution containing the microspheres, in which the solvent is allowed to evaporate slowly in suitable conditions. In this case, the driving force is capillary interaction, which originates from the deformation of the surface of a liquid caused by the presence of floating or submerged particles.[12]

These two properties—weak interactions and thermodynamic stability—can be recalled to rationalise another property often found in self-assembled systems: the *sensitivity to perturbations* exerted by the external environment. These are small fluctuations that alter thermodynamic variables that might lead to marked changes in the structure and even compromise it, either during or after SA. The weak nature of interactions is responsible for the flexibility of the architecture and allows for rearrangements of the structure in the direction determined by thermodynamics. If fluctuations bring the thermodynamic variables back to the starting condition, the structure is likely to go back to its initial configuration. This leads us to identify one more property of SA, which is generally not observed in materials synthesised by other techniques: *reversibility*.

SA is a process which is easily influenced by external parameters: if this can make synthesis more problematic due to the many free parameters that require control, on the other hand it has the exciting advantage that a large variety of shapes and functions on many length scales can be obtained.[13]

Generally speaking, the fundamental condition needed for NBBs to self-assemble into an ordered structure is the simultaneous presence of long-range repulsive and short-range attractive forces.[14]

By choosing precursors with suitable physicochemical properties, it is possible to exert a fine control on the formation processes that produce complex structures. Clearly, the most important tool when it comes to designing a synthesis strategy for a material, is the knowledge of the chemistry of the building units. For example, it was demonstrated that it was possible to use diblock copolymers with different block reactivities in order to selectively embed maghemite nanoparticles and generate periodic materials with potential use as waveguides.[15]

In 2008, *Advances in Colloid and Interface Science* published a study in which it was concluded that every self-assembly process in reality presents a co-assembly, which makes the former term a misnomer of a kind.[16] The thesis is built on the concept of mutual ordering of the self-assembling system and its environment.

6.2 Self-assembly at the macroscopic scale

Self-assembly processes can be observed in systems of macroscopic building blocks. These building blocks can be externally propelled[17] or self-propelled.[18] Since the 1950s, scientists have built self-assembly systems exhibiting centimeter-sized components ranging from passive mechanical parts to mobile robots.[19] For systems at this scale, the component design can be precisely controlled. For some systems, the components' interaction preferences are programmable. The self-assembly processes can be easily monitored and analyzed by the components themselves or by external observers.

In April 2014, Skylar Tibbits of the Massachusetts Institute of Technology, demonstrated a combination of 3D printed plastic with a "smart material" that self-assembles in water.[20] Tibbits refers to this as "4D printing".[21]

6.3 Consistent concepts of self-organization and self-assembly

People regularly use the terms "self-organization" and "self-assembly" interchangeably. As complex system science becomes more popular though, there is a higher need to clearly distinguish the differences between the two mechanisms to understand their significance in physical and biological systems. Both processes explain how collective order develops from "dynamic small-scale interactions", according to an article in a November/December 2008 issue of the journal *Complexity*.[22] Self-organization is a non-equilibrium process where self-assembly is a spontaneous process that leads toward equilibrium. Self-assembly requires components to remain essentially unchanged throughout the process. Besides the thermodynamic difference between the two, there is also a difference in formation. The first difference is what "encodes the global order of the whole" in self-assembly whereas in self-organization these initial encodings are not necessary. Another slight contrast refers to the minimum number of units needed to make an order. Self-organization appears to have a minimum number of units whereas self-assembly does not. The concepts may have particular application in connection with natural selection.[23] Eventually, these patterns may form one theory of pattern formation in nature.[24]

6.4 See also

- Crystal engineering
- Autopoiesis
- Langmuir–Blodgett film
- Nanotechnology
- Pick-and-place machine
- Self-assembly of nanoparticles

6.5 References

[1] Wetterskog, Erik; Agthe, Michael; Mayence, Arnaud; Grins, Jekabs; Wang, Dong; Rana, Subhasis; Ahniyaz, Anwar; Salazar-Alvarez, German; Bergström, Lennart (2014). "Precise control over shape and size of iron oxide nanocrystals suitable for assembly into ordered particle arrays". *Science and Technology of Advanced Materials* **15** (5): 055010. Bibcode:2014STAdM..15e5010W. doi:10.1088/1468-6996/15/5/055010.

[2] Pham, Tuan Anh; Song, Fei; Nguyen, Manh-Thuong; Stöhr, Meike (2014). "Self-assembly of pyrene derivatives on Au(111): Substituent effects on intermolecular interactions". *Chem. Commun* **50** (91): 14089–14092. doi:10.1039/C4CC02753A.

[3] Strong, M. (2004). "Protein Nanomachines". *PLoS Biol.* **2** (3): e73–e74. doi:10.1371/journal.pbio.0020073. PMC 368168. PMID 15024422.

[4] "Structural Diversity and the Role of Particle Shape and Dense Fluid Behavior in Assemblies of Hard Polyhedra" (PDF). Archived from the original on 2012-02-10. Retrieved 23 June 2012.

[5] Solem, J. C. (2002). "Self-assembling micrites based on the Platonic solids". *Robotics and Autonomous Systems* **38** (2): 69–92. doi:10.1016/s0921-8890(01)00167-1.

[6] Trewhella, J.; Solem, J. C. (1998). "Future Research Directions for Los Alamos: A Perspective from the Los Alamos Fellows" (PDF). *Los Alamos National Laboratory Report LA-UR-02-7722*: 9.

[7] Whitesides, G.M.; Boncheva, M. (2002). "Beyond molecules: Self-assembly of mesoscopic and macroscopic components". *PNAS* **99** (8): 4769–74. doi:10.1073/pnas.082065899. PMC 122665. PMID 11959929.

[8] Whitesides, George M.; Kriebel, Jennah K.; Love, J. Christopher (2005). "Molecular engineering of surfaces using self-assembled monolayers". *Science Progress* **88** (Pt 1): 17–48. doi:10.3184/003685005783238462. PMID 16372593.

[9] Berillo, Dmitriy; Mattiasson, Bo; Galaev, Igor Yu.; Kirsebom, Harald (2012). "Formation of macroporous self-assembled hydrogels through cryogelation of Fmoc–Phe–Phe". *Journal of Colloid and Interface Science* **368** (1): 226–230. doi:10.1016/j.jcis.2011.11.006. PMID 22129632.

[10] Chen, Pu; Luo, Zhengyuan; Güven, Sinan; Tasoglu, Savas; Ganesan, Adarsh Venkataraman; Weng, Andrew; Demirci, Utkan (2014). "Microscale Assembly Directed by Liquid-Based Template". *Advanced Materials* **26** (34): 5936–5941. doi:10.1002/adma.201402079. PMID 24956442.

[11] Ozin, Geoffrey A.; Arsenault, André C. (2005). *Nanochemistry: a chemical approach to nanomaterials*. Cambridge: Royal Society of Chemistry. ISBN 0-85404-664-X.

[12] Denkov, N.; Velev, O.; Kralchevski, P.; Ivanov, I.; Yoshimura, H.; Nagayama, K. (1992). "Mechanism of formation of two-dimensional crystals from latex particles on substrates". *Langmuir* **8** (12): 3183–3190. doi:10.1021/la00048a054.

[13] Lehn, Jm (Mar 2002). "Toward self-organization and complex matter". *Science* **295** (5564): 2400–3. doi:10.1126/science.1071063. PMID 11923524.

[14] Forster, Paul M.; Cheetham, Anthony K. (2002). "Open-Framework Nickel Succinate, [Ni$_7$(C$_4$H$_4$O$_4$)$_6$(OH)$_2$(H$_2$O)$_2$]·2H$_2$O: A New Hybrid Material with Three-Dimensional Ni–O–Ni Connectivity". *Angewandte Chemie International Edition* **41** (3): 457–459. doi:10.1002/1521-3773(20020201)41:3<457::AID-ANIE457>3.0.CO;2-W.

[15] Gazit, Oz; Khalfin, Rafail; Cohen, Yachin; Tannenbaum, Rina (2009). "Self-Assembled Diblock Copolymer "Nanoreactors" as "Catalysts" for Metal Nanoparticle Synthesis". *The Journal of Physical Chemistry C* **113** (2): 576–583. doi:10.1021/jp807668h.

[16] Uskoković, Vuk (2008). "Isn't self-assembly a misnomer? Multi-disciplinary arguments in favor of co-assembly". *Advances in Colloid and Interface Science* **141** (1–2): 37–47. doi:10.1016/j.cis.2008.02.004. PMID 18406396.

[17] Hosokawa K.; Shimoyama, I.; Miura, H. (1994). "Dynamics of self-assembling systems: Analogy with chemical kinetics". *Artificial Life* **1** (4): 413–427. doi:10.1162/artl.1994.1.413.

[18] Groß R.; Dorigo, M.; Mondada, Francesco; Dorigo, Marco (2006). "Autonomous self-assembly in swarm-bots". *IEEE Transactions on Robotics* **22** (6): 1115–1130. doi:10.1109/TRO.2006.882919.

[19] Groß R.; Dorigo, M. (2008). "Self-assembly at the macroscopic scale". *Proceedings of the IEEE* **96** (9): 1490–1508. doi:10.1109/JPROC.2008.927352.

[20] D'Monte, Leslie (7 May 2014) Indian market sees promise in 3D printers. livemint.com

[21] The emergence of "4D printing". ted.com (2013)

[22] Halley, J. D.; Winkler, D.A. (2008). "Consistent Concepts of Self-organization and Self-assembly". *Complexity* **14** (2): 10–17. doi:10.1002/cplx.20235.

[23] Compare: Halley, J.D.; Winkler, D.A. (2008). "Critical-like self-organization and natural selection: Two facets of a single evolutionary process?". *Bio-Systems* **92** (2): 148–158. doi:10.1016/j.biosystems.2008.01.005. PMID 18353531. Retrieved 2016-04-04. We argue that critical-like dynamics self-organize relatively easily in non-equilibrium systems, and that in biological systems such dynamics serve as templates upon which natural selection builds further elaborations. These critical-like states can be modified by natural selection in two fundamental ways, reflecting the selective advantage (if any) of heritable variations either among avalanche participants or among whole systems.

[24] Halley, J. D.; Winkler, D.A. (2008). "Consistent Concepts of Self-organization and Self-assembly". *Complexity* **14** (2): 15. doi:10.1002/cplx.20235. [...] it may one day even be possible to integrate these pattern forming mechanisms into the one general theory of pattern formation in nature.

6.6 External links and further reading

- Ariga, K.; Hill, J. P.; Lee, M. V.; Vinu, A.; Charvet, R.; Acharya, S. (2008). "Challenges and breakthroughs in recent research on self-assembly". *Science and Technology of Advanced Materials* **9**: 014109. Bibcode:2008STAdM...9a4109A. doi:10.1088/1468-6996/9/1/014109.

- Kuniaki Nagayama, *Freeview Video 'Self-Assembly: Nature's Way To Do It*, A Royal Institution Lecture by the Vega Science Trust.

- Paper Molecular Self-Assembly

- Stephens, A. D. (1977). "The management of cystinuria in 1976". *Proceedings of the Royal Society of Medicine*. 70 Suppl 3: 24–6. PMC 1543588. PMID 122665.

- Whitesides, G. M.; Grzybowski, Bartosz (2002). "Self-Assembly at All Scales". *Science* **295** (5564): 2418–21. Bibcode:2002Sci...295.2418W. doi:10.1126/science.1070821. PMID 11923529.

- Damasceno, P. F.; Engel, M.; Glotzer, S. C. (2012). "Predictive Self-Assembly of Polyhedra into Complex Structures". *Science* **337** (6093): 453–7. Bibcode:2012Sci...337..453D. doi:10.1126/science.1220869. PMID 22837525.

- Rothemund, P. W. K.; Papadakis, N.; Winfree, E. (2004). "Algorithmic Self-Assembly of DNA Sierpinski Triangles". *PLoS Biology* **2** (12): e424. doi:10.1371/journal.pbio.0020424. PMC 534809. PMID 15583715.

- Wiki: *C2 Self Assembly from a computer programming perspective.*

- Pelesko, J.A., (2007) *Self Assembly: The Science of Things That Put Themselves Together*, Chapman & Hall/CRC Press.

- A brief page on self-assembly at the University of Delaware *Self Assembly*

- Mohammadzadegan R, Sheikhi MH (2007) DNA Nano-Gears *Molecular Simulation* 33(13); 1071–1081.

- Structure and Dynamics of Organic Nanostructures

- Metal organic coordination networks of oligopyridines and Cu on graphite

- Varga, M.; Korkmaz N. S-layer proteins as Self-assembly Tool in Nano Bio Technology In: *Bio and Nano Packaging Techniques for Electron Devices (G. Gerlach and K.-J. Wolter, eds.) Springer, Heidelberg, ISBN 978-3-642-28521-9* DOI 10.1007/978-3-642-28522-6

- M. Varga, G. Roedel and W. Pompe, "Engineering of self-assembling proteins for biosensing applications," *Nanotechnology (IEEE-NANO), 2011 11th IEEE Conference on*, Portland, OR, 2011, pp. 1602-1606. doi: 10.1109/NANO.2011.6144303 http://ieeexplore.ieee.org/stamp/stamp.jsp?tp=&arnumber=6144303&isnumber=6144287

Chapter 7

Cellular automaton

Gosper's Glider Gun creating "gliders" in the cellular automaton Conway's Game of Life[1]

A **cellular automaton** (pl. **cellular automata**, abbrev. **CA**) is a discrete model studied in computability theory, mathematics, physics, complexity science, theoretical biology and microstructure modeling. Cellular automata are also called **cellular spaces**, **tessellation automata**, **homogeneous structures**, **cellular structures**, **tessellation structures**, and **iterative arrays**.[2]

A cellular automaton consists of a regular grid of *cells*, each in one of a finite number of *states*, such as *on* and *off* (in contrast to a coupled map lattice). The grid can be in any finite number of dimensions. For each cell, a set of cells called its *neighborhood* is defined relative to the specified cell. An initial state (time $t = 0$) is selected by assigning a state for each cell. A new *generation* is created (advancing t by 1), according to some fixed *rule* (generally, a mathematical function) that determines the new state of each cell in terms of the current state of the cell and the states of the cells in its neighborhood. Typically, the rule for updating the state of cells is the same for each cell and does not change over time, and is applied to the whole grid simultaneously, though exceptions are known, such as the stochastic cellular automaton and asynchronous cellular automaton.

The concept was originally discovered in the 1940s by Stanislaw Ulam and John von Neumann while they were contemporaries at Los Alamos National Laboratory. While studied by some throughout the 1950s and 1960s, it was not until the 1970s and Conway's Game of Life, a two-dimensional cellular automaton, that interest in the subject expanded beyond academia. In the 1980s, Stephen Wolfram engaged in a systematic study of one-dimensional cellular automata, or what he calls elementary cellular automata; his research assistant Matthew Cook showed that one of these rules is Turing-complete. Wolfram published *A New Kind of Science* in 2002, claiming that cellular automata have applications in many fields of science. These include computer processors and cryptography.

The primary classifications of cellular automata, as outlined by Wolfram, are numbered one to four. They are, in order, automata in which patterns generally stabilize into homogeneity, automata in which patterns evolve into mostly stable or oscillating structures, automata in which patterns evolve in a seemingly chaotic fashion, and automata in which patterns become extremely complex and may last for a long time, with stable local structures. This last class are thought to be computationally universal, or capable of simulating a Turing machine. Special types of cellular automata are *reversible*, where only a single configuration leads directly to a subsequent one, and *totalistic*, in which the future value of individual cells only depends on the total value of a group of neighboring cells. Cellular automata can simulate a variety of real-world systems, including biological and chemical ones.

7.1 Overview

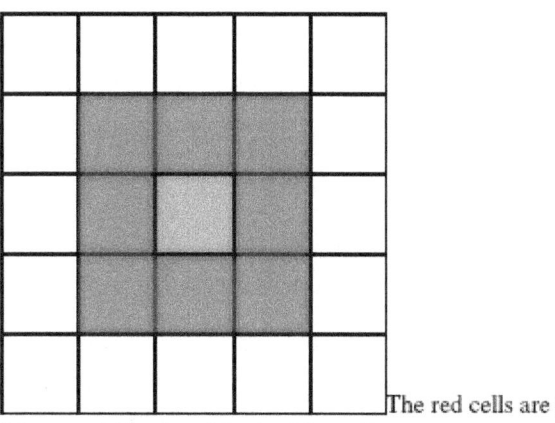
The red cells are the Moore neighborhood for the blue cell.

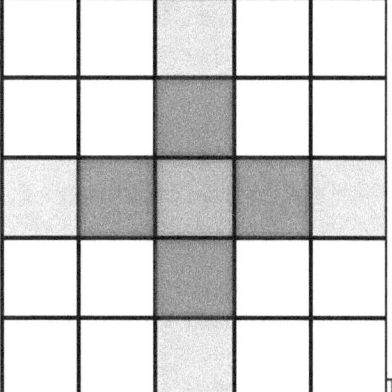
The red cells are the von Neumann neighborhood for the blue cell. The extended neighborhood includes the pink cells as well.

One way to simulate a two-dimensional cellular automaton is with an infinite sheet of graph paper along with a set of rules for the cells to follow. Each square is called a "cell" and each cell has two possible states, black and white. The *neighborhood* of a cell is the nearby, usually adjacent, cells. The two most common types of neighborhoods are the *von Neumann neighborhood* and the *Moore neighborhood*.[3] The former, named after the founding cellular automaton theorist, consists of the four orthogonally adjacent cells.[3] The latter includes the von Neumann neighborhood as well as the four remaining cells surrounding the cell whose state is to be calculated.[3] For such a cell and its Moore neighborhood, there are 512 (= 2^9) possible patterns. For each of the 512 possible patterns, the rule table would state whether the center cell will be black or white on the next time interval. Conway's Game of Life is a popular version of this model. Another common neighborhood type is the *extended von Neumann neighborhood*, which includes the two closest cells in each orthogonal direction, for a total of eight.[3] The general equation for such a system of rules is k^{k^s}, where k is the number of possible states for a cell, and s is the number of neighboring cells (including the cell to be calculated itself) used to determine the cell's next state.[4] Thus, in the two dimensional system with a Moore neighborhood, the total number of automata possible would be 2^{2^9}, or 1.34×10^{154}.

It is usually assumed that every cell in the universe starts in the same state, except for a finite number of cells in other states; the assignment of state values is called a *configuration*.[5] More generally, it is sometimes assumed that the universe starts out covered with a periodic pattern, and only a finite number of cells violate that pattern. The latter assumption is common in one-dimensional cellular automata.

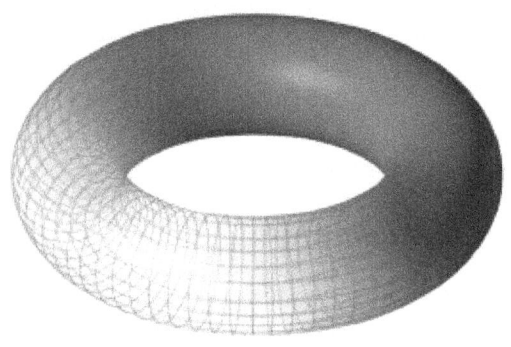
A torus, a toroidal shape

Cellular automata are often simulated on a finite grid rather than an infinite one. In two dimensions, the universe would be a rectangle instead of an infinite plane. The obvious problem with finite grids is how to handle the cells on the edges. How they are handled will affect the values of all the cells in the grid. One possible method is to allow the values in those cells to remain constant. Another method is to define neighborhoods differently for these cells. One could say that they have fewer neighbors, but then one would also have to define new rules for the cells located on the edges. These cells are usually handled with a *toroidal* arrangement: when one goes off the top, one comes in at the corresponding position on the bottom, and when one goes off the left, one comes in on the right. (This essentially simulates an infinite periodic tiling, and in the field of partial differential equations is sometimes referred to as *periodic* boundary conditions.) This can be visualized as taping the left and right edges of the rectangle to form a tube, then taping the top and bottom edges of the tube to form a torus (doughnut shape). Universes of other dimensions are handled similarly. This solves boundary problems with neighborhoods, but another advantage is that it is easily programmable using modular arithmetic functions. For example, in a 1-dimensional cellular automaton like the examples below, the neighborhood of a cell xi^t is $\{xi_{-1}^{t-1}, xi^{t-1}, xi_{+1}^{t-1}\}$, where t is the time step (vertical), and i is the index (horizontal) in one generation.

7.2 History

Stanislaw Ulam, while working at the Los Alamos National Laboratory in the 1940s, studied the growth of crystals, using a simple lattice network as his model.[6] At the same time, John von Neumann, Ulam's colleague at Los Alamos, was working on the problem of self-replicating systems.[7] Von Neumann's initial design was founded upon the notion of one robot building another robot. This design is known as the kinematic model.[8][9] As he developed this design, von Neumann came to realize the great difficulty of building a self-replicating robot, and of the great cost in providing the robot with a "sea of parts" from which to build its replicant. Neumann read a paper entitled "The general and logical theory of automata" at the Hixon Symposium in 1948.[7] Ulam was the one who suggested using a *discrete* system for creating a reductionist model of self-replication.[10][11] Nils Aall Barricelli performed many of the earliest explorations of these models of artificial life.

John von Neumann, Los Alamos ID badge

Ulam and von Neumann created a method for calculating liquid motion in the late 1950s. The driving concept of the method was to consider a liquid as a group of discrete units and calculate the motion of each based on its neighbors' behaviors.[12] Thus was born the first system of cellular automata. Like Ulam's lattice network, von Neumann's cellular automata are two-dimensional, with his self-replicator implemented algorithmically. The result was a universal copier and constructor working within a cellular automaton with a small neighborhood (only those cells that touch are neighbors; for von Neumann's cellular automata, only orthogonal cells), and with 29 states per cell.[13] Von Neumann gave an existence proof that a particular pattern would make endless copies of itself within the given cellular universe by designing a 200,000 cell configuration that could do so.[13] This design is known as the tessellation model, and is called a von Neumann universal constructor.[14]

Also in the 1940s, Norbert Wiener and Arturo Rosenblueth developed a model of excitable media with some of the characteristics of a cellular automaton.[15] Their specific motivation was the mathematical description of impulse conduction in cardiac systems. However their model is not a cellular automaton because the medium in which signals propagate is continuous, and wave fronts are curves.[15][16] A true cellular automaton model of excitable media was developed and studied by J. M. Greenberg and S. P. Hastings in 1978; see Greenberg-Hastings cellular automaton. The original work of Wiener and Rosenblueth contains many insights and continues to be cited in modern research publications on cardiac arrhythmia and excitable systems.[17]

In the 1960s, cellular automata were studied as a particular type of dynamical system and the connection with the mathematical field of symbolic dynamics was established for the first time. In 1969, Gustav A. Hedlund compiled many results following this point of view[18] in what is still considered as a seminal paper for the mathematical study of cellular automata. The most fundamental result is the characterization in the Curtis–Hedlund–Lyndon theorem of the set of global rules of cellular automata as the set of continuous endomorphisms of shift spaces.

In 1969, German computer pioneer Konrad Zuse published his book *Calculating Space*, proposing that the physical laws of the universe are discrete by nature, and that the entire universe is the output of a deterministic computation on a single cellular automaton; "Zuse's Theory" became the foundation of the field of study called *digital physics*.[19]

Also in 1969 computer scientist Alvy Ray Smith completed a Stanford PhD dissertation on Cellular Automata Theory, the first mathematical treatment of CA as a general class of computers. Many papers came from this dissertation: He showed the equivalence of neighborhoods of various shapes, how to reduce a Moore to a von Neumann neighborhood or how to reduce any neighborhood to a von Neumann neighborhood.[20] He proved that two-dimensional CA are computation universal, introduced 1-dimensional CA, and showed that they too are computation universal, even with simple neighborhoods.[21] He showed how to subsume the complex von Neumann proof of construction universality (and hence self-reproducing machines) into a consequence of computation universality in a 1-dimensional CA.[22] Intended as the introduction to the German edition of von Neumann's book on CA, he wrote a survey of the field with dozens of references to papers, by many authors in many countries over a decade or so of work, often overlooked by modern CA researchers.[23]

In the 1970s a two-state, two-dimensional cellular automaton named Game of Life became widely known, particularly among the early computing community. Invented by John Conway and popularized by Martin Gardner in a *Scientific American* article,[24] its rules are as follows: If a cell has two black neighbors, it stays the same. If it has three black neighbors, it becomes black. In all other situations it becomes white. Despite its simplicity, the system achieves an impressive diversity of behavior, fluctuating between apparent randomness and order. One of the most apparent features of the Game of Life is the frequent occurrence of *gliders*, arrangements of cells that essentially move themselves across the grid. It is possible to arrange the automaton so that the gliders interact to perform computations, and after much effort it has been shown that the Game of Life can emulate a universal Turing machine.[25] It was viewed as a largely recreational topic, and little follow-up work was done outside of investigating the particularities of the Game of Life and a few related rules in the early 1970s.[26]

Stephen Wolfram independently began working on cellular automata in mid 1981 after considering how complex patterns seemed formed in nature in violation of the Second Law of Thermodynamics.[27] His investigations were initially spurred by an interest in modelling systems such as neural networks.[27] He published his first paper in *Reviews of Modern Physics* investigating *elementary cellular automata* (Rule 30 in particular) in June 1983.[2][27] The unexpected complexity of the behavior of these simple rules led Wolfram to suspect that complexity in nature may be due to similar mechanisms.[27] His investigations, however, led him to realize that cellular automata were poor at modelling neural networks.[27] Additionally, during this period Wolfram formulated the concepts of intrinsic randomness and computational irreducibility,[28] and suggested that rule 110 may be universal—a fact proved later by Wolfram's research assistant Matthew Cook in the 1990s.[29]

In 2002 Wolfram published a 1280-page text *A New Kind of Science*, which extensively argues that the discoveries about cellular automata are not isolated facts but are robust and have significance for all disciplines of science.[30] Despite confusion in the press,[31][32] the book did not argue for a fundamental theory of physics based on cellular automata,[33] and although it did describe a few specific physical models based on cellular automata,[34] it also provided models based on qualitatively different abstract systems.[35]

7.3 Classification

Wolfram, in *A New Kind of Science* and several papers dating from the mid-1980s, defined four classes into which cellular automata and several other simple computational models can be divided depending on their behavior. While earlier studies in cellular automata tended to try to identify type of patterns for specific rules, Wolfram's classification was the first attempt to classify the rules themselves. In order of complexity the classes are:

- Class 1: Nearly all initial patterns evolve quickly into a stable, homogeneous state. Any randomness in the initial pattern disappears.[36]

- Class 2: Nearly all initial patterns evolve quickly into stable or oscillating structures. Some of the randomness in the initial pattern may filter out, but some remains. Local changes to the initial pattern tend to remain local.[36]

- Class 3: Nearly all initial patterns evolve in a pseudo-random or chaotic manner. Any stable structures that appear are quickly destroyed by the surrounding noise. Local changes to the initial pattern tend to spread indefinitely.[36]

- Class 4: Nearly all initial patterns evolve into structures that interact in complex and interesting ways, with the formation of local structures that are able to survive for long periods of time.[37] Class 2 type stable or oscillating structures may be the eventual outcome, but the number of steps required to reach this state may be very large, even when the initial pattern is relatively simple. Local changes to the initial pattern may spread indefinitely. Wolfram has conjectured that many, if not all class 4 cellular automata are capable of universal computation. This has been proven for Rule 110 and Conway's game of Life.

These definitions are qualitative in nature and there is some room for interpretation. According to Wolfram, "...with almost any general classification scheme there are inevitably cases which get assigned to one class by one definition and another class by another definition. And so it is with cellular automata: there are occasionally rules...that show some features of one class and some of another."[38] Wolfram's classification has been empirically matched to a clustering of the compressed lengths of the outputs of cellular automata.[39]

There have been several attempts to classify cellular automata in formally rigorous classes, inspired by the Wolfram's classification. For instance, Culik and Yu proposed three well-defined classes (and a fourth one for the automata not matching any of these), which are sometimes called Culik-Yu classes; membership in these proved undecidable.[40][41][42] Wolfram's class 2 can be partitioned into two subgroups of stable (fixed-point) and oscillating (periodic) rules.[43]

7.3.1 Reversible

Main article: Reversible cellular automaton

A cellular automaton is *reversible* if, for every current configuration of the cellular automaton, there is exactly one past configuration (preimage).[44] If one thinks of a cellular automaton as a function mapping configurations to configurations, reversibility implies that this function is bijective.[44] If a cellular automaton is reversible, its time-reversed behavior can also be described as a cellular automaton; this fact is a consequence of the Curtis–Hedlund–Lyndon theorem, a topological characterization of cellular automata.[45][46] For cellular automata in which not every configuration has a preimage, the configurations without preimages are called *Garden of Eden* patterns.[47]

For one-dimensional cellular automata there are known algorithms for deciding whether a rule is reversible or irreversible.[48][49] However, for cellular automata of two or more dimensions reversibility is undecidable; that is, there is no algorithm that takes as input an automaton rule and is guaranteed to determine correctly whether the automaton is reversible. The proof by Jarkko Kari is related to the tiling problem by Wang tiles.[50]

Reversible cellular automata are often used to simulate such physical phenomena as gas and fluid dynamics, since they obey the laws of thermodynamics. Such cellular automata have rules specially constructed to be reversible. Such systems have been studied by Tommaso Toffoli, Norman Margolus and others. Several techniques can be used to explicitly construct reversible cellular automata with known inverses. Two common ones are the second order cellular automaton and the block cellular automaton, both of which involve modifying the definition of a cellular automaton in some way. Although such automata do not strictly satisfy the definition given above, it can be shown that they can be emulated by conventional cellular automata with sufficiently large neighborhoods and numbers of states, and can therefore be considered a subset of conventional cellular automata. Conversely, it has been shown that every reversible cellular automaton can be emulated by a block cellular automaton.[51][52]

7.3.2 Totalistic

A special class of cellular automata are *totalistic* cellular automata. The state of each cell in a totalistic cellular automaton is represented by a number (usually an integer value drawn from a finite set), and the value of a cell at time t depends only on the *sum* of the values of the cells in its neighborhood (possibly including the cell itself) at time $t-1$.[53][54] If the state of the cell at time t depends on both its own state and the total of its neighbors at time $t-1$ then the cellular automaton is properly called *outer totalistic*.[54] Conway's Game of Life is an example of an outer totalistic cellular automaton with cell values 0 and 1; outer totalistic cellular automata with the same Moore neighborhood structure as Life are sometimes called life-like cellular automata.[55][56]

7.3.3 Related automata

There are many possible generalizations of the cellular automaton concept.

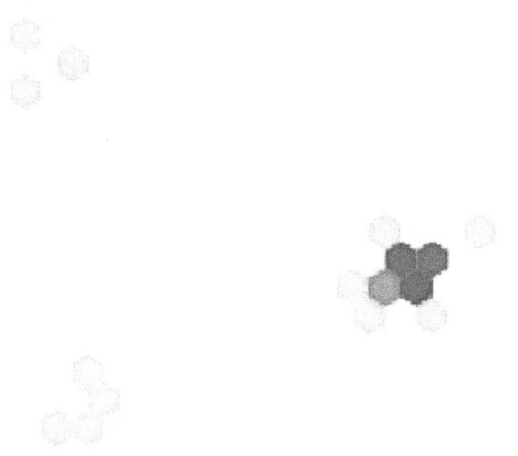

A cellular automaton based on hexagonal cells instead of squares (rule 34/2)

One way is by using something other than a rectangular (cubic, *etc.*) grid. For example, if a plane is tiled with regular hexagons, those hexagons could be used as cells. In many cases the resulting cellular automata are equivalent to those with rectangular grids with specially designed neighborhoods and rules. Another variation would be to make the grid itself irregular, such as with Penrose tiles.[57]

Also, rules can be probabilistic rather than deterministic. Such cellular automata are called probabilistic cellular automata. A probabilistic rule gives, for each pattern at time t, the probabilities that the central cell will transition to each possible state at time $t + 1$. Sometimes a simpler rule is used; for example: "The rule is the Game of Life, but on each time step there is a 0.001% probability that each cell will transition to the opposite color."

The neighborhood or rules could change over time or space. For example, initially the new state of a cell could be determined by the horizontally adjacent cells, but for the next generation the vertical cells would be used.

In cellular automata, the new state of a cell is not affected by the new state of other cells. This could be changed so that, for instance, a 2 by 2 block of cells can be determined by itself and the cells adjacent to itself.

There are *continuous automata*. These are like totalistic cellular automata, but instead of the rule and states being discrete (*e.g.* a table, using states $\{0,1,2\}$), continuous functions are used, and the states become continuous (usually values in $[0,1]$). The state of a location is a finite number of real numbers. Certain cellular automata can yield diffusion in liquid patterns in this way.

Continuous spatial automata have a continuum of locations. The state of a location is a finite number of real numbers. Time is also continuous, and the state evolves according to differential equations. One important example is reaction-diffusion textures, differential equations proposed by Alan Turing to explain how chemical reactions could create the stripes on zebras and spots on leopards.[58] When these are approximated by cellular automata, they often yield similar patterns. MacLennan considers continuous spatial automata as a model of computation.

There are known examples of continuous spatial automata, which exhibit propagating phenomena analogous to gliders in the Game of Life.[59]

7.4 Elementary cellular automata

Main article: Elementary cellular automaton

The simplest nontrivial cellular automaton would be one-dimensional, with two possible states per cell, and a cell's neighbors defined as the adjacent cells on either side of it. A cell and its two neighbors form a neighborhood of 3 cells, so there are $2^3 = 8$ possible patterns for a neighborhood. A rule consists of deciding, for each pattern, whether the cell will be a 1 or a 0 in the next generation. There are then $2^8 = 256$ possible rules.[4] These 256 cellular automata are generally referred to by their Wolfram code, a standard naming convention invented by Wolfram that gives each rule a number from 0 to 255. A number of papers have analyzed and compared these 256 cellular automata. The rule 30 and rule 110 cellular automata are particularly interesting. The images below show the history of each when the starting configuration consists of a 1 (at the top of each image) surrounded by 0s. Each row of pixels represents a generation in the history of the automaton, with $t=0$ being the top row. Each pixel is colored white for 0 and black for 1.

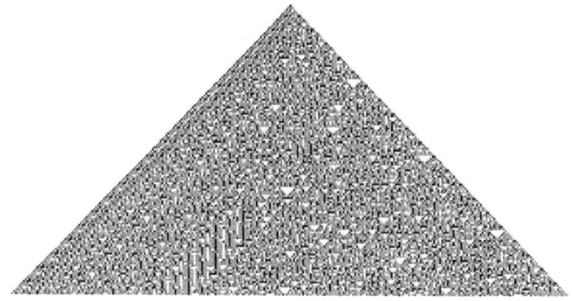

Rule 30

Rule 30 cellular automaton

Rule 30 exhibits *class 3* behavior, meaning even simple input patterns such as that shown lead to chaotic, seemingly random histories.

Rule 110

Rule 110 cellular automaton

Rule 110, like the Game of Life, exhibits what Wolfram calls *class 4* behavior, which is neither completely random nor completely repetitive. Localized structures appear and interact in various complicated-looking ways. In the course of the development of *A New Kind of Science*, as a research assistant to Wolfram in 1994, Matthew Cook proved that some of these structures were rich enough to support universality. This result is interesting because rule 110 is an extremely simple one-dimensional system, and difficult to engineer to perform specific behavior. This result therefore provides significant support for Wolfram's view that class 4 systems are inherently likely to be universal. Cook presented his proof at a Santa Fe Institute conference on Cellular Automata in 1998, but Wolfram blocked the proof from being included in the conference proceedings, as Wolfram did not want the proof announced before the publication of *A New Kind of Science*.[60] In 2004, Cook's proof was finally published in Wolfram's journal *Complex Systems* (Vol. 15, No. 1), over ten years after Cook came up with it. Rule 110 has been the basis for some of the smallest universal Turing machines.[61]

7.5 Rule space

An elementary cellular automaton rule is specified by 8 bits, and all elementary cellular automaton rules can be considered to sit on the vertices of the 8-dimensional unit hypercube. This unit hypercube is the cellular automaton rule space. For next-nearest-neighbor cellular automata, a rule is specified by $2^5 = 32$ bits, and the cellular automaton rule space is a 32-dimensional unit hypercube. A distance between two rules can be defined by the number of steps required to move from one vertex, which represents the first rule, and another vertex, representing another rule, along the edge of the hypercube. This rule-to-rule distance is also called the Hamming distance.

Cellular automaton rule space allows us to ask the question concerning whether rules with similar dynamical behavior are "close" to each other. Graphically drawing a high dimensional hypercube on the 2-dimensional plane remains a difficult task, and one crude locator of a rule in the hypercube is the number of bit-1 in the 8-bit string for elementary rules (or 32-bit string for the next-nearest-neighbor rules). Drawing the rules in different Wolfram classes in these slices of the rule space show that class 1 rules tend to have lower number of bit-1's, thus located in one region of the space, whereas class 3 rules tend to have higher proportion (50%) of bit-1's.[43]

For larger cellular automaton rule space, it is shown that class 4 rules are located between the class 1 and class 3 rules.[62] This observation is the foundation for the phrase edge of chaos, and is reminiscent of the phase transition in thermodynamics.

7.6 Biology

Further information: Patterns in nature

Some biological processes occur—or can be simulated—by

Conus textile *exhibits a cellular automaton pattern on its shell.*[63]

cellular automata.

Patterns of some seashells, like the ones in *Conus* and *Cymbiola* genus, are generated by natural cellular automata. The pigment cells reside in a narrow band along the shell's lip. Each cell secretes pigments according to the activating and inhibiting activity of its neighbor pigment cells, obeying a natural version of a mathematical rule.[63] The cell band leaves the colored pattern on the shell as it grows slowly. For example, the widespread species *Conus textile* bears a pattern resembling Wolfram's rule 30 cellular automaton.[63]

Plants regulate their intake and loss of gases via a cellular automaton mechanism. Each stoma on the leaf acts as a cell.[64]

Moving wave patterns on the skin of cephalopods can be simulated with a two-state, two-dimensional cellular automata, each state corresponding to either an expanded or retracted chromatophore.[65]

Threshold automata have been invented to simulate neurons, and complex behaviors such as recognition and learning can be simulated.[66]

Fibroblasts bear similarities to cellular automata, as each fibroblast only interacts with its neighbors.[67]

7.7 Chemical types

The Belousov–Zhabotinsky reaction is a spatio-temporal chemical oscillator that can be simulated by means of a cellular automaton. In the 1950s A. M. Zhabotinsky (extending the work of B. P. Belousov) discovered that when a thin, homogenous layer of a mixture of malonic acid, acidified bromate, and a ceric salt were mixed together and left undisturbed, fascinating geometric patterns such as concentric circles and spirals propagate across the medium. In the

"Computer Recreations" section of the August 1988 issue of *Scientific American*,[68] A. K. Dewdney discussed a cellular automaton[69] developed by Martin Gerhardt and Heike Schuster of the University of Bielefeld (Germany). This automaton produces wave patterns that resemble those in the Belousov-Zhabotinsky reaction.

7.8 Applications

7.8.1 Computer processors

Cellular automaton processors are physical implementations of CA concepts, which can process information computationally. Processing elements are arranged in a regular grid of identical cells. The grid is usually a square tiling, or tessellation, of two or three dimensions; other tilings are possible, but not yet used. Cell states are determined only by interactions with adjacent neighbor cells. No means exists to communicate directly with cells farther away.[70] One such cellular automaton processor array configuration is the systolic array. Cell interaction can be via electric charge, magnetism, vibration (phonons at quantum scales), or any other physically useful means. This can be done in several ways so no wires are needed between any elements. This is very unlike processors used in most computers today, von Neumann designs, which are divided into sections with elements that can communicate with distant elements over wires.

7.8.2 Cryptography

Rule 30 was originally suggested as a possible block cipher for use in cryptography. Two dimensional cellular automata are used for random number generation.[71]

Cellular automata have been proposed for public key cryptography. The one-way function is the evolution of a finite CA whose inverse is believed to be hard to find. Given the rule, anyone can easily calculate future states, but it appears to be very difficult to calculate previous states.

7.8.3 Error correction coding

CA have been applied to design error correction codes in a paper by D. Roy Chowdhury, S. Basu, I. Sen Gupta, and P. Pal Chaudhuri. The paper defines a new scheme of building single bit error correction and double bit error detection (SEC-DED) codes using CA, and also reports a fast hardware decoder for the code.[72]

7.9 Modeling physical reality

Main articles: Digital physics and digital philosophy

As Andrew Ilachinski points out in his *Cellular Automata*, many scholars have raised the question of whether the universe is a cellular automaton.[73] Ilachinski argues that the importance of this question may be better appreciated with a simple observation, which can be stated as follows. Consider the evolution of rule 110: if it were some kind of "alien physics", what would be a reasonable description of the observed patterns?[74] If an observer did not know how the images were generated, that observer might end up conjecturing about the movement of some particle-like objects. Indeed, physicist James Crutchfield has constructed a rigorous mathematical theory out of this idea, proving the statistical emergence of "particles" from cellular automata.[75] Then, as the argument goes, one might wonder if *our* world, which is currently well described by physics with particle-like objects, could be a CA at its most fundamental level.

While a complete theory along this line has not been developed, entertaining and developing this hypothesis led scholars to interesting speculation and fruitful intuitions on how can we make sense of our world within a discrete framework. Marvin Minsky, the AI pioneer, investigated how to understand particle interaction with a four-dimensional CA lattice;[76] Konrad Zuse—the inventor of the first working computer, the Z3—developed an irregularly organized lattice to address the question of the information content of particles.[77] More recently, Edward Fredkin exposed what he terms the "finite nature hypothesis", i.e., the idea that "ultimately every quantity of physics, including space and time, will turn out to be discrete and finite."[78] Fredkin and Wolfram are strong proponents of a CA-based physics.

In recent years, other suggestions along these lines have emerged from literature in non-standard computation. Wolfram's *A New Kind of Science* considers CA the key to understanding a variety of subjects, physics included. The *Mathematics of the Models of Reference*—created by iLabs[79] founder Gabriele Rossi and developed with Francesco Berto and Jacopo Tagliabue—features an original 2D/3D universe based on a new "rhombic dodecahedron-based" lattice and a unique rule. This model satisfies universality (it is equivalent to a Turing Machine) and perfect reversibility (a *desideratum* if one wants to conserve various quantities easily and never lose information), and it comes embedded in a first-order theory, allowing computable, qualitative statements on the universe evolution.[80]

7.10 Specific rules

Specific types of cellular automata include:

- Brian's Brain
- Langton's ant
- Wireworld
- Rule 90
- Rule 184
- von Neumann cellular automata
- Nobili cellular automata
- Codd's cellular automaton
- Langton's loops
- CoDi

7.11 Problems solved

Problems that can be solved with cellular automata include:

- Firing squad synchronization problem
- Majority problem

7.12 See also

- Automata theory
- Bidirectional traffic
- Cellular automata in popular culture
- Cyclic cellular automaton
- Excitable medium
- Mirek's Cellberation
- Movable cellular automaton
- Quantum cellular automata
- Spatial decision support system
- Turmites

7.13 Reference notes

[1] Daniel Dennett (1995), *Darwin's Dangerous Idea*, Penguin Books, London, ISBN 978-0-14-016734-4, ISBN 0-14-016734-X

[2] Wolfram, Stephen (1983). "Statistical Mechanics of Cellular Automata". *Reviews of Modern Physics* **55** (3): 601–644. Bibcode:1983RvMP...55..601W. doi:10.1103/RevModPhys.55.601.

[3] Kier, Seybold, Cheng 2005, p. 15

[4] Bialynicki-Birula, Bialynicka-Birula 2004, p. 9

[5] Schiff 2011, p. 41

[6] Pickover, Clifford A. (2009). *The Math Book: From Pythagoras to the 57th Dimension, 250 Milestones in the History of Mathematics*. Sterling Publishing Company, Inc. p. 406. ISBN 978-1402757969.

[7] Schiff 2011, p. 1

[8] John von Neumann, "The general and logical theory of automata," in L.A. Jeffress, ed., Cerebral Mechanisms in Behavior – The Hixon Symposium, John Wiley & Sons, New York, 1951, pp. 1–31.

[9] Kemeny, John G. (1955). "Man viewed as a machine". *Sci. Amer.* **192**: 58–67. doi:10.1038/scientificamerican0455-58.; *Sci. Amer.* 1955; 192:6 (errata).

[10] Schiff 2011, p. 3

[11] Ilachinski 2001, p. xxix

[12] Bialynicki-Birula, Bialynicka-Birula 2004, p. 8

[13] Wolfram 2002, p. 876

[14] von Neumann, John; Burks, Arthur W. (1966). *Theory of Self-Reproducing Automata*. University of Illinois Press.

[15] Wiener, N.; Rosenblueth, A. (1946). "The mathematical formulation of the problem of conduction of impulses in a network of connected excitable elements, specifically in cardiac muscle". *Arch. Inst. Cardiol. México* **16**: 205.

[16] Letichevskii, A. A.; Reshodko, L. V. (1974). "N. Wiener's theory of the activity of excitable media". *Cybernetics* **8**: 856–864. doi:10.1007/bf01068458.

[17] Davidenko, J. M.; Pertsov, A. V.; Salomonsz, R.; Baxter, W.; Jalife, J. (1992). "Stationary and drifting spiral waves of excitation in isolated cardiac muscle". *Nature* **355** (6358): 349–351. Bibcode:1992Natur.355..349D. doi:10.1038/355349a0. PMID 1731248.

[18] Hedlund, G. A. (1969). "Endomorphisms and automorphisms of the shift dynamical system". *Math. Systems Theory* **3** (4): 320–3751. doi:10.1007/BF01691062.

[19] Schiff 2011, p. 182

[20] Smith, Alvy Ray. "Cellular Automata Complexity Trade-Offs" (PDF).

[21] Smith, Alvy Ray. "Simple Computation-Universal Cellular Spaces" (PDF).

[22] Smith, Alvy Ray. "Simple Nontrivial Self-Reproducing Machines" (PDF).

[23] Smith, Alvy Ray. "Introduction to and Survey of Cellular Automata or Polyautomata Theory" (PDF).

[24] Gardner, Martin (1970). "Mathematical Games: The fantastic combinations of John Conway's new solitaire game "life"". *Scientific American* (223): 120–123.

[25] Paul Chapman. Life universal computer. http://www.igblan.free-online.co.uk/igblan/ca/ November 2002

[26] Wainwright 2010, p. 16

[27] Wolfram 2002, p. 880

[28] Wolfram 2002, p. 881

[29] Mitchell, Melanie (4 October 2002). "Is the Universe a Universal Computer?". *Science* **298** (5591): 65–68. doi:10.1126/science.1075073.

[30] Wolfram 2002, pp. 1–7

[31] Johnson, George (9 June 2002). "'A New Kind of Science': You Know That Space-Time Thing? Never Mind". *The New York Times* (The New York Times Company). Retrieved 22 January 2013.

[32] "The Science of Everything". *The Economist*. 30 May 2002. Retrieved 22 January 2013.

[33] Wolfram 2002, pp. 433–546

[34] Wolfram 2002, pp. 51–114

[35] Wolfram 2002, pp. 115–168

[36] Ilachinsky 2001, p. 12

[37] Ilachinsky 2001, p. 13

[38] Wolfram 2002, p. 231

[39] Zenil, Hector (2010). "Compression-based investigation of the dynamical properties of cellular automata and other systems" (PDF). *Complex Systems* **19** (1).

[40] G. Cattaneo; E. Formenti; L. Margara (1998). "Topological chaos and CA". In M. Delorme; J. Mazoyer. *Cellular automata: a parallel model*. Springer. p. 239. ISBN 978-0-7923-5493-2.

[41] Burton H. Voorhees (1996). *Computational analysis of one-dimensional cellular automata*. World Scientific. p. 8. ISBN 978-981-02-2221-5.

[42] Max Garzon (1995). *Models of massive parallelism: analysis of cellular automata and neural networks*. Springer. p. 149. ISBN 978-3-540-56149-1.

[43] Li, Wentian; Packard, Norman (1990). "The structure of the elementary cellular automata rule space" (PDF). *Complex Systems* **4**: 281–297. Retrieved January 25, 2013.

[44] Kari, Jarrko 1991, p. 379

[45] Richardson, D. (1972). "Tessellations with local transformations". *J. Computer System Sci.* **6** (5): 373–388. doi:10.1016/S0022-0000(72)80009-6.

[46] Margenstern, Maurice (2007). *Cellular Automata in Hyperbolic Spaces – Tome I, Volume 1*. Archives contemporaines. p. 134. ISBN 978-2-84703-033-4.

[47] Schiff 2011, p. 103

[48] Amoroso, Serafino; Patt, Yale N. (1972). "Decision Procedures for Surjectivity and Injectivity of Parallel Maps for Tessellation Structures". *J. Comput. Syst. Sci.* **6** (5): 448–464. doi:10.1016/s0022-0000(72)80013-8.

[49] Sutner, Klaus (1991). "De Bruijn Graphs and Linear Cellular Automata" (PDF). *Complex Systems* **5**: 19–30.

[50] Kari, Jarkko (1990). "Reversibility of 2D cellular automata is undecidable". *Physica D* **45**: 379–385. Bibcode:1990PhyD...45..379K. doi:10.1016/0167-2789(90)90195-U.

[51] Kari, Jarkko (1999). "On the circuit depth of structurally reversible cellular automata". *Fundamenta Informaticae* **38**: 93–107.

[52] Durand-Lose, Jérôme (2001). "Representing reversible cellular automata with reversible block cellular automata". *Discrete Mathematics and Theoretical Computer Science* **AA**: 145–154.

[53] Wolfram 2002, p. 60

[54] Ilachinski, Andrew (2001). *Cellular automata: a discrete universe*. World Scientific. pp. 44–45. ISBN 978-981-238-183-5.

[55] The phrase "life-like cellular automaton" dates back at least to Barral, Chaté & Manneville (1992), who used it in a broader sense to refer to outer totalistic automata, not necessarily of two dimensions. The more specific meaning given here was used e.g. in several chapters of Adamatzky (2010). See: Barral, Bernard; Chaté, Hugues; Manneville, Paul (1992). "Collective behaviors in a family of high-dimensional cellular automata". *Physics Letters A* **163** (4): 279–285. Bibcode:1992PhLA..163..279B. doi:10.1016/0375-9601(92)91013-H.

[56] Eppstein 2010, pp. 72–73

[57] Jacob Aron. "First gliders navigate ever-changing Penrose universe". *New Scientist*.

[58] Murray, J. "Mathematical Biology II". Springer.

[59] Pivato, M: "RealLife: The continuum limit of Larger than Life cellular automata", *Theoretical Computer Science*, 372 (1), March 2007, pp. 46–68

[60] Giles, Jim (2002). "What Kind of Science is This?". *Nature* (417): 216–218.

[61] Weinberg, Steven (October 24, 2002). "Is the Universe a Computer?". *The New York Review of Books* (Rea S. Hederman). Retrieved October 12, 2012.

[62] Wentian Li; Norman Packard; Chris G Langton (1990). "Transition phenomena in cellular automata rule space". *Physica D* 45 (1-3): 77–94. Bibcode:1990PhyD...45...77L. doi:10.1016/0167-2789(90)90175-O.

[63] Coombs, Stephen (February 15, 2009), *The Geometry and Pigmentation of Seashells* (PDF), pp. 3-4, retrieved September 2, 2012

[64] Peak, West; Messinger, Mott (2004). "Evidence for complex, collective dynamics and emergent, distributed computation in plants". *Proceedings of the National Institute of Science of the USA* 101 (4): 918–922. Bibcode:2004PNAS..101..918P. doi:10.1073/pnas.0307811100. PMC 327117. PMID 14732685.

[65] http://gilly.stanford.edu/past_research_files/APackardneuralnet.pdf

[66] Ilachinsky 2001, p. 275

[67] Yves Bouligand (1986). *Disordered Systems and Biological Organization*. pp. 374–375.

[68] A. K. Dewdney, The hodgepodge machine makes waves, Scientific American, p. 104, August 1988.

[69] Gerhardt, M.; Schuster, H. (1989). "A cellular automaton describing the formation of spatially ordered structures in chemical systems". *Physica D* 36: 209–221. doi:10.1016/0167-2789(89)90081-x.

[70] Muhtaroglu, Ali (August 1996). "4.1 Cellular Automaton Processor (CAP)". *Cellular Automaton Processor Based Systems for Genetic Sequence Comparison/Database Searching*. Cornell University. pp. 62–74.

[71] Tomassini, M.; Sipper, M.; Perrenoud, M. (2000). "On the generation of high-quality random numbers by two-dimensional cellular automata". *IEEE Transactions on Computers* 49 (10): 1146–1151. doi:10.1109/12.888056.

[72] Chowdhury, D. Roy; Basu, S.; Gupta, I. Sen; Chaudhuri, P. Pal (June 1994). "Design of CAECC - cellular automata based error correcting code". *IEEE Transactions on Computers* 43 (6). doi:10.1109/12.286310.

[73] Ilachinsky 2001, p. 660

[74] Ilachinsky 2001, pp. 661–662

[75] J. P. Crutchfield, "The Calculi of Emergence: Computation, Dynamics, and Induction", Physica D 75, 11–54, 1994.

[76] Minsky, M. "Cellular Vacuum". *International Journal of Theoretical Physics* 21 (537–551): 1982.

[77] K. Zuse, "The Computing Universe", Int. Jour. of Theo. Phy. 21, 589–600, 1982.

[78] E. Fredkin, "Digital mechanics: an informational process based on reversible universal cellular automata", Physica D 45, 254–270, 1990

[79] "Ilabs".

[80] F. Berto, G. Rossi, J. Tagliabue, The Mathematics of the Models of Reference, College Publications, 2010

7.14 References

- Adamatzky, Andrew, ed. (2010). *Game of Life Cellular Automata*. Springer. ISBN 978-1-84996-216-2.

- Bialynicki-Birula, Iwo; Bialynicka-Birula, Iwona (2004). *Modeling Reality: How Computers Mirror Life*. Oxford University Press. ISBN 0198531001.

- Chopard, Bastien; Droz, Michel (2005). *Cellular Automata Modeling of Physical Systems*. Cambridge University Press. ISBN 0-521-46168-5.

- Gutowitz, Howard, ed. (1991). *Cellular Automata: Theory and Experiment*. MIT Press. ISBN 9780262570862.

- Ilachinski, Andrew (2001). *Cellular Automata: A Discrete Universe*. World Scientific. ISBN 9789812381835.

- Kier, Lemont B.; Seybold, Paul G.; Cheng, Chao-Kun (2005). *Modeling Chemical Systems using Cellular Automata*. Springer. ISBN 9781402036576.

- Schiff, Joel L. (2011). *Cellular Automata: A Discrete View of the World*. Wiley & Sons, Inc. ISBN 9781118030639.

- Wolfram, Stephen (2002). *A New Kind of Science*. Wolfram Media. ISBN 978-1579550080.

- Cellular automaton FAQ from the newsgroup comp.theory.cell-automata

- "Neighbourhood Survey" (includes discussion on triangular grids, and larger neighborhood CAs)

- von Neumann, John, 1966, *The Theory of Self-reproducing Automata*, A. Burks, ed., Univ. of Illinois Press, Urbana, IL.

- Cosma Shalizi's Cellular Automata Notebook contains an extensive list of academic and professional reference material.

- Wolfram's papers on CAs

- A.M. Turing. 1952. The Chemical Basis of Morphogenesis. *Phil. Trans. Royal Society*, vol. B237, pp. 37–72. (proposes reaction-diffusion, a type of continuous automaton).

- Evolving Cellular Automata with Genetic Algorithms: A Review of Recent Work, Melanie Mitchell, James P. Crutchfeld, Rajarshi Das (In Proceedings of the First International Conference on Evolutionary Computation and Its Applications (EvCA'96). Moscow, Russia: Russian Academy of Sciences, 1996.)

- The Evolutionary Design of Collective Computation in Cellular Automata, James P. Crutchfeld, Melanie Mitchell, Rajarshi Das (In J. P. Crutch˜eld and P. K. Schuster (editors), Evolutionary Dynamics|Exploring the Interplay of Selection, Neutrality, Accident, and Function. New York: Oxford University Press, 2002.)

- The Evolution of Emergent Computation, James P. Crutchfield and Melanie Mitchell (SFI Technical Report 94-03-012)

- Ganguly, Sikdar, Deutsch and Chaudhuri "A Survey on Cellular Automata"

7.15 External links

- Berto, Francesco; Tagliabue, Jacopo. "Cellular Automata". *Stanford Encyclopedia of Philosophy*.

- Mirek's Cellebration – Home to free MCell and MJCell cellular automata explorer software and rule libraries. The software supports a large number of 1D and 2D rules. The site provides both an extensive rules lexicon and many image galleries loaded with examples of rules. MCell is a Windows application, while MJCell is a Java applet. Source code is available.

- Modern Cellular Automata – Easy to use interactive exhibits of live color 2D cellular automata, powered by Java applet. Included are exhibits of traditional, reversible, hexagonal, multiple step, fractal generating, and pattern generating rules. Thousands of rules are provided for viewing. Free software is available.

- Self-replication loops in Cellular Space – Java applet powered exhibits of self replication loops.

- A collection of over 10 different cellular automata applets (in Monash University's Virtual Lab)

- Golly supports von Neumann, Nobili, GOL, and a great many other systems of cellular automata. Developed by Tomas Rokicki and Andrew Trevorrow. This is the only simulator currently available that can demonstrate von Neumann type self-replication.

- Fourier Life - A collection of rules that demonstrate self-replicating patterns which form from a starting field of random cells. Most of the rules were found using an algorithm that uses a Fourier transform to detect self-replication.

- Wolfram Atlas – An atlas of various types of one-dimensional cellular automata.

- Conway Life

- First replicating creature spawned in life simulator

- *The Mathematics of the Models of Reference*, featuring a general tutorial on CA, interactive applet, free code and resources on CA as model of fundamental physics

- Fourmilab Cellular Automata Laboratory

- Busy Boxes, a 3-D, reversible, SALT-architecture CA

- Cellular Automata Repository (CA researchers, historic links, free software, books and beyond)

Chapter 8

Emergence

For other uses, see Emergence (disambiguation).
See also: Emergent (disambiguation), Spontaneous order, and Self-organization
In philosophy, systems theory, science, and art, **emergence**

A termite "cathedral" mound produced by a termite colony offers a classic example of emergence in nature.

The formation of complex symmetrical and fractal patterns in snowflakes exemplifies emergence in a physical system.

is a process whereby larger entities, patterns, and regularities arise through interactions among smaller or simpler entities that themselves do not exhibit such properties.

Emergence is central in theories of integrative levels and of complex systems. For instance, the phenomenon of *life* as studied in biology is commonly perceived as an emergent property of interacting molecules as studied in chemistry, whose phenomena reflect interactions among elementary particles, modeled in particle physics, that at such higher mass—via substantial conglomeration—exhibit motion as modeled in gravitational physics. Neurobiological phenomena are often presumed to suffice as the underlying basis of psychological phenomena, whereby economic phenomena are in turn presumed to principally emerge.

In philosophy, emergence typically refers to emergentism. Almost all accounts of emergentism include a form of epistemic or ontological irreducibility to the lower levels.[1]

8.1 In philosophy

Main article: Emergentism

In philosophy, emergence is often understood to be a claim about the etiology of a system's properties. An emergent property of a system, in this context, is one that is not a property of any component of that system, but is still a feature of the system as a whole. Nicolai Hartmann, one of the first modern philosophers to write on emergence, termed this *categorial novum* (new category).

8.1.1 Definitions

This idea of emergence has been around since at least the time of Aristotle.[2] John Stuart Mill[3] and Julian Huxley[4] are two of many scientists and philosophers who have written on the concept.

The term "emergent" was coined by philosopher G. H. Lewes, who wrote:

> "Every resultant is either a sum or a difference of the co-operant forces; their sum, when their directions are the same -- their difference, when their directions are contrary. Further, every resultant is clearly traceable in its components, because these are homogeneous and commensurable. It is otherwise with emergents, when, instead of adding measurable motion to measurable motion, or things of one kind to other individuals of their kind, there is a co-operation of things of unlike kinds. The emergent is unlike its components insofar as these are incommensurable, and it cannot be reduced to their sum or their difference."[5][6]

Economist Jeffrey Goldstein provided a current definition of emergence in the journal *Emergence*.[7] Goldstein initially defined emergence as: "the arising of novel and coherent structures, patterns and properties during the process of self-organization in complex systems".

Goldstein's definition can be further elaborated to describe the qualities of this definition in more detail:

> "The common characteristics are: (1) radical novelty (features not previously observed in systems); (2) coherence or correlation (meaning integrated wholes that maintain themselves over some period of time); (3) A global or macro "level" (i.e. there is some property of "wholeness"); (4) it is the product of a dynamical process (it evolves); and (5) it is "ostensive" (it can be perceived)." For good measure, Goldstein throws in supervenience.[8]

Systems scientist Peter Corning also says that living systems cannot be reduced to underlying laws of physics:

> Rules, or laws, have no causal efficacy; they do not in fact "generate" anything. They serve merely to describe regularities and consistent relationships in nature. These patterns may be very illuminating and important, but the underlying causal agencies must be separately specified (though often they are not). But that aside, the game of chess illustrates ... why any laws or rules of emergence and evolution are insufficient. Even in a chess game, you cannot use the rules to predict "history" — i.e., the course of any given game. Indeed, you cannot even reliably predict the next move in a chess game. Why? Because the "system" involves more than the rules of the game. It also includes the players and their unfolding, moment-by-moment decisions among a very large number of available options at each choice point. The game of chess is inescapably historical, even though it is also constrained and shaped by a set of rules, not to mention the laws of physics. Moreover, and this is a key point, the game of chess is also shaped by teleonomic, cybernetic, feedback-driven influences. It is not simply a self-ordered process; it involves an organized, "purposeful" activity.[8]

8.1.2 Strong and weak emergence

Usage of the notion "emergence" may generally be subdivided into two perspectives, that of "weak emergence" and "strong emergence". In terms of physical systems, weak emergence is a type of emergence in which the emergent property is amenable to computer simulation. This is opposed to the older notion of strong emergence, in which the emergent property cannot be simulated by a computer.

Some common points between the two notions are that emergence concerns new properties produced as the system grows, which is to say ones which are not shared with its components or prior states. Also, it is assumed that the properties are supervenient rather than metaphysically primitive (Bedau 1997).

Weak emergence describes new properties arising in systems as a result of the interactions at an elemental level. However, it is stipulated that the properties can be determined by observing or simulating the system, and not by any process of a priori analysis.

Bedau notes that weak emergence is not a universal metaphysical solvent, as weak emergence leads to the conclusion that matter itself contains elements of awareness to it. However, Bedau concludes that adopting this view would provide a precise notion that emergence is involved in consciousness, and second, the notion of weak emergence is metaphysically benign (Bedau 1997).

Strong emergence describes the direct causal action of a high-level system upon its components; qualities produced this way are irreducible to the system's constituent parts (Laughlin 2005). The whole is other than the sum of its parts. An example from physics of such emergence is water, being seemingly unpredictable even after an exhaustive study of the properties of its constituent atoms of hydrogen and oxygen.[9] It follows then that no simulation of the system can exist, for such a simulation would itself constitute a reduction of the system to its constituent parts (Bedau 1997).

However, "the debate about whether or not the whole can be predicted from the properties of the parts misses the point. Wholes produce unique combined effects, but many of these effects may be co-determined by the context and the interactions between the whole and its environment(s)" (Corning 2002). In accordance with his Synergism Hypothesis (Corning 1983 2005), Corning also stated, "It is the synergistic effects produced by wholes that are the very cause of the evolution of complexity in nature." Novelist Arthur Koestler used the metaphor of Janus (a symbol of the unity underlying complements like open/shut, peace/war) to illustrate how the two perspectives (strong vs. weak or holistic vs. reductionistic) should be treated as non-exclusive, and should work together to address the issues of emergence (Koestler 1969). Further,

> The ability to reduce everything to simple fundamental laws does not imply the ability to start from those laws and reconstruct the universe. The constructionist hypothesis breaks down when confronted with the twin difficulties of scale and complexity. At each level of complexity entirely new properties appear. Psychology is not applied biology, nor is biology applied chemistry. We can now see that the whole becomes not merely more, but very different from the sum of its parts. (Anderson 1972)

The plausibility of strong emergence is questioned by some as contravening our usual understanding of physics. Mark A. Bedau observes:

> Although strong emergence is logically possible, it is uncomfortably like magic. How does an irreducible but supervenient downward causal power arise, since by definition it cannot be due to the aggregation of the micro-level potentialities? Such causal powers would be quite unlike anything within our scientific ken. This not only indicates how they will discomfort reasonable forms of materialism. Their mysteriousness will only heighten the traditional worry that emergence entails illegitimately getting something from nothing.[10]

Meanwhile, others have worked towards developing analytical evidence of strong emergence. In 2009, Gu et al. presented a class of physical systems that exhibits non-computable macroscopic properties.[11][12] More precisely, if one could compute certain macroscopic properties of these systems from the microscopic description of these systems, then one would be able to solve computational problems known to be undecidable in computer science. They concluded that

> Although macroscopic concepts are essential for understanding our world, much of fundamental physics has been devoted to the search for a 'theory of everything', a set of equations that perfectly describe the behavior of all fundamental particles. The view that this is the goal of science rests in part on the rationale that such a theory would allow us to derive the behavior of all macroscopic concepts, at least in principle. The evidence we have presented suggests that this view may be overly optimistic. A 'theory of everything' is one of many components necessary for complete understanding of the universe, but is not necessarily the only one. The development of macroscopic laws from first principles may involve more than just systematic logic, and could require conjectures suggested by experiments, simulations or insight.[11]

Emergent structures are patterns that emerge via collective actions of many individual entities. To explain such patterns, one might conclude, per Aristotle,[2] that emergent structures are other than the sum of their parts on the assumption that the emergent order will not arise if the various parts simply interact independently of one another. However, there are those who disagree.[13] According to this argument, the interaction of each part with its immediate surroundings causes a complex chain of processes that can lead to order in some form. In fact, some systems in nature are observed to exhibit emergence based upon the interactions of autonomous parts, and some others exhibit emergence that at least at present cannot be reduced in this way.

8.1.3 Objective or subjective quality

The properties of complexity and organization of any system are considered by Crutchfield to be subjective qualities determined by the observer.

> "Defining structure and detecting the emergence of complexity in nature are inherently subjective, though essential, scientific activities. Despite the difficulties, these problems can be analysed in terms of how model-building observers infer from measurements the computational capabilities embedded in non-linear processes. An observer's notion of what is ordered, what is random, and what is complex in its environment depends directly on its computational resources: the amount of raw measurement data, of memory, and of time available for estimation and inference. The discovery of structure in an environment depends more critically and subtly, though, on how those resources are organized. The descriptive power of the observer's chosen (or implicit) computational model class, for example, can be an overwhelming determinant in finding regularity in data."(Crutchfield 1994)

On the other hand, Peter Corning argues "Must the synergies be perceived/observed in order to qualify as emergent effects, as some theorists claim? Most emphatically not. The synergies associated with emergence are real and measurable, even if nobody is there to observe them." (Corning 2002)

8.2 In religion, art and humanities

In religion, emergence grounds expressions of religious naturalism and syntheism in which a sense of the sacred is perceived in the workings of entirely naturalistic processes by which more complex forms arise or evolve from simpler forms. Examples are detailed in *The Sacred Emergence of Nature* by Ursula Goodenough & Terrence Deacon and *Beyond Reductionism: Reinventing the Sacred* by Stuart Kauffman, both from 2006, and in *Syntheism - Creating God in The Internet Age* by Alexander Bard & Jan Söderqvist from 2014. An early argument (1904-5) for the emergence of social formations, in part stemming from religion, can be found in Max Weber's most famous work, *The Protestant Ethic and the Spirit of Capitalism* [14]

In art, emergence is used to explore the origins of novelty, creativity, and authorship. Some art/literary theorists (Wheeler, 2006;[15] Alexander, 2011[16] have proposed alternatives to postmodern understandings of "authorship" using the complexity sciences and emergence theory. They contend that artistic selfhood and meaning are emergent, relatively objective phenomena. Michael J. Pearce has used emergence to describe the experience of works of art in relation to contemporary neuroscience.[17]) The concept of emergence has also been applied to the theory of literature and art, history, linguistics, cognitive sciences, etc. by the teachings of Jean-Marie Grassin at the University of Limoges (v. esp.: J. Fontanille, B. Westphal, J. Vion-Dury, éds. L'Émergence—Poétique de l'Émergence, en réponse aux travaux de Jean-Marie Grassin, Bern, Berlin, etc., 2011; and: the article "Emergence" in the *International Dictionary of Literary Terms (DITL)*.

In international development, concepts of emergence have been used within a theory of social change termed SEED-SCALE to show how standard principles interact to bring forward socio-economic development fitted to cultural values, community economics, and natural environment (local solutions emerging from the larger socio-econo-biosphere). These principles can be implemented utilizing a sequence of standardized tasks that self-assemble in individually specific ways utilizing recursive evaluative criteria.[18]

In postcolonial studies, the term "Emerging Literature" refers to a contemporary body of texts that is gaining momentum in the global literary landscape (v. esp.: J.M. Grassin, ed. *Emerging Literatures*, Bern, Berlin, etc. : Peter Lang, 1996). By opposition, "emergent literature" is rather a concept used in the theory of literature.

8.3 Emergent properties and processes

An emergent behavior or emergent property can appear when a number of simple entities (agents) operate in an environment, forming more complex behaviors as a collective. If emergence happens over disparate size scales, then the reason is usually a causal relation across different scales. In other words, there is often a form of top-down feedback in systems with emergent properties.[19] The processes from which emergent properties result may occur in either the observed or observing system, and can commonly be identified by their patterns of accumulating change, most generally called 'growth'. Emergent behaviours can occur because of intricate causal relations across different scales and feedback, known as interconnectivity. The emergent property itself may be either very predictable or unpredictable and unprecedented, and represent a new level of the system's evolution. The complex behaviour or properties are not a property of any single such entity, nor can they easily be predicted or deduced from behaviour in the lower-level entities, and might in fact be irreducible to such behavior.

The shape and behaviour of a flock of birds or school of fish are good examples of emergent properties.

One reason why emergent behaviour is hard to predict is that the number of interactions between components of a system increases exponentially with the number of components, thus potentially allowing for many new and subtle types of behaviour to emerge. Emergence is often a product of particular patterns of interaction. Negative feedback introduces constraints that serve to fix structures or behaviours. In contrast, positive feedback promotes change, allowing local variations to grow into global patterns. Another way in which interactions leads to emergent properties is dual-phase evolution. This occurs where interactions are applied intermittently, leading to two phases: one in which patterns form or grow, the other in which they are refined or removed.

On the other hand, merely having a large number of interactions is not enough by itself to guarantee emergent behaviour; many of the interactions may be negligible or irrelevant, or may cancel each other out. In some cases, a large number of interactions can in fact work against the emergence of interesting behaviour, by creating a lot of "noise" to drown out any emerging "signal"; the emergent behaviour may need to be temporarily isolated from other interactions before it reaches enough critical mass to be self-supporting. Thus it is not just the sheer number of connections between components which encourages emergence; it is also how these connections are organised. A hierarchical organisation is one example that can generate emergent behaviour (a bureaucracy may behave in a way quite different from that of the individual humans in that bureaucracy); but perhaps more interestingly, emergent behaviour can also arise from more decentralized organisational structures, such as a marketplace. In some cases, the system has to reach a combined threshold of diversity, organisation, and connectivity before emergent behaviour appears.

Unintended consequences and side effects are closely related to emergent properties. Luc Steels writes: *"A component has a particular functionality but this is not recognizable as a subfunction of the global functionality. Instead a component implements a behaviour whose side effect contributes to the global functionality [...] Each behaviour has a side effect and the sum of the side effects gives the desired functionality"* (Steels 1990). In other words, the global or macroscopic functionality of a system with "emergent functionality" is the sum of all "side effects", of all emergent properties and functionalities.

Systems with emergent properties or emergent structures may appear to defy entropic principles and the second law of thermodynamics, because they form and increase order despite the lack of command and central control. This is possible because open systems can extract information and order out of the environment.

Emergence helps to explain why the fallacy of division is a fallacy.

8.4 Emergent structures in nature

Main article: Patterns in nature

Emergent structures can be found in many natural phenom-

Ripple patterns in a sand dune created by wind or water is an example of an emergent structure in nature.

Giant's Causeway in Northern Ireland is an example of a complex emergent structure created by natural processes.

ena, from the physical to the biological domain. For example, the shape of weather phenomena such as hurricanes are emergent structures. The development and growth of complex, orderly crystals, as driven by the random motion of water molecules within a conducive natural environment, is another example of an emergent process, where randomness can give rise to complex and deeply attractive, orderly structures.

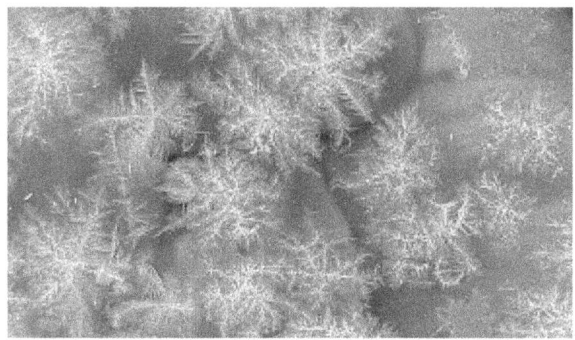

Water crystals forming on glass demonstrate an emergent, fractal natural process occurring under appropriate conditions of temperature and humidity.

However, crystalline structure and hurricanes are said to have a self-organizing phase.

Symphony of the Stones carved by the Goght River at Garni Gorge in Armenia is an example of an emergent natural structure.

It is useful to distinguish three forms of emergent structures. A *first-order* emergent structure occurs as a result of shape interactions (for example, hydrogen bonds in water molecules lead to surface tension). A *second-order* emergent structure involves shape interactions played out sequentially over time (for example, changing atmospheric conditions as a snowflake falls to the ground build upon and alter its form). Finally, a *third-order* emergent structure is a consequence of shape, time, and heritable instructions. For example, an organism's genetic code sets boundary conditions on the interaction of biological systems in space and time.

8.4.1 Non-living, physical systems

In physics, emergence is used to describe a property, law, or phenomenon which occurs at macroscopic scales (in space or time) but not at microscopic scales, despite the fact that a macroscopic system can be viewed as a very large ensemble of microscopic systems.

An emergent property need not be more complicated than the underlying non-emergent properties which generate it. For instance, the laws of thermodynamics are remarkably simple, even if the laws which govern the interactions between component particles are complex. The term emergence in physics is thus used not to signify complexity, but rather to distinguish which laws and concepts apply to macroscopic scales, and which ones apply to microscopic scales.

Some examples include:

- Classical mechanics: The laws of classical mechanics can be said to emerge as a limiting case from the rules of quantum mechanics applied to large enough masses. This is particularly strange since quantum mechanics is generally thought of as *more* complicated than classical mechanics.

- Friction: Forces between elementary particles are conservative. However, friction emerges when considering more complex structures of matter, whose surfaces can convert mechanical energy into heat energy when rubbed against each other. Similar considerations apply to other emergent concepts in continuum mechanics such as viscosity, elasticity, tensile strength, etc.

- Patterned ground: the distinct, and often symmetrical geometric shapes formed by ground material in periglacial regions.

- Statistical mechanics was initially derived using the concept of a large enough ensemble that fluctuations about the most likely distribution can be all but ignored. However, small clusters do not exhibit sharp first order phase transitions such as melting, and at the boundary it is not possible to completely categorize the cluster as a liquid or solid, since these concepts are (without extra definitions) only applicable to macroscopic systems. Describing a system using statistical mechanics methods is much simpler than using a low-level atomistic approach.

- Electrical networks: The bulk conductive response of binary (RC) electrical networks with random arrangements can be seen as emergent properties of such physical systems. Such arrangements can be used as simple physical prototypes for deriving mathematical formulae for the emergent responses of complex systems.[20]

- Weather.

Temperature is sometimes used as an example of an emergent macroscopic behaviour. In classical dynamics, a *snapshot* of the instantaneous momenta of a large number of particles at equilibrium is sufficient to find the average kinetic energy per degree of freedom which is proportional to the temperature. For a small number of particles the instantaneous momenta at a given time are not statistically sufficient to determine the temperature of the system. However, using the ergodic hypothesis, the temperature can still be obtained to arbitrary precision by further averaging the momenta over a long enough time.

Convection in a liquid or gas is another example of emergent macroscopic behaviour that makes sense only when considering differentials of temperature. Convection cells, particularly Bénard cells, are an example of a self-organizing system (more specifically, a dissipative system) whose structure is determined both by the constraints of the system and by random perturbations: the possible realizations of the shape and size of the cells depends on the temperature gradient as well as the nature of the fluid and shape of the container, but which configurations are actually realized is due to random perturbations (thus these systems exhibit a form of symmetry breaking).

In some theories of particle physics, even such basic structures as mass, space, and time are viewed as emergent phenomena, arising from more fundamental concepts such as the Higgs boson or strings. In some interpretations of quantum mechanics, the perception of a deterministic reality, in which all objects have a definite position, momentum, and so forth, is actually an emergent phenomenon, with the true state of matter being described instead by a wavefunction which need not have a single position or momentum. Most of the laws of physics themselves as we experience them today appear to have emerged during the course of time making emergence the most fundamental principle in the universe and raising the question of what might be the most fundamental law of physics from which all others emerged. Chemistry can in turn be viewed as an emergent property of the laws of physics. Biology (including biological evolution) can be viewed as an emergent property of the laws of chemistry. Similarly, psychology could be understood as an emergent property of neurobiological laws. Finally, free-market theories understand economy as an emergent feature of psychology.

According to Laughlin (2005), for many particle systems, nothing can be calculated exactly from the microscopic equations, and macroscopic systems are characterised by broken symmetry: the symmetry present in the microscopic equations is not present in the macroscopic system, due to phase transitions. As a result, these macroscopic systems are described in their own terminology, and have properties that do not depend on many microscopic details. This does not mean that the microscopic interactions are irrelevant, but simply that you do not see them anymore — you only see a renormalized effect of them. Laughlin is a pragmatic theoretical physicist: if you cannot, possibly ever, calculate the broken symmetry macroscopic properties from the microscopic equations, then what is the point of talking about reducibility?

8.4.2 Living, biological systems

Emergence and evolution

See also: Abiogenesis

Life is a major source of complexity, and evolution is the major process behind the varying forms of life. In this view, evolution is the process describing the growth of complexity in the natural world and in speaking of the emergence of complex living beings and life-forms, this view refers therefore to processes of sudden changes in evolution.

Life is thought to have emerged in the early RNA world when RNA chains began to express the basic conditions necessary for natural selection to operate as conceived by Darwin: heritability, variation of type, and competition for limited resources. Fitness of an RNA replicator (its per capita rate of increase) would likely be a function of adaptive capacities that were intrinsic (in the sense that they were determined by the nucleotide sequence) and the availability of resources.[21][22] The three primary adaptive capacities may have been (1) the capacity to replicate with moderate fidelity (giving rise to both heritability and variation of type); (2) the capacity to avoid decay; and (3) the capacity to acquire and process resources.[21][22] These capacities would have been determined initially by the folded configurations of the RNA replicators (see "Ribozyme") that, in turn, would be encoded in their individual nucleotide sequences. Competitive success among different replicators would have depended on the relative values of these adaptive capacities.

Regarding causality in evolution Peter Corning observes:

> "Synergistic effects of various kinds have played a major causal role in the evolutionary process generally and in the evolution of cooperation and complexity in particular... Natural selection is often portrayed as a "mechanism", or is personified as a causal agency... In reality, the differential "selection" of a trait, or an adaptation, is a consequence of the functional effects it produces in relation to the survival and reproductive success of a given organism in a given environment. It is these functional effects that are ultimately responsible for the trans-generational continuities and changes in nature." (Corning 2002)

Per his definition of emergence, Corning also addresses emergence and evolution:

> "[In] evolutionary processes, causation is iterative; effects are also causes. And this is equally true of the synergistic effects produced by emergent systems. In other words, emergence itself... has been the underlying cause of the evolution of emergent phenomena in biological evolution; it is the synergies produced by organized systems that are the key." (Corning 2002)

Swarming is a well-known behaviour in many animal species from marching locusts to schooling fish to flocking birds. Emergent structures are a common strategy found in many animal groups: colonies of ants, mounds built by termites, swarms of bees, shoals/schools of fish, flocks of birds, and herds/packs of mammals.

An example to consider in detail is an ant colony. The queen does not give direct orders and does not tell the ants what to do. Instead, each ant reacts to stimuli in the form of chemical scent from larvae, other ants, intruders, food and buildup of waste, and leaves behind a chemical trail, which, in turn, provides a stimulus to other ants. Here each ant is an autonomous unit that reacts depending only on its local environment and the genetically encoded rules for its variety of ant. Despite the lack of centralized decision making, ant colonies exhibit complex behavior and have even been able to demonstrate the ability to solve geometric problems. For example, colonies routinely find the maximum distance from all colony entrances to dispose of dead bodies.[23]

It appears that environmental factors may play a role in influencing emergence. Research suggests induced emergence of the bee species Perdita portalis. In this species, the bees emerge in a pattern consistent with rainfall. Specifically, the pattern of emergence is consistent with southwestern deserts' late summer rains and lack of activity in the spring.[24]

Organization of life

A broader example of emergent properties in biology is viewed in the biological organisation of life, ranging from the subatomic level to the entire biosphere. For example, individual atoms can be combined to form molecules such as polypeptide chains, which in turn fold and refold to form proteins, which in turn create even more complex structures. These proteins, assuming their functional status from their spatial conformation, interact together and with other molecules to achieve higher biological functions and eventually create an organism. Another example is how cascade phenotype reactions, as detailed in chaos theory, arise from individual genes mutating respective positioning.[25] At the highest level, all the biological communities in the world form the biosphere, where its human participants form societies, and the complex interactions of meta-social systems such as the stock market.

8.5 In humanity

8.5.1 Spontaneous order

See also: Spontaneous order and Self-organization

Groups of human beings, left free to each regulate themselves, tend to produce spontaneous order, rather than the meaningless chaos often feared. This has been observed in society at least since Chuang Tzu in ancient China. A classic traffic roundabout is a good example, with cars moving in and out with such effective organization that some modern cities have begun replacing stoplights at problem intersections with traffic circles , and getting better results. Open-source software and Wiki projects form an even more compelling illustration.

Emergent processes or behaviours can be seen in many other places, such as cities, cabal and market-dominant minority phenomena in economics, organizational phenomena in computer simulations and cellular automata. Whenever you have a multitude of individuals interacting with one another, there often comes a moment when disorder gives way to order and something new emerges: a pattern, a decision, a structure, or a change in direction (Miller 2010, 29).[26]

Economics

The stock market (or any market for that matter) is an example of emergence on a grand scale. As a whole it precisely regulates the relative security prices of companies across the world, yet it has no leader; when no central planning is in place, there is no one entity which controls the workings of the entire market. Agents, or investors, have knowledge of only a limited number of companies within their portfolio, and must follow the regulatory rules of the market and analyse the transactions individually or in large groupings. Trends and patterns emerge which are studied intensively by technical analysts.

World Wide Web and the Internet

The World Wide Web is a popular example of a decentralized system exhibiting emergent properties. There is no central organization rationing the number of links, yet the number of links pointing to each page follows a power law in which a few pages are linked to many times and most pages are seldom linked to. A related property of the network of links in the World Wide Web is that almost any pair of pages can be connected to each other through a relatively short chain of links. Although relatively well known now, this property was initially unexpected in an unregulated network. It is shared with many other types of networks called small-world networks (Barabasi, Jeong, & Albert 1999, pp. 130–131).

Internet traffic can also exhibit some seemingly emergent properties. In the congestion control mechanism, TCP flows can become globally synchronized at bottlenecks, simultaneously increasing and then decreasing throughput in coordination. Congestion, widely regarded as a nuisance, is possibly an emergent property of the spreading of bottlenecks across a network in high traffic flows which can be considered as a phase transition [see review of related research in (Smith 2008, pp. 1–31)].

Another important example of emergence in web-based systems is social bookmarking (also called collaborative tagging). In social bookmarking systems, users assign tags to resources shared with other users, which gives rise to a type of information organisation that emerges from this crowdsourcing process. Recent research which analyzes empirically the complex dynamics of such systems[27] has shown that consensus on stable distributions and a simple form of shared vocabularies does indeed emerge, even in the absence of a central controlled vocabulary. Some believe that this could be because users who contribute tags all use the same language, and they share similar semantic structures underlying the choice of words. The convergence in social tags may therefore be interpreted as the emergence of structures as people who have similar semantic interpretation collaboratively index online information, a process called semantic imitation.[28] [29]

Architecture and cities

Emergent structures appear at many different levels of organization or as spontaneous order. Emergent self-organization appears frequently in cities where no planning or zoning entity predetermines the layout of the city. (Krugman 1996, pp. 9–29) The interdisciplinary study of emergent behaviors is not generally considered a homogeneous field, but divided across its application or problem domains.

Architects and Landscape Architects may not design all the pathways of a complex of buildings. Instead they might let usage patterns emerge and then place pavement where pathways have become worn in.

The on-course action and vehicle progression of the 2007

Traffic patterns in cities can be seen as an example of spontaneous order

Urban Challenge could possibly be regarded as an example of cybernetic emergence. Patterns of road use, indeterministic obstacle clearance times, etc. will work together to form a complex emergent pattern that can not be deterministically planned in advance.

The architectural school of Christopher Alexander takes a deeper approach to emergence attempting to rewrite the process of urban growth itself in order to affect form, establishing a new methodology of planning and design tied to traditional practices, an Emergent Urbanism. Urban emergence has also been linked to theories of urban complexity (Batty 2005) and urban evolution (Marshall 2009).

Building ecology is a conceptual framework for understanding architecture and the built environment as the interface between the dynamically interdependent elements of buildings, their occupants, and the larger environment. Rather than viewing buildings as inanimate or static objects, building ecologist Hal Levin views them as interfaces or intersecting domains of living and non-living systems.[30] The microbial ecology of the indoor environment is strongly dependent on the building materials, occupants, contents, environmental context and the indoor and outdoor climate. The strong relationship between atmospheric chemistry and indoor air quality and the chemical reactions occurring indoors. The chemicals may be nutrients, neutral or biocides for the microbial organisms. The microbes produce chemicals that affect the building materials and occupant health and well being. Humans manipulate the ventilation, temperature and humidity to achieve comfort with the concomitant effects on the microbes that populate and evolve.[30][31][32]

Eric Bonabeau's attempt to define emergent phenomena is through traffic: "traffic jams are actually very complicated and mysterious. On an individual level, each driver is trying to get somewhere and is following (or breaking) certain rules, some legal (the speed limit) and others societal or personal (slow down to let another driver change into your lane). But a traffic jam is a separate and distinct entity that emerges from those individual behaviors. Gridlock on a highway, for example, can travel backward for no apparent reason, even as the cars are moving forward." He has also likened emergent phenomena to the analysis of market trends and employee behavior.[33]

Computational emergent phenomena have also been utilized in architectural design processes, for example for formal explorations and experiments in digital materiality.[34]

8.5.2 Computer AI

Some artificially intelligent computer applications utilize emergent behavior for animation. One example is Boids, which mimics the swarming behavior of birds.

8.5.3 Language

It has been argued that the structure and regularity of language grammar, or at least language change, is an emergent phenomenon (Hopper 1998). While each speaker merely tries to reach his or her own communicative goals, he or she uses language in a particular way. If enough speakers behave in that way, language is changed (Keller 1994). In a wider sense, the norms of a language, i.e. the linguistic conventions of its speech society, can be seen as a system emerging from long-time participation in communicative problem-solving in various social circumstances. (Määttä 2000)

8.5.4 Emergent change processes

Within the field of group facilitation and organization development, there have been a number of new group processes that are designed to maximize emergence and self-organization, by offering a minimal set of effective initial conditions. Examples of these processes include SEED-SCALE, Appreciative Inquiry, Future Search, the World Cafe or Knowledge Cafe, Open Space Technology, and others. (Holman, 2010)

8.6 See also

- Abstraction
- Agent-based model
- Anthropic principle
- Big History
- Connectionism
- Consilience
- Constructal theory
- Dynamical system
- Deus ex machina
- Dual-phase evolution
- Emergenesis
- Emergent algorithm
- Emergent evolution
- Emergent gameplay
- Emergent organization
- Epiphenomenon
- Externality
- Free will
- Generative sciences
- Innovation butterfly
- Interconnectedness
- Irreducible complexity
- Langton's ant
- Law of Complexity-Consciousness
- Mass action (sociology)
- Neural networks
- Organic Wholes of G.E. Moore
- Polytely
- Society of Mind theory
- Structuralism
- Swarm intelligence
- System of systems
- Teleology
- Synergetics (Fuller)
- Synergetics (Haken)

8.7 References

[1] O'Connor, Timothy; Wong, Hong Yu (February 28, 2012). Edward N. Zalta, ed. "Emergent Properties". *The Stanford Encyclopedia of Philosophy (Spring 2012 Edition).*

[2] Aristotle, *Metaphysics*, Book H 1045a 8–10: "... the totality is not, as it were, a mere heap, but the whole is something besides the parts ...", i.e., the whole is other than the sum of the parts.

[3] "The chemical combination of two substances produces, as is well known, a third substance with properties different from those of either of the two substances separately, or of both of them taken together" (Mill 1843)

[4] Julian Huxley: "now and again there is a sudden rapid passage to a totally new and more comprehensive type of order or organization, with quite new emergent properties, and involving quite new methods of further evolution" (Huxley & Huxley 1947)

[5] (Lewes 1875, p. 412)

[6] (Blitz 1992)

[7] (Goldstein 1999)

[8] Corning, Peter A. (2002), "The Re-Emergence of "Emergence": A Venerable Concept in Search of a Theory" (PDF), *Complexity* 7 (6): 18–30, doi:10.1002/cplx.10043

[9] Luisi, Pier L. (2006). *The Emergence of Life: From Chemical Origins to Synthetic Biology*. Cambridge, England: Cambridge University Press. p. 119. ISBN 0521821177.

[10] (Bedau 1997)

[11] Gu, Mile, et al. "More really is different." Physica D: Nonlinear Phenomena 238.9 (2009): 835-839.

[12] Binder, P-M. "Computation: The edge of reductionism." Nature 459.7245 (2009): 332-334.

[13] Steven Weinberg. "A Designer Universe?". Retrieved 2008-07-14. A version of the original quote from address at the Conference on Cosmic Design, American Association for the Advancement of Science, Washington, D.C. in April 1999

[14] McKinnon, AM. (2010). 'Elective affinities of the Protestant ethic: Weber and the chemistry of capitalism'. Sociological Theory, vol 28, no. 1, pp. 108-126.

[15] Wheeler, Wendy (2006). *The Whole Creature: Complexity, Biosemiotics and the Evolution of Culture*. London: Lawrence & Wishart. p. 192. ISBN 1-905007-30-2.

[16] Alexander, Victoria N. (2011). *The Biologist's Mistress: Rethinking Self-Organization in Art, Literature, and Nature*. Litchfield Park, AZ: Emergent Publications. ISBN 0-9842165-5-3.

[17] Pearce, Michael J. (2015). *Art in the Age of Emergence*. Manchester, England: Cambridge Scholars Publishing. ISBN 1443870579.

[18] Daniel C. Taylor, Carl E. Taylor, Jesse O. Taylor, "Empowerment on an Unstable Planet: From Seeds of Human Energy to a Scale of Global Change" (New York: Oxford University Press, 2012)

[19] See, e.g., Korotayev, A.; Malkov, A.; Khaltourina, D. (2006), *Introduction to Social Macrodynamics: Compact Macromodels of the World System Growth*, Moscow: URSS, ISBN 5-484-00414-4

[20] "The origin of power-law emergent scaling in large binary networks" D. P. Almond, C. J. Budd, M. A. Freitag, G. W. Hunt, N. J. McCullen and N. D. Smith. Physica A: Statistical Mechanics and its Applications, Volume 392, Issue 4, 15 February 2013

[21] Bernstein, H; Byerly, HC; Hopf, FA; Michod, RA; Vemulapalli, GK (1983). "The Darwinian Dynamic". *Quarterly Review of Biology* 58: 185–207. doi:10.1086/413216.

[22] Michod RE. (2000) Darwinian Dynamics: Evolutionary Transitions in Fitness and Individuality. Princeton University Press, Princeton, New Jersey ISBN 0691050112 ISBN 978-0691050119

[23] Steven Johnson. 2001. Emergence: The Connected Lives of Ants, Brains, Cities, and Software

[24] Danforth, Bryan (1991). "Female Foraging and Intranest Behavior of a Communal Bee, Perdita portalis (Hymenoptera: Andrenidae)". *Annals of the Entomological Society of America* 84 (5): 537–548. doi:10.1093/aesa/84.5.537.

[25] Campbell, Neil A., and Jane B. Reece. *Biology*. 6th ed. San Francisco: Benjamin Cummings, 2002.

[26] Miller, Peter. 2010. The Smart Swarm: How understanding flocks, schools, and colonies can make us better at communicating, decision making, and getting things done. New York: Avery.

[27] Valentin Robu, Harry Halpin, Hana Shepherd Emergence of consensus and shared vocabularies in collaborative tagging systems, ACM Transactions on the Web (TWEB), Vol. 3(4), article 14, ACM Press, September 2009.

[28] Fu, Wai-Tat; Kannampallil, Thomas George; Kang, Ruogu (August 2009), "A Semantic Imitation Model of Social Tagging", *Proceedings of the IEEE conference on Social Computing*: 66–72, doi:10.1109/CSE.2009.382, ISBN 978-1-4244-5334-4

[29] Fu, Wai-Tat; Kannampallil, Thomas; Kang, Ruogu; He, Jibo (2010), "Semantic Imitation in Social Tagging", *ACM Transactions on Computer-Human Interaction* 17 (3): 1–37, doi:10.1145/1806923.1806926

[30] http://www.microbe.net/fact-sheet-building-ecology/

[31] http://www.microbe.net

[32] http://buildingecology.com

[33] Bonabeau E. Predicting the Unpredictable. Harvard Business Review [serial online]. March 2002;80(3):109-116. Available from: Business Source Complete, Ipswich, MA. Accessed February 1, 2012.

[34] Roudavski, Stanislav and Gwyllim Jahn (2012). 'Emergent Materiality though an Embedded Multi-Agent System', in 15th Generative Art Conference, ed. by Celestino Soddu (Lucca, Italy: Domus Argenia), pp. 348-363

8.8 Bibliography

- Anderson, P.W. (1972), "More is Different: Broken Symmetry and the Nature of the Hierarchical Structure of Science", *Science* **177** (4047): 393–396, Bibcode:1972Sci...177..393A, doi:10.1126/science.177.4047.393, PMID 17796623

- Bedau, Mark A. (1997), *Weak Emergence* (PDF)

- Corning, Peter A. (1983), *The Synergism Hypothesis: A Theory of Progressive Evolution*, New York: McGraw-Hill

- Koestler, Arthur (1969), A. Koestler; J. R. Smythies, eds., *Beyond Reductionism: New Perspectives in the Life Sciences*, London: Hutchinson

- Laughlin, Robert (2005), *A Different Universe: Reinventing Physics from the Bottom Down*, Basic Books, ISBN 0-465-03828-X

8.9 Further reading

- Alexander, V. N. (2011). *The Biologist's Mistress: Rethinking Self-Organization in Art, Literature and Nature*. Litchfield Park AZ: Emergent Publications.

- Anderson, P.W. (1972), "More is Different: Broken Symmetry and the Nature of the Hierarchical Structure of Science" (PDF), *Science* **177** (4047): 393–396, Bibcode:1972Sci...177..393A, doi:10.1126/science.177.4047.393, PMID 17796623

- Barabási, Albert-László; Jeong, Hawoong; Albert, Réka (1999), "The Diameter of the World Wide Web", *Nature* **401** (6749): 130–131, arXiv:cond-mat/9907038, Bibcode:1999Natur.401..130A, doi:10.1038/43601

- Bar-Yam, Yaneer (2004), "A Mathematical Theory of Strong Emergence using Multiscale Variety" (PDF), *Complexity* **9** (6): 15–24, doi:10.1002/cplx.20029

- Bateson, Gregory (1972), *Steps to an Ecology of Mind*, Ballantine Books, ISBN 0-226-03905-6

- Batty, Michael (2005), *Cities and Complexity*, MIT Press, ISBN 0-262-52479-1

- Bedau, Mark A. (1997)."Weak Emergence".

- Blitz, David. (1992). *Emergent Evolution: Qualitative Novelty and the Levels of Reality*. Dordrecht: Kluwer Academic.

- Bunge, Mario Augusto (2003), *Emergence and Convergence: Qualitiative Novelty and the Unity of Knowledge*, Toronto: University of Toronto Press

- Chalmers, David J. (2002). "Strong and Weak Emergence" http://consc.net/papers/emergence.pdf Republished in P. Clayton and P. Davies, eds. (2006) *The Re-Emergence of Emergence*. Oxford: Oxford University Press.

- Philip Clayton (2005). *Mind and Emergence: From Quantum to Consciousness* Oxford: OUP, ISBN 978-0-19-927252-5

- Philip Clayton & Paul Davies (eds.) (2006). *The Re-Emergence of Emergence: The Emergentist Hypothesis from Science to Religion* Oxford: Oxford University Press.

- Corning, Peter A. (2005). "Holistic Darwinism: Synergy, Cybernetics and the Bioeconomics of Evolution." Chicago: University of Chicago Press.

- Crutchfield, James P. (1994), "Special issue on the Proceedings of the Oji International Seminar: Complex *Systems — from Complex Dynamics to Artificial Reality*" (PDF), *Physica D* **75**: 11–54, Bibcode:1994PhyD...75...11C, doi:10.1016/0167-2789(94)90273-9 |contribution= ignored (help)

- Felipe Cucker and Stephen Smale (2007), The Japanese Journal of Mathematics, *The Mathematics of Emergence*

- Delsemme, Armand (1998), *Our Cosmic Origins: From the Big Bang to the Emergence of Life and Intelligence*, Cambridge University Press

- De Wolf, Tom; Holvoet, Tom (2005), "Emergence Versus Self-Organisation: Different Concepts but Promising When Combined", *Engineering Self Organising Systems: Methodologies and Applications, Lecture Notes in Computer Science:* 3464, pp. 1–15, doi:10.1007/11494676_1

- Fromm, Jochen (2004), *The Emergence of Complexity*, Kassel University Press, ISBN 3-89958-069-9*
  Fromm, Jochen (2005a), *Types and Forms of Emergence*, arXiv:nlin.AO/0506028

- Fromm, Jochen (2005b), *Ten Questions about Emergence*, arXiv:nlin.AO/0509049

- Goodwin, Brian (2001), *How the Leopard Changed Its Spots: The Evolution of Complexity*, Princeton University Press

- Goldstein, Jeffrey (1999), "Emergence as a Construct: History and Issues" (PDF), *Emergence: Complexity and Organization* 1 (1): 49–72, doi:10.1207/s15327000em0101_4

- Haag, James W. (2008). *Emergent Freedom: Naturalizing Free Will* Goettingen: Vandenhoeck & Ruprecht, ISBN 978-3-525-56988-7

- Hayek, Friedrich (1973), *Law, Legislation and Liberty*, ISBN 0-226-32086-3

- Hofstadter, Douglas R. (1979), *Gödel, Escher, Bach: an Eternal Golden Braid*, Harvester Press

- Holland, John H. (1998), *Emergence from Chaos to Order*, Oxford University Press, ISBN 0-7382-0142-1

- Holman, Peggy. (2010). Engaging Emergence: Turning upheaval into opportunity. San Francisco: Barrett-Koehler. ISBN 978-1-60509-521-9

- Hopfield, John J. (1982), "Neural networks and physical systems with emergent collective computational abilities", *Proc. Natl. Acad. Sci. USA* **79** (8): 2554–2558, Bibcode:1982PNAS...79.2554H, doi:10.1073/pnas.79.8.2554, PMC 346238, PMID 6953413

- Hopper, P. 1998. Emergent Grammar. In: Tomasello, M. eds. 1998. The new psychology of language: Cognitive and functional approaches to language structure. Mahwah, NJ: Earlbaum, pp. 155–176.

- Huxley, Julian S.; Huxley, Thomas Henry (1947), *Evolution and Ethics: 1893-1943*, London, 1947: The Pilot Press, p. 120

- Johnson, Steven Berlin (2001), *Emergence: The Connected Lives of Ants, Brains, Cities, and Software*, Scribner's, ISBN 0-684-86876-8

- Kauffman, Stuart (1993), *The Origins of Order: Self-Organization and Selection in Evolution*, Oxford University Press, ISBN 0-19-507951-5

- Keller, Rudi (1994), *On Language Change: The Invisible Hand in Language*, London/New York: Routledge, ISBN 0-415-07671-4

- Kauffman, Stuart (1995), *At Home in the Universe*, New York: Oxford University Press

- Kelly, Kevin (1994), *Out of Control: The New Biology of Machines, Social Systems, and the Economic World*, Perseus Books, ISBN 0-201-48340-8

- Koestler, Arthur (1969), A. Koestler & J. R. Smythies, ed., *Beyond Reductionism: New Perspectives in the Life Sciences*, London: Hutchinson

- Korotayev, A.; Malkov, A.; Khaltourina, D. (2006), *Introduction to Social Macrodynamics: Compact Macromodels of the World System Growth*, Moscow: URSS, ISBN 5-484-00414-4

- Krugman, Paul (1996), *The Self-organizing Economy*, Oxford: Blackwell, ISBN 1-55786-698-8, ISBN 0-87609-177-X

- Laughlin, Robert (2005), *A Different Universe: Reinventing Physics from the Bottom Down*, Basic Books, ISBN 0-465-03828-X

- Leland, W.E.; Willinger, M.S.; Taqqu, M.S.; Wilson, D.V. (1994), "On the self-similar nature of Ethernet traffic (extended version)", *IEEE/ACM Transactions on Networking* **2**: 1–15, doi:10.1109/90.282603

- Lewes, G. H. (1875), *Problems of Life and Mind (First Series)* 2, London: Trübner, ISBN 1-4255-5578-0

- Lewin, Roger (2000), *Complexity - Life at the Edge of Chaos* (second ed.), University of Chicago Press, ISBN 0-226-47654-5, ISBN 0-226-47655-3

- Ignazio Licata & Ammar Sakaji (eds) (2008). *Physics of Emergence and Organization*, ISBN 978-981-277-994-6, World Scientific and Imperial College Press.

- Marshall, Stephen (2009), *Cities Design and Evolution*, Routledge, ISBN 978-0-415-42329-8, ISBN 0-415-42329-5

- Mill, John Stuart (1843), "On the Composition of Causes", *A System of Logic, Ratiocinative and Inductive* (1872 ed.), London: John W. Parker and Son, p. 371

- Morowitz, Harold J. (2002), *The Emergence of Everything: How the World Became Complex*, Oxford University Press, ISBN 0-19-513513-X

- Pearce, Michael J. (2015), *Art in the Age of Emergence.*, Cambridge Scholars Publishing, ISBN 1-443-87057-9, ISBN 1-443-87057-9

- Ritchey, Tom (2014), "On a Morphology of Theories of Emergence" (PDF), *Acta Morphologica Generalis* 3 (3)

- Ryan, Alex J. (2006), "Emergence is Coupled to Scope, not Level", *Complexity*, (to be submitted): 67–77, arXiv:nlin.AO/0609011, doi:10.1002/cplx.20203

- Schelling, Thomas C. (1978), *Micromotives and Macrobehaviour*, W. W. Norton, ISBN 0-393-05701-1

- Jackie (Jianhong) Shen (2008), *Cucker–Smale Flocking Emergence under Hierarchical Leadership* In: SIAM J. Applied Math., 68:3,

- Smith, John Maynard; Szathmáry, Eörs (1997), *The Major Transitions in Evolution*, Oxford University Press, ISBN 0-19-850294-X

- Smith, Reginald D. (2008), *The Dynamics of Internet Traffic: Self-Similarity, Self-Organization, and Complex Phenomena*, arXiv:0807.3374, Bibcode:2008arXiv0807.3374S

- Solé, Ricard and Goodwin, Brian (2000) Signs of life: how complexity pervades biology, Basic Books, New York.

- Steels, Luc (1990), "Towards a Theory of Emergent Functionality", in Jean-Arcady Meyer; Stewart W. Wilson, *From Animals to Animats (Proceedings of the First International Conference on Simulation of Adaptive behaviour)*, Cambridge, MA & London, England: Bradford Books (MIT Press), pp. 451–461

- Wan, Poe Yu-ze (2011), "Emergence a la Systems Theory: Epistemological *Totalausschluss* or Ontological Novelty?", *Philosophy of the Social Sciences, 41(2)*, pp. 178–210

- Wan, Poe Yu-ze (2011), *Reframing the Social: Emergentist Systemism and Social Theory*, Ashgate Publishing

- Weinstock, Michael (2010), *The Architecture of Emergence - the evolution of form in Nature and Civilisation*, John Wiley and Sons, ISBN 0-470-06633-4

- Wolfram, Stephen (2002), *A New Kind of Science*, ISBN 1-57955-008-8

- Young, Louise B. (2002), *The Unfinished Universe*, ISBN 0-19-508039-4

8.10 External links

- "Emergence". *Internet Encyclopedia of Philosophy.*

- "Emergent Properties". *Stanford Encyclopedia of Philosophy.*

- Emergence at PhilPapers

- Emergence at the Indiana Philosophy Ontology Project

- The Emergent Universe: An interactive introduction to emergent phenomena, from ant colonies to Alzheimer's.

- Exploring Emergence: An introduction to emergence using CA and Conway's Game of Life from the MIT Media Lab

- ISCE group: Institute for the Study of Coherence and Emergence.

- Towards modeling of emergence: lecture slides from Helsinki University of Technology

- Biomimetic Architecture - Emergence applied to building and construction

- Studies in Emergent Order: Studies in Emergent Order (SIEO) is an open-access journal

- Emergence

Chapter 9

Gauss's principle of least constraint

The **principle of least constraint** is another formulation of classical mechanics enunciated by Carl Friedrich Gauss in 1829.

The principle of least constraint is a least squares principle stating that the true motion of a mechanical system of N masses is the minimum of the quantity

$$Z \stackrel{\text{def}}{=} \sum_{k=1}^{N} m_k \left| \frac{d^2 \mathbf{r}_k}{dt^2} - \frac{\mathbf{F}_k}{m_k} \right|^2$$

for all trajectories satisfying any imposed constraints, where m_k, $\mathbf{r}_k$ and $\mathbf{F}_k$ represent the mass, position and applied forces of the kth mass.

Gauss's principle is equivalent to D'Alembert's principle.

The principle of least constraint is qualitatively similar to Hamilton's principle, which states that the true path taken by a mechanical system is an extremum of the action. However, Gauss's principle is a true (local) *minimal* principle, whereas the other is an *extremal* principle.

9.1 Hertz's principle of least curvature

Hertz's principle of least curvature is a special case of Gauss's principle, restricted by the two conditions that there be no applied forces and that all masses are identical. (Without loss of generality, the masses may be set equal to one.) Under these conditions, Gauss's minimized quantity can be written

$$Z = \sum_{k=1}^{N} \left| \frac{d^2 \mathbf{r}_k}{dt^2} \right|^2$$

The kinetic energy T is also conserved under these conditions

$$T \stackrel{\text{def}}{=} \frac{1}{2} \sum_{k=1}^{N} \left| \frac{d \mathbf{r}_k}{dt} \right|^2$$

Since the line element ds^2 in the $3N$-dimensional space of the coordinates is defined

$$ds^2 \stackrel{\text{def}}{=} \sum_{k=1}^{N} |d\mathbf{r}_k|^2$$

the conservation of energy may also be written

$$\left(\frac{ds}{dt} \right)^2 = 2T$$

Dividing Z by $2T$ yields another minimal quantity

$$K \stackrel{\text{def}}{=} \sum_{k=1}^{N} \left| \frac{d^2 \mathbf{r}_k}{ds^2} \right|^2$$

Since $\sqrt{K}$ is the local curvature of the trajectory in the $3N$-dimensional space of the coordinates, minimization of K is equivalent to finding the trajectory of least curvature (a geodesic) that is consistent with the constraints. Hertz's principle is also a special case of Jacobi's formulation of the least-action principle.

9.2 See also

- Appell's equation of motion

9.3 References

- Gauss, C. F. (1829). "Über ein neues allgemeines Grundgesetz der Mechanik". *Crelle's Journal* 4: 232. doi:10.1515/crll.1829.4.232.

- Gauss, C. F. *Werke* **5**. p. 23.
- Hertz, H. (1896). *Principles of Mechanics*. Miscellaneous Papers **III**. Macmillan.

9.4 External links

- Gauss's principle of least constraint
- Hertz's principle of least curvature

Chapter 10

Self-organized criticality

In physics, **self-organized criticality** (SOC) is a property of (classes of) dynamical systems that have a critical point as an attractor. Their macroscopic behaviour thus displays the spatial and/or temporal scale-invariance characteristic of the critical point of a phase transition, but without the need to tune control parameters to precise values.

The concept was put forward by Per Bak, Chao Tang and Kurt Wiesenfeld ("BTW") in a paper[1] published in 1987 in *Physical Review Letters*, and is considered to be one of the mechanisms by which complexity [2] arises in nature. Its concepts have been enthusiastically applied across fields as diverse as geophysics, physical cosmology, evolutionary biology and ecology, bio-inspired computing and optimization (mathematics), economics, quantum gravity, sociology, solar physics, plasma physics, neurobiology [3][4][5][6][7][8][9][10][11][12] [13] [14] [15] and others.

SOC is typically observed in slowly driven non-equilibrium systems with extended degrees of freedom and a high level of nonlinearity. Many individual examples have been identified since BTW's original paper, but to date there is no known set of general characteristics that *guarantee* a system will display SOC.

10.1 Overview

Self-organized criticality is one of a number of important discoveries made in statistical physics and related fields over the latter half of the 20th century, discoveries which relate particularly to the study of complexity in nature. For example, the study of cellular automata, from the early discoveries of Stanislaw Ulam and John von Neumann through to John Conway's Game of Life and the extensive work of Stephen Wolfram, made it clear that complexity could be generated as an emergent feature of extended systems with simple local interactions. Over a similar period of time, Benoît Mandelbrot's large body of work on fractals showed that much complexity in nature could be described by certain ubiquitous mathematical laws, while the extensive study of phase transitions carried out in the 1960s and 1970s showed how scale invariant phenomena such as fractals and power laws emerged at the critical point between phases. However, the term Self-Organized Criticality was firstly introduced by Bak, Tang and Wiesenfeld's 1987 paper which clearly linked together these factors: a simple cellular automaton was shown to produce several characteristic features observed in natural complexity (fractal geometry, pink (1/f) noise and power laws) in a way that could be linked to critical-point phenomena. Crucially, however, the paper emphasized that the complexity observed emerged in a robust manner that did not depend on finely tuned details of the system: variable parameters in the model could be changed widely without affecting the emergence of critical behaviour (hence, *self-organized* criticality). Thus, the key result of BTW's paper was its discovery of a mechanism by which the emergence of complexity from simple local interactions could be *spontaneous* — and therefore plausible as a source of natural complexity — rather than something that was only possible in the lab (or lab computer) where it was possible to tune control parameters to precise values. The publication of this research sparked considerable interest from both theoreticians and experimentalists, and important papers on the subject are among the most cited papers in the scientific literature.

Due to BTW's metaphorical visualization of their model as a "sandpile" on which new sand grains were being slowly sprinkled to cause "avalanches", much of the initial experimental work tended to focus on examining real avalanches in granular matter, the most famous and extensive such study probably being the Oslo ricepile experiment. Other experiments include those carried out on magnetic-domain patterns, the Barkhausen effect and vortices in superconductors. Early theoretical work included the development of a variety of alternative SOC-generating dynamics distinct from the BTW model, attempts to prove model properties analytically (including calculating the critical exponents[16][17]), and examination of the necessary conditions for SOC to emerge. One of the important issues for the latter investigation was whether conservation of energy was required in the local dynamical exchanges of models: the answer in general is no, but with (minor) reservations, as some exchange dynamics (such as those of BTW) do require local conservation at least on average. In the long term, key theoretical issues yet to be resolved include the calculation of the possible universality classes of SOC behaviour and the question of whether it is possible to derive a general rule for determining if an arbitrary algorithm displays SOC.

Alongside these largely lab-based approaches, many other investigations have centered around large-scale natural or social systems that are known (or suspected) to display scale-invariant behavior. Although these approaches were not always welcomed (at least initially) by specialists in the subjects examined, SOC has nevertheless become established as a strong candidate for explaining a number of natural phenomena, including: earthquakes (which, long before SOC was discovered, were known as a source of scale-invariant behavior such as the Gutenberg–Richter law describing the statistical distribution of earthquake sizes and the Omori law describing the frequency of aftershocks, and where models that displayed SOC were proposed and analyzed prior to the BTW 87 paper;[3][4]); solar flares; fluctuations in economic systems such as financial markets (references to SOC are common in econophysics); landscape formation; forest fires; landslides; epidemics; neuronal avalanches in cortex;[6][11] 1/f noise in the amplitude envelope of electrophysiological signals;[5] and biological evolution (where SOC has been invoked, for example, as the dynamical mechanism behind the theory of "punctuated equilibria" put forward by Niles Eldredge and Stephen Jay Gould). These "applied" investigations of SOC have included both attempts at modelling (either developing new models or adapting existing ones to the specifics of a given natural system), and extensive data analysis to determine the existence and/or characteristics of natural scaling laws.

The recent excitement generated by scale-free networks has raised some interesting new questions for SOC-related research: a number of different SOC models have been shown to generate such networks as an emergent phenomenon, as opposed to the simpler models proposed by network researchers where the network tends to be assumed to exist independently of any physical space or dynamics.

Despite the considerable interest and research output generated from the SOC hypothesis there remains no general agreement with regards to its mathematical mechanisms. Bak Tang and Wiesenfeld based their hypothesis on the behavior of their sandpile model.[1] However, this model was subsequently shown to actually generate $1/f^2$ noise rather than 1/f noise.[18] Other simulation models were proposed later that could produce true 1/f noise,[19], and experimental sandpile models were observed to yield 1/f noise.[20] In addition to the nonconservative theoretical model mentioned above, other theoretical models for SOC have been based upon information theory[21] and mean field theory,.[22]

10.2 Examples of self-organized critical dynamics

In chronological order of development:

- Bak–Tang–Wiesenfeld sandpile
- Forest-fire model
- Olami–Feder–Christensen model
- Bak–Sneppen model

10.3 See also

- 1/f noise
- Complex systems
- Detrended fluctuation analysis, a method to detect power-law scaling in time series.
- Dual-phase evolution, another process that contributes to self-organization in complex systems.
- Fractals
- Power laws
- Scale invariance
- Self-organization
- Critical exponents
- Ilya Prigogine, a systems scientist who helped formalize dissipative system behavior in general terms.
- Red Queen hypothesis
- Self-organized criticality control

10.4 References

[1] Bak, P., Tang, C. and Wiesenfeld, K. (1987). "Self-organized criticality: an explanation of 1/f noise". *Physical Review Letters* **59** (4): 381–384. Bibcode:1987PhRvL..59..381B. doi:10.1103/PhysRevLett.59.381. Papercore summary: http://papercore.org/Bak1987.

[2] Bak, P., and Paczuski, M. (1995). "Complexity, contingency, and criticality". *Proc Natl Acad Sci U S A.* **92** (15): 6689–6696. Bibcode:1995PNAS...92.6689B. doi:10.1073/pnas.92.15.6689. PMC 41396. PMID 11607561.

[3] Turcotte, D. L.; Smalley, R. F., Jr.; Solla, S. A. (1985). "Collapse of loaded fractal trees". *Nature* **313** (6004): 671–672. Bibcode:1985Natur.313..671T. doi:10.1038/313671a0.

[4] Smalley, R. F., Jr.; Turcotte, D. L.; Solla, S. A. (1985). "A renormalization group approach to the stick-slip behavior of faults". *Journal of Geophysical Research* **90** (B2): 1894. Bibcode:1985JGR....90.1894S. doi:10.1029/JB090iB02p01894.

[5] K. Linkenkaer-Hansen; V. V. Nikouline; J. M. Palva & R. J. Ilmoniemi. (2001). "Long-Range Temporal Correlations and Scaling Behavior in Human Brain Oscillations". *J. Neurosci.* **21** (4): 1370–1377. PMID 11160408.

[6] J. M. Beggs & D. Plenz (2006). "Neuronal Avalanches in Neocortical Circuits". *J. Neurosci 23.*

[7] Chialvo, D. R. (2004). "Critical brain networks". *Physica A* **340** (4): 756–765. arXiv:cond-mat/0402538. Bibcode:2004PhyA..340..756C. doi:10.1016/j.physa.2004.05.064.

[8] Stefan Boettcher (1999). "Extremal optimization of graph partitioning at the percolation threshold". *J. Phys. A: Math. Gen* **32** (28): 5201–5211. arXiv:cond-mat/9901353. Bibcode:1999JPhA...32.5201B. doi:10.1088/0305-4470/32/28/302.

[9] D. Fraiman, P. Balenzuela, J. Foss and D. R. Chialvo (2004). "Ising-like dynamics in large scale brain functional networks". *Physical Review E* **79** (6): 061922. arXiv:0811.3721. Bibcode:2009PhRvE..79f1922F. doi:10.1103/PhysRevE.79.061922.

[10] L. de Arcangelis; C. Perrone-Capano & H. J. Herrmann (2006). "Self-organized criticality model for brain plasticity". *Phys. Rev. Lett.* **96**. arXiv:q-bio/0602014. Bibcode:2006PhRvL..96b8107D. doi:10.1103/physrevlett.96.028107.

[11] Poil, SS; Hardstone, R; Mansvelder, HD; Linkenkaer-Hansen, K (Jul 2012). "Critical-state dynamics of avalanches and oscillations jointly emerge from balanced excitation/inhibition in neuronal networks". *Journal of Neuroscience* **32** (29): 9817–23. doi:10.1523/JNEUROSCI.5990-11. PMC 3553543. PMID 22815496.

[12] Manfred G. Kitzbichler, Marie L. Smith, Søren R. Christensen, Ed Bullmore1 (2009). Behrens, Tim, ed. "Broadband Criticality of Human Brain Network Synchronization". *PLoS Comput Biol* **5** (3): e1000314. Bibcode:2009PLSCB...5E0314K. doi:10.1371/journal.pcbi.1000314. PMC 2647739. PMID 19300473.

[13] Chialvo, D. R. (2010). "Emergent complex neural dynamics". *Nature Physics* **6**: 744–750. arXiv:1010.2530. Bibcode:2010NatPh...6..744C. doi:10.1038/nphys1803.

[14] Tagliazucchi E, Balenzuela P, Fraiman D and Chialvo DR. (2012). "Criticality in large-scale brain fMRI dynamics unveiled by a novel point process analysis". *Front. Physiol.* **3**: 15. doi:10.3389/fphys.2012.00015.

[15] Haimovici A, Tagliazucchi E, Balenzuela P and Chialvo DR. (2013). "Brain Organization into Resting State Networks Emerges at Criticality on a Model of the Human Connectome". *Physical Review Letters* **110**: 178101. arXiv:1209.5353. Bibcode:2013PhRvL.110q8101H. doi:10.1103/PhysRevLett.110.178101.

[16] Tang, C. and Bak, P. (1988). "Critical exponents and scaling relations for self-organized critical phenomena". *Physical Review Letters* **60** (23): 2347–2350. Bibcode:1988PhRvL..60.2347T. doi:10.1103/PhysRevLett.60.2347.

[17] Tang, C. and Bak, P. (1988). "Mean field theory of self-organized critical phenomena". *Journal of Statistical Physics* **51** (5-6): 797–802. Bibcode:1988JSP....51..797T. doi:10.1007/BF01014884.

[18] Jensen, H. J., Christensen, K. and Fogedby, H. C. (1989). "1/f noise, distribution of lifetimes, and a pile of sand". *Phys. Rev. B* **40**: 7425–7427. Bibcode:1989PhRvB..40.7425J. doi:10.1103/physrevb.40.7425.

[19] Maslov, S., Tang, C. and Zhang, Y. - C. (1999). "1/f noise in Bak-Tang-Wiesenfeld models on narrow stripes". *Phys. Rev. Lett.* **83**: 2449–2452. arXiv:cond-mat/9902074. Bibcode:1999PhRvL..83.2449M. doi:10.1103/physrevlett.83.2449.

[20] Frette, V., Christinasen, K., Malthe-Sørenssen,A., Feder, J, Jøssang, T and Meaken, P (1996). "Avalanche dynamics in a pile of rice". *Nature* **379**: 49–52. Bibcode:1996Natur.379...49F. doi:10.1038/379049a0.

[21] Dewar, R. (2003). "Information theory explanation of the fluctuation theorem, maximum entropy production and self-organized criticality in non-equilibrium stationary states". *J. Phys. A: Math. Gen.* **36**: 631–641. arXiv:cond-mat/0005382. Bibcode:2003JPhA...36..631D. doi:10.1088/0305-4470/36/3/303.

[22] Vespignani, A., and Zapperi,S. (1998). "How self-organized criticality works: a unified mean-field picture". *Phys. Rev. E* **57**: 6345–6362. arXiv:cond-mat/9709192. Bibcode:1998PhRvE..57.6345V. doi:10.1103/physreve.57.6345.

10.5 Further reading

- Adami, C. (1995). "Self-organized criticality in living systems". *Physics Letters A* **203** (1): 29–32.

Bibcode:1995PhLA..203...29A. doi:10.1016/0375-9601(95)00372-A.

- Bak, P. (1996). *How Nature Works: The Science of Self-Organized Criticality*. New York: Copernicus. ISBN 0-387-94791-4.

- Bak, P. and Paczuski, M. (1995). "Complexity, contingency, and criticality". *Proceedings of the National Academy of Sciences of the USA* **92** (15): 6689–6696. Bibcode:1995PNAS...92.6689B. doi:10.1073/pnas.92.15.6689. PMC 41396. PMID 11607561.

- Bak, P. and Sneppen, K. (1993). "Punctuated equilibrium and criticality in a simple model of evolution". *Physical Review Letters* **71** (24): 4083–4086. Bibcode:1993PhRvL..71.4083B. doi:10.1103/PhysRevLett.71.4083. PMID 10055149.

- Bak, P., Tang, C. and Wiesenfeld, K. (1987). "Self-organized criticality: an explanation of $1/f$ noise". *Physical Review Letters* **59** (4): 381–384. Bibcode:1987PhRvL..59..381B. doi:10.1103/PhysRevLett.59.381.

- Bak, P., Tang, C. and Wiesenfeld, K. (1988). "Self-organized criticality". *Physical Review A* **38** (1): 364–374. Bibcode:1988PhRvA..38..364B. doi:10.1103/PhysRevA.38.364. Papercore summary.

- Buchanan, M. (2000). *Ubiquity*. London: Weidenfeld & Nicolson. ISBN 0-7538-1297-5.

- Jensen, H. J. (1998). *Self-Organized Criticality*. Cambridge: Cambridge University Press. ISBN 0-521-48371-9.

- Katz, J. I. (1986). "A model of propagating brittle failure in heterogeneous media". *Journal of Geophysical Research* **91** (B10): 10412. Bibcode:1986JGR....9110412K. doi:10.1029/JB091iB10p10412.

- Kron, T./Grund, T. (2009). "Society as a Selforganized Critical System". *Cybernetics and Human Knowing* **16**: 65–82.

- Paczuski, M. (2005). "Networks as renormalized models for emergent behavior in physical systems". *The Science and Culture Series – Physics*. arXiv:physics/0502028. Bibcode:2005cmn..conf..363P. doi:10.1142/9789812701558_0042. ISBN 978-981-256-525-9.

- Turcotte, D. L. (1997). *Fractals and Chaos in Geology and Geophysics*. Cambridge: Cambridge University Press. ISBN 0-521-56733-5.

- Turcotte, D. L. (1999). "Self-organized criticality". *Reports on Progress in Physics* **62** (10): 1377–1429. Bibcode:1999RPPh...62.1377T. doi:10.1088/0034-4885/62/10/201.

- Md. Nurujjaman/A. N. Sekar Iyengar (2007). "Realization of {SOC} behavior in a dc glow discharge plasma". *Physics Letters A* **360**: 717–721. arXiv:physics/0611069. Bibcode:2007PhLA..360..717N. doi:10.1016/j.physleta.2006.09.005.

- Self-organized criticality on arxiv.org

Chapter 11

Spontaneous order

See also: Emergence and Self-organization

Spontaneous order, also named "self-organization", is the spontaneous emergence of order out of seeming chaos. It is a process found in physical, biological, and social networks including economics, though the term "self-organization" is more often used for physical and biological processes, while "spontaneous order" is typically used to describe the emergence of various kinds of social orders from a combination of self-interested individuals who are not intentionally trying to create order through planning. The evolution of life on Earth, language, crystal structure, the Internet and a free market economy have all been proposed as examples of systems which evolved through spontaneous order.[1] Naturalists often point to the inherent "watch-like" precision of uncultivated ecosystems and to the universe itself as ultimate examples of this phenomenon.

Spontaneous orders are to be distinguished from organizations. Spontaneous orders are distinguished by being scale-free networks, while organizations are hierarchical networks. Further, organizations can be and often are a part of spontaneous social orders, but the reverse is not true. Further, while organizations are created and controlled by humans, spontaneous orders are created, controlled, *and controllable* by no one. In economics and the social sciences, spontaneous order is defined as "the result of human actions, not of human design."

Spontaneous order is also used as a synonym for any emergent behavior of which self-interested spontaneous order is just an instance.

11.1 History

According to Murray Rothbard, Zhuangzi (369–286 BCE) was the first to work out the idea of spontaneous order. The philosopher rejected the authoritarianism of Confucianism, writing that there "has been such a thing as letting mankind alone; there has never been such a thing as governing mankind [with success]." He articulated an early form of spontaneous order, asserting that "good order results spontaneously when things are let alone", a concept later "developed particularly by Proudhon in the nineteenth" century.[2]

The thinkers of the Scottish Enlightenment were the first to seriously develop and inquire into the idea of the market as a spontaneous order. In 1767, the sociologist and historian Adam Ferguson described the phenomenon of spontaneous order in society as the "result of human action, but not the execution of any human design".[3][4]

The Austrian School of Economics, led by Carl Menger, Ludwig von Mises and Friedrich Hayek, would later refine the concept and make it a centerpiece in its social and economic thought.

11.2 Examples

11.2.1 Markets

Many economic classical liberals, such as Hayek, have argued that market economies are a spontaneous order, "a more efficient allocation of societal resources than any design could achieve."[5] They claim this spontaneous order (referred to as the extended order in Hayek's "The Fatal Conceit") is superior to any order a human mind can design due to the specifics of the information required.[6] Centralized statistical data cannot convey this information because the statistics are created by abstracting away from the particulars of the situation.[7]

In a market economy, price is the aggregation of information acquired when people are free to use their individual knowledge. Price then allows everyone dealing in a commodity or its substitutes to make decisions based on more information than he or she could personally acquire, information not statistically conveyable to a centralized authority. Interference from a central authority which affects price will have consequences they could not foresee because they do not know all of the particulars involved.

This is illustrated in the concept of the invisible hand proposed by Adam Smith in *The Wealth of Nations*.[1] Thus in this view by acting on information with greater detail and accuracy than possible for any centralized authority, a more efficient economy is created to the benefit of a whole society.

Lawrence Reed, president of the Foundation for Economic Education, describes spontaneous order as follows:

> Spontaneous order is what happens when you leave people alone—when entrepreneurs... see the desires of people... and then provide for them.
> They respond to market signals, to prices. Prices tell them what's needed and how urgently and where. And it's infinitely better and more productive than relying on a handful of elites in some distant bureaucracy.[8]

11.2.2 Game studies

The concept of spontaneous order is closely related with modern game studies. As early as in the 1940s, historian Johan Huizinga wrote that "in myth and ritual the great instinctive forces of civilized life have their origin: law and order, commerce and profit, craft and art, poetry, wisdom and science. All are rooted in the primeval soil of play." Following on this in his book *The Fatal Conceit*, Hayek notably wrote that "a game is indeed a clear instance of a process wherein obedience to common rules by elements pursuing different and even conflicting purposes results in overall order."

11.2.3 Anarchism

Anarchists argue that the state is in fact an artificial creation of the ruling elite, and that true spontaneous order would arise if it was eliminated. Construed by some but not all as the ushering in of organization by anarchist law. In the anarchist view, such spontaneous order would involve the voluntary cooperation of individuals. According to the *Oxford Dictionary of Sociology*, "the work of many symbolic interactionists is largely compatible with the anarchist vision, since it harbours a view of society as spontaneous order."[9]

11.2.4 Sobornost

The concept of spontaneous order can also be seen in the works of the Russian Slavophile movements and specifically in the works of Fyodor Dostoyevsky. The concept of an organic social manifestation as a concept in Russia expressed under the idea of sobornost. Sobornost was also used by Leo Tolstoy as an underpinning to the ideology of Christian anarchism. The concept was used to describe the uniting force behind the peasant or serf Obshchina in pre-Soviet Russia.[10]

11.2.5 Recent developments

Perhaps the most famous theorist of social spontaneous orders is Friedrich Hayek. In addition to arguing the economy is a spontaneous order, which he termed a catallaxy, he argued that common law[11] and the brain[12] are also types of spontaneous orders. In "The Republic of Science,"[13] Michael Polanyi also argued that science is a spontaneous order, a theory further developed by Bill Butos and Thomas McQuade in a variety of papers. Gus DiZerega has argued that democracy is the spontaneous order form of government,[14] David Emmanuel Andersson has argued that religion in places like the United States is a spontaneous order,[15] and Troy Camplin argues that artistic and literary production are spontaneous orders.[16] Paul Krugman too has contributed to spontaneous order theory in his book *The Self-Organizing Economy*,[17] in which he claims that cities are self-organizing systems.

11.3 See also

- Anonymous
- Deregulation
- Extended order
- Free price system
- "I, Pencil" by Leonard Read
- Invisible hand
- Mutual aid
- Natural law
- Natural order
- Organised order
- Revolutionary spontaneity
- Stigmergy
- Tragedy of the commons

11.4 References

[1] Norman Barry, The Tradition of Spontaneous Order, *Literature of Liberty: A Review of Contemporary Liberal Thought*, Library of Economics and Liberty, 1982, accessed 2010-12-12

[2] Rothbard, Murray. *Concepts of the Role of Intellectuals in Social Change Toward Laissez Faire*, The Journal of Libertarian Studies, Vol IX No. 2 (Fall 1990)

[3] Adam Ferguson on The History of Economic Thought Website

[4] Ferguson, Adam (1767). *An Essay on the History of Civil Society*. The Online Library of Liberty: T. Cadell, London. p. 205.

[5] Hayek cited. Petsoulas, Christian. Hayek's Liberalism and Its Origins: His Idea of Spontaneous Order and the Scottish Enlightenment. Routledge. 2001. p. 2

[6] Hayek, F.A. *The Fatal Conceit: The Errors of Socialism*. The University of Chicago Press. 1991. Page 6.

[7] Hayek cited. Boaz, David. The Libertarian Reader. The Free Press. 1997. p. 220

[8] Stossel, John (2011-02-10) Spontaneous Order, *Reason*

[9] Marshall, Gordon; et al. (1998) [1994]. *Oxford Dictionary of Sociology* (2 ed.). Oxford: Oxford University Press. pp. 19–20. ISBN 0-19-280081-7.

[10] Faith and Order: The Reconciliation of Law and Religion By Harold Joseph pg 388 Berman Wm. B. Eerdmans Publishing Religion and law ISBN 0-8028-4852-4 https://books.google.com/books?id=j1208xA7F_0C&lpg=PA388&ots=p0N6U4zWbf&pg=PA388

[11] The Constitution of Liberty; Law, Legislation and Liberty

[12] The Sensory Order

[13] http://fiesta.bren.ucsb.edu/~{}gsd/595e/docs/41.%20Polanyi_Republic_of_Science.pdf

[14] Persuasion, Power, and Polity

[15] http://www.amazon.com/Persuasion-Power-Polity-Democratic-Self-Organization/dp/1572732571/ref=sr_1_4?ie=UTF8&s=books&qid=1302773406&sr=1-4

[16] http://studiesinemergentorder.org/current-issue/sieo3-195/

[17] The Self-Organizing Economy

11.5 External links

- The Tradition of Spontaneous Order, by Norman Barry, Library of Economics and Liberty

Chapter 12

Metastability

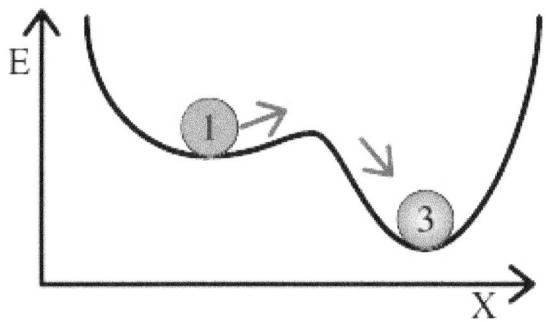

A metastable state of weaker bond (1), a transitional 'saddle' configuration (2) and a stable state of stronger bond (3).

Metastability denotes the phenomenon when a system spends an extended time in a configuration other than the system's state of least energy. During a metastable state of finite lifetime all state-describing parameters reach and hold stationary values. While in isolation:

- the state of least energy is the only one the system will inhabit for an indefinite length of time, until more external energy is added to the system (unique "absolutely stable" state);
- the system will spontaneously leave any other state (of higher energy) to eventually return (after a sequence of transitions) to the least energetic state.

A single particle analogy may be drawn with a ball resting in a hollow on a slope. With some perturbation the ball may start rolling again to lower levels. Isomerisation is another common example, where higher energy isomers are long lived as they are prevented from rearranging to the ground state by (possibly large) barriers in the potential energy.

The metastability concept originates in the physics of first-order phase transitions later to acquire new meanings in the study of aggregated subatomic particles (in atomic nuclei or in atoms) or in molecules, macromolecules or clusters of atoms and molecules. Later on it was borrowed for the study of decision-making and information transmitting systems.

Many complex natural and man-made systems can demonstrate metastability.

- Metastability is common in physics and chemistry - from an atom (many-body assembly) to statistical ensembles of molecules (viscous fluids, amorphous solids, liquid crystals etc.) at molecular levels or as a whole (see metastable phases of matter and grain piles below). The abundance of states is more prevalent as the systems grow larger and/or if the forces of their mutual interaction are spatially less uniform or more diverse.

- In dynamic systems (with feedback) like electronic circuits, signal trafficking, decisional systems and neuroscience - it is the time-invariance of the active or reactive patterns with respect to the external influences that defines stability and metastability (see brain metastability below). Here the equivalent of the thermal fluctuations is the "white noise" affecting the signal propagation and the decision-making.

12.1 Statistical physics and thermodynamics

Non-equilibrium thermodynamics is a branch of physics that studies the dynamics of statistical ensembles of molecules via unstable states. Being "stuck" in a thermodynamic trough without being at the lowest energy state is known as having **kinetic stability** or being *kinetically persistent*. The particular motion or kinetics of the atoms involved has resulted in getting stuck, despite there being preferable (lower-energy) alternatives.

12.1.1 States of matter

Metastable states of matter range from **melting solids** (or **freezing liquids**), **boiling liquids** (or **condensing gases**) and sublimating solids to supercooled liquids or superheated liquid-gas mixtures. Extremely pure, supercooled water stays liquid below 0 °C and remains so until applied vibrations or condensing seed doping initiates crystallization centers. This is a common situation for the droplets of atmospheric clouds.

12.1.2 Condensed matter and macromolecules

Metastable phases are common in condensed matter. For example, diamond is a metastable form of carbon at standard temperature and pressure. It can be converted to graphite (plus leftover kinetic energy), but only after overcoming an activation energy - an intervening hill. Martensite is a metastable phase used to control the hardness of most steel. The bonds between the building blocks of polymers such as DNA, RNA and proteins are also metastable. Metastable polymorphs of silica are commonly observed. In some cases, such as in the allotropes of solid boron, acquiring a sample of the stable phase is difficult.[1] Generally speaking, emulsions/colloidal systems and glasses are metastable.

Small avalanches demonstrate metastability at Mount Baker Ski Area.

Sandpiles are one system which can exhibit metastability if a steep slope or tunnel is present. Sand grains form a pile due to friction. It is possible for an entire large sand pile to reach a point where it is stable, but the addition of a single grain causes large parts of it to collapse.

The avalanche is a well-known problem with large piles of snow and ice crystals on steep slopes. In dry conditions, snow slopes act similarly to sandpiles. An entire mountainside of snow can suddenly slide due to the presence of a skier, or even a loud noise or vibration.

12.2 Quantum mechanics

Aggregated systems of subatomic particles described by quantum mechanics (quarks inside nucleons, nucleons inside atomic nuclei, electrons inside atoms, molecules or atomic clusters) are found to have many distinguishable states. Of these, one (or a small degenerate set) is indefinitely stable: the ground state or global minimum.

All other states besides the ground state (or those degenerate with it) have higher energies. Of all these other states, the **metastable** states are the ones having lifetimes lasting at least 10^2 to 10^3 times longer than the shortest lived states of the set.

A *metastable state* is then long-lived (locally stable with respect to configurations of 'neighbouring' energies) but not eternal (as the global minimum is). Being excited - of an energy above the ground state - it will eventually decay to a more stable state, releasing energy. Indeed, above absolute zero, all states of a system have a non-zero probability to decay; that is, to spontaneously fall into another state (usually lower in energy). One mechanism for this to happen is through tunnelling.

12.2.1 Nuclear physics

Some energetic states of an atomic nucleus (having distinct spatial mass, charge, spin, isospin distributions) are much longer-lived than others (nuclear isomers of the same isotope). E.g. Technetium-99m.[2]

12.2.2 Atomic and molecular physics

Some atomic energy levels are metastable. Rydberg atoms are an example of metastable excited atomic states. Transitions from metastable excited levels are typically those forbidden by electric dipole selection rules. This means that any transitions from this level are relatively unlikely to occur. In a sense, an electron that happens to find itself in a metastable configuration is trapped there. Of course, since transitions away from a metastable state are not impossible (merely unlikely), the electron will eventually be able to decay to a less energetic state by spontaneous emission.

This property of the electron is used in lasers. When light of suitable wavelength falls on atoms, their electrons jump to a higher energy state. When the incoming radiations are removed, the excited electron goes back to its original level, typically within a duration of around 10^{-8} seconds. However, when an electron goes to a metastable state, it remains there for a relatively longer duration of 10^{-3} seconds. This phenomenon leads to accumulation of electrons in the metastable state, since the rate of addition of electrons to the metastable state is higher than the rate of their de-excitation. This leads to the phenomenon called population inversion, which forms the basis of lasing action of lasers.

12.2.3 Chemistry

In chemical systems, a system of atoms or molecules involving a change in chemical bond can be in a metastable state, which lasts for a relatively long period of time. Molecular vibrations and thermal motion make chemical species at the energetic equivalent of the top of a round hill very short-lived. Metastable states that persist for many seconds (or years) are found in energetic *valleys* which are not the lowest possible valley (point 1 in illustration). A common type of metastability is isomerism.

The stability or metastability of a given chemical system depends its environment, particularly temperature and pressure. The difference between producing a stable vs. metastable entity can have important consequences. For instances, having the wrong crystal polymorph can result in failure of a drug while in storage between manufacture and administration.[3] The map of which state is the most stable as a function of pressure, temperature and/or composition is known as a phase diagram. In regions where a particular state is not the most stable, it may still be metastable.

Reaction intermediates are relatively short-lived, and are usually thermodynamically unstable rather than metastable. The IUPAC recommends referring to these as *transient* rather than metastable.[4]

Metastability is also used to refer to specific situations in mass spectrometry[5] and spectrochemistry.[6]

See also: Chemical stability and Chemical equilibrium § Metastable mixtures

12.2.4 Electron systems in biochemistry

The evolution of a many-body quantum system between its characteristic set of states may be influenced by the following external actions:

- The environment may act chaotically onto the system and add uncertainty to all state energies (while decreasing their lifetimes) — as in the spectral line broadening.

- Just as well, resonant exterior actions may nudge the system into a **lower** cohesive **energy state** while making it release an **intrinsic amount** or **quanta** of its energy — as in the stimulated emissions.

- Alternatively, external catalytic fields of forces may **briefly flatten** some of the **barriers** (ridges separating adjacent valleys) in the potential landscape of the system and help it **tunnel** through to lower energy states (see image above).

- Under the impact of thermal or directional external actions, some systems (see macromolecule complexes involving enzyme-cofactor association) may wander for extremely long periods of time among a certain sub-group of their states (all having distinct configurations but energy differences within the thermal fluctuation range). As such the enzymes will enter a biochemical reaction sequence with an **initial** configuration, perform through its many steps as catalysts while continuously contorting, and eventually leave that reaction sequence in the **same** configuration as they have entered it, ready to perform again.

12.3 Electronic circuits

Metastability in electronics is usually seen as a problem. A changing circuit is supposed to settle into one of a small number of desired states, but if the circuit is vulnerable to metastability, it can get stuck in an undesirable state.

12.4 Computational neuroscience

Metastability in the brain is a phenomenon studied in computational neuroscience to elucidate how the human brain recognizes patterns. Here, the term metastability is used rather loosely. There is no lower-energy state, but there are semi-transient signals in the brain that persist for a while and are different than the usual equilibrium state.

12.5 See also

- False vacuum
- Hysteresis

12.6 References

[1] van Setten, Uijttewaal, de Wijs and de Groot (2007). *JACS* **129**: 2458–2465. Missing or empty |title= (help)

[2] Hyperphysics

[3] Process Chemistry in the Pharmaceutical Industry. Kumar G. Gadamasetti, editor. 1999, pp. 375–378

12.6. REFERENCES

[4] IUPAC Gold Book

[5] IUPAC Gold Book - metastable ion inmass spectrometry

[6] IUPAC Gold Book - metastable state inspectrochemistry

Chapter 13

Butterfly effect

For other uses, see Butterfly effect (disambiguation).

The **butterfly effect** is the concept that small causes can

A plot of Lorenz's strange attractor for values ϱ=28, σ = 10, β = 8/3. The butterfly effect or sensitive dependence on initial conditions is the property of a dynamical system that, starting from any of various arbitrarily close alternative initial conditions on the attractor, the iterated points will become arbitrarily spread out from each other.

have large effects. Initially, it was used with weather prediction but later the term became a metaphor used in and out of science.[1]

In chaos theory, the **butterfly effect** is the sensitive dependence on initial conditions in which a small change in one state of a deterministic nonlinear system can result in large differences in a later state. The name, coined by Edward Lorenz for the effect which had been known long before, is derived from the metaphorical example of the details of a hurricane (exact time of formation, exact path taken) being influenced by minor perturbations such as the flapping of the wings of a distant butterfly several weeks earlier.

Lorenz discovered the effect when he observed that runs of his weather model with initial condition data that was rounded in a seemingly inconsequential manner would fail to reproduce the results of runs with the unrounded initial condition data. A very small change in initial conditions had created a significantly different outcome.

The idea, that small causes may have large effects in general and in weather specifically, was used from Henri Poincaré to Norbert Wiener. Edward Lorenz's work placed the concept of instability of the atmosphere onto a quantitative base and linked the concept of instability to the properties of large classes of dynamic systems which are undergoing nonlinear dynamics and deterministic chaos.[1]

The butterfly effect is exhibited by very simple systems. For example, the randomness of the outcomes of throwing dice depends on this characteristic to amplify small differences in initial conditions—the precise direction, thrust, and orientation of the throw—into significantly different dice paths and outcomes, which makes it virtually impossible to throw dice exactly the same way twice.

13.1 History

In The Vocation of Man (1800), Fichte says that "you could not remove a single grain of sand from its place without thereby ... changing something throughout all parts of the immeasurable whole".

Chaos theory and the sensitive dependence on initial conditions were described in the literature in a particular case of the three-body problem by Henri Poincaré in 1890.[2] He later proposed that such phenomena could be common, for example, in meteorology.[3]

In 1898,[2] Jacques Hadamard noted general divergence of trajectories in spaces of negative curvature. Pierre Duhem discussed the possible general significance of this in 1908.[2] The idea that one butterfly could eventually have a far-reaching ripple effect on subsequent historic events made its earliest known appearance in "A Sound of Thunder", a 1952 short story by Ray Bradbury about time travel.[4]

In 1961, Lorenz was running a numerical computer model to redo a weather prediction from the middle of the previous run as a shortcut. He entered the initial condition 0.506 from the printout instead of entering the full precision 0.506127 value. The result was a completely different weather scenario.[5]

Lorenz wrote:

In 1963 Lorenz published a theoretical study of this effect in a highly cited, seminal paper called *Deterministic Nonperiodic Flow*[7][8] (the calculations were performed on a Royal McBee LGP-30 computer).[9][10] Elsewhere he stated:

> One meteorologist remarked that if the theory were correct, one flap of a sea gull's wings would be enough to alter the course of the weather forever. The controversy has not yet been settled, but the most recent evidence seems to favor the sea gulls.[10]

Following suggestions from colleagues, in later speeches and papers Lorenz used the more poetic butterfly. According to Lorenz, when he failed to provide a title for a talk he was to present at the 139th meeting of the American Association for the Advancement of Science in 1972, Philip Merilees concocted *Does the flap of a butterfly's wings in Brazil set off a tornado in Texas?* as a title.[11] Although a butterfly flapping its wings has remained constant in the expression of this concept, the location of the butterfly, the consequences, and the location of the consequences have varied widely.[12]

The phrase refers to the idea that a butterfly's wings might create tiny changes in the atmosphere that may ultimately alter the path of a tornado or delay, accelerate or even prevent the occurrence of a tornado in another location. The butterfly does not power or directly create the tornado, but the term is intended to imply that the flap of the butterfly's wings can *cause* the tornado: in the sense that the flap of the wings is a part of the initial conditions; one set of conditions leads to a tornado while the other set of conditions doesn't. The flapping wing represents a small change in the initial condition of the system, which cascades to large-scale alterations of events (compare: domino effect). Had the butterfly not flapped its wings, the trajectory of the system might have been vastly different—but it's also equally possible that the set of conditions without the butterfly flapping its wings is the set that leads to a tornado.

The butterfly effect presents an obvious challenge to prediction, since initial conditions for a system such as the weather can never be known to complete accuracy. This problem motivated the development of ensemble forecasting, in which a number of forecasts are made from perturbed initial conditions.[13]

Some scientists have since argued that the weather system is not as sensitive to initial condition as previously believed.[14] David Orrell argues that the major contributor to weather forecast error is model error, with sensitivity to initial conditions playing a relatively small role.[15][16] Stephen Wolfram also notes that the Lorenz equations are highly simplified and do not contain terms that represent viscous effects; he believes that these terms would tend to damp out small perturbations.[17]

13.2 Illustration

13.3 Theory and mathematical definition

Recurrence, the approximate return of a system towards its initial conditions, together with sensitive dependence on initial conditions, are the two main ingredients for chaotic motion. They have the practical consequence of making complex systems, such as the weather, difficult to predict past a certain time range (approximately a week in the case of weather) since it is impossible to measure the starting atmospheric conditions completely accurately.

A dynamical system displays sensitive dependence on initial conditions if points arbitrarily close together separate over time at an exponential rate. The definition is not topological, but essentially metrical.

If M is the state space for the map f^t, then f^t displays sensitive dependence to initial conditions if for any x in M and any $\delta > 0$, there are y in M, with distance $d(.,.)$ such that $0 < d(x,y) < \delta$ and such that

$$d(f^\tau(x), f^\tau(y)) > e^{a\tau} d(x,y)$$

for some positive parameter a. The definition does not require that all points from a neighborhood separate from the base point x, but it requires one positive Lyapunov exponent.

The simplest mathematical framework exhibiting sensitive dependence on initial conditions is provided by a particular parametrization of the logistic map:

$$x_{n+1} = 4x_n(1 - x_n), \quad 0 \leq x_0 \leq 1,$$

which, unlike most chaotic maps, has a closed-form solution:

$$x_n = \sin^2(2^n \theta \pi)$$

where the initial condition parameter θ is given by $\theta = \frac{1}{\pi}\sin^{-1}(x_0^{1/2})$. For rational θ, after a finite number of iterations x_n maps into a periodic sequence. But almost all θ are irrational, and, for irrational θ, x_n never repeats itself – it is non-periodic. This solution equation clearly demonstrates the two key features of chaos – stretching and folding: the factor 2^n shows the exponential growth of stretching, which results in sensitive dependence on initial conditions (the butterfly effect), while the squared sine function keeps x_n folded within the range [0, 1].

13.4 In physical systems

13.4.1 In weather

The butterfly effect is most familiar in terms of weather; it can easily be demonstrated in standard weather prediction models, for example. The climate scientists James Annan and William Connolley explain that chaos is important in the development of weather prediction methods; models are sensitive to initial conditions. They add the caveat: "Of course the existence of an unknown butterfly flapping its wings has no direct bearing on weather forecasts, since it will take far too long for such a small perturbation to grow to a significant size, and we have many more immediate uncertainties to worry about. So the direct impact of this phenomenon on weather prediction is often somewhat overstated."[18]

13.4.2 In quantum mechanics

The potential for sensitive dependence on initial conditions (the butterfly effect) has been studied in a number of cases in semiclassical and quantum physics including atoms in strong fields and the anisotropic Kepler problem.[19][20] Some authors have argued that extreme (exponential) dependence on initial conditions is not expected in pure quantum treatments,[21][22] however, the sensitive dependence on initial conditions demonstrated in classical motion is included in the semiclassical treatments developed by Martin Gutzwiller[23] and Delos and co-workers.[24]

Other authors suggest that the butterfly effect can be observed in quantum systems. Karkuszewski et al. consider the time evolution of quantum systems which have slightly different Hamiltonians. They investigate the level of sensitivity of quantum systems to small changes in their given Hamiltonians.[25] Poulin et al. presented a quantum algorithm to measure fidelity decay, which "measures the rate at which identical initial states diverge when subjected to slightly different dynamics". They consider fidelity decay to be "the closest quantum analog to the (purely classical) butterfly effect".[26] Whereas the classical butterfly effect considers the effect of a small change in the position and/or velocity of an object in a given Hamiltonian system, the quantum butterfly effect considers the effect of a small change in the Hamiltonian system with a given initial position and velocity.[27][28] This quantum butterfly effect has been demonstrated experimentally.[29] Quantum and semiclassical treatments of system sensitivity to initial conditions are known as quantum chaos.[21][27]

13.5 In popular culture

Main article: Butterfly effect in popular culture

The journalist Peter Dizikes, writing in *The Boston Globe* in 2008, notes that popular culture likes the idea of the butterfly effect, but gets it wrong. Whereas Lorenz suggested correctly with his butterfly metaphor that predictability "is inherently limited", popular culture supposes that each event can be explained by finding the small reasons that caused it. Dizikes explains: "It speaks to our larger expectation that the world should be comprehensible - that everything happens for a reason, and that we can pinpoint all those reasons, however small they may be. But nature itself defies this expectation."[30]

13.6 See also

- Actuality and potentiality
- Avalanche effect
- Behavioral cusp
- Butterfly effect in popular culture
- Cascading failure
- Causality
- Chain reaction
- Clapotis

- Determinism
- Domino effect
- Dynamical systems
- Fractal
- Great Stirrup Controversy
- Innovation butterfly
- Kessler syndrome
- Law of unintended consequences
- Point of divergence
- Positive feedback
- Representativeness heuristic
- Ripple effect
- Snowball effect
- Traffic congestion
- Tropical cyclogenesis

13.7 References

[1] "Butterfly effect - Scholarpedia". *www.scholarpedia.org*. Retrieved 2016-01-02.

[2] Some Historical Notes: History of Chaos Theory

[3] Steves, Bonnie; Maciejewski, AJ (September 2001). *The Restless Universe Applications of Gravitational N-Body Dynamics to Planetary Stellar and Galactic Systems*. USA: CRC Press. ISBN 0750308222. Retrieved January 6, 2014.

[4] Flam, Faye (2012-06-15). "The Physics of Ray Bradbury's "A Sound of Thunder"". *Philadelphia Inquirer*. Retrieved 2015-09-02.

[5] Gleick, James (1987). *Chaos: Making a New Science*. Viking. p. 16. ISBN 0-8133-4085-3.

[6] "Chaos at 50".

[7] Lorenz, Edward N. (March 1963). "Deterministic Nonperiodic Flow". *Journal of the Atmospheric Sciences* **20** (2): 130–141. Bibcode:1963JAtS...20..130L. doi:10.1175/1520-0469(1963)020<0130:DNF>2.0.CO;2. ISSN 1520-0469. Retrieved 3 June 2010.

[8] Google Scholar citation record

[9] "Part19". Cs.ualberta.ca. 1960-11-22. Retrieved 2014-06-08.

[10] Lorenz, Edward N. (1963). "The Predictability of Hydrodynamic Flow" (PDF). *Transactions of the New York Academy of Sciences* **25** (4): 409–432. Retrieved 1 September 2014.

[11] Lorenz: "Predictability", AAAS 139th meeting, 1972 Retrieved May 22, 2015

[12] "The Butterfly Effects: Variations on a Meme". *AP42 ...and everything*. Retrieved 3 August 2011.

[13] Woods, Austin (2005). *Medium-range weather prediction: The European approach; The story of the European Centre for Medium-Range Weather Forecasts*. New York: Springer. p. 118. ISBN 978-0387269283.

[14] Orrell, David; Smith, Leonard; Barkmeijer, Jan; Palmer, Tim (2001). "Model error in weather forecasting". *Nonlinear Processes in Geophysics* **9**: 357–371.

[15] Orrell, David (2002). "Role of the metric in forecast error growth: How chaotic is the weather?". *Tellus* **54A**: 350–362. doi:10.3402/tellusa.v54i4.12159.

[16] Orrell, David (2012). *Truth or Beauty: Science and the Quest for Order*. New Haven: Yale University Press. p. 208. ISBN 978-0300186611.

[17] Wolfram, Stephen (2002). *A New Kind of Science*. Wolfram Media. p. 998. ISBN 978-1579550080.

[18] "Chaos and Climate". RealClimate. Retrieved 2014-06-08.

[19] Heller, E. J.; Tomsovic, S. (July 1993). "Postmodern Quantum Mechanics". *Physics Today* **46**: 38. Bibcode:1993PhT....46g..38H. doi:10.1063/1.881358.

[20] Gutzwiller, Martin C. (1990). *Chaos in Classical and Quantum Mechanics*. New York: Springer-Verlag. ISBN 0-387-97173-4.

[21] Rudnick, Ze'ev (January 2008). "What is...Quantum Chaos" (PDF). *Notices of the American Mathematical Society*.

[22] Berry, Michael (1989). "Quantum chaology, not quantum chaos". *Physica Scripta* **40** (3): 335–336. Bibcode:1989PhyS...40..335B. doi:10.1088/0031-8949/40/3/013.

[23] Gutzwiller, Martin C. (1971). "Periodic Orbits and Classical Quantization Conditions". *Journal of Mathematical Physics* **12** (3): 343. Bibcode:1971JMP....12..343G. doi:10.1063/1.1665596.

[24] Gao, J. & Delos, J. B. (1992). "Closed-orbit theory of oscillations in atomic photoabsorption cross sections in a strong electric field. II. Derivation of formulas". *Physical Review A* **46** (3): 1455–1467. Bibcode:1992PhRvA..46.1455G. doi:10.1103/PhysRevA.46.1455.

[25] Karkuszewski, Zbyszek P.; Jarzynski, Christopher; Zurek, Wojciech H. (2002). "Quantum Chaotic Environments, the Butterfly Effect, and Decoherence". *Physical Review Letters* **89** (17): 170405. arXiv:quant-ph/0111002. Bibcode:2002PhRvL..89q0405K. doi:10.1103/PhysRevLett.89.170405.

[26] Poulin, David; Blume-Kohout, Robin; Laflamme, Raymond & Ollivier, Harold (2004). "Exponential Speedup with a Single Bit of Quantum Information: Measuring the Average Fidelity Decay". *Physical Review Letters* **92** (17): 177906. arXiv:quant-ph/0310038. Bibcode:2004PhRvL..92q7906P. doi:10.1103/PhysRevLett.92.177906. PMID 15169196.

[27] Poulin, David. "A Rough Guide to Quantum Chaos" (PDF).

[28] Peres, A. (1995). *Quantum Theory: Concepts and Methods*. Dordrecht: Kluwer Academic.

[29] Lee, Jae-Seung & Khitrin, A. K. (2004). "Quantum amplifier: Measurement with entangled spins". *Journal of Chemical Physics* **121** (9): 3949. Bibcode:2004JChPh.121.3949L. doi:10.1063/1.1788661.

[30] Dizikes, Petyer (8 June 2008). "The meaning of the butterfly". The Boston Globe. Retrieved 8 June 2016.

13.8 Further reading

- James Gleick, *Chaos: Making a New Science*, New York: Viking, 1987. 368 pp.

- Devaney, Robert L. (2003). *Introduction to Chaotic Dynamical Systems*. Westview Press. ISBN 0670811785.

- Hilborn, Robert C. (2004). "Sea gulls, butterflies, and grasshoppers: A brief history of the butterfly effect in nonlinear dynamics". *American Journal of Physics* **72** (4): 425–427. Bibcode:2004AmJPh..72..425H. doi:10.1119/1.1636492.

13.9 External links

- The meaning of the butterfly: Why pop culture loves the 'butterfly effect,' and gets it totally wrong, Peter Dizikes, *Boston Globe*, June 8, 2008

- New England Complex Systems Institute - Concepts: Butterfly Effect

- The Chaos Hypertextbook. An introductory primer on chaos and fractals

- ChaosBook.org. Advanced graduate textbook on chaos (no fractals)

- Weisstein, Eric W., "Butterfly Effect", *MathWorld*.

Chapter 14

Self-organized criticality control

In applied physics, the concept of **controlling self-organized criticality** refers to the control of processes by which a self-organized system dissipates energy. The objective of the control is to reduce the probability of occurrence of and size of energy dissipation bursts, often called *avalanches*, of self-organized systems. Dissipation of energy in a self-organized critical system into a lower energy state can be costly for society, since it depends on avalanches of all sizes usually following a kind of power law distribution and large avalanches can be damaging and disruptive.[1][2][3]

14.1 Self-organized criticality control schemes

Several strategies have been proposed to deal with the issue of controlling self-organized criticality:

1. *The design of controlled avalanches.* Daniel O. Cajueiro and Roberto F. S. Andrade show that if well-formulated small and medium avalanches are exogenously triggered in the system, the energy of the system is released in a way that large avalanches are rarer.[1][2][3]

2. *The modification of the degree of interdependence of the network where the avalanche spreads.* Charles D. Brummitt, Raissa M. D'Souza and E. A. Leicht show that the dynamics of self-organized critical systems on complex networks depend on connectivity of the complex network. They find that while some connectivity is beneficial (since it suppresses the largest cascades in the system), too much connectivity gives space for the development of very large cascades and increases the size of capacity of the system.[4]

3. *The modification of the deposition process of the self-organized system.* Pierre-Andre Noel, Charles D. Brummitt and Raissa M. D'Souza show that it is possible to control the self-organized system by modifying the natural deposition process of the self-organized system adjusting the place where the avalanche starts.[5]

4. *Dynamically modifying the local thresholds of cascading failures.* In a model of an electric transmission network, Heiko Hoffmann and David W. Payton demonstrated that either randomly upgrading lines (sort of like preventive maintenance) or upgrading broken lines to a random breakage threshold suppresses self-organized criticality.[6] Apparently, these strategies undermine the self-organization of large critical clusters. Here, a critical cluster is a collection of transmission lines that are near the failure threshold and that collapse entirely if triggered.

14.2 Applications

There are at least four different types of events that may arise in nature or society, where these ideas of control may help us to avoid them:[1][2][3][4][5]

1. Flood caused by systems of dams and reservoirs or interconnected valleys.

2. Snow avalanches that take place in snow hills.

3. Forest fires in areas susceptible to a lightning bolt or a match lighting.

4. Cascades of load shedding that take place in power grids.

14.3 See also

- Self-organized criticality
- Complex Networks
- Abelian sandpile model

14.4 References

[1] D. O. Cajueiro and R. F. S. Andrade (2010). "Controlling self-organized criticality in sandpile models". *Physical Review E* **81**: 015102#R. arXiv:1305.6648. Bibcode:2010PhRvE..81a5102C. doi:10.1103/physreve.81.015102.

[2] D. O. Cajueiro and R. F. S. Andrade (2010). "Controlling self-organized criticality in complex networks". *European Physical Journal B* **77**: 291–296. arXiv:1305.6656. Bibcode:2010EPJB...77..291C. doi:10.1140/epjb/e2010-00229-8.

[3] D. O. Cajueiro and R. F. S. Andrade (2010). "Dynamical programming approach for controlling the directed Abelian Dhar-Ramaswamy model". *Physical Review E* **82**: 031108. arXiv:1305.6668. Bibcode:2010PhRvE..82c1108C. doi:10.1103/physreve.82.031108.

[4] C. D. Brummitt, R. M. D'Souza and E. A. Leicht (2012). "Suppressing cascades of load in interdependent networks". *PNAS* **109**: E680–E689. arXiv:1106.4499. Bibcode:2012PNAS..109E.680B. doi:10.1073/pnas.1110586109.

[5] P. A. Noel, C. D. Brummitt and R. M. D'Souza (2013). "Controlling self-organized criticality on networks using models that self-organize". *Physical Review Letters* **111**: 078701. arXiv:1305.1877. Bibcode:2013PhRvL.111g8701N. doi:10.1103/physrevlett.111.078701.

[6] H. Hoffmann and D. W. Payton (2014). "Suppressing cascades in a self-organized-critical model with non-contiguous spread of failures". *Chaos, Solitons and Fractals* **67**: 87–93. doi:10.1016/j.chaos.2014.06.011.

Chapter 15

Spontaneous magnetization

Spontaneous magnetization is the appearance of an ordered spin state (magnetization) at zero applied magnetic field in a ferromagnetic or ferrimagnetic material below a critical point called the Curie temperature or TC.

15.1 Overview

Heated to temperatures above TC, ferromagnetic materials become paramagnetic and their magnetic behavior is dominated by spin waves or magnons, which are boson collective excitations with energies in the meV range. The magnetization that occurs below TC is a famous example of the "spontaneous" breaking of a global symmetry, a phenomenon that is described by Goldstone's theorem. The term "symmetry breaking" refers to the choice of a magnetization direction by the spins, which have spherical symmetry above TC, but a preferred axis (the magnetization direction) below TC.

15.2 Temperature dependence

To first order, the temperature dependence of spontaneous magnetization at low temperatures is given by Bloch's Law:[1]

$$M(T) = M(0)\left(1 - (T/T_c)^{3/2}\right),$$

where M(0) is the spontaneous magnetization at absolute zero. The decrease in spontaneous magnetization at higher temperatures is caused by the increasing excitation of spin waves. In a particle description, the spin waves correspond to magnons, which are the massless Goldstone bosons corresponding to the broken symmetry. This is exactly true for an isotropic magnet.

Magnetic anisotropy, that is the existence of an easy direction along which the moments align spontaneously in the crystal, corresponds however to "massive" magnons. This is a way of saying that they cost a minimum amount of energy to excite, hence they are very unlikely to be excited as $T \to 0$. Hence the magnetization of an anisotropic magnet is harder to destroy at low temperature and the temperature dependence of the magnetization deviates accordingly from the Bloch's law. All real magnets are anisotropic to some extent.

Near the Curie temperature,

$$M(T) \propto (T - T_c)^\beta,$$

where β is a critical exponent that depends on composition. The exponent is 0.34 for Fe and 0.51 for Ni.[2]

An empirical interpolation of the two regimes is given by

$$\frac{M(T)}{M(0)} = (1 - (T/T_c)^\alpha)^\beta,$$

it is easy to check two limits of this interpolation that follow laws similar to the Bloch law, for $T \to 0$, and the critical behavior, for $T \to T_C$, respectively.

15.3 Notes and references

[1] Ashcroft & Mermin 1976, p. 708

[2] Chikazumi 1997, pp. 128–129

15.4 Further reading

- Ashcroft, Neil W.; Mermin, N. David (1976). *Solid State Physics*. Holt, Rinehart and Winston. ISBN 0-03-083993-9.

- Chikazumi, Sōshin (1997). *Physics of Ferromagnetism*. Clarendon Press. ISBN 0-19-851776-9.

Chapter 16

Crystallization

"Crystallizing" and "Crystallized" redirect here. For the song, see Crystalised. For other uses, see Crystallization (disambiguation).

Crystallization is the (natural or artificial) process where a solid forms where the atoms or molecules are highly organized in a structure known as a crystal. Some of the ways which crystals form are through precipitating from a solution, melt or more rarely deposited directly from a gas. Crystallization is also a chemical solid–liquid separation technique, in which mass transfer of a solute from the liquid solution to a pure solid crystalline phase occurs. In chemical engineering crystallization occurs in a crystallizer. Crystallization is therefore related to precipitation, although the result is not amorphous or disordered, but a crystal.

16.1 Process

See also: Crystallization § Crystallization dynamics
The crystallization process consists of two major events,

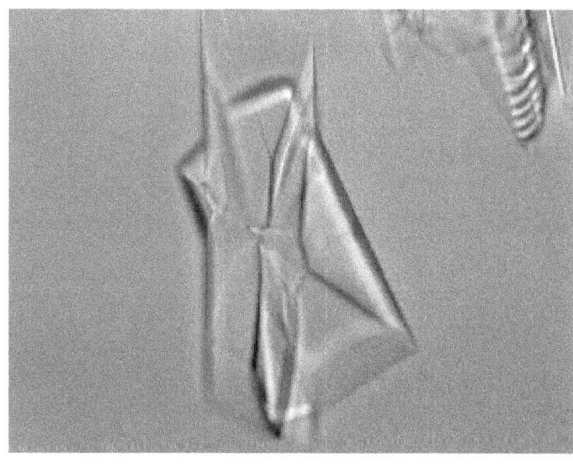

Time-lapse of growth of a citric acid crystal. The video covers an area of 2.0 by 1.5 mm and was captured over 7.2 min.

nucleation and *crystal growth* which are driven by thermodynamic properties as well as chemical properties. In crystal growth *Nucleation* is the step where the solute molecules or atoms dispersed in the solvent start to gather into clusters, on the microscopic scale (elevating solute concentration in a small region), that become stable under the current operating conditions. These stable clusters constitute the nuclei. Therefore, the clusters need to reach a critical size in order to become stable nuclei. Such critical size is dictated by many different factors (temperature, supersaturation, etc.). It is at the stage of nucleation that the atoms or molecules arrange in a defined and periodic manner that defines the crystal structure — note that "crystal structure" is a special term that refers to the relative arrangement of the atoms or molecules, not the macroscopic properties of the crystal (size and shape), although those are a result of the internal crystal structure.

The *crystal growth* is the subsequent size increase of the nuclei that succeed in achieving the critical cluster size. Crystal growth is a dynamic process occurring in equilibrium where solute molecules or atoms precipitate out of solution, and dissolve back into solution. Supersaturation is one of the driving forces of crystallization, as the solubility of a species is an equilibrium process quantified by Ksp. Depending upon the conditions, either nucleation or growth may be predominant over the other, dictating crystal size.

Many compounds have the ability to crystallize with some having different crystal structures, a phenomenon called polymorphism. Each polymorph is in fact a different thermodynamic solid state and crystal polymorphs of the same compound exhibit different physical properties, such as dissolution rate, shape (angles between facets and facet growth rates), melting point, etc. For this reason, polymorphism is of major importance in industrial manufacture of crystalline products. Additionally, crystal phases can sometimes be interconverted by varying factors such as temperature.

16.2 Crystallization in nature

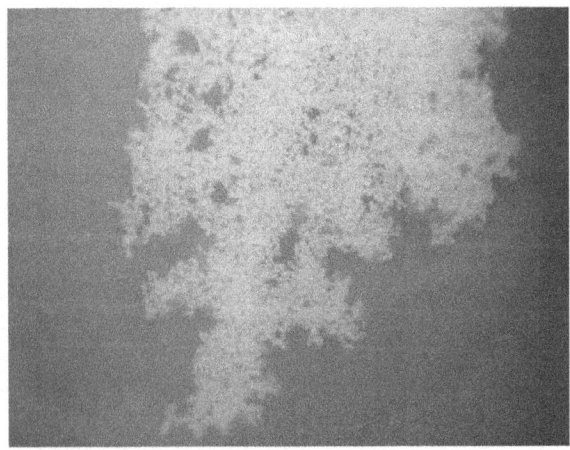

Snowflakes are a very well known example, where subtle differences in crystal growth conditions result in different geometries.

Crystallized honey

There are many examples of natural process that involve crystallization.

Geological time scale process examples include:

- Natural (mineral) crystal formation (see also gemstone);
- Stalactite/stalagmite, rings formation.

Usual time scale process examples include:

- Snow flakes formation;
- Honey crystallization (nearly all types of honey crystallize).

16.3 Methods

Crystal formation can be divided into two types, where the first type of crystals are composed of a cation and anion, also known as a salt, such as sodium acetate. The second type of crystals are composed of uncharged species, for example menthol.

Crystal formation can be achieved by various methods, such as: cooling, evaporation, addition of a second solvent to reduce the solubility of the solute (technique known as antisolvent or drown-out), solvent layering, sublimation, changing the cation or anion, as well as other methods.

The formation of a supersaturated solution does not guarantee crystal formation, and often a seed crystal or scratching the glass is required to form nucleation sites.

A typical laboratory technique for crystal formation is to dissolve the solid in a solution in which it is partially soluble, usually at high temperatures to obtain supersaturation. The hot mixture is then filtered to remove any insoluble impurities. The filtrate is allowed to slowly cool. Crystals that form are then filtered and washed with a solvent in which they are not soluble, but is miscible with the mother liquor. The process is then repeated to increase the purity in a technique known as recrystallization.

16.3.1 Typical equipment

Equipment for the main industrial processes for crystallization.

1. *Tank crystallizers.* Tank crystallization is an old method still used in some specialized cases. Saturated solutions, in tank crystallization, are allowed to cool in open tanks. After a period of time the mother liquor is drained and the crystals removed. Nucleation and size of crystals are difficult to control. Typically, labor costs are very high.

16.4 Thermodynamic view

The nature of a crystallization process is governed by both thermodynamic and kinetic factors, which can make it highly variable and difficult to control. Factors such as impurity level, mixing regime, vessel design, and cooling profile can have a major impact on the size, number, and shape of crystals produced.

Now put yourself in the place of a molecule within a pure and *perfect crystal*, being heated by an external source. At some sharply defined temperature, the complicated architecture of the crystal collapses to that of a liquid. Textbook thermodynamics says that melting occurs because the entropy, S, gain in your system by spatial randomization of the molecules has overcome the enthalpy, H, loss due to breaking the crystal packing forces:

$$T(S_{liquid} - S_{solid}) > H_{liquid} - H_{solid}$$
$$G_{liquid} < G_{solid}$$

This rule suffers no exceptions when the temperature is rising. By the same token, on cooling the melt, at the very same temperature the bell should ring again, and molecules should click back into the very same crystalline form. The entropy decrease due to the ordering of molecules within the system is overcompensated by the thermal randomization of the surroundings, due to the release of the heat of fusion; the entropy of the universe increases.

But liquids that behave in this way on cooling are the exception rather than the rule; in spite of the second principle of thermodynamics, crystallization usually occurs at lower temperatures (supercooling). This can only mean that a crystal is more easily destroyed than it is formed. Similarly, it is usually much easier to dissolve a perfect crystal in a solvent than to grow again a good crystal from the resulting solution. The nucleation and growth of a crystal are under kinetic, rather than thermodynamic, control.

16.5 Crystallization dynamics

As mentioned above, a crystal is formed following a well-defined pattern, or structure, dictated by forces acting at the molecular level. As a consequence, during its formation process the crystal is in an environment where the solute concentration reaches a certain critical value, before changing status. Solid formation, impossible below the solubility threshold at the given temperature and pressure conditions, may then take place at a concentration higher than the theoretical solubility level. The difference between the actual value of the solute concentration at the crystallization limit and the theoretical (static) solubility threshold is called supersaturation and is a fundamental factor in crystallization.

16.5.1 Nucleation

Main article: Nucleation

Nucleation is the initiation of a phase change in a small region, such as the formation of a solid crystal from a liquid solution. It is a consequence of rapid local fluctuations on a molecular scale in a homogeneous phase that is in a state of metastable equilibrium. Total nucleation is the sum effect of two categories of nucleation – primary and secondary.

Primary nucleation

Primary nucleation is the initial formation of a crystal where there are no other crystals present or where, if there are crystals present in the system, they do not have any influence on the process. This can occur in two conditions. The first is homogeneous nucleation, which is nucleation that is not influenced in any way by solids. These solids include the walls of the crystallizer vessel and particles of any foreign substance. The second category, then, is heterogeneous nucleation. This occurs when solid particles of foreign substances cause an increase in the rate of nucleation that would otherwise not be seen without the existence of these foreign particles. Homogeneous nucleation rarely occurs in practice due to the high energy necessary to begin nucleation without a solid surface to catalyse the nucleation.

Primary nucleation (both homogeneous and heterogeneous) has been modelled with the following:[1]

$$B = \frac{dN}{dt} = k_n(c - c^*)^n$$

- B is the number of nuclei formed per unit volume per unit time.
- N is the number of nuclei per unit volume.
- k_n is a rate constant.
- c is the instantaneous solute concentration.
- c^* is the solute concentration at saturation.
- $(c-c^*)$ is also known as supersaturation.
- n is an empirical exponent that can be as large as 10, but generally ranges between 3 and 4.

Secondary nucleation

Secondary nucleation is the formation of nuclei attributable to the influence of the existing microscopic crystals in the magma.[2] The first type of known secondary crystallization is attributable to fluid shear, the other due to collisions between already existing crystals with either a solid surface of the crystallizer or with other crystals themselves. Fluid shear nucleation occurs when liquid travels across a Crystal at a high speed, sweeping away nuclei that would otherwise be incorporated into a Crystal, causing the swept-away nuclei to become new crystals. Contact nucleation has been found to be the most effective and common method for nucleation. The benefits include the following [2]

- Low kinetic order and rate-proportional to supersaturation, allowing easy control without unstable operation.
- Occurs at low supersaturation, where growth rate is optimum for good quality.
- Low necessary energy at which crystals strike avoids the breaking of existing crystals into new crystals.
- The quantitative fundamentals have already been isolated and are being incorporated into practice.

The following model, although somewhat simplified, is often used to model secondary nucleation:[1]

$$B = \frac{dN}{dt} = k_1 M_T^j (c - c^*)^b$$

- k_1 is a rate constant.
- M_T is the suspension density.
- j is an empirical exponent that can range up to 1.5, but is generally 1.
- b is an empirical exponent that can range up to 5, but is generally 2.

16.5.2 Crystal growth

Once the first small crystal, the nucleus, forms it acts as a convergence point (if unstable due to supersaturation) for molecules of solute touching – or adjacent to – the crystal so that it increases its own dimension in successive layers. The pattern of growth resembles the rings of an onion, as shown in the picture, where each colour indicates the same mass of solute; this mass creates increasingly thin layers due to the increasing surface area of the growing crystal. The supersaturated solute mass the original nucleus may *capture* in a time unit is called the *growth rate* expressed in kg/(m^2*h), and is a constant specific to the process. Growth rate is influenced by several physical factors, such as surface tension of solution, pressure, temperature, relative crystal velocity in the solution, Reynolds number, and so forth.

The main values to control are therefore:

- Supersaturation value, as an index of the quantity of solute available for the growth of the crystal;
- Total crystal surface in unit fluid mass, as an index of the capability of the solute to fix onto the crystal;
- Retention time, as an index of the probability of a molecule of solute to come into contact with an existing crystal;
- Flow pattern, again as an index of the probability of a molecule of solute to come into contact with an existing crystal (higher in laminar flow, lower in turbulent flow, but the reverse applies to the probability of contact).

The first value is a consequence of the physical characteristics of the solution, while the others define a difference between a well- and poorly designed crystallizer.

16.5.3 Crystal size distribution

The appearance and size range of a crystalline product is extremely important in crystallization. If further processing of the crystals is desired, large crystals with uniform size are important for washing, filtering, transportation, and storage. The importance lies in the fact that large crystals are easier to filter out of a solution than small crystals. Also, larger crystals have a smaller surface area to volume ratio, leading to a higher purity. This higher purity is due to less retention of mother liquor which contains impurities, and a smaller loss of yield when the crystals are washed to remove the mother liquor. The theoretical crystal size distribution can be estimated as a function of operating conditions with a fairly complicated mathematical process called population balance theory (using population balance equations).

16.6 Main crystallization processes

Some of the important factors influencing solubility are:

- Concentration
- Temperature
- Polarity

- Ionic Strength

So we may identify two main families of crystallization processes:

- Cooling crystallization
- Evaporative crystallization

This division is not really clear-cut, since hybrid systems exist, where cooling is performed through evaporation, thus obtaining at the same time a concentration of the solution.

A crystallization process often referred to in chemical engineering is the fractional crystallization. This is not a different process, rather a special application of one (or both) of the above.

16.6.1 Cooling crystallization

Application

Most chemical compounds, dissolved in most solvents, show the so-called *direct* solubility that is, the solubility threshold increases with temperature.

So, whenever the conditions are favourable, crystal formation results from simply cooling the solution. Here *cooling* is a relative term: austenite crystals in a steel form well above 1000 °C. An example of this crystallization process is the production of Glauber's salt, a crystalline form of sodium sulfate. In the diagram, where equilibrium temperature is on the x-axis and equilibrium concentration (as mass percent of solute in saturated solution) in y-axis, it is clear that sulfate solubility quickly decreases below 32.5 °C. Assuming a saturated solution at 30 °C, by cooling it to 0 °C (note that this is possible thanks to the freezing-point depression), the precipitation of a mass of sulfate occurs corresponding to the change in solubility from 29% (equilibrium value at 30 °C) to approximately 4.5% (at 0 °C) – actually a larger crystal mass is precipitated, since sulfate entrains hydration water, and this has the side effect of increasing the final concentration.

There are of course limitation in the use of cooling crystallization:

- Many solutes precipitate in hydrate form at low temperatures: in the previous example this is acceptable, and even useful, but it may be detrimental when, for example, the mass of water of hydration to reach a stable hydrate crystallization form is more than the available water: a single block of hydrate solute will be formed – this occurs in the case of calcium chloride);

- Maximum supersaturation will take place in the coldest points. These may be the heat exchanger tubes which are sensitive to scaling, and heat exchange may be greatly reduced or discontinued;

- A decrease in temperature usually implies an increase of the viscosity of a solution. Too high a viscosity may give hydraulic problems, and the laminar flow thus created may affect the crystallization dynamics.

- It is of course not applicable to compounds having *reverse* solubility, a term to indicate that solubility increases with temperature decrease (an example occurs with sodium sulfate where solubility is reversed above 32.5 °C).

Cooling crystallizers

The simplest cooling crystallizers are tanks provided with a mixer for internal circulation, where temperature decrease is obtained by heat exchange with an intermediate fluid circulating in a jacket. These simple machines are used in batch processes, as in processing of pharmaceuticals and are prone to scaling. Batch processes normally provide a relatively variable quality of product along the batch.

The *Swenson-Walker* crystallizer is a model, specifically conceived by Swenson Co. around 1920, having a semi-cylindric horizontal hollow trough in which a hollow screw conveyor or some hollow discs, in which a refrigerating fluid is circulated, plunge during rotation on a longitudinal axis. The refrigerating fluid is sometimes also circulated in a jacket around the trough. Crystals precipitate on the cold surfaces of the screw/discs, from which they are removed by scrapers and settle on the bottom of the trough. The screw, if provided, pushes the slurry towards a discharge port.

A common practice is to cool the solutions by flash evaporation: when a liquid at a given T_0 temperature is transferred in a chamber at a pressure P_1 such that the liquid saturation temperature T_1 at P_1 is lower than T_0, the liquid will release heat according to the temperature difference and a quantity of solvent, whose total latent heat of vaporization equals the difference in enthalpy. In simple words, the liquid is cooled by evaporating a part of it.

In the sugar industry vertical cooling crystallizers are used to exhaust the molasses in the last crystallization stage downstream of vacuum pans, prior to centrifugation. The massecuite enters the crystallizers at the top, and cooling water is pumped through pipes in counterflow.

16.6.2 Evaporative crystallization

Another option is to obtain, at an approximately constant temperature, the precipitation of the crystals by increasing the solute concentration above the solubility threshold. To obtain this, the solute/solvent mass ratio is increased using the technique of evaporation. This process is of course insensitive to change in temperature (as long as hydration state remains unchanged).

All considerations on control of crystallization parameters are the same as for the cooling models.

Evaporative crystallizers

Most industrial crystallizers are of the evaporative type, such as the very large sodium chloride and sucrose units, whose production accounts for more than 50% of the total world production of crystals. The most common type is the *forced circulation* (FC) model (see evaporator). A pumping device (a pump or an axial flow mixer) keeps the crystal slurry in homogeneous suspension throughout the tank, including the exchange surfaces; by controlling pump flow, control of the contact time of the crystal mass with the supersaturated solution is achieved, together with reasonable velocities at the exchange surfaces. The Oslo, mentioned above, is a refining of the evaporative forced circulation crystallizer, now equipped with a large crystals settling zone to increase the retention time (usually low in the FC) and to roughly separate heavy slurry zones from clear liquid.

16.6.3 DTB crystallizer

Whichever the form of the crystallizer, to achieve an effective process control it is important to control the retention time and the crystal mass, to obtain the optimum conditions in terms of crystal specific surface and the fastest possible growth. This is achieved by a separation – to put it simply – of the crystals from the liquid mass, in order to manage the two flows in a different way. The practical way is to perform a gravity settling to be able to extract (and possibly recycle separately) the (almost) clear liquid, while managing the mass flow around the crystallizer to obtain a precise slurry density elsewhere. A typical example is the DTB (*Draft Tube and Baffle*) crystallizer, an idea of Richard Chisum Bennett (a Swenson engineer and later President of Swenson) at the end of the 1950s. The DTB crystallizer (see images) has an internal circulator, typically an axial flow mixer – yellow – pushing upwards in a draft tube while outside the crystallizer there is a settling area in an annulus; in it the exhaust solution moves upwards at a very low velocity, so that large crystals settle – and return to the main circulation – while only the fines, below a given grain size are extracted and eventually destroyed by increasing or decreasing temperature, thus creating additional supersaturation. A quasi-perfect control of all parameters is achieved. This crystallizer, and the derivative models (Krystal, CSC, etc.) could be the ultimate solution if not for a major limitation in the evaporative capacity, due to the limited diameter of the vapour head and the relatively low external circulation not allowing large amounts of energy to be supplied to the system.

16.7 See also

- Abnormal grain growth
- Chiral resolution by crystallization
- Crystal
- Crystal habit
- Crystal structure
- Crystallite
- Crystal growth
- Fractional crystallization (chemistry)
- Igneous differentiation
- Laser heated pedestal growth
- Micro-pulling-down
- Pumpable ice technology
- Quasicrystal
- Recrystallization (chemistry)
- Recrystallization (metallurgy)
- Seed crystal
- Single crystal
- Symplectite
- Vitrification
- X-Ray Crystallography

16.8 References

[1] Tavare, N.S. (1995). *Industrial Crystallization* Plenum Press, New York

[2] McCabe & Smith (2000). *Unit Operations of Chemical Engineering* McGraw-Hill, New York

16.9 Further reading

- A. Mersmann, *Crystallization Technology Handbook* (2001) CRC; 2nd ed. ISBN 0-8247-0528-9

- Tine Arkenbout-de Vroome, *Melt Crystallization Technology* (1995) CRC ISBN 1-56676-181-6

- "Small Molecule Crystallization" (PDF) at Illinois Institute of Technology website

- Glynn P.D. and Reardon E.J. (1990) "Solid-solution aqueous-solution equilibria: thermodynamic theory and representation". Amer. J. Sci. 290, 164–201.

- Geankoplis, C.J. (2003) "Transport Processes and Separation Process Principles". 4th Ed. Prentice-Hall Inc.

16.10 External links

- Batch Crystallization
- Industrial Crystallization
- Time-lapse video with salt crystallization
- Evaporator Crystallizer Treatment of Landfill Leachate

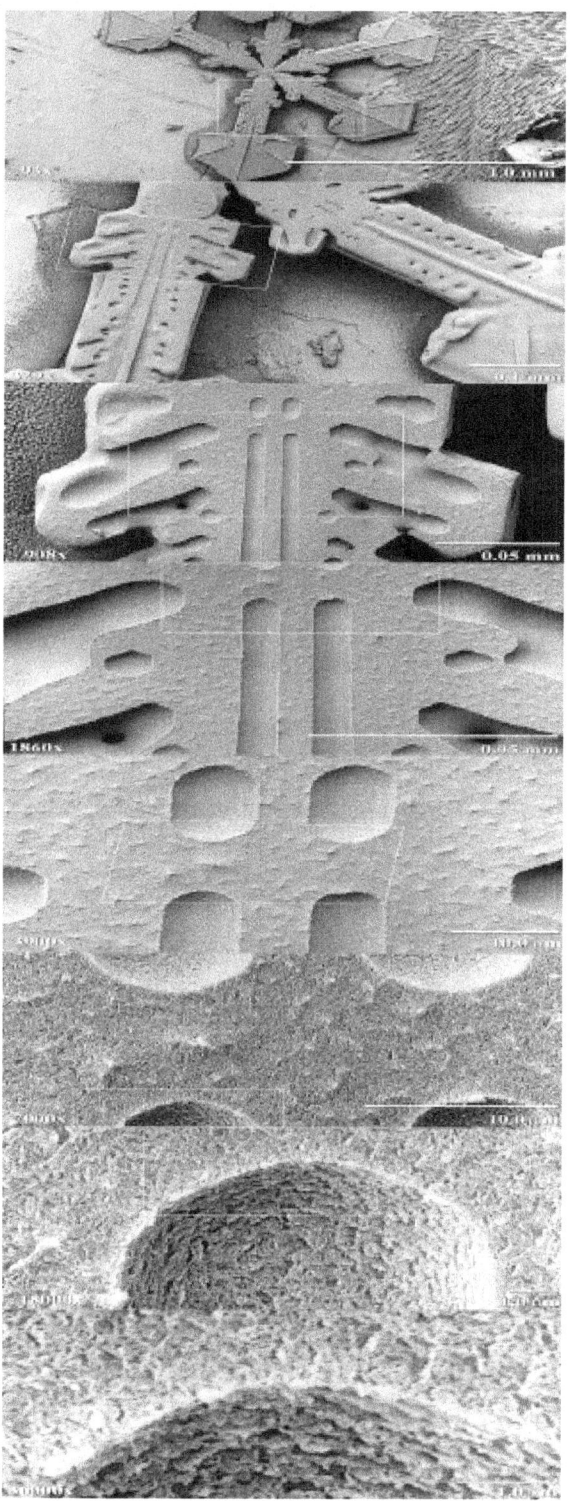

Low-temperature SEM magnification series for a snow crystal. The crystals are captured, stored, and sputter coated with platinum at cryo-temperatures for imaging.

16.10. EXTERNAL LINKS

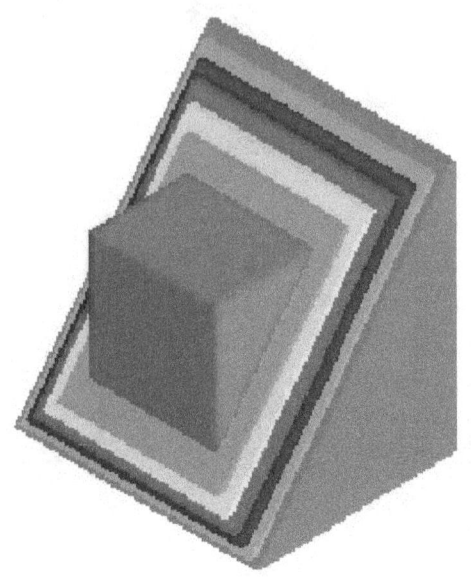

Crystal growth

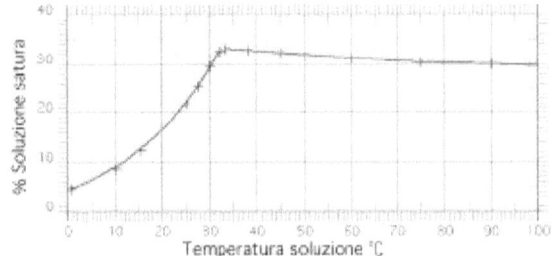

Vertical cooling crystallizer in a beet sugar factory

Solubility of the system $Na_2SO_4 - H_2O$

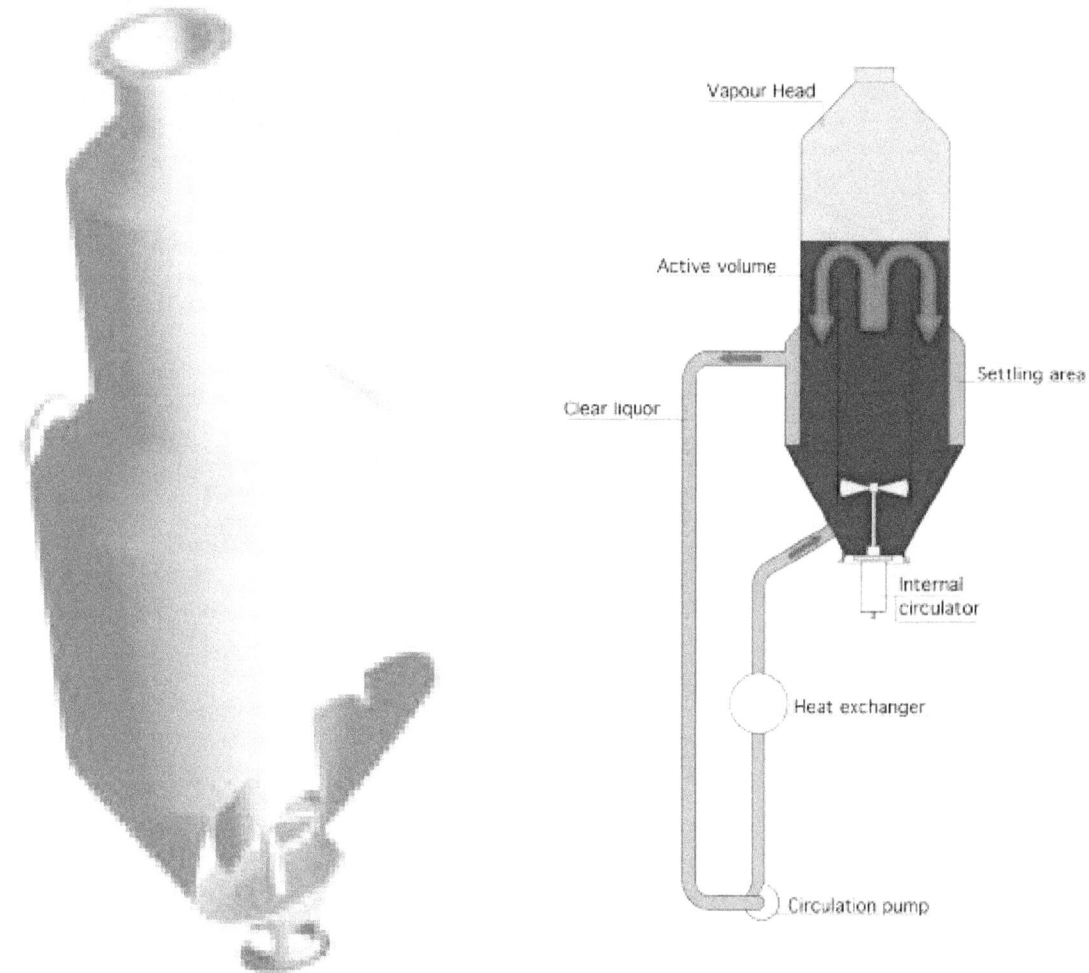

DTB Crystallizer

Schematic of DTB

Chapter 17

Phase transition

17.1 Types of phase transition

Examples of phase transitions include:

- The transitions between the solid, liquid, and gaseous phases of a single component, due to the effects of temperature and/or pressure:

    - (see also vapor pressure and phase diagram)

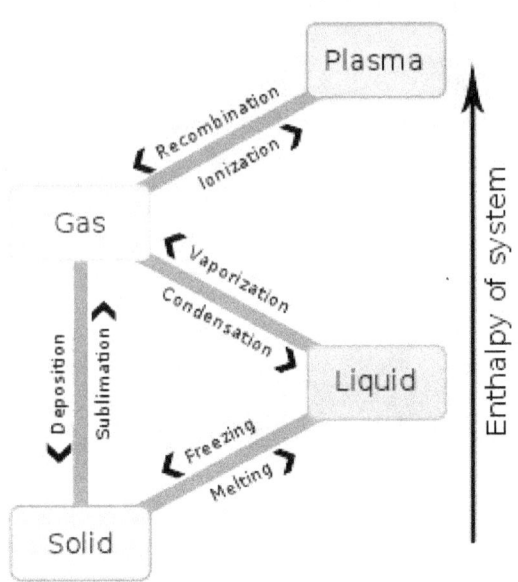

This diagram shows the nomenclature for the different phase transitions.

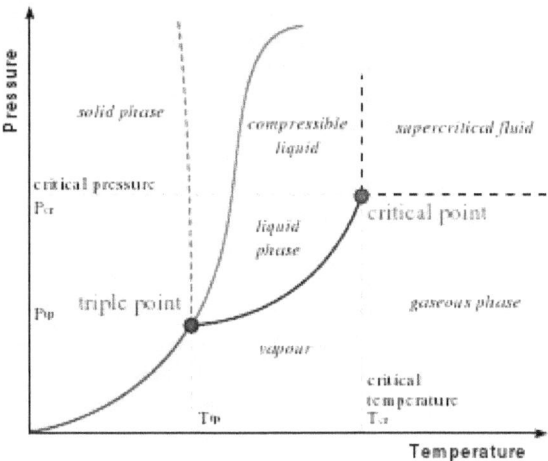

A typical phase diagram. The dotted line gives the anomalous behavior of water.

The term **phase transition** is most commonly used to describe transitions between solid, liquid and gaseous states of matter, and, in rare cases, plasma. A phase of a thermodynamic system and the states of matter have uniform physical properties. During a phase transition of a given medium certain properties of the medium change, often discontinuously, as a result of the change of some external condition, such as temperature, pressure, or others. For example, a liquid may become gas upon heating to the boiling point, resulting in an abrupt change in volume. The measurement of the external conditions at which the transformation occurs is termed the phase transition. Phase transitions are common in nature and used today in many technologies.

- A eutectic transformation, in which a two component single phase liquid is cooled and transforms into two solid phases. The same process, but beginning with a solid instead of a liquid is called a eutectoid transformation.

- A peritectic transformation, in which a two component single phase solid is heated and transforms into a solid phase and a liquid phase.

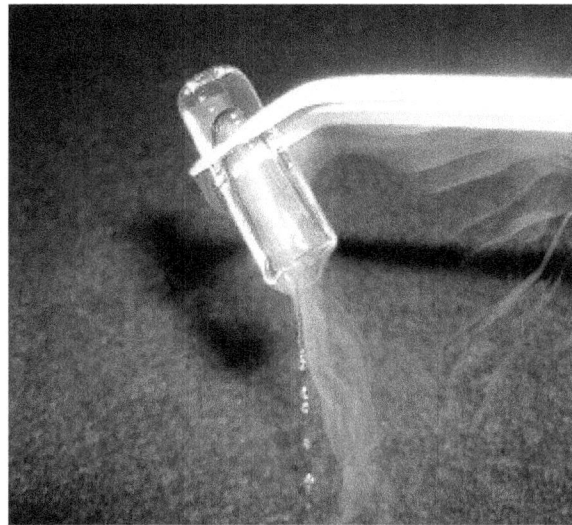

A small piece of rapidly melting solid argon simultaneously shows the transitions from solid to liquid and liquid to gas.

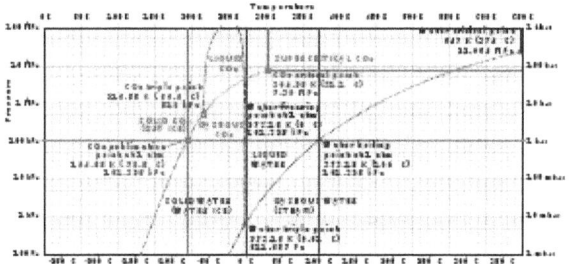

Comparison of phase diagrams of carbon dioxide (red) and water (blue) explaining their different phase transitions at 1 atmosphere

- A spinodal decomposition, in which a single phase is cooled and separates into two different compositions of that same phase.

- Transition to a mesophase between solid and liquid, such as one of the "liquid crystal" phases.

- The transition between the ferromagnetic and paramagnetic phases of magnetic materials at the Curie point.

- The transition between differently ordered, commensurate or incommensurate, magnetic structures, such as in cerium antimonide.

- The martensitic transformation which occurs as one of the many phase transformations in carbon steel and stands as a model for displacive phase transformations.

- Changes in the crystallographic structure such as between ferrite and austenite of iron.

- Order-disorder transitions such as in alpha-titanium aluminides.

- The dependence of the adsorption geometry on coverage and temperature, such as for hydrogen on iron (110).

- The emergence of superconductivity in certain metals and ceramics when cooled below a critical temperature.

- The transition between different molecular structures (polymorphs, allotropes or polyamorphs), especially of solids, such as between an amorphous structure and a crystal structure, between two different crystal structures, or between two amorphous structures.

- Quantum condensation of bosonic fluids (Bose–Einstein condensation). The superfluid transition in liquid helium is an example of this.

- The breaking of symmetries in the laws of physics during the early history of the universe as its temperature cooled.

- Isotope fractionation occurs during a phase transition, the ratio of light to heavy isotopes in the involved molecules changes. When water vapor condenses (an equilibrium fractionation), the heavier water isotopes ($18O$ and $2H$) become enriched in the liquid phase while the lighter isotopes ($16O$ and $1H$) tend toward the vapor phase.[1]

Phase transitions occur when the thermodynamic free energy of a system is non-analytic for some choice of thermodynamic variables (cf. phases). This condition generally stems from the interactions of a large number of particles in a system, and does not appear in systems that are too small. It is important to note that phase transitions can occur and are defined for non-thermodynamic systems, where temperature is not a parameter. Examples include: quantum phase transitions, dynamic phase transitions, and topological (structural) phase transitions. In these types of systems other parameters take the place of temperature. For instance, connection probability replaces temperature for percolating networks.

At the phase transition point (for instance, boiling point) the two phases of a substance, liquid and vapor, have identical free energies and therefore are equally likely to exist. Below the boiling point, the liquid is the more stable state of the two, whereas above the gaseous form is preferred.

It is sometimes possible to change the state of a system diabatically (as opposed to adiabatically) in such a way that it can be brought past a phase transition point without undergoing a phase transition. The resulting state is metastable, i.e., less stable than the phase to which the transition would have occurred, but not unstable either. This occurs in superheating, supercooling, and supersaturation, for example.

17.2 Classifications

17.2.1 Ehrenfest classification

Paul Ehrenfest classified phase transitions based on the behavior of the thermodynamic free energy as a function of other thermodynamic variables.[2] Under this scheme, phase transitions were labeled by the lowest derivative of the free energy that is discontinuous at the transition. *First-order phase transitions* exhibit a discontinuity in the first derivative of the free energy with respect to some thermodynamic variable.[3] The various solid/liquid/gas transitions are classified as first-order transitions because they involve a discontinuous change in density, which is the (inverse of the) first derivative of the free energy with respect to pressure. *Second-order phase transitions* are continuous in the first derivative (the order parameter, which is the first derivative of the free energy with respect to the external field, is continuous across the transition) but exhibit discontinuity in a second derivative of the free energy.[3] These include the ferromagnetic phase transition in materials such as iron, where the magnetization, which is the first derivative of the free energy with respect to the applied magnetic field strength, increases continuously from zero as the temperature is lowered below the Curie temperature. The magnetic susceptibility, the second derivative of the free energy with the field, changes discontinuously. Under the Ehrenfest classification scheme, there could in principle be third, fourth, and higher-order phase transitions.

Though useful, Ehrenfest's classification has been found to be an incomplete method of classifying phase transitions, for it does not take into account the case where a derivative of free energy diverges (which is only possible in the thermodynamic limit). For instance, in the ferromagnetic transition, the heat capacity diverges to infinity.

17.2.2 Modern classifications

In the modern classification scheme, phase transitions are divided into two broad categories, named similarly to the Ehrenfest classes:[2]

First-order phase transitions are those that involve a latent heat. During such a transition, a system either absorbs or releases a fixed (and typically large) amount of energy per volume. During this process, the temperature of the system will stay constant as heat is added: the system is in a "mixed-phase regime" in which some parts of the system have completed the transition and others have not. Familiar examples are the melting of ice or the boiling of water (the water does not instantly turn into vapor, but forms a turbulent mixture of liquid water and vapor bubbles). Imry and Wortis showed that quenched disorder can broaden a first-order transition in that the transformation is completed over a finite range of temperatures, but phenomena like supercooling and superheating survive and hysteresis is observed on thermal cycling.[4][5][6]

Second-order phase transitions are also called *continuous phase transitions*. They are characterized by a divergent susceptibility, an infinite correlation length, and a power-law decay of correlations near criticality. Examples of second-order phase transitions are the ferromagnetic transition, superconducting transition (for a Type-I superconductor the phase transition is second-order at zero external field and for a Type-II superconductor the phase transition is second-order for both normal state-mixed state and mixed state-superconducting state transitions) and the superfluid transition. In contrast to viscosity, thermal expansion and heat capacity of amorphous materials show a relatively sudden change at the glass transition temperature[7] which enables accurate detection using differential scanning calorimetry measurements. Lev Landau gave a phenomenological theory of second-order phase transitions.

Apart from isolated, simple phase transitions, there exist transition lines as well as multicritical points, when varying external parameters like the magnetic field or composition.

Several transitions are known as the *infinite-order phase transitions*. They are continuous but break no symmetries. The most famous example is the Kosterlitz–Thouless transition in the two-dimensional XY model. Many quantum phase transitions, e.g., in two-dimensional electron gases, belong to this class.

The liquid–glass transition is observed in many polymers and other liquids that can be supercooled far below the melting point of the crystalline phase. This is atypical in several respects. It is not a transition between thermodynamic ground states: it is widely believed that the true ground state is always crystalline. Glass is a *quenched disorder* state, and its entropy, density, and so on, depend on the thermal history. Therefore, the glass transition is primarily a dynamic phenomenon: on cooling a liquid, internal degrees of freedom successively fall out of equilibrium. Some theoretical methods predict an underlying phase transition in the hypothetical limit of infinitely long relaxation times.[8][9] No direct experimental evidence supports the existence of these transitions.

17.3 Characteristic properties

17.3.1 Phase coexistence

A disorder-broadened first-order transition occurs over a finite range of temperatures where the fraction of the low-temperature equilibrium phase grows from zero to one

(100%) as the temperature is lowered. This continuous variation of the coexisting fractions with temperature raised interesting possibilities. On cooling, some liquids vitrify into a glass rather than transform to the equilibrium crystal phase. This happens if the cooling rate is faster than a critical cooling rate, and is attributed to the molecular motions becoming so slow that the molecules cannot rearrange into the crystal positions.[10] This slowing down happens below a glass-formation temperature Tg, which may depend on the applied pressure.,[7][11] If the first-order freezing transition occurs over a range of temperatures, and Tg falls within this range, then there is an interesting possibility that the transition is arrested when it is partial and incomplete. Extending these ideas to first-order magnetic transitions being arrested at low temperatures, resulted in the observation of incomplete magnetic transitions, with two magnetic phases coexisting, down to the lowest temperature. First reported in the case of a ferromagnetic to anti-ferromagnetic transition,[12] such persistent phase coexistence has now been reported across a variety of first-order magnetic transitions. These include colossal-magnetoresistance manganite materials,[13][14] magnetocaloric materials,[15] magnetic shape memory materials,[16] and other materials.[17] The interesting feature of these observations of Tg falling within the temperature range over which the transition occurs is that the first-order magnetic transition is influenced by magnetic field, just like the structural transition is influenced by pressure. The relative ease with which magnetic field can be controlled, in contrast to pressure, raises the possibility that one can study the interplay between Tg and Tc in an exhaustive way. Phase coexistence across first-order magnetic transitions will then enable the resolution of outstanding issues in understanding glasses.

17.3.2 Critical points

In any system containing liquid and gaseous phases, there exists a special combination of pressure and temperature, known as the critical point, at which the transition between liquid and gas becomes a second-order transition. Near the critical point, the fluid is sufficiently hot and compressed that the distinction between the liquid and gaseous phases is almost non-existent. This is associated with the phenomenon of critical opalescence, a milky appearance of the liquid due to density fluctuations at all possible wavelengths (including those of visible light).

17.3.3 Symmetry

Phase transitions often involve a symmetry breaking process. For instance, the cooling of a fluid into a crystalline solid breaks continuous translation symmetry: each point in the fluid has the same properties, but each point in a crystal does not have the same properties (unless the points are chosen from the lattice points of the crystal lattice). Typically, the high-temperature phase contains more symmetries than the low-temperature phase due to spontaneous symmetry breaking, with the exception of certain accidental symmetries (e.g. the formation of heavy virtual particles, which only occurs at low temperatures).[18]

17.3.4 Order parameters

An order parameter is a measure of the degree of order across the boundaries in a phase transition system; it normally ranges between zero in one phase (usually above the critical point) and nonzero in the other.[19] At the critical point, the order parameter susceptibility will usually diverge.

An example of an order parameter is the net magnetization in a ferromagnetic system undergoing a phase transition. For liquid/gas transitions, the order parameter is the difference of the densities.

From a theoretical perspective, order parameters arise from symmetry breaking. When this happens, one needs to introduce one or more extra variables to describe the state of the system. For example, in the ferromagnetic phase, one must provide the net magnetization, whose direction was spontaneously chosen when the system cooled below the Curie point. However, note that order parameters can also be defined for non-symmetry-breaking transitions. Some phase transitions, such as superconducting and ferromagnetic, can have order parameters for more than one degree of freedom. In such phases, the order parameter may take the form of a complex number, a vector, or even a tensor, the magnitude of which goes to zero at the phase transition.

There also exist dual descriptions of phase transitions in terms of disorder parameters. These indicate the presence of line-like excitations such as vortex- or defect lines.

17.3.5 Relevance in cosmology

Symmetry-breaking phase transitions play an important role in cosmology. It has been speculated by Lee Smolin and Benjamin and Jeremy Bernstein that, in the hot early universe, the vacuum (i.e. the various quantum fields that fill space) possessed a large number of symmetries. As the universe expanded and cooled, the vacuum underwent a series of symmetry-breaking phase transitions. For example, the electroweak transition broke the SU(2)×U(1) symmetry of the electroweak field into the U(1) symmetry of the present-day electromagnetic field. This transition is important to understanding the asymmetry between the amount of matter and antimatter in the present-day universe (see electroweak baryogenesis.)

Progressive phase transitions in an expanding universe are implicated in the development of order in the universe, as is illustrated by the work of Eric Chaisson[20] and David Layzer.[21] See also Relational order theories.

See also: Order-disorder

17.3.6 Critical exponents and universality classes

Main article: critical exponent

Continuous phase transitions are easier to study than first-order transitions due to the absence of latent heat, and they have been discovered to have many interesting properties. The phenomena associated with continuous phase transitions are called critical phenomena, due to their association with critical points.

It turns out that continuous phase transitions can be characterized by parameters known as critical exponents. The most important one is perhaps the exponent describing the divergence of the thermal correlation length by approaching the transition. For instance, let us examine the behavior of the heat capacity near such a transition. We vary the temperature T of the system while keeping all the other thermodynamic variables fixed, and find that the transition occurs at some critical temperature T_c. When T is near T_c, the heat capacity C typically has a power law behavior,

$$C \propto |T_c - T|^{-\alpha}.$$

Such a behaviour has the heat capacity of amorphous materials near the glass transition temperature where the universal critical exponent $\alpha = 0.59$[22] A similar behavior, but with the exponent ν instead of α, applies for the correlation length.

The exponent ν is positive. This is different with α. Its actual value depends on the type of phase transition we are considering.

It is widely believed that the critical exponents are the same above and below the critical temperature. It has now been shown that this is not necessarily true: When a continuous symmetry is explicitly broken down to a discrete symmetry by irrelevant (in the renormalization group sense) anisotropies, then some exponents (such as γ, the exponent of the susceptibility) are not identical.[23]

For $-1 < \alpha < 0$, the heat capacity has a "kink" at the transition temperature. This is the behavior of liquid helium at the lambda transition from a normal state to the superfluid state, for which experiments have found $\alpha = -0.013 \pm 0.003$. At least one experiment was performed in the zero-gravity conditions of an orbiting satellite to minimize pressure differences in the sample.[24] This experimental value of α agrees with theoretical predictions based on variational perturbation theory.[25]

For $0 < \alpha < 1$, the heat capacity diverges at the transition temperature (though, since $\alpha < 1$, the enthalpy stays finite). An example of such behavior is the 3D ferromagnetic phase transition. In the three-dimensional Ising model for uniaxial magnets, detailed theoretical studies have yielded the exponent $\alpha \sim +0.110$.

Some model systems do not obey a power-law behavior. For example, mean field theory predicts a finite discontinuity of the heat capacity at the transition temperature, and the two-dimensional Ising model has a logarithmic divergence. However, these systems are limiting cases and an exception to the rule. Real phase transitions exhibit power-law behavior.

Several other critical exponents, β, γ, δ, ν, and η, are defined, examining the power law behavior of a measurable physical quantity near the phase transition. Exponents are related by scaling relations, such as

$$\beta = \gamma/(\delta - 1), \qquad \nu = \gamma/(2 - \eta)$$

It can be shown that there are only two independent exponents, e.g. ν and η.

It is a remarkable fact that phase transitions arising in different systems often possess the same set of critical exponents. This phenomenon is known as *universality*. For example, the critical exponents at the liquid–gas critical point have been found to be independent of the chemical composition of the fluid.

More impressively, but understandably from above, they are an exact match for the critical exponents of the ferromagnetic phase transition in uniaxial magnets. Such systems are said to be in the same universality class. Universality is a prediction of the renormalization group theory of phase transitions, which states that the thermodynamic properties of a system near a phase transition depend only on a small number of features, such as dimensionality and symmetry, and are insensitive to the underlying microscopic properties of the system. Again, the divergence of the correlation length is the essential point.

17.3.7 Critical slowing down and other phenomena

There are also other critical phenomena; e.g., besides *static functions* there is also *critical dynamics*. As a consequence, at a phase transition one may observe critical slowing down or *speeding up*. The large *static universality classes* of a continuous phase transition split into smaller *dynamic universality* classes. In addition to the critical exponents, there are also universal relations for certain static or dynamic functions of the magnetic fields and temperature differences from the critical value.

17.3.8 Percolation theory

Another phenomenon which shows phase transitions and critical exponents is percolation. The simplest example is perhaps percolation in a two dimensional square lattice. Sites are randomly occupied with probability p. For small values of p the occupied sites form only small clusters. At a certain threshold p_c a giant cluster is formed and we have a second-order phase transition.[26] The behavior of P∞ near p_c is, P∞~$(p-p_c)^\beta$, where β is a critical exponent.

17.3.9 Phase transitions in biological systems

Phase transitions play many important roles in biological systems. Examples include the lipid bilayer formation, the coil-globule transition in the process of protein folding and DNA melting, liquid crystal-like transitions in the process of DNA condensation, and cooperative ligand binding to DNA and proteins with the character of phase transition.[27]

In *biological membranes*, gel to liquid crystalline phase transitions play a very critical role in physiological functioning of biomembranes. In gel phase, due to low fluidity of membrane lipid fatty-acyl chains, membrane proteins have restricted movement and thus are restrained in exercise of their physiological role. Plants depend critically on photosynthesis by chloroplast thylakoid membranes which are exposed cold environmental temperatures. Thylakoid membranes retain innate fluidity even at relatively low temperatures because of high degree of fatty-acyl disorder allowed by their high content of linolenic acid, 18-carbon chain with 3-double bonds.[28] Gel-to-liquid crystalline phase transition temperature of biological membranes can be determined by many techniques including calorimetry, flouorescence, spin label electron paramagnetic resonance and NMR by recording measurements of the concerned parameter by at series of sample temperatures. A simple method for its determination from 13-C NMR line intensities has also been proposed.[29]

It has been proposed that some biological systems might lie near critical points. Examples include neural networks in the salamander retina,[30] bird flocks[31] gene expression networks in Drosophila,[32] and protein folding.[33] However, it is not clear whether or not alternative reasons could explain some of the phenomena supporting arguments for criticality.[34]

17.4 See also

- Allotropy
- Autocatalytic reactions and order creation
- Crystal growth
    - Abnormal grain growth
- Differential scanning calorimetry
- Diffusionless transformations
- Ehrenfest equations
- Jamming (physics)
- Kelvin probe force microscope
- Landau theory of second order phase transitions
- Laser-heated pedestal growth
- List of states of matter
- Micro-Pulling-Down
- Percolation theory
    - Continuum percolation theory
- Superfluid film
- Superradiant phase transition

17.5 References

[1] Carol Kendall (2004). "Fundamentals of Stable Isotope Geochemistry". USGS. Retrieved 10 April 2014.

[2] Jaeger, Gregg (1 May 1998). "The Ehrenfest Classification of Phase Transitions: Introduction and Evolution". *Archive for History of Exact Sciences* 53 (1): 51–81. doi:10.1007/s004070050021.

[3] Blundell, Stephen J.; Katherine M. Blundell (2008). *Concepts in Thermal Physics*. Oxford University Press. ISBN 978-0-19-856770-7.

17.5. REFERENCES

[4] Imry, Y.; Wortis, M. (1979). "Influence of quenched impurities on first-order phase transitions". *Phys. Rev. B* **19** (7): 3580–3585. Bibcode:1979PhRvB..19.3580I. doi:10.1103/physrevb.19.3580.

[5] Kumar, K.; et al. (2006). "Relating supercooling and glass-like arrest of kinetics for phase separated systems: DopedCeFe2and(La,Pr,Ca)MnO3". *Phys. Rev. B* **73** (18): 184435. arXiv:cond-mat/0602627. Bibcode:2006PhRvB..73r4435K. doi:10.1103/PhysRevB.73.184435.

[6] Pasquini, G.; et al. (2008). "Single-Qubit Lasing and Cooling at the Rabi Frequency". *Phys. Rev. Lett* **100** (3): 247003. arXiv:cond-mat/0701041. Bibcode:2008PhRvL.100c7003H. doi:10.1103/PhysRevLett.100.037003.

[7] Ojovan, M.I. (2013). "Ordering and structural changes at the glass-liquid transition". *J. Non-Cryst. Solids* **382**: 79–86. Bibcode:2013JNCS..382...79O. doi:10.1016/j.jnoncrysol.2013.10.016.

[8] Gotze, Wolfgang. "Complex Dynamics of Glass-Forming Liquids: A Mode-Coupling Theory."

[9] Lubchenko, V. Wolynes; Wolynes, Peter G. (2007). "Theory of Structural Glasses and Supercooled Liquids". *Annual Review of Physical Chemistry* **58**: 235–266. arXiv:cond-mat/0607349. Bibcode:2007ARPC...58..235L. doi:10.1146/annurev.physchem.58.032806.104653. PMID 17067282.

[10] Greer, A. L. (1995). "Metallic Glasses". *Science* **267** (5206): 1947–1953. Bibcode:1995Sci...267.1947G. doi:10.1126/science.267.5206.1947.

[11] Tarjus, G. (2007). "Materials science: Metal turned to glass". *Nature* **448** (7155): 758–759. Bibcode:2007Natur.448..758T. doi:10.1038/448758a. PMID 17700684.

[12] Manekar, M. A.; et al. (2001). "Nonequilibrium relaxation study of Ising spin glass models". *Physical Review B* **64** (2): 104416. Bibcode:2001PhRvB..64b4416O. doi:10.1103/PhysRevB.64.024416. |first2= missing |last2= in Authors list (help)

[13] Banerjee, A; Pramanik, A K; Kumar, Kranti; Chaddah, P (2006). "Coexisting tunable fractions of glassy and equilibrium long-range-order phases in manganites". *Journal of Physics: Condensed Matter* **18** (49): L605. arXiv:cond-mat/0611152. Bibcode:2006JPCM...18L.605B. doi:10.1088/0953-8984/18/49/L02.

[14] Wu, W.; et al. (2006). "Magnetic imaging of a supercooling glass transition in a weakly disordered ferromagnet". *Nature Materials* **5** (11): 881–886. Bibcode:2006NatMa...5..881W. doi:10.1038/nmat1743.

[15] Roy, S. B.; et al. (2006). "Evidence of a magnetic glass state in the magnetocaloric materialGd5Ge4". *Physical Review B* **74**: 012403. Bibcode:2006PhRvB..74a2403R. doi:10.1103/PhysRevB.74.012403.

[16] Lakhani, A.; et al. (2012). "Magnetic glass in shape memory alloy: Ni45Co5Mn38Sn12". *J. Phys. Condens. Matter* **24** (38): 386004. arXiv:1206.2024. Bibcode:2012JPCM...24L6004L. doi:10.1088/0953-8984/24/38/386004.

[17] Kushwaha, P.; et al. (2009). "Non-Korringa nuclear relaxation in the ferromagnetic phase of the bilayered manganiteLa1.2Sr1.8Mn2O7". *Physical Review B* **80** (2): 174413. Bibcode:2009PhRvB..80b4413H. doi:10.1103/PhysRevB.80.024413.

[18] Ivancevic, Vladimir G.; Ivancevic, Tijiana, T. (2008). *Complex Nonlinearity*. Berlin: Springer. pp. 176–177. ISBN 978-3-540-79357-1. Retrieved 12 October 2014.

[19] A. D. McNaught and A. Wilkinson (ed.). "Compendium of Chemical Terminology". IUPAC. ISBN 0-86542-684-8. Retrieved 2007-10-23.

[20] Chaisson, *Cosmic Evolution*, Harvard, 2001

[21] David Layzer, *Cosmogenesis, The Development of Order in the Universe*, Oxford Univ. Press, 1991

[22] Ojovan, Michael I; Lee, William E (2006). "Topologically disordered systems at the glass transition" (PDF). *Journal of Physics: Condensed Matter* **18** (50): 11507–11520. Bibcode:2006JPCM...1811507O. doi:10.1088/0953-8984/18/50/007.

[23] Leonard, F.; Delamotte, B. (2015). "Critical exponents can be different on the two sides of a transition". *Phys. Rev. Lett* **115**: 200601. arXiv:1508.07852. Bibcode:2015PhRvL.115t0601L. doi:10.1103/PhysRevLett.115.200601.

[24] Lipa, J.; Nissen, J.; Stricker, D.; Swanson, D.; Chui, T. (2003). "Specific heat of liquid helium in zero gravity very near the lambda point". *Physical Review B* **68** (17): 174518. arXiv:cond-mat/0310163. Bibcode:2003PhRvB..68q4518L. doi:10.1103/PhysRevB.68.174518.

[25] Kleinert, Hagen (1999). "Critical exponents from seven-loop strong-coupling φ4 theory in three dimensions". *Physical Review D* **60** (8): 085001. arXiv:hep-th/9812197. Bibcode:1999PhRvD..60h5001K. doi:10.1103/PhysRevD.60.085001.

[26] Armin Bunde and Shlomo Havlin (1996). *Fractals and Disordered Systems*. Springer.

[27] D.Y. Lando and V.B. Teif (2000). "Long-range interactions between ligands bound to a DNA molecule give rise to adsorption with the character of phase transition of the first kind". *J. Biomol. Struct. Dynam.* **17** (5): 903–911. doi:10.1080/07391102.2000.10506578.

[28] YashRoy, R.C. (1987). "13-C NMR studies of lipid fatty acyl chains of chloroplast membranes". *Indian Journal of Biochemistry and Biophysics* 24 (6): 177–178.

[29] YashRoy, R C (1990). "Determination of membrane lipid phase transition temperature from 13-C NMR intensities". *Journal of Biochemical and Biophysical Methods* 20 (4): 353–356. doi:10.1016/0165-022X(90)90097-V. PMID 2365951.

[30] Tkacik, Gasper; Mora, Thierry; Marre, Olivier; Amodei, Dario; Berry II, Michael J.; Bialek, William (2014). "Thermodynamics for a network of neurons: Signatures of criticality". arXiv:1407.5946 [q-bio.NC].

[31] Bialek, W; Cavagna, A; Giardina, I (2014). "Social interactions dominate speed control in poising natural flocks near criticality". *PNAS*.

[32] Krotov, D; Dubuis, J O; Gregor, T; Bialek, W (2014). "Morphogenesis at criticality". *PNAS*.

[33] Mora, Thierry; Bialek, William (2010). "Are biological systems poised at criticality?". *arXiv*.

[34] Schwab, David J; Nemenman, Ilya; Mehta, Pankaj (2013). "Zipf's law and criticality in multivariate data without fine-tuning". *arXiv*.

17.5.1 Further reading

- Anderson, P.W., *Basic Notions of Condensed Matter Physics*, Perseus Publishing (1997).

- Fisher, M.E. (1974). "The renormalization group in the theory of critical behavior". *Rev. Mod. Phys.* 46: 597–616. Bibcode:1974RvMP...46..597F. doi:10.1103/revmodphys.46.597.

- Goldenfeld, N., *Lectures on Phase Transitions and the Renormalization Group*, Perseus Publishing (1992).

- Ivancevic, Vladimir G; Ivancevic, Tijana T (2008), *Chaos, Phase Transitions, Topology Change and Path Integrals*, Berlin: Springer, ISBN 978-3-540-79356-4, retrieved 14 March 2013 e-ISBN 978-3-540-79357-1

- Kogut, J.; Wilson, K (1974). "The Renormalization Group and the epsilon-Expansion". *Phys. Rep.* 12: 75. Bibcode:1974PhR....12...75W. doi:10.1016/0370-1573(74)90023-4.

- Krieger, Martin H., *Constitutions of matter : mathematically modelling the most everyday of physical phenomena*, University of Chicago Press, 1996. Contains a detailed pedagogical discussion of Onsager's solution of the 2-D Ising Model.

- Landau, L.D. and Lifshitz, E.M., *Statistical Physics Part 1*, vol. 5 of *Course of Theoretical Physics*, Pergamon Press, 3rd Ed. (1994).

- Kleinert, H., *Gauge Fields in Condensed Matter*, Vol. I, "Superfluid and Vortex lines; Disorder Fields, Phase Transitions,", pp. 1–742, World Scientific (Singapore, 1989); Paperback ISBN 9971-5-0210-0 *(readable online physik.fu-berlin.de)*

- Kleinert, H. and Verena Schulte-Frohlinde, *Critical Properties of φ^4-Theories*, World Scientific (Singapore, 2001); Paperback ISBN 981-02-4659-5 *(readable online here)*.

- Mussardo G., "Statistical Field Theory. An Introduction to Exactly Solved Models of Statistical Physics", Oxford University Press, 2010.

- Schroeder, Manfred R., *Fractals, chaos, power laws : minutes from an infinite paradise*, New York: W. H. Freeman, 1991. Very well-written book in "semi-popular" style—not a textbook—aimed at an audience with some training in mathematics and the physical sciences. Explains what scaling in phase transitions is all about, among other things.

- Yeomans J. M., *Statistical Mechanics of Phase Transitions*, Oxford University Press, 1992.

- H. E. Stanley, *Introduction to Phase Transitions and Critical Phenomena* (Oxford University Press, Oxford and New York 1971).

17.6 External links

- Interactive Phase Transitions on lattices with Java applets

- Universality classes from Sklogwiki

Chapter 18

Critical opalescence

Left-to-right sequence of heating a mass of ethane in a constant volume. In the center panel, critical opalescence is seen.

18.1 External links

- A time-lapse video of critical opalescence in a binary mixture
- Account of Einstein's work on daylight and critical opalescence, with references

More-detailed experimental demonstrations of critical opalescence may be found at

- http://physicsofmatter.com/NotTheBook/CriticalOpal/Explanation.html
- http://www.msm.cam.ac.uk/doitpoms/tlplib/solid-solutions/demo.php

Critical opalescence is a phenomenon which arises in the region of a continuous, or second-order, phase transition. Originally reported by Charles Cagniard de la Tour in 1823 in mixtures of alcohol and water, its importance was recognised by Thomas Andrews in 1869 following his experiments on the liquid-gas transition in carbon dioxide, many other examples have been discovered since. The phenomenon is most commonly demonstrated in binary fluid mixtures, such as methanol and cyclohexane. As the critical point is approached, the sizes of the gas and liquid region begin to fluctuate over increasingly large length scales. As the density fluctuations become of a size comparable to the wavelength of light, the light is scattered and causes the normally transparent liquid to appear cloudy. Tellingly, the opalescence does not diminish as one gets closer to the critical point, where the largest fluctuations can reach even centimetre proportions, confirming the physical relevance of smaller fluctuations.

In 1908 the Polish physicist Marian Smoluchowski became the first to ascribe the phenomenon of critical opalescence to large density fluctuations. In 1910 Albert Einstein showed that the link between critical opalescence and Rayleigh scattering is quantitative .

Chapter 19

Percolation

In coffee percolation, soluble compounds leave the coffee grounds and join the water to form coffee. Insoluble compounds (and granulates) remain within the coffee filter.

In physics, chemistry and materials science, **percolation** (from Latin *percōlāre*, "to filter" or "trickle through") refers to the movement and filtering of fluids through porous materials.

19.1 Background

During the last decades, percolation theory, an extensive mathematical studies model of percolation, has brought new understanding and techniques to a broad range of topics in physics, materials science, complex networks, epidemiology, and other fields . For example, in geology, percolation refers to filtration of water through soil and permeable rocks. The water flows to groundwater storage (aquifer).

Percolation typically exhibits universality. Statistical physics concepts such as scaling theory, renormalization, phase transition, critical phenomena and fractals are used to characterize percolation properties. Combinatorics is commonly employed to study percolation thresholds.

Due to the complexity involved in obtaining exact results from analytical models of percolation, computer simulations are typically used. The current fastest algorithm for percolation was published in 2000 by Mark Newman and Robert Ziff.[1]

19.2 Examples

- Coffee percolation, where the solvent is water, the permeable substance is the coffee grounds, and the soluble constituents are the chemical compounds that give coffee its color, taste, and aroma
- Movement of weathered material down on a slope under the earth's surface
- Cracking of trees with the presence of two conditions, sunlight and under the influence of pressure
- Robustness of networks to random and targeted attacks
- Transport in porous media
- Epidemic spreading[2][3]
- Surface roughening
- Dental Percolation, increase rate of decay under crowns because of a conducive environment for strep mutans and lactobacillus

19.3 See also

- Branched polymer
- Conductance
- Critical exponents

- Fragmentation
- Gelation
- Groundwater recharge
- Immunization
- Network theory
- Percolation critical exponents
- Percolation theory
- Polymerization
- Self-organization
- Self-organized criticality
- Septic tank
- Supercooled water
- Water pipe percolator

19.4 References

[1] Newman, Mark; Ziff, Robert (2000). "Efficient Monte Carlo Algorithm and High-Precision Results for Percolation". *Physical Review Letters* (American Physical Society) **85** (19): 4104–4107. arXiv:cond-mat/0005264. Bibcode:2000PhRvL..85.4104N. doi:10.1103/PhysRevLett.85.4104. PMID 11056635. Retrieved 19 November 2013.

[2] Parshani, Roni; Carmi, Shai; Havlin, Shlomo (2010). "Epidemic Threshold for the Susceptible-Infectious-Susceptible Model on Random Networks". *Physical Review Letters* **104** (25). arXiv:0909.3811. Bibcode:2010PhRvL.104y8701P. doi:10.1103/PhysRevLett.104.258701. ISSN 0031-9007.

[3] Grassberger, P. "On the Critical Behavior of the General Epidemic Process and Dynamic Percolation". *Mathematical Biosciences*.

19.5 Further reading

- Harry Kesten, What is percolation? *Notices of the AMS*, May 2006.
- Muhammad Sahimi. *Applications of Percolation Theory*. Taylor & Francis, 1994. ISBN 0-7484-0075-3 (cloth), ISBN 0-7484-0076-1 (paper)
- Geoffrey Grimmett. *Percolation (2. ed)*. Springer Verlag, 1999.

- D.Stauffer and A.Aharony. *Introduction to Percolation Theory*
- A. Bunde, S. Havlin (Editors) *Fractals and Disordered Systems*, Springer, 1996
- S. Kirkpatrick *Percolation and conduction* Rev. Mod. Phys. 45, 574, 1973
- D. Ben-Avraham, S. Havlin *Diffusion and Reactions in Fractals and Disordered Systems*, Cambridge University Press, 2000
- Edouard Rodrigues, Remarkable properties of pawns on a hexboard
- R. Cohen and S. Havlin *Complex Networks: Structure, Robustness and Function*
- Bollobás, Béla; Riordan, Oliver (2006), Percolation, Cambridge University Press, ISBN 0521872324
- Grimmett, Geoffrey (1999), Percolation, Springer

19.6 External links

- Introduction to Percolation Theory: short course by Shlomo Havlin

Chapter 20

Structure formation

In physical cosmology, **structure formation** refers to the formation of galaxies, galaxy clusters and larger structures from small early density fluctuations. The universe, as is now known from observations of the cosmic microwave background radiation, began in a hot, dense, nearly uniform state approximately 13.8 billion years ago.[1] However, looking in the sky today, we see structures on all scales, from stars and planets to galaxies and, on still larger scales still, galaxy clusters and sheet-like structures of galaxies separated by enormous voids containing few galaxies. Structure formation attempts to model how these structures formed by gravitational instability of small early density ripples.[2][3][4][5]

The modern Lambda-CDM model is successful at predicting the observed large-scale distribution of galaxies, clusters and voids; but on the scale of individual galaxies there are many complications due to highly nonlinear processes involving baryonic physics, gas heating and cooling, star formation and feedback. Understanding the processes of galaxy formation is a major topic of modern cosmology research, both via observations such as the Hubble Ultra-Deep Field and via large computer simulations.

20.1 Overview

Under present models, the structure of the visible universe was formed in the following stages:

20.1.1 Very early universe

In this stage, some mechanism, such as cosmic inflation, was responsible for establishing the initial conditions of the universe: homogeneity, isotropy, and flatness.[3][6] Cosmic inflation also would have amplified minute quantum fluctuations (pre-inflation) into slight density ripples of overdensity and underdensity (post-inflation).

20.1.2 Growth of structure

The early universe was dominated by radiation; in this case density fluctuations larger than the cosmic horizon grow proportional to the scale factor, as the gravitational potential fluctuations remain constant. Structures smaller than the horizon remained essentially frozen due to radiation domination impeding growth. As the universe expanded, the density of radiation drops faster than matter (due to redshifting of photon energy); this led to a crossover called matter-radiation equality at ~ 50,000 years after the Big Bang. After this all dark matter ripples could grow freely, forming seeds into which the baryons could later fall. The size of the universe at this epoch forms a turnover in the matter power spectrum which can be measured in large redshift surveys.

20.1.3 Recombination

The universe was dominated by radiation for most of this stage, and due to the intense heat and radiation, the primordial hydrogen and helium were fully ionized into nuclei and free electrons. In this hot and dense situation, the radiation (photons) could not travel far before Thomson scattering off an electron. The universe was very hot and dense, but expanding rapidly and therefore cooling. Finally, at a little less than 400,000 years after the 'bang', it become cool enough (around 3000 K) for the protons to capture negatively charged electrons, forming neutral hydrogen atoms. (Helium atoms formed somewhat earlier due to their larger binding energy). Once nearly all the charged particles were bound in neutral atoms, the photons no longer interacted with them and were free to propagate for the next 13.8 billion years; we currently detect those photons redshifted by a factor 1090 down to 2.725 K as the Cosmic Microwave Background Radiation (CMB) filling today's universe. Several remarkable space-based missions (COBE, WMAP, Planck), have detected very slight variations in the density and temperature of the CMB. These variations were subtle, and the CMB appears very nearly uniformly the same in every direction. However, the slight temperature variations of order a few parts in 100,000 are of enormous importance, for they essentially were early "seeds" from which all subsequent complex structures in the universe ultimately developed.

The theory of what happened after the universe's first 400,000 years is one of hierarchical structure formation: the smaller gravitationally bound structures such as matter peaks containing the first stars and stellar clusters formed first, and these subsequently merged with gas and dark matter to form galaxies, followed by groups, clusters and superclusters of galaxies.

20.2 Very early universe

The very early universe is still a poorly understood epoch, from the viewpoint of fundamental physics. The prevailing theory, cosmic inflation, does a good job explaining the observed flatness, homogeneity and isotropy of the universe, as well as the absence of exotic relic particles (such as magnetic monopoles). Another prediction borne out by observation is that tiny perturbations in the primordial universe seed the later formation of structure. These fluctuations, while they form the foundation for all structure, appear most clearly as tiny temperature fluctuations at one part in 100,000. (To put this in perspective, the same level of fluctuations on a topographic map of the United States would show no feature taller than a few centimeters.) These fluctuations are critical, because they provide the seeds from which the largest structures can grow and eventually collapse to form galaxies and stars. COBE (Cosmic Background Explorer) provided the first detection of the intrinsic fluctuations in the cosmic microwave background radiation in the 1990s.

These perturbations are thought to have a very specific character: they form a Gaussian random field whose covariance function is diagonal and nearly scale-invariant. Observed fluctuations appear to have exactly this form, and in addition the *spectral index* measured by WMAP—the spectral index measures the deviation from a scale-invariant (or Harrison-Zel'dovich) spectrum—is very nearly the value predicted by the simplest and most robust models of inflation. Another important property of the primordial perturbations, that they are adiabatic (or isentropic between the various kinds of matter that compose the universe), is predicted by cosmic inflation and has been confirmed by observations.

Other theories of the very early universe have been proposed that are claimed to make similar predictions, such as the brane gas cosmology, cyclic model, pre-big bang model and holographic universe, but they remain nascent and are not widely accepted. Some theories, such as cosmic strings, have largely been refuted by increasingly precise data.

20.2.1 The horizon problem

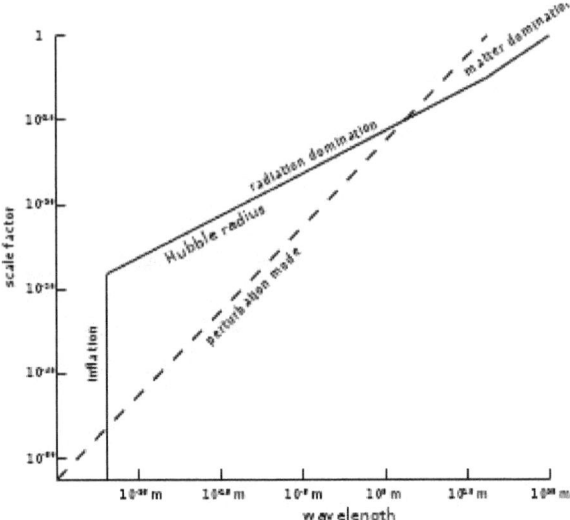

The physical size of the Hubble radius (solid line) as a function of the scale factor of the universe. The physical wavelength of a perturbation mode (dashed line) is shown as well. The plot illustrates how the perturbation mode exits the horizon during cosmic inflation in order to reenter during radiation domination. If cosmic inflation never happened, and radiation domination continued back until a gravitational singularity, then the mode would never have exited the horizon in the very early universe.

An important concept in structure formation is the notion of the Hubble radius, often called simply the *horizon*, as it is closely related to the particle horizon. The Hubble radius, which is related to the Hubble parameter H as $R = c/H$, where c is the speed of light, defines, roughly speaking, the volume of the nearby universe that has recently (in the last expansion time) been in causal contact with an observer. Since the universe is continually expanding, its energy density is continually decreasing (in the absence of truly exotic matter such as phantom energy). The Friedmann equation relates the energy density of the universe to the Hubble parameter and shows that the Hubble radius is continually increasing.

The horizon problem of big bang cosmology says that, without inflation, perturbations were never in causal contact before they entered the horizon and thus the homogeneity and isotropy of, for example, the large scale galaxy distributions cannot be explained. This is because, in an ordinary Friedmann–Lemaître–Robertson–Walker cosmology, the Hubble radius increases more rapidly than space expands, so perturbations only enter the Hubble radius, and are not pushed out by the expansion. This paradox is resolved by cosmic inflation, which suggests that during a phase of rapid expansion in the early universe the Hubble radius was nearly constant. Thus, large scale isotropy is due to quantum fluctuations produced during cosmic inflation that are pushed outside the horizon.

20.3 Primordial plasma

The end of inflation is called reheating, when the inflation particles decay into a hot, thermal plasma of other particles. In this epoch, the energy content of the universe is entirely radiation, with standard model particles having relativistic velocities. As the plasma cools, baryogenesis and leptogenesis are thought to occur, as the quark–gluon plasma cools, electroweak symmetry breaking occurs and the universe becomes principally composed of ordinary protons, neutrons and electrons. As the universe cools further, big bang nucleosynthesis occurs and small quantities of deuterium, helium and lithium nuclei are created. As the universe cools and expands, the energy in photons begins to redshift away, particles become non-relativistic and ordinary matter begins to dominate the universe. Eventually, atoms begin to form as free electrons bind to nuclei. This suppresses Thomson scattering of photons. Combined with the rarefaction of the universe (and consequent increase in the mean free path of photons), this makes the universe transparent and the cosmic microwave background is emitted at recombination (the *surface of last scattering*).

20.3.1 Acoustic oscillations

Main article: baryon acoustic oscillations

The primordial plasma would have had very slight overdensities of matter, thought to have derived from the enlargement of quantum fluctuations during inflation. Whatever the source, these overdensities gravitationally attract matter. But the intense heat of the near constant photon-matter interactions of this epoch rather forcefully seeks thermal equilibrium, which creates a large amount of outward pressure. These counteracting forces of gravity and pressure create oscillations, analogous to sound waves created in air by pressure differences.

These perturbations are important, as they are responsible for the subtle physics that result in the cosmic microwave background anisotropy. In this epoch, the amplitude of perturbations that enter the horizon oscillate sinusoidally, with dense regions becoming more rarefied and then becoming dense again, with a frequency which is related to the size of the perturbation. If the perturbation oscillates an integral or half-integral number of times between coming into the horizon and recombination, it appears as an acoustic peak of the cosmic microwave background anisotropy. (A half-oscillation, in which a dense region becomes a rarefied region or vice versa, appears as a peak because the anisotropy is displayed as a *power spectrum*, so underdensities contribute to the power just as much as overdensities.) The physics that determines the detailed peak structure of the microwave background is complicated, but these oscillations provide the essence.[7][8][9][10][11]

20.4 Linear structure

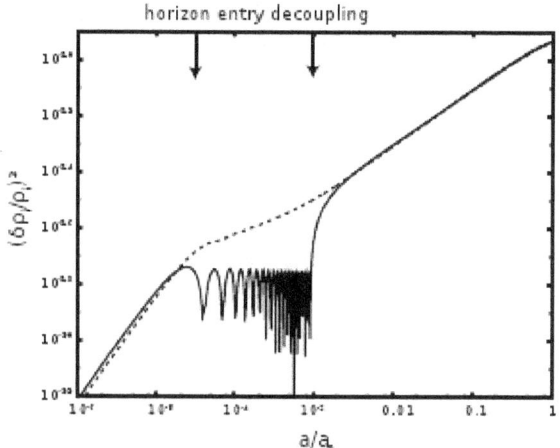

Evolution of two perturbations to the ΛCDM homogeneous big bang model. Between entering the horizon and decoupling, the dark matter perturbation (dashed line) grows logarithmically, before the growth accelerates in matter domination. On the other hand, between entering the horizon and decoupling, the perturbation in the baryon-photon fluid (solid line) oscillates rapidly. After decoupling, it grows rapidly to match the dominant matter perturbation, the dark matter mode.

One of the key realizations made by cosmologists in the 1970s and 1980s was that the majority of the matter content of the universe was composed not of atoms, but rather a mysterious form of matter known as dark matter. Dark matter interacts through the force of gravity, but it is not composed of baryons and it is known with very high accuracy that it does not emit or absorb radiation. It may be composed of particles that interact through the weak interaction, such as neutrinos, but it cannot be composed entirely of the three known kinds of neutrinos (although some have suggested it is a sterile neutrino). Recent evidence suggests that there is about five times as much dark matter as baryonic matter, and thus the dynamics of the universe in this epoch are dominated by dark matter.

Dark matter plays a key role in structure formation because it feels only the force of gravity: the gravitational Jeans instability which allows compact structures to form is not opposed by any force, such as radiation pressure. As a result, dark matter begins to collapse into a complex network of dark matter halos well before ordinary matter, which is impeded by pressure forces. Without dark matter, the epoch of galaxy formation would occur substantially later in the universe than is observed.

The physics of structure formation in this epoch is particularly simple, as dark matter perturbations with different wavelengths evolve independently. As the Hubble radius grows in the expanding universe, it encompasses larger and larger perturbations. During matter domination, all causal dark matter perturbations grow through gravitational clustering. However, the shorter-wavelength perturbations that are encompassed during radiation domination have their growth retarded until matter domination. At this stage, luminous, baryonic matter is expected to simply mirror the evolution of the dark matter, and their distributions should closely trace one another.

It is a simple matter to calculate this "linear power spectrum" and, as a tool for cosmology, it is of comparable importance to the cosmic microwave background. The power spectrum has been measured by galaxy surveys, such as the Sloan Digital Sky Survey, and by surveys of the Lyman-α forest. Since these surveys observe radiation emitted from galaxies and quasars, they do not directly measure the dark matter, but the large scale distribution of galaxies (and of absorption lines in the Lyman-α forest) is expected to closely mirror the distribution of dark matter. This depends on the fact that galaxies will be larger and more numerous in denser parts of the universe, whereas they will be comparatively scarce in rarefied regions.

20.5 Nonlinear structure

When the perturbations have grown sufficiently, a small region might become substantially denser than the mean density of the universe. At this point, the physics involved becomes substantially more complicated. When the deviations from homogeneity are small, the dark matter may be treated as a pressureless fluid and evolves by very simple equations. In regions which are significantly denser than the background, the full Newtonian theory of gravity must be included. (The Newtonian theory is appropriate because the masses involved are much less than those required to form a black hole, and the speed of gravity may be ignored as the light-crossing time for the structure is still smaller than the characteristic dynamical time.) One sign that the linear and fluid approximations become invalid is that dark matter starts to form caustics in which the trajectories of adjacent particles cross, or particles start to form orbits. These dynamics are generally best understood using N-body simulations (although a variety of semi-analytic schemes, such as the Press–Schechter formalism, can be used in some cases). While in principle these simulations are quite simple, in practice they are very difficult to implement, as they require simulating millions or even billions of particles. Moreover, despite the large number of particles, each particle typically weighs 10^9 solar masses and discretization effects may become significant. The largest such simulation as of 2005 is the Millennium simulation.[12]

The result of N-body simulations suggests that the universe is composed largely of voids, whose densities might be as low as one tenth the cosmological mean. The matter condenses in large filaments and haloes which have an intricate web-like structure. These form galaxy groups, clusters and superclusters. While the simulations appear to agree broadly with observations, their interpretation is complicated by the understanding of how dense accumulations of dark matter spur galaxy formation. In particular, many more small haloes form than we see in astronomical observations as dwarf galaxies and globular clusters. This is known as the galaxy bias problem, and a variety of explanations have been proposed. Most account for it as an effect in the complicated physics of galaxy formation, but some have suggested that it is a problem with our model of dark matter and that some effect, such as warm dark matter, prevents the formation of the smallest haloes.

20.6 Gas evolution

See also: galaxy formation and evolution and stellar evolution

The final stage in evolution comes when baryons condense in the centres of galaxy haloes to form galaxies, stars and quasars. A paradoxical aspect of structure formation is that while dark matter greatly accelerates the formation of dense haloes, because dark matter does not have radiation pressure, the formation of smaller structures from dark matter is impossible because dark matter cannot dissipate angular momentum, whereas ordinary baryonic matter can collapse to form dense objects by dissipating angular momentum through radiative cooling. Understanding these processes is an enormously difficult computational problem, because they can involve the physics of gravity, magnetohydrodynamics, atomic physics, nuclear reactions, turbulence and even general relativity. In most cases, it is not yet possible to perform simulations that can be compared quantitatively with observations, and the best that can be achieved are approximate simulations that illustrate the main qualitative features of a process such as star formation.

20.7 Modelling structure formation

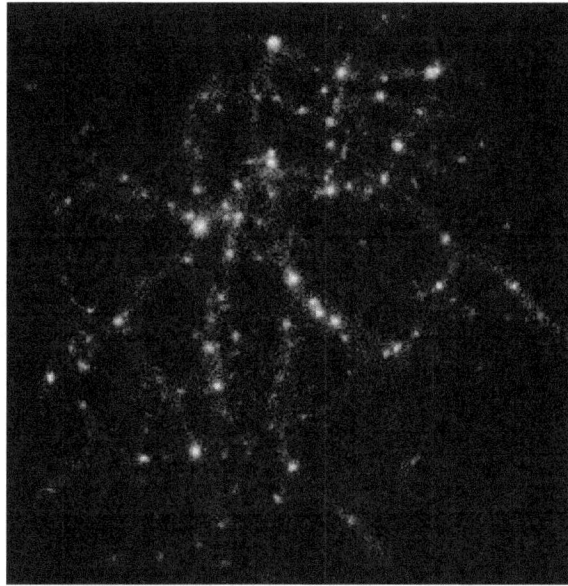

Snapshot from a computer simulation of large scale structure formation in a Lambda-CDM universe.

20.7.1 Cosmological perturbations

Main article: cosmological perturbation theory

Much of the difficulty, and many of the disputes, in understanding the large-scale structure of the universe can be resolved by better understanding the choice of gauge in general relativity. By the scalar-vector-tensor decomposition, the metric includes four scalar perturbations, two vector perturbations, and one tensor perturbation. Only the scalar perturbations are significant: the vectors are exponentially suppressed in the early universe, and the tensor mode makes only a small (but important) contribution in the form of primordial gravitational radiation and the B-modes of the cosmic microwave background polarization. Two of the four scalar modes may be removed by a physically meaningless coordinate transformation. Which modes are eliminated determine the infinite number of possible gauge fixings. The most popular gauge is Newtonian gauge (and the closely related conformal Newtonian gauge), in which the retained scalars are the Newtonian potentials Φ and Ψ, which correspond exactly to the Newtonian potential energy from Newtonian gravity. Many other gauges are used, including synchronous gauge, which can be an efficient gauge for numerical computation (it is used by CMBFAST). Each gauge still includes some unphysical degrees of freedom. There is a so-called gauge-invariant formalism, in which only gauge invariant combinations of variables are considered.

20.7.2 Inflation and initial conditions

The initial conditions for the universe are thought to arise from the scale invariant quantum mechanical fluctuations of cosmic inflation. The perturbation of the background energy density at a given point $\rho(\mathbf{x}, t)$ in space is then given by an isotropic, homogeneous Gaussian random field of mean zero. This means that the spatial Fourier transform of ρ – $\hat{\rho}(\mathbf{k}, t)$ has the following correlation functions

$$\langle \hat{\rho}(\mathbf{k}, t) \hat{\rho}(\mathbf{k}', t) \rangle = f(k) \delta^{(3)}(\mathbf{k} - \mathbf{k}')$$

where $\delta^{(3)}$ is the three-dimensional Dirac delta function and $k = |\mathbf{k}|$ is the length of $\mathbf{k}$. Moreover, the spectrum predicted by inflation is nearly scale invariant, which means

$$\langle \hat{\rho}(\mathbf{k}, t) \hat{\rho}(\mathbf{k}', t) \rangle = k^{n_s - 1} \delta^{(3)}(\mathbf{k} - \mathbf{k}')$$

where $n_s - 1$ is a small number. Finally, the initial conditions are adiabatic or isentropic, which means that the fractional perturbation in the entropy of each species of particle is equal.

20.8 See also

- Big Bang
- Chronology of the universe
- Galaxy formation and evolution
- Illustris project
- Stellar evolution
- Timeline of the Big Bang

20.9 References

[1] "Cosmic Detectives". The European Space Agency (ESA). 2013-04-02. Retrieved 2013-04-15.

[2] Dodelson, Scott (2003). *Modern Cosmology*. Academic Press. ISBN 0-12-219141-2.

[3] Liddle, Andrew; David Lyth (2000). *Cosmological Inflation and Large-Scale Structure*. Cambridge. ISBN 0-521-57598-2.

[4] Padmanabhan, T. (1993). *Structure formation in the universe*. Cambridge University Press. ISBN 0-521-42486-0.

[5] Peebles, P. J. E. (1980). *The Large-Scale Structure of the Universe*. Princeton University Press. ISBN 0-691-08240-5.

[6] Kolb, Edward; Michael Turner (1988). *The Early Universe*. Addison-Wesley. ISBN 0-201-11604-9.

[7] Harrison, E. R. (1970). "Fluctuations at the threshold of classical cosmology". *Phys. Rev.* **D1** (10): 2726. Bibcode:1970PhRvD...1.2726H. doi:10.1103/PhysRevD.1.2726.

[8] Peebles, P. J. E.; Yu, J. T. (1970). "Primeval adiabatic perturbation in an expanding universe". *Astrophysical Journal* **162**: 815. Bibcode:1970ApJ...162..815P. doi:10.1086/150713.

[9] Zel'dovich, Yaa B. (1972). "A hypothesis, unifying the structure and entropy of the Universe". *Monthly Notices of the Royal Astronomical Society* **160**: 1P. Bibcode:1972MNRAS.160P...1Z. doi:10.1093/mnras/160.1.1p.

[10] R. A. Sunyaev, "Fluctuations of the microwave background radiation", in *Large Scale Structure of the Universe* ed. M. S. Longair and J. Einasto, 393. Dordrecht: Reidel 1978.

[11] U. Seljak & M. Zaldarriaga (1996). "A line-of-sight integration approach to cosmic microwave background anisotropies". *Astrophys. J.* **469**: 437–444. arXiv:astro-ph/9603033. Bibcode:1996ApJ...469..437S. doi:10.1086/177793.

[12] Springel, V.; et al. (2005). "Simulations of the formation, evolution and clustering of galaxies and quasars". *Nature* **435** (7042): 629–636. arXiv:astro-ph/0504097. Bibcode:2005Natur.435..629S. doi:10.1038/nature03597. PMID 15931216.

Chapter 21

De Sitter universe

A **de Sitter universe** is a cosmological solution to the Einstein field equations of general relativity, named after Willem de Sitter. It models the universe as spatially flat and neglects ordinary matter, so the dynamics of the universe are dominated by the cosmological constant, thought to correspond to dark energy in our universe or the inflaton field in the early universe. According to the models of inflation and current observations of the accelerating universe, the concordance models of physical cosmology are converging on a consistent model where our universe was best described as a de Sitter universe at about a time $t = 10^{-33}$ seconds after the fiducial Big Bang singularity, and far into the future.

21.1 Mathematical expression

A de Sitter universe has no ordinary matter content but with a positive cosmological constant (Λ) that sets the expansion rate, H. A larger cosmological constant leads to a larger expansion rate:

$$H \propto \sqrt{\Lambda},$$

where the constants of proportionality depend on conventions.

It is common to describe a patch of this solution as an expanding universe of the FLRW form where the scale factor is given by[1]

$$a(t) = e^{Ht},$$

where the constant H is the Hubble expansion rate and t is time. As in all FLRW spaces, $a(t)$, the scale factor, describes the expansion of physical spatial distances.

Unique to universes described by the FLRW metric, a de Sitter universe has a Hubble Law that is not only consistent through all space, but also through all time (since the deceleration parameter is $q = -1$), thus satisfying the perfect cosmological principle that assumes isotropy and homogeneity throughout space and time. As a class of models with different values of the Hubble constant, the static universe that Einstein developed, and for which he invented the cosmological constant, can be considered a special case of the de Sitter universe where the expansion is finely tuned to exactly cancel out the collapse associated with the positive curvature associated with a nonzero matter density. There are ways to cast de Sitter space with static coordinates (see de Sitter space), so unlike other FLRW models, de Sitter space can be thought of as a static solution to Einstein's equations even though the geodesics followed by observers necessarily diverge as expected from the expansion of physical spatial dimensions. As a model for the universe, de Sitter's solution was not considered viable for the observed universe until models for inflation and dark energy were developed. Before then, it was assumed that the Big Bang implied only an acceptance of the weaker cosmological principle, which holds that isotropy applies spatially but not temporally.[2]

21.2 Potential for the Universe

Because our Universe entered the Dark Energy Dominated Era about five billion years ago, our universe is probably approaching a de Sitter universe in the infinite future. If the current acceleration of our universe is due to a cosmological constant then as the universe continues to expand all of the matter and radiation will be diluted. Eventually there will be almost nothing left but the vacuum energy, tiny thermal fluctuations, quantum fluctuations and our universe will have become a de Sitter universe.

21.3 Relative expansion

The exponential expansion of the scale factor means that the physical distance between any two non-accelerating observers will eventually be growing faster than the speed of light. At this point those two observers will no longer be able to make contact. Therefore, any observer in a de Sitter universe would see event horizons beyond which that observer can never see nor learn any information. If our universe is approaching a de Sitter universe then eventually we will not be able to observe any galaxies other than our own Milky Way (and any others in the gravitationally bound Local Group, assuming they were to somehow survive to that time without merging).

21.4 Modelling cosmic inflation

Another application of de Sitter space is in the early universe during cosmic inflation. Many inflationary models are approximately de Sitter space and can be modelled by giving the Hubble parameter a mild time dependence. For simplicity, some calculations involving inflation in the early universe can be performed in de Sitter space rather than a more realistic inflationary universe. By using the de Sitter universe instead, where the expansion is truly exponential, there are many simplifications.

21.5 See also

- Cosmic inflation
- De Sitter space for more mathematical properties
- Deceleration parameter
- Causal patch

21.6 References

[1] Adler, Ronald; Maurice Bazin; Menahem Schiffer (1965). *Introduction to General Relativity*. NY: McGraw-Hill. p. 468.

[2] Dodelson, Scott (2003). *Modern Cosmology* (4. [print.]. ed.). San Diego: Academic Press. ISBN 978-0-12-219141-1.

Chapter 22

Diffusion-limited aggregation

A DLA cluster grown from a copper sulfate solution in an electrodeposition cell

A Brownian tree resulting from a computer simulation

A DLA obtained by allowing random walkers to adhere to a straight line. Different colors indicate different arrival time of the random walkers.

Diffusion-limited aggregation (DLA) is the process whereby particles undergoing a random walk due to Brownian motion cluster together to form aggregates of such particles. This theory, proposed by T.A. Witten Jr. and L.M. Sander in 1981,[1] is applicable to aggregation in any system where diffusion is the primary means of transport in the system. DLA can be observed in many systems such as electrodeposition, Hele-Shaw flow, mineral deposits, and dielectric breakdown.

The clusters formed in DLA processes are referred to as Brownian trees. These clusters are an example of a fractal. In 2-D these fractals exhibit a dimension of approximately 1.71 for free particles that are unrestricted by a lattice, however computer simulation of DLA on a lattice will change the fractal dimension slightly for a DLA in the same embedding dimension. Some variations are also observed depending on the geometry of the growth, whether it be from a single point radially outward or from a plane or line for example. Two examples of aggregates generated using a microcomputer by allowing random walkers to adhere to an aggregate (originally (i) a straight line consisting 1300 particles and (ii) one particle at center) are shown on the right.

Computer simulation of DLA is one of the primary means of studying this model. Several methods are available to accomplish this. Simulations can be done on a lattice of any desired geometry of embedding dimension, in fact this has been done in up to 8 dimensions,[2] or the simulation can be done more along the lines of a standard molecular dynamics simulation where a particle is allowed to freely random walk until it gets within a certain critical range at which time

A DLA consisting about 33,000 particles obtained by allowing random walkers to adhere to a seed at the center. Different colors indicate different arrival time of the random walkers.

it is pulled onto the cluster. Of critical importance is that the number of particles undergoing Brownian motion in the system is kept very low so that only the diffusive nature of the system is present.

22.1 Artwork based on diffusion-limited aggregation

High-voltage dielectric breakdown within a block of plexiglas creates a fractal pattern called a Lichtenberg figure. The branching discharges ultimately become hairlike, but are thought to extend down to the molecular level.[3]

Sunflow rendered image of a point cloud created using toxiclibs/simutils with the DLA process applied to a spiral curve

The intricate and organic forms that can be generated with diffusion-limited aggregation algorithms have been explored by artists. Simutils, part of the toxiclibs open source library for the Java programming language developed by Karsten Schmidt, allows users to apply the DLA process to pre-defined guidelines or curves in the simulation space and via various other parameters dynamically direct the growth of 3D forms.[4]

22.2 See also

- Eden growth model

22.3 References

[1] T. A. Witten Jr, L. M. Sander, Phys. Rev. Lett. 47, 1400 (1981)

[2] R. Ball, M. Nauenberg, T. A. Witten, Phys. Rev. A 29, 2017 (1984)

[3] Bert Hickman What are Lichtenberg figures, and how do we make them? http://lichdesc.teslamania.com

[4] Schmidt, K. simutils-0001: Diffusion-limited aggregation

22.4 External links

- JavaScript based DLA
- Diffusion-Limited Aggregation: A Model for Pattern Formation
- A Java applet demonstration of DLA from Hong Kong University
- Another DLA java applet

- Free, open source program for generating DLAs using freely available ImageJ software

Chapter 23

Reaction–diffusion system

Reaction–diffusion systems are mathematical models which correspond to several physical phenomena: the most common is the change in space and time of the concentration of one or more chemical substances: local chemical reactions in which the substances are transformed into each other, and diffusion which causes the substances to spread out over a surface in space.

Reaction–diffusion systems are naturally applied in chemistry. However, the system can also describe dynamical processes of non-chemical nature. Examples are found in biology, geology and physics (neutron diffusion theory) and ecology. Mathematically, reaction–diffusion systems take the form of semi-linear parabolic partial differential equations. They can be represented in the general form

$$\partial_t q = \underline{\underline{D}} \nabla^2 q + R(q),$$

where $q(x, t)$ represents the unknown vector function, D is a diagonal matrix of diffusion coefficients, and R accounts for all local reactions. The solutions of reaction–diffusion equations display a wide range of behaviours, including the formation of travelling waves and wave-like phenomena as well as other self-organized patterns like stripes, hexagons or more intricate structure like dissipative solitons. Each function, for which a reaction diffusion differential equation holds, represents in fact a *concentration variable*.

23.1 One-component reaction–diffusion equations

The simplest reaction–diffusion equation is in one spatial dimension in plane geometry,

$$\partial_t u = D \partial_x^2 u + R(u),$$

is also referred to as the KPP (Kolmogorov-Petrovsky-Piskounov) equation.[1] If the reaction term vanishes, then the equation represents a pure diffusion process. The corresponding equation is Fick's second law. The choice $R(u) = u(1 - u)$ yields Fisher's equation that was originally used to describe the spreading of biological populations,[2] the Newell-Whitehead-Segel equation with $R(u) = u(1 - u^2)$ to describe Rayleigh-Benard convection,[3][4] the more general Zeldovich equation with $R(u) = u(1 - u)(u - \alpha)$ and $0 < \alpha < 1$ that arises in combustion theory,[5] and its particular degenerate case with $R(u) = u^2 - u^3$ that is sometimes referred to as the Zeldovich equation as well.[6]

The dynamics of one-component systems is subject to certain restrictions as the evolution equation can also be written in the variational form

$$\partial_t u = -\frac{\delta \mathcal{L}}{\delta u}$$

and therefore describes a permanent decrease of the "free energy" $\mathcal{L}$ given by the functional

$$\mathcal{L} = \int_{-\infty}^{\infty} \left[\frac{D}{2} (\partial_x u)^2 - V(u) \right] dx$$

with a potential $V(u)$ such that $R(u) = dV(u)/du$.

In systems with more than one stationary homogeneous solution, a typical solution is given by travelling fronts connecting the homogeneous states. These solutions move with constant speed without changing their shape and are of the form $u(x, t) = \hat{u}(\xi)$ with $\xi = x - ct$, where c is the speed of the travelling wave. Note that while travelling waves are generically stable structures, all non-monotonous stationary solutions (e.g. localized domains composed of a front-antifront pair) are unstable. For $c = 0$, there is a simple proof for this statement:[7] if $u_0(x)$ is a stationary solution and $u = u_0(x) + \tilde{u}(x, t)$ is an infinitesimally perturbed solution, linear stability analysis yields the equation

$$\partial_t \tilde{u} = D \partial_x^2 \tilde{u} - U(x) \tilde{u}, \qquad U(x) = -R'(u)|_{u=u_0(x)}.$$

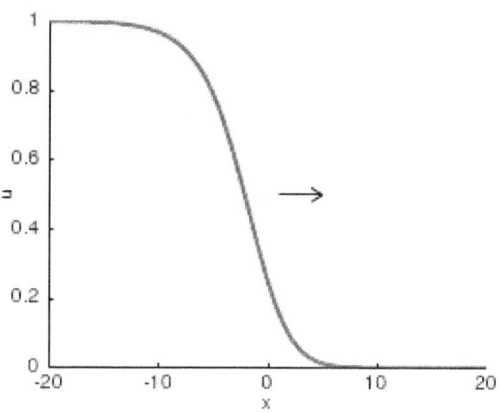

A travelling wave front solution for Fisher's equation.

With the ansatz $\bar{u} = \psi(x)\exp(-\lambda t)$ we arrive at the eigenvalue problem

$$\hat{H}\psi = \lambda\psi, \qquad \hat{H} = -D\partial_x^2 + U(x),$$

of Schrödinger type where negative eigenvalues result in the instability of the solution. Due to translational invariance $\psi = \partial x\, u_0(x)$ is a neutral eigenfunction with the eigenvalue $\lambda = 0$, and all other eigenfunctions can be sorted according to an increasing number of knots with the magnitude of the corresponding real eigenvalue increases monotonically with the number of zeros. The eigenfunction $\psi = \partial x\, u_0(x)$ should have at least one zero, and for a non-monotonic stationary solution the corresponding eigenvalue $\lambda = 0$ cannot be the lowest one, thereby implying instability.

To determine the velocity c of a moving front, one may go to a moving coordinate system and look at stationary solutions:

$$D\partial_\xi^2 \hat{u}(\xi) + c\partial_\xi \hat{u}(\xi) + R(\hat{u}(\xi)) = 0.$$

This equation has a nice mechanical analogue as the motion of a mass D with position $\hat{u}$ in the course of the "time" ξ under the force R with the damping coefficient c which allows for a rather illustrative access to the construction of different types of solutions and the determination of c.

When going from one to more space dimensions, a number of statements from one-dimensional systems can still be applied. Planar or curved wave fronts are typical structures, and a new effect arises as the local velocity of a curved front becomes dependent on the local radius of curvature (this can be seen by going to polar coordinates). This phenomenon leads to the so-called curvature-driven instability.[8]

23.2 Two-component reaction–diffusion equations

Two-component systems allow for a much larger range of possible phenomena than their one-component counterparts. An important idea that was first proposed by Alan Turing is that a state that is stable in the local system can become unstable in the presence of diffusion.[9]

A linear stability analysis however shows that when linearizing the general two-component system

$$\begin{pmatrix}\partial_t u & \partial_t v\end{pmatrix} = \begin{pmatrix}D_u & 0 \\ 0 & D_v\end{pmatrix}\begin{pmatrix}\partial_{xx}u \\ \partial_{xx}v\end{pmatrix} + \begin{pmatrix}F(u,v) \\ G(u,v)\end{pmatrix}$$

a plane wave perturbation

$$\tilde{q}_k(x,t) = \begin{pmatrix}\tilde{u}(t) \\ \tilde{v}(t)\end{pmatrix} e^{i k\cdot x}$$

of the stationary homogeneous solution will satisfy

$$\begin{pmatrix}\partial_t \tilde{u}_k(t) \\ \partial_t \tilde{v}_k(t)\end{pmatrix} = -k^2 \begin{pmatrix}D_u \tilde{u}_k(t) \\ D_v \tilde{v}_k(t)\end{pmatrix} + R'\begin{pmatrix}\tilde{u}_k(t) \\ \tilde{v}_k(t)\end{pmatrix}.$$

Turing's idea can only be realized in four equivalence classes of systems characterized by the signs of the Jacobian R' of the reaction function. In particular, if a finite wave vector k is supposed to be the most unstable one, the Jacobian must have the signs

$$\begin{pmatrix}+ & - \\ + & -\end{pmatrix}, \quad \begin{pmatrix}+ & + \\ - & -\end{pmatrix}, \quad \begin{pmatrix}- & + \\ - & +\end{pmatrix}, \quad \begin{pmatrix}- & - \\ + & +\end{pmatrix}.$$

This class of systems is named *activator-inhibitor system* after its first representative: close to the ground state, one component stimulates the production of both components while the other one inhibits their growth. Its most prominent representative is the FitzHugh–Nagumo equation

$$\partial_t u = d_u^2 \nabla^2 u + f(u) - \sigma v,$$
$$\tau \partial_t v = d_v^2 \nabla^2 v + u - v$$

with $f(u) = \lambda u - u^3 - \kappa$ which describes how an action potential travels through a nerve.[10][11] Here, d_u, d_v, τ, σ and λ are positive constants.

When an activator-inhibitor system undergoes a change of parameters, one may pass from conditions under which a homogeneous ground state is stable to conditions under which it is linearly unstable. The corresponding bifurcation may be either a Hopf bifurcation to a globally oscillating homogeneous state with a dominant wave number $k = 0$ or a *Turing bifurcation* to a globally patterned state with a dominant finite wave number. The latter in two spatial dimensions typically leads to stripe or hexagonal patterns.

- Subcritical Turing bifurcation: formation of a hexagonal pattern from noisy initial conditions in the above two-component reaction-diffusion system of Fitzhugh-Nagumo type.
- Noisy initial conditions at $t = 0$.
- State of the system at $t = 10$.
- Almost converged state at $t = 100$.

For the Fitzhugh-Nagumo example, the neutral stability curves marking the boundary of the linearly stable region for the Turing and Hopf bifurcation are given by

$$q_n^H(k): \quad \frac{1}{\tau} + \left(d_u^2 + \frac{1}{\tau}d_v^2\right)k^2 = f'(u_h),$$

$$q_n^T(k): \quad \frac{\kappa}{1 + d_v^2 k^2} + d_u^2 k^2 = f'(u_h).$$

If the bifurcation is subcritical, often localized structures (dissipative solitons) can be observed in the hysteretic region where the pattern coexists with the ground state. Other frequently encountered structures comprise pulse trains (also known as periodic travelling waves), spiral waves and target patterns. These three solution types are also generic features of two- (or more-) component reaction-diffusion equations in which the local dynamics have a stable limit cycle[12]

- Other patterns found in the above two-component reaction-diffusion system of Fitzhugh-Nagumo type.
- Rotating spiral.
- Target pattern.
- Stationary localized pulse (dissipative soliton).

23.3 Three- and more-component reaction–diffusion equations

For a variety of systems, reaction-diffusion equations with more than two components have been proposed, e.g. as models for the Belousov-Zhabotinsky reaction,[13] for blood clotting[14] or planar gas discharge systems.[15]

It is known that systems with more components allow for a variety of phenomena not possible in systems with one or two components (e.g. stable running pulses in more than one spatial dimension without global feedback),.[16] An introduction and systematic overview of the possible phenomena in dependence on the properties of the underlying system is given in.[17]

23.4 Applications and universality

In recent times, reaction–diffusion systems have attracted much interest as a prototype model for pattern formation. The above-mentioned patterns (fronts, spirals, targets, hexagons, stripes and dissipative solitons) can be found in various types of reaction-diffusion systems in spite of large discrepancies e.g. in the local reaction terms. It has also been argued that reaction-diffusion processes are an essential basis for processes connected to morphogenesis in biology[18] and may even be related to animal coats and skin pigmentation.[19][20]

See also: The chemical basis of morphogenesis

Other applications of reaction-diffusion equations include ecological invasions,[21] spread of epidemics,[22] tumour growth[23][24][25] and wound healing.[26] Another reason for the interest in reaction-diffusion systems is that although they are nonlinear partial differential equations, there are often possibilities for an analytical treatment.[7][8][27][28][29]

23.5 Experiments

Well-controllable experiments in chemical reaction-diffusion systems have up to now been realized in three ways. First, gel reactors[30] or filled capillary tubes[31] may be used. Second, temperature pulses on catalytic surfaces have been investigated.[32][33] Third, the propagation of running nerve pulses is modelled using reaction-diffusion systems.[10][34]

Aside from these generic examples, it has turned out that under appropriate circumstances electric transport systems like plasmas[35] or semiconductors[36] can be described in a reaction-diffusion approach. For these systems various experiments on pattern formation have been carried out.

23.6 Numerical treatments

A reaction-diffusion system can be solved by using methods of numerical mathematics. There are existing several numerical treatments in research literature.[37] Also for complex geometries numerical solution methods are proposed.[38][39]

23.7 See also

- Autowave
- Diffusion-controlled reaction
- Chemical kinetics
- Phase space method
- Autocatalytic reactions and order creation
- Pattern formation
- Patterns in nature
- Periodic travelling wave
- Stochastic geometry
- MClone

23.8 Some examples of reaction-diffusion equations

- Fisher equation
- Fisher-Kolmogorov equation

23.9 References

[1] A. Kolmogorov et al., Moscow Univ. Bull. Math. A 1 (1937): 1

[2] R. A. Fisher, Ann. Eug. 7 (1937): 355

[3] A. C. Newell and J. A. Whitehead, J. Fluid Mech. 38 (1969): 279

[4] L. A. Segel, J. Fluid Mech. 38 (1969): 203

[5] Y. B. Zeldovich and D. A. Frank-Kamenetsky, Acta Physicochim. 9 (1938): 341

[6] B. H. Gilding and R. Kersner, Travelling Waves in Nonlinear Diffusion Convection Reaction, Birkhäuser (2004)

[7] P. C. Fife, Mathematical Aspects of Reacting and Diffusing Systems, Springer (1979)

[8] A. S. Mikhailov, Foundations of Synergetics I. Distributed Active Systems, Springer (1990)

[9] A. M. Turing, Phil. Transact. Royal Soc. B 237 (1952): 37

[10] R. FitzHugh, Biophys. J. 1 (1961): 445

[11] J. Nagumo et al., Proc. Inst. Radio Engin. Electr. 50 (1962): 2061

[12] N. Kopell and L.N. Howard, Stud. Appl. Math. 52 (1973): 291

[13] V. K. Vanag and I. R. Epstein, Phys. Rev. Lett. 92 (2004): 128301

[14] E. S. Lobanova and F. I. Ataullakhanov, Phys. Rev. Lett. 93 (2004): 098303

[15] H.-G. Purwins et al. in: Dissipative Solitons, Lectures Notes in Physics, Ed. N. Akhmediev and A. Ankiewicz, Springer (2005)

[16] C. P. Schenk et al., Phys. Rev. Lett. 78 (1997): 3781

[17] A. W. Liehr: *Dissipative Solitons in Reaction Diffusion Systems. Mechanism, Dynamics, Interaction.* Volume 70 of Springer Series in Synergetics, Springer, Berlin Heidelberg 2013, ISBN 978-3-642-31250-2

[18] L.G. Harrison, Kinetic Theory of Living Pattern, Cambridge University Press (1993)

[19] H. Meinhardt, Models of Biological Pattern Formation, Academic Press (1982)

[20] Murray, James D. (9 March 2013). *Mathematical Biology*. Springer Science & Business Media. pp. 436–450. ISBN 978-3-662-08539-4.

[21] E.E. Holmes et al, Ecology 75 (1994): 17

[22] J.D. Murray et al, Proc. R. Soc. Lond. B 229 (1986: 111

[23] M.A.J. Chaplain J. Bio. Systems 3 (1995): 929

[24] J.A. Sherratt and M.A. Nowak, Proc. R. Soc. Lond. B 248 (1992): 261

[25] R.A. Gatenby and E.T. Gawlinski, Cancer Res. 56 (1996): 5745

[26] J.A. Sherratt and J.D. Murray, Proc. R. Soc. Lond. B 241 (1990): 29

[27] P. Grindrod, Patterns and Waves: The Theory and Applications of Reaction-Diffusion Equations, Clarendon Press (1991)

[28] J. Smoller, Shock Waves and Reaction Diffusion Equations, Springer (1994)

[29] B. S. Kerner and V. V. Osipov, Autosolitons. A New Approach to Problems of Self-Organization and Turbulence, Kluwer Academic Publishers (1994)

[30] K.-J. Lee et al., Nature 369 (1994): 215

[31] C. T. Hamik and O. Steinbock, New J. Phys. 5 (2003): 58

[32] H. H. Rotermund et al., Phys. Rev. Lett. 66 (1991): 3083

[33] M. D. Graham et al., J. Phys. Chem. 97 (1993): 7564

[34] A. L. Hodgkin and A. F. Huxley, J. Physiol. 117 (1952): 500

[35] M. Bode and H.-G. Purwins, Physica D 86 (1995): 53

[36] E. Schöll, Nonlinear Spatio-Temporal Dynamics and Chaos in Semiconductors, Cambridge University Press (2001)

[37] S.Tang et al., J.Austral.Math.Soc. Ser.B 35(1993): 223-243

[38] Isaacson, Samuel A.; Peskin, Charles S. (2006). "Incorporating Diffusion in Complex Geometries into Stochastic Chemical Kinetics Simulations". *SIAM J. Sci. Comput.* **28** (1): 47–74. doi:10.1137/040605060.

[39] Linker, Patrick (2016). "Numerical methods for solving the reactive diffusion equation in complex geometries". *The Winnower*.

23.10 External links

- Java applet showing a reaction–diffusion simulation

- Another applet showing Gray-Scott reaction-diffusion.

- Java applet Uses reaction-diffusion to simulate pattern formation in several snake species.

- TexRD software random texture generator based on reaction-diffusion for graphists and scientific use

- Reaction-Diffusion by the Gray-Scott Model: Pearson's parameterization a visual map of the parameter space of Gray-Scott reaction diffusion.

- A Thesis on reaction-diffusion patterns with an overview of the field

- ReDiLab - Reaction Diffusion Laboratory Flash & GPU based application simulating Belousov-Zhabotinsky, Gray Scott, Willamowski–Rössler and FitzHugh-Nagumo with full source code.

- ReDiLab:Node Node based UI for coupling multiple reaction diffusion systems together supporting Belousov-Zhabotinsky, Gray Scott, Willamowski–Rössler and FitzHugh-Nagumo.

- Turing, biology and pattern formation

Chapter 24

Molecular self-assembly

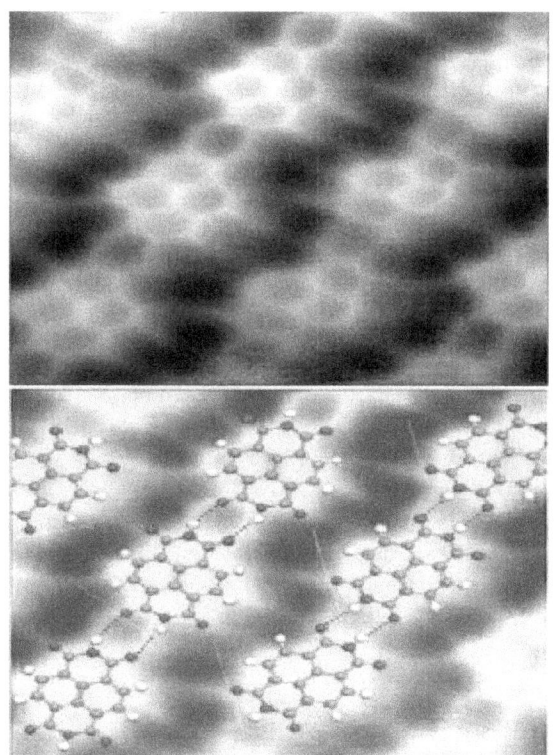

AFM image of napthalenetetracarboxylic diimide molecules on silver interacting via hydrogen bonding (77 K).

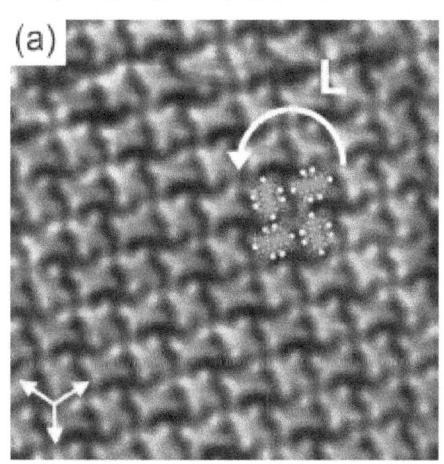

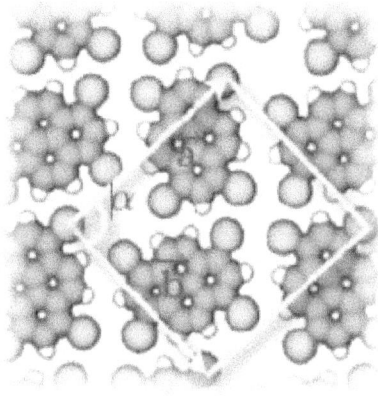

image of self-assembled Br_4-pyrene molecules on Au(111) surface (top) and its model (bottom; pink spheres are Br atoms).[2]

Molecular self-assembly is the process by which molecules adopt a defined arrangement without guidance or management from an outside source. There are two types of self-assembly. These are **intramolecular** self-assembly and **intermolecular** self-assembly. Commonly, the term molecular self-assembly refers to intermolecular self-assembly, while the intramolecular analog is more commonly called folding.

24.1 Supramolecular systems

Molecular self-assembly is a key concept in supramolecular chemistry.[3][4][5] This is because assembly of molecules in such systems is directed through noncovalent interactions (e.g., hydrogen bonding, metal coordination, hydrophobic forces, van der Waals forces, π-π interactions, and/or electrostatic) as well as electromagnetic interactions. Common examples include the formation of micelles, vesicles, liquid crystal phases, and Langmuir monolayers by surfactant molecules.[6] Further examples of supramolecular assemblies demonstrate that a variety of different shapes and sizes can be obtained using molecular self-assembly.[7]

Molecular self-assembly allows the construction of challenging molecular topologies. One example is Borromean rings, interlocking rings wherein removal of one ring unlocks each of the other rings. DNA has been used to prepare a molecular analog of Borromean rings.[8] More recently, a similar structure has been prepared using non-biological building blocks.[9] While a mechanistic understanding of how supramolecular self-assembly occurs remains largely unknown, both experimental and theoretic work has been pursued on this topic. [10] [11]

24.2 Biological systems

Molecular self-assembly underlies the construction of biologic macromolecular assemblies in living organisms, and so is crucial to the function of cells. It is exhibited in the self-assembly of lipids to form the membrane, the formation of double helical DNA through hydrogen bonding of the individual strands, and the assembly of proteins to form quaternary structures. Molecular self-assembly of incorrectly folded proteins into insoluble amyloid fibers is responsible for infectious prion-related neurodegenerative diseases. Molecular self-assembly of nanoscale structures plays a role in the growth of the remarkable β-keratin lamellae/setae/spatulae structures used to give geckos the ability to climb walls and adhere to ceilings and rock overhangs.[12][13]

24.3 Nanotechnology

Molecular self-assembly is an important aspect of bottom-up approaches to nanotechnology. Using molecular self-assembly the final (desired) structure is programmed in the shape and functional groups of the molecules. Self-assembly is referred to as a 'bottom-up' manufacturing technique in contrast to a 'top-down' technique such as lithography where the desired final structure is carved from a larger block of matter. In the speculative vision of molecular nanotechnology, microchips of the future might be made by molecular self-assembly. An advantage to constructing nanostructure using molecular self-assembly for biological materials is that they will degrade back into individual molecules that can be broken down by the body.

24.3.1 DNA nanotechnology

Main article: DNA nanotechnology

DNA nanotechnology is an area of current research that uses the bottom-up, self-assembly approach for nanotechnological goals. DNA nanotechnology uses the unique molecular recognition properties of DNA and other nucleic acids to create self-assembling branched DNA complexes with useful properties.[14] DNA is thus used as a structural material rather than as a carrier of biological information, to make structures such as two-dimensional periodic lattices (both tile-based as well as using the "DNA origami" method) and three-dimensional structures in the shapes of polyhedra.[15] These DNA structures have also been used as templates in the assembly of other molecules such as gold nanoparticles[16] and streptavidin proteins.[17]

24.4 Two-dimensional monolayers

Main article: Monolayer

The spontaneous assembly of a single layer of molecules at interfaces is usually referred to as two-dimensional self-assembly. Early direct proofs showing that molecules can assembly into higher-order architectures at solid interfaces came with the development of scanning tunneling microscopy and shortly thereafter.[18] Eventually two strategies became popular for the self-assembly of 2D architectures, namely self-assembly following ultra-high-vacuum deposition and annealing and self-assembly at the solid-liquid interface.[19] The design of molecules and conditions leading to the formation of highly-crystalline architectures is considered today a form of 2D crystal engineering at the nanoscopic scale.

24.5 See also

- Supramolecular assembly
- Foldamer
- Macromolecular assembly
- Ice-nine
- Self-assembly of nanoparticles

24.6 References

[1] Sweetman, A. M.; Jarvis, S. P.; Sang, Hongqian; Lekkas, I.; Rahe, P.; Wang, Yu; Wang, Jianbo; Champness, N.R.; Kantorovich, L.; Moriarty, P. (2014). "Mapping the force field of a hydrogen-bonded assembly". *Nature Communications* 5. doi:10.1038/ncomms4931.

[2] Pham, Tuan Anh; Song, Fei; Nguyen, Manh-Thuong; Stöhr, Meike (2014). "Self-assembly of pyrene derivatives on Au(111): Substituent effects on intermolecular interactions". *Chem. Commun* **50** (91): 14089. doi:10.1039/C4CC02753A.

[3] Lehn, J.-M. (1988). "Perspectives in Supramolecular Chemistry-From Molecular Recognition towards Molecular Information Processing and Self-Organization". *Angew. Chem. Int. Ed. Engl.* **27** (11): 89–121. doi:10.1002/anie.198800891.

[4] Lehn, J.-M. (1990). "Supramolecular Chemistry-Scope and Perspectives: Molecules, Supermolecules, and Molecular Devices (Nobel Lecture)". *Angew. Chem. Int. Ed. Engl.* **29** (11): 1304–1319. doi:10.1002/anie.199013041.

[5] Lehn, J.-M. *Supramolecular Chemistry: Concepts and Perspectives*. Wiley-VCH. ISBN 978-3-527-29311-7.

[6] Rosen, Milton J. (2004). *Surfactants and interfacial phenomena*. Hoboken, NJ: Wiley-Interscience. ISBN 978-0-471-47818-8.

[7] Ariga, Katsuhiko; Hill, Jonathan P; Lee, Michael V; Vinu, Ajayan; Charvet, Richard; Acharya, Somobrata (2008). "Challenges and breakthroughs in recent research on self-assembly". *Science and Technology of Advanced Materials* (free-download review) **9**: 014109. Bibcode:2008STAdM...9a4109A. doi:10.1088/1468-6996/9/1/014109.

[8] Mao, C; Sun, W; Seeman, N. C. (1997). "Assembly of Borromean rings from DNA". *Nature* **386** (6621): 137–138. Bibcode:1997Natur.386..137M. doi:10.1038/386137b0. PMID 9062186.

[9] Chichak, K. S.; Cantrill, S. J.; Pease, A. R.; Chiu, S. H.; Cave, G. W.; Atwood, J. L.; Stoddart, J. F. (2004). "Molecular Borromean Rings". *Science* **304** (5675): 1308–1312. Bibcode:2004Sci...304.1308C. doi:10.1126/science.1096914. PMID 15166376.

[10] Larsen, Randy (2008). "How fast do metal organic polyhedra form in solution? Kinetics of [Cu2(5-OH-bdc)2L2]12 formation in methanol". *Journal of the American Chemical Society* **130** (34): 11246–11247. doi:10.1021/ja802605v.

[11] Alkordi, Mohamed; Belof, Jonathan; Rivera, Edwin; Wojtas, Lukasz; Eddaoudi, Mohamed (2011). "Insight into the construction of metal–organic polyhedra: metal–organic cubes as a case study". *Chemical Science* **2**: 1695–1705. doi:10.1039/C1SC00269D.

[12] Min, Younjin; et al. (2008). "The role of interparticle and external forces in nanoparticle assembly". *Nature Materials* **7** (7): 527–38. Bibcode:2008NatMa...7..527M. doi:10.1038/nmat2206. PMID 18574482.

[13] Santos, Daniel; Spenko, Matthew; Parness, Aaron; Kim, Sangbae; Cutkosky, Mark (2007). "Directional adhesion for climbing: theoretical and practical considerations". *Journal of Adhesion Science and Technology* **21** (12–13): 1317–1341. doi:10.1163/156856107782328399. Gecko "feet and toes are a hierarchical system of complex structures consisting of lamellae, setae, and spatulae. The distinguishing characteristics of the gecko adhesion system have been described [as] (1) anisotropic attachment, (2) high pulloff force to preload ratio, (3) low detachment force, (4) material independence, (5) self-cleaning, (6) anti-self sticking and (7) non-sticky default state. ... The gecko's adhesive structures are made from β-keratin (modulus of elasticity [approx.] 2 GPa). Such a stiff material is not inherently sticky; however, because of the gecko adhesive's hierarchical nature and extremely small distal features (spatulae are [approx.] 200 nm in size), the gecko's foot is able to intimately conform to the surface and generate significant attraction using van der Waals forces.

[14] Seeman, N. C. (2003). "DNA in a material world". *Nature* **421** (6921): 427–431. Bibcode:2003Natur.421..427S. doi:10.1038/nature01406. PMID 12540916.

[15] Chen, J. & Seeman, N. C. (1991). "Synthesis from DNA of a molecule with the connectivity of a cube". *Nature* **350** (6319): 631–633. Bibcode:1991Natur.350..631C. doi:10.1038/350631a0. PMID 2017259.

[16] Mirkin, C. A.; Letsinger, R. L.; Mucic, R. C.; Storhoff, J. J. (1996). "A DNA-based method for rationally assembling nanoparticles into macroscopic materials". *Nature* **382** (6592): 607–609. Bibcode:1996Natur.382..607M. doi:10.1038/382607a0. PMID 8757129.

[17] Yan, H; Park, S. H.; Finkelstein, G; Reif, J. H.; Labean, T. H. (2003). "DNA-Templated Self-Assembly of Protein Arrays and Highly Conductive Nanowires". *Science* **301** (5641): 1882–1884. Bibcode:2003Sci...301.1882Y. doi:10.1126/science.1089389. PMID 14512621.

[18] Foster, J. S. & Frommer, J. E. (1988). "Imaging of liquid crystals using a tunnelling microscope". *Nature* **333** (6173): 542–545. Bibcode:1988Natur.333..542F. doi:10.1038/333542a0.

[19] Rabe, J.P. & Buchholz, S. (1991). "Commensurability and Mobility in Two-Dimensional Molecular Patterns on Graphite". *Science* **353** (5018): 424–427. Bibcode:1991Sci...253..424R. doi:10.1126/science.253.5018.424. JSTOR 2878886.

24.7 External and further reading

- "Challenges and breakthroughs in recent research on self-assembly" Sci. Technol. Adv. Mater. 9 No 1(2008) 014109 (96 pages) free download

- G Kurth, Dirk (2008). "Metallo-supramolecular modules as a paradigm for materials science". *Science and*

24.7. EXTERNAL AND FURTHER READING

- *Technology of Advanced Materials* (free-download review) **9**: 014103. Bibcode:2008STAdM...9a4103G. doi:10.1088/1468-6996/9/1/014103.

- Bureekaew, Sareeya; Shimomura, Satoru; Kitagawa, Susumu (2008). "Chemistry and application of flexible porous coordination polymers". *Science and Technology of Advanced Materials* (free-download review) **9**: 014108. Bibcode:2008STAdM...9a4108B. doi:10.1088/1468-6996/9/1/014108.

- H.E. Hoster, M. Roos, A. Breitruck, C. Meier, K. Tonigold, T. Waldmann, U. Ziener, K. Landfester, R.J. Behm, *Structure Formation in Bis(terpyridine)Derivative Adlayers – Molecule-Substrate vs. Molecule-Molecule Interactions*, Langmuir 23 (2007) 11570

- Molecular Self-Assembly papers

- Beyond molecules: Self-assembly of mesoscopic and macroscopic components

- Whitesides, G. M. & Grzyboski, B. (2002) Science 295, 2418–2421.

- Rothemund P, Papadakis N, Winfree E (2004). "Algorithmic Self-Assembly of DNA Sierpinski Triangles". *PLoS Biol* **2** (12): 12. doi:10.1371/journal.pbio.0020424. PMC 534809. PMID 15583715.

- C2 Wiki: Self Assembly from a computer programming perspective.

- Structure and Dynamics of Organic Nanostructures

- Metal organic coordination networks of oligopyridines and Cu on graphite

Chapter 25

Autocatalysis

A single chemical reaction is said to have undergone **autocatalysis**, or be **autocatalytic**, if one of the reaction products is also a reactant and therefore a catalyst in the same or a coupled reaction.[1] The reaction is called an **autocatalytic reaction**.

The rate equations for autocatalytic reactions are fundamentally nonlinear. This nonlinearity can lead to the spontaneous generation of order. A dramatic example of this order is that which is found in living systems. The spontaneous order creation corresponds to a decrease in the entropy of the system, which must be compensated by a larger increase in the entropy of the surroundings in order to satisfy the Second Law of Thermodynamics.

A *set* of chemical reactions can be said to be "collectively autocatalytic" if a number of those reactions produce, as reaction products, catalysts for enough of the other reactions that the entire set of chemical reactions is self-sustaining given an input of energy and food molecules (see autocatalytic set).

25.1 Chemical reactions

Main articles: Chemical reaction and Chemical kinetics

A chemical reaction of two reactants and two products can be written as

$$\alpha A + \beta B \rightleftharpoons \sigma S + \tau T$$

where the Greek letters are stoichiometric coefficients and the capital Latin letters represent chemical species. The chemical reaction proceeds in both the forward and reverse direction. This equation is easily generalized to any number of reactants, products, and reactions.

25.1.1 Chemical equilibrium

In chemical equilibrium the forward and reverse reaction rates are such that each chemical species is being created at the same rate it is being destroyed. In other words, the rate of the forward reaction is equal to the rate of the reverse reaction.

$$k_+[A]^\alpha[B]^\beta = k_-[S]^\sigma[T]^\tau$$

Here, the brackets indicate the concentration of the chemical species, in moles per liter, and k_+ and k_- are rate constants.

25.1.2 Far from equilibrium

Far from equilibrium, the forward and reverse reaction rates no longer balance and the concentration of reactants and products is no longer constant. For every forward reaction α molecules of A are destroyed. For every reverse reaction α molecules of A are created. In the case of an elementary reaction step the reaction order in each direction equals the molecularity, so that the rate of change in number of moles of A is then

$$\frac{d}{dt}[A] = -\alpha k_+[A]^\alpha[B]^\beta + \alpha k_-[S]^\sigma[T]^\tau$$

$$\frac{d}{dt}[B] = -\beta k_+[A]^\alpha[B]^\beta + \beta k_-[S]^\sigma[T]^\tau$$

$$\frac{d}{dt}[S] = \sigma k_+[A]^\alpha[B]^\beta - \sigma k_-[S]^\sigma[T]^\tau$$

$$\frac{d}{dt}[T] = \tau k_+[A]^\alpha[B]^\beta - \tau k_-[S]^\sigma[T]^\tau$$

This system of equations has a single stable fixed point when the forward rates and the reverse rates are equal. This means that the system evolves to the equilibrium state, and this is the only state to which it evolves.

25.1.3 Autocatalytic reactions

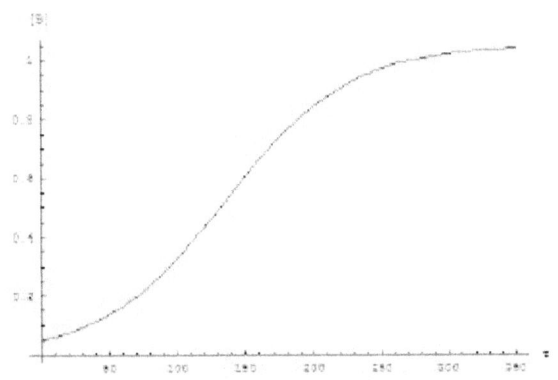

Sigmoid variation of product concentration in autocatalytic reactions

Autocatalytic reactions are those in which at least one of the products is a reactant. Perhaps the simplest autocatalytic reaction can be written[1]

$$A + B \rightleftharpoons 2B$$

with the rate equations (for an elementary reaction)

$$\frac{d}{dt}[A] = -k_+[A][B] + k_-[B]^2$$

$$\frac{d}{dt}[B] = +k_+[A][B] - k_-[B]^2$$

This reaction is one in which a molecule of species A interacts with a molecule of species B. The A molecule is converted into a B molecule. The final product consists of the original B molecule plus the B molecule created in the reaction.

The key feature of these rate equations is that they are nonlinear; the second term on the right varies as the square of the concentration of B. This feature can lead to multiple fixed points of the system, much like a quadratic equation can have multiple roots. Multiple fixed points allow for multiple states of the system. A system existing in multiple macroscopic states is more orderly (has lower entropy) than a system in a single state.

The concentrations of A and B vary in time according to[1][2]

$$[A] = \frac{[A]_0 + [B]_0}{1 + \frac{[B]_0}{[A]_0} e^{([A]_0+[B]_0)kt}}$$

and

$$[B] = \frac{[A]_0 + [B]_0}{1 + \frac{[A]_0}{[B]_0} e^{-([A]_0+[B]_0)kt}}$$

The graph for these equations is a sigmoid curve, which is typical for autocatalytic reactions: these chemical reactions proceed slowly at the start (the induction period) because there is little catalyst present, the rate of reaction increases progressively as the reaction proceeds as the amount of catalyst increases and then it again slows down as the reactant concentration decreases. If the concentration of a reactant or product in an experiment follows a sigmoid curve, the reaction may be autocatalytic.

These kinetic equations apply for example to the acid-catalyzed hydrolysis of some esters to carboxylic acids and alcohols.[2] There must be at least some acid present initially to start the catalyzed mechanism; if not the reaction must start by an alternate uncatalyzed path which is usually slower. The above equations for the catalyzed mechanism would imply that the concentration of acid product remains zero forever.[2]

25.2 Creation of order

25.2.1 Background

The Second Law of Thermodynamics states that the disorder (entropy) of a physical or chemical system and its surroundings (a closed system) must increase with time. Systems left to themselves become increasingly random, and orderly energy of a system like uniform motion degrades eventually to the random motion of particles in a heat bath.

There are, however, many instances in which physical systems spontaneously become emergent or orderly. For example, despite the destruction they cause, hurricanes have a very orderly vortex motion when compared to the random motion of the air molecules in a closed room. Even more spectacular is the order created by chemical systems; the most dramatic being the order associated with life.

This is consistent with the Second Law, which requires that the total disorder of a system *and its surroundings* must increase with time. Order can be created in a system by an even greater decrease in order of the systems surroundings.[3] In the hurricane example, hurricanes are formed from unequal heating within the atmosphere. The Earth's atmosphere is then far from thermal equilibrium. The order of the Earth's atmosphere increases, but at the expense of the order of the sun. The sun is becoming more disorderly as it ages and throws off light and material to the rest of the universe. The total disorder of the sun and the earth increases despite the fact that orderly hurricanes are generated on earth.

A similar example exists for living chemical systems. The sun provides energy to green plants. The green plants are food for other living chemical systems. The energy absorbed by plants and converted into chemical energy generates a system on earth that is orderly and far from chemical equilibrium. Here, the difference from chemical equilibrium is determined by an excess of reactants over the equilibrium amount. Once again, order on earth is generated at the expense of entropy increase of the sun. The total entropy of the earth and the rest of the universe increases, consistent with the Second Law.

Some autocatalytic reactions also generate order in a system at the expense of its surroundings. For example, (clock reactions) have intermediates whose concentrations oscillate in time, corresponding to temporal order. Other reactions generate spatial separation of chemical species corresponding to spatial order. More complex reactions are involved in metabolic pathways and metabolic networks in biological systems.

The transition to order as the distance from equilibrium increases is not usually continuous. Order typically appears abruptly. The threshold between the disorder of chemical equilibrium and order is known as a phase transition. The conditions for a phase transition can be determined with the mathematical machinery of non-equilibrium thermodynamics.

25.2.2 Temporal order

A chemical reaction cannot oscillate about a position of final equilibrium because the second law of thermodynamics requires that a thermodynamic system approach equilibrium and not recede from it. For a closed system at constant temperature and pressure, the Gibbs free energy must decrease continuously and not oscillate. However it is possible that the concentrations of some reaction intermediates oscillate, and also that the *rate* of formation of products oscillates.[4]

Idealized example: Lotka-Volterra equation

Consider a coupled set of two autocatalytic reactions in which the concentration of one of the reactants A is much larger than its equilibrium value. In this case the forward reaction rate is so much larger than the reverse rates that we can neglect the reverse rates.

$A + X \to 2X$

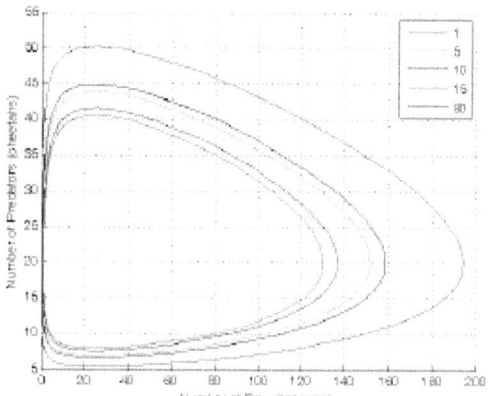

The Lotka-Volterra equation is isomorphic with the predator prey model and the two reaction autocatalytic model. In this example baboons and cheetahs are equivalent to two different chemical species X and Y in autocatalytic reactions.

$X + Y \to 2Y$

$Y \to E$

with the rate equations

$$\frac{d}{dt}[X] = k_1[A][X] - k_2[X][Y]$$

$$\frac{d}{dt}[Y] = k_2[X][Y] - k_3[Y]$$

Here, we have neglected the depletion of the reactant A, since its concentration is so large. The rate constants for the three reactions are k_1, k_2, and k_3, respectively.

This system of rate equations is known as the Lotka-Volterra equation and is most closely associated with population dynamics in predator-prey relationships. This system of equations can yield oscillating concentrations of the reaction intermediates X and Y. The amplitude of the oscillations depends on the concentration of A (which decreases without oscillation). Such oscillations are a form of emergent temporal order that is not present in equilibrium.

Another idealized example: Brusselator

Another example of a system that demonstrates temporal order is the Brusselator (see Prigogine reference). It is characterized by the reactions

$A \to X$

$2X + Y \to 3X$

25.3. BIOLOGICAL EXAMPLE

$B + X \rightarrow Y + D$

$X \rightarrow E$

with the rate equations

$$\frac{d}{dt}[X] = [A] + [X]^2[Y] - [B][X] - [X]$$

$$\frac{d}{dt}[Y] = [B][X] - [X]^2[Y]$$

where, for convenience, the rate constants have been set to 1.

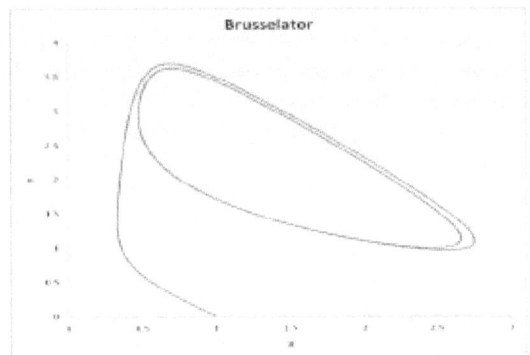

The Brusselator in the unstable regime . A=1. B=2.5. X(0)=1.Y(0)=0.The system approaches a limit cycle . For B <1+A the sys-tem is stable and approaches a fixed point.

The Brusselator has a fixed point at

$[X] = A$

$[Y] = \frac{B}{A}$

The fixed point becomes unstable when

$B > 1 + A^2$

leading to an oscillation of the system. Unlike the Lotka-Volterra equation, the oscillations of the Brusselator do not depend on the amount of reactant present initially. Instead, after sufficient time, the oscillations approach a limit cycle.[5]

Real examples

Real examples of clock reactions are the Belousov-Zhabotinsky reaction (BZ reaction), the Briggs-Rauscher reaction, the Bray-Liebhafsky reaction and the iodine clock reaction. These are oscillatory reactions, and the concentration of products and reactants can be approximated in terms of damped oscillations.

The best-known reaction, the BZ reaction, can be created with a mixture of potassium bromate, malonic acid, and manganese sulfate prepared in a heated solution of sulfuric acid.[6]

25.2.3 Spatial order

An idealized example of spatial spontaneous symmetry breaking is the case in which we have two boxes of material separated by a permeable membrane so that material can diffuse between the two boxes. It is assumed that identical Brusselators are in each box with nearly identical initial conditions. (see Prigogine reference)

$$\frac{d}{dt}[X_1] = [A] + [X_1]^2[Y_1] - [B][X_1] - [X_1] + D_x(X_2 - X_1)$$

$$\frac{d}{dt}[Y_1] = [B][X_1] - [X_1]^2[Y_1] + D_y(Y_2 - Y_1)$$

$$\frac{d}{dt}[X_2] = [A] + [X_2]^2[Y_2] - [B][X_2] - [X_2] + D_x(X_1 - X_2)$$

$$\frac{d}{dt}[Y_2] = [B][X_2] - [X_2]^2[Y_2] + D_y(Y_1 - Y_2)$$

Here, the numerical subscripts indicate which box the material is in. There are additional terms proportional to the diffusion coefficient D that account for the exchange of material between boxes.

If the system is initiated with the same conditions in each box, then a small fluctuation will lead to separation of materials between the two boxes. One box will have a predominance of X, and the other will have a predominance of Y.

25.3 Biological example

It is known that an important metabolic cycle, glycolysis, displays temporal order.[7] Glycolysis consists of the degradation of one molecule of glucose and the overall production of two molecules of ATP. The process is therefore of great importance to the energetics of living cells. The global glycolysis reaction involves glucose, ADP, NAD, pyruvate, ATP, and NADH.

glucose+2ADP+2P$_i$+2NAD $\rightarrow$ 2(pyruvate)+2ATP+2NADH

The details of the process are quite involved, however, a section of the process is autocatalyzed by phosphofructokinase (PFK). This portion of the process is responsible for oscillations in the pathway that lead to the process oscillating between an active and an inactive form. Thus, the autocatalytic reaction can modulate the process.

25.4 Phase transitions

The initial amounts of reactants determine the distance from chemical equilibrium of the system. The greater the initial concentrations the further the system is from equilibrium. As the initial concentration increases, an abrupt change in order occurs. This abrupt change is known as phase transition. At the phase transition, fluctuations in macroscopic quantities, such as chemical concentrations, increase as the system oscillates between the more ordered state (lower entropy, such as ice) and the more disordered state (higher entropy, such as liquid water). Also, at the phase transition, macroscopic equations, such as the rate equations, fail. Rate equations can be derived from microscopic considerations. The derivations typically rely on a mean field theory approximation to microscopic dynamical equations. Mean field theory breaks down in the presence of large fluctuations (see Mean field theory article for a discussion). Therefore, since large fluctuations occur in the neighborhood of a phase transition, macroscopic equations, such as rate equations, fail. As the initial concentration increases further, the system settles into an ordered state in which fluctuations are again small. (see Prigogine reference)

25.5 Asymmetric autocatalysis

Asymmetric autocatalysis occurs when the reaction product is chiral and thus acts as a chiral catalyst for its own production. Reactions of this type, such as the Soai reaction, have the property that they can amplify a very small enantiomeric excess into a large one. This has been proposed as an important step in the origin of biological homochirality.[8]

25.6 Role in origin of life

Main article: Abiogenesis

In 1995 Stuart Kauffman proposed that life initially arose as autocatalytic chemical networks.[9]

British ethologist Richard Dawkins wrote about autocatalysis as a potential explanation for abiogenesis in his 2004 book *The Ancestor's Tale*. He cites experiments performed by Julius Rebek and his colleagues at the Scripps Research Institute in California in which they combined amino adenosine and pentafluorophenyl ester with the autocatalyst amino adenosine triacid ester (AATE). One system from the experiment contained variants of AATE which catalysed the synthesis of themselves. This experiment demonstrated the possibility that autocatalysts could exhibit competition within a population of entities with heredity, which could be interpreted as a rudimentary form of natural selection, and that certain environmental changes (such as irradiation) could alter the chemical structure of some of these self-replicating molecules (an analogue for mutation) in such ways that could either boost or interfere with its ability to react, thus boosting or interfering with its ability to replicate and spread in the population.[10]

Autocatalysis plays a major role in the processes of life. Two researchers who have emphasised its role in the origins of life are Robert Ulanowicz[11] and Stuart Kauffman.[12]

Autocatalysis occurs in the initial transcripts of rRNA. The introns are capable of excising themselves by the process of two nucleophilic transesterification reactions. The RNA able to do this is sometimes referred to as a ribozyme. Additionally, the citric acid cycle is an autocatalytic cycle run in reverse.

Ultimately, biological metabolism itself can be seen as a vast autocatalytic set, in that all of the molecular constituents of a biological cell are produced by reactions involving this same set of molecules.

25.7 Examples of autocatalytic reactions

- DNA replication
- Haloform reaction
- Tin pest
- Reaction of Permanganate with Oxalic Acid
- The mechanism of the above reaction [13]
- Vinegar syndrome
- Binding of oxygen by hemoglobin
- The spontaneous degradation of aspirin into salicylic acid and acetic acid, causing very old aspirin in sealed containers to smell mildly of vinegar.
- The α-bromination of acetophenone with bromine.
- Liesegang rings

25.8 See also

- Autocatalytic reactions and order creation
- Catalytic cycle
- Reaction–diffusion system
- Abiogenesis
- Stuart Kauffman
- Morphogenesis

25.9 References

[1] Steinfeld J.I., Francisco J.S. and Hase W.L. *Chemical Kinetics and Dynamics* (2nd ed., Prentice-Hall 1999) p.151-2 ISBN 0-13-737123-3

[2] Moore J.W. and Pearson R.G. *Kinetics and Mechanism* (John Wiley 1981) p.26 ISBN 0-471-03558-0

[3] Ilya Prigogine (1980). *From Being to Becoming: Time and Complexity in the Physical Sciences*. San Francisco: W. H. Freeman. ISBN 0-7167-1107-9.

[4] Espenson, J.H. *Chemical Kinetics and Reaction Mechanisms* (2nd ed., McGraw-Hill 2002) p.190 ISBN 0-07-288362-6

[5] http://www.math.ohio-state.edu/~{ }ault/Papers/Brusselator.pdf Dynamics of the Brusselator

[6] Peterson, Gabriel. "The Belousov-Zhabotinsky Reaction". Archived from the original on December 31, 2012.

[7] G. Nicolis and Ilya Prigogine (1977). *Self-Organization in Nonequilibrium Systems*. New York: John Wiley and Sons. ISBN 0-471-02401-5.

[8] Soai K, Sato I, Shibata T (2001). "Asymmetric autocatalysis and the origin of chiral homogeneity in organic compounds.". *Chem Rec* 1 (4): 321–32. doi:10.1002/tcr.1017. PMID 11893072.

[9] Stuart Kauffman (1995). *At Home in the Universe: The Search for the Laws of Self-Organization and Complexity*. Oxford University Press. ISBN 0-19-509599-5.

[10] Rebeck, Julius (July 1994). "Synthetic Self-Replicating Molecules". *Scientific American*: 48–55.

[11] Ecology, the Ascendent Perspective", Robert Ulanowicz, Columbia Univ. Press 1997

[12] Investigations, Stuart Kauffman.

[13] Kovacs KA, Grof P, Burai L, Riedel M (2004). "Revising the Mechanism of the Permanganate/Oxalate Reaction". *J. Phys. Chem. A* 108 (50): 11026. doi:10.1021/jp047061u.

25.10 External links

- Some Remarks on Autocatalysis and Autopoiesis (Barry McMullin)
- Jain, Sanjay; Krishna, Sandeep (21 December 1998). "Autocatalytic Sets and the Growth of Complexity in an Evolutionary Model". *Physical Review Letters* 81 (25): 5684–5687. doi:10.1103/PhysRevLett.81.5684.

Chapter 26

Liquid crystal

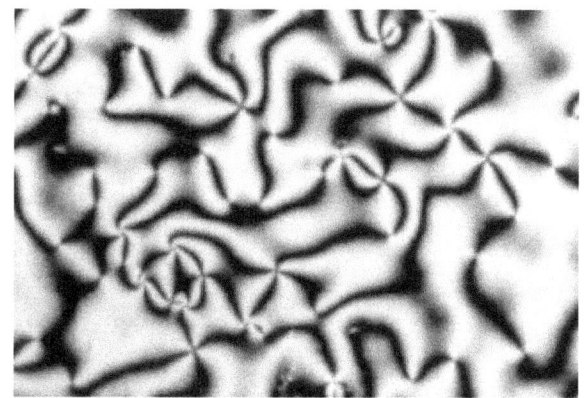

Schlieren texture of liquid crystal nematic phase

Liquid crystals (LCs) are matter in a state that has properties between those of the conventional liquid and those of solid crystal.[1] For instance, a liquid crystal may flow like a liquid, but its molecules may be oriented in a crystal-like way. There are many different types of liquid-crystal phases, which can be distinguished by their different optical properties (such as birefringence). When viewed under a microscope using a polarized light source, different liquid crystal phases will appear to have distinct textures. The contrasting areas in the textures correspond to domains where the liquid-crystal molecules are oriented in different directions. Within a domain, however, the molecules are well ordered. LC materials may not always be in a liquid-crystal phase (just as water may turn into ice or steam).

Liquid crystals can be divided into thermotropic, lyotropic and metallotropic phases. Thermotropic and lyotropic liquid crystals consist mostly of organic molecules although few minerals are also known. Thermotropic LCs exhibit a phase transition into the liquid-crystal phase as temperature is changed. Lyotropic LCs exhibit phase transitions as a function of both temperature and concentration of the liquid-crystal molecules in a solvent (typically water). Metallotropic LCs are composed of both organic and inorganic molecules; their liquid-crystal transition depends not only on temperature and concentration, but also on the inorganic-organic composition ratio.

Examples of liquid crystals can be found both in the natural world and in technological applications. Most contemporary electronic displays use liquid crystals. Lyotropic liquid-crystalline phases are abundant in living systems but can also be found in the mineral world. For example, many proteins and cell membranes are liquid crystals. Other well-known examples of liquid crystals are solutions of soap and various related detergents, as well as the tobacco mosaic virus, and some clays.

26.1 History

In 1888, Austrian botanical physiologist Friedrich Reinitzer, working at the Karl-Ferdinands-Universität, examined the physico-chemical properties of various derivatives of cholesterol which now belong to the class of materials known as cholesteric liquid crystals. Previously, other researchers had observed distinct color effects when cooling cholesterol derivatives just above the freezing point, but had not associated it with a new phenomenon. Reinitzer perceived that color changes in a derivative cholesteryl benzoate were not the most peculiar feature.

Chemical structure of cholesteryl benzoate molecule

He found that cholesteryl benzoate does not melt in the same manner as other compounds, but has two melting points. At 145.5 °C (293.9 °F) it melts into a cloudy liquid, and at 178.5 °C (353.3 °F) it melts again and the cloudy liquid becomes clear. The phenomenon is reversible. Seeking help from a physicist, on March 14, 1888, he wrote to Otto Lehmann, at that time a *Privatdozent* in Aachen. They exchanged letters and samples. Lehmann examined the intermediate cloudy fluid, and reported seeing crystallites. Reinitzer's Viennese colleague von Zepharovich also indicated that the intermediate "fluid" was crystalline. The exchange of letters with Lehmann ended on April 24, with many questions unanswered. Reinitzer presented his results, with credits to Lehmann and von Zepharovich, at a meeting of the Vienna Chemical Society on May 3, 1888.[2]

By that time, Reinitzer had discovered and described three important features of cholesteric liquid crystals (the name coined by Otto Lehmann in 1904): the existence of two melting points, the reflection of circularly polarized light, and the ability to rotate the polarization direction of light.

After his accidental discovery, Reinitzer did not pursue studying liquid crystals further. The research was continued by Lehmann, who realized that he had encountered a new phenomenon and was in a position to investigate it: In his postdoctoral years he had acquired expertise in crystallography and microscopy. Lehmann started a systematic study, first of cholesteryl benzoate, and then of related compounds which exhibited the double-melting phenomenon. He was able to make observations in polarized light, and his microscope was equipped with a hot stage (sample holder equipped with a heater) enabling high temperature observations. The intermediate cloudy phase clearly sustained flow, but other features, particularly the signature under a microscope, convinced Lehmann that he was dealing with a solid. By the end of August 1889 he had published his results in the Zeitschrift für Physikalische Chemie.[3]

Lehmann's work was continued and significantly expanded by the German chemist Daniel Vorländer, who from the beginning of 20th century until his retirement in 1935, had synthesized most of the liquid crystals known. However, liquid crystals were not popular among scientists and the material remained a pure scientific curiosity for about 80 years.[4]

After World War II work on the synthesis of liquid crystals was restarted at university research laboratories in Europe. George William Gray, a prominent researcher of liquid crystals, began investigating these materials in England in the late 1940s. His group synthesized many new materials that exhibited the liquid crystalline state and developed a better understanding of how to design molecules that exhibit the state. His book *Molecular Structure and the Properties of Liquid Crystals*[5] became a guidebook on the subject. One of the first U.S. chemists to study liquid crystals was Glenn H. Brown, starting in 1953 at the University of Cincinnati and later at Kent State University. In 1965, he organized the first international conference on liquid crystals, in Kent,

Otto Lehmann

Ohio, with about 100 of the world's top liquid crystal scientists in attendance. This conference marked the beginning of a worldwide effort to perform research in this field, which soon led to the development of practical applications for these unique materials.[6][7]

Liquid crystal materials became a focus of research in the development of flat panel electronic displays beginning in 1962 at RCA Laboratories.[8] When physical chemist Richard Williams applied an electric field to a thin layer of a nematic liquid crystal at 125 °C, he observed the formation of a regular pattern that he called domains (now known as Williams Domains). This led his colleague George H. Heilmeier to perform research on a liquid crystal-based flat panel display to replace the cathode ray vacuum tube used in televisions. But the para-Azoxyanisole that Williams and Heilmeier used exhibits the nematic liquid crystal state only above 116 °C, which made it impractical to use in a commercial display product. A material that could be operated at room temperature was clearly needed.

In 1966, Joel E. Goldmacher and Joseph A. Castellano, research chemists in Heilmeier group at RCA, discovered that mixtures made exclusively of nematic compounds that differed only in the number of carbon atoms in the terminal side chains could yield room-temperature nematic liquid crystals. A ternary mixture of Schiff base compounds resulted in a material that had a nematic range of

22–105 °C.[9] Operation at room temperature enabled the first practical display device to be made.[10] The team then proceeded to prepare numerous mixtures of nematic compounds many of which had much lower melting points. This technique of mixing nematic compounds to obtain wide operating temperature range eventually became the industry standard and is used to this very day to tailor materials to meet specific applications.

Chemical structure of N-(4-Methoxybenzylidene)–4-butylaniline (MBBA) molecule

In 1969, Hans Kelker succeeded in synthesizing a substance that had a nematic phase at room temperature, MBBA, which is one of the most popular subjects of liquid crystal research.[11] The next step to commercialization of liquid crystal displays was the synthesis of further chemically stable substances (cyanobiphenyls) with low melting temperatures by George Gray.[12] That work with Ken Harrison and the UK MOD (RRE Malvern), in 1973, led to design of new materials resulting in rapid adoption of small area LCDs within electronic products.

These molecules are rod-shaped; some are created in the lab and some could appear spontaneously in nature. Since then, two new types of LC molecules were discovered, both are *man-make*: disc-shaped (created by S. Chandrasekhar's group in India, 1977) and bowl-shaped (invented by Lui Lam[13] in China, 1982, and synthesized in Europe three years later).

In 1991, when liquid crystal displays were already well established, Pierre-Gilles de Gennes working at the Université Paris-Sud received the Nobel Prize in physics "for discovering that methods developed for studying order phenomena in simple systems can be generalized to more complex forms of matter, in particular to liquid crystals and polymers".[14]

26.2 Design of liquid crystalline materials

A large number of chemical compounds are known to exhibit one or several liquid crystalline phases. Despite significant differences in chemical composition, these molecules have some common features in chemical and physical properties. There are three types of thermotropic liquid crystals: discotics, bowlics and rod-shaped molecules. Discotics are flat disc-like molecules consisting of a core of adjacent aromatic rings; the core in a bowlic is not flat but like a rice bowl (a three-dimensional object).[15][16] This allows for two dimensional columnar ordering, for both discotics and bowlics. Rod-shaped molecules have an elongated, anisotropic geometry which allows for preferential alignment along one spatial direction.

• The molecular shape should be relatively thin, flat or bowl-like, especially within rigid molecular frameworks.
• The molecular length should be at least 1.3 nm, consistent with the presence of long alkyl group on many room-temperature liquid crystals.
• The structure should not be branched or angular, except for the bowlics.
• A low melting point is preferable in order to avoid metastable, monotropic liquid crystalline phases. Low-temperature mesomorphic behavior in general is technologically more useful, and alkyl terminal groups promote this.

An extended, structurally rigid, highly anisotropic shape seems to be the main criterion for liquid crystalline behavior, and as a result many liquid crystalline materials are based on benzene rings.[17]

26.3 Liquid-crystal phases

The various liquid-crystal phases (called mesophases) can be characterized by the type of ordering. One can distinguish positional order (whether molecules are arranged in any sort of ordered lattice) and orientational order (whether molecules are mostly pointing in the same direction), and moreover order can be either short-range (only between molecules close to each other) or long-range (extending to larger, sometimes macroscopic, dimensions). Most thermotropic LCs will have an isotropic phase at high temperature. That is that heating will eventually drive them into a conventional liquid phase characterized by random and isotropic molecular ordering (little to no long-range order), and fluid-like flow behavior. Under other conditions (for instance, lower temperature), a LC might inhabit one or more phases with significant anisotropic orientational structure and short-range orientational order while still having an ability to flow.[1][18]

26.3. LIQUID-CRYSTAL PHASES

The ordering of liquid crystalline phases is extensive on the molecular scale. This order extends up to the entire domain size, which may be on the order of micrometers, but usually does not extend to the macroscopic scale as often occurs in classical crystalline solids. However some techniques, such as the use of boundaries or an applied electric field, can be used to enforce a single ordered domain in a macroscopic liquid crystal sample. The ordering in a liquid crystal might extend along only one dimension, with the material being essentially disordered in the other two directions.[19][20]

26.3.1 Thermotropic liquid crystals

See also: Thermotropic crystal

Thermotropic phases are those that occur in a certain temperature range. If the temperature rise is too high, thermal motion will destroy the delicate cooperative ordering of the LC phase, pushing the material into a conventional isotropic liquid phase. At too low temperature, most LC materials will form a conventional crystal.[1][18] Many thermotropic LCs exhibit a variety of phases as temperature is changed. For instance, on heating a particular type of LC molecule (called mesogen) may exhibit various smectic phases followed by the nematic phase and finally the isotropic phase as temperature is increased. An example of a compound displaying thermotropic LC behavior is para-azoxyanisole.[21]

Nematic phase

See also: Biaxial nematic and Twisted nematic field effect

One of the most common LC phases is the nematic. The word *nematic* comes from the Greek νήμα (*Greek: nema*), which means "thread". This term originates from the thread-like topological defects observed in nematics, which are formally called 'disclinations'. Nematics also exhibit so-called "hedgehog" topological defects. In a nematic phase, the *calamitic* or rod-shaped organic molecules have no positional order, but they self-align to have long-range directional order with their long axes roughly parallel.[22] Thus, the molecules are free to flow and their center of mass positions are randomly distributed as in a liquid, but still maintain their long-range directional order. Most nematics are uniaxial: they have one axis that is longer and preferred, with the other two being equivalent (can be approximated as cylinders or rods). However, some liquid crystals are biaxial nematics, meaning that in addition to orienting their long axis, they also orient along a secondary axis.[23] Nematics have fluidity similar to that of ordinary (isotropic) liquids but they can be easily aligned by an external magnetic or electric field. Aligned nematics have the optical properties of uniaxial crystals and this makes them extremely useful in liquid crystal displays (LCD).[8]

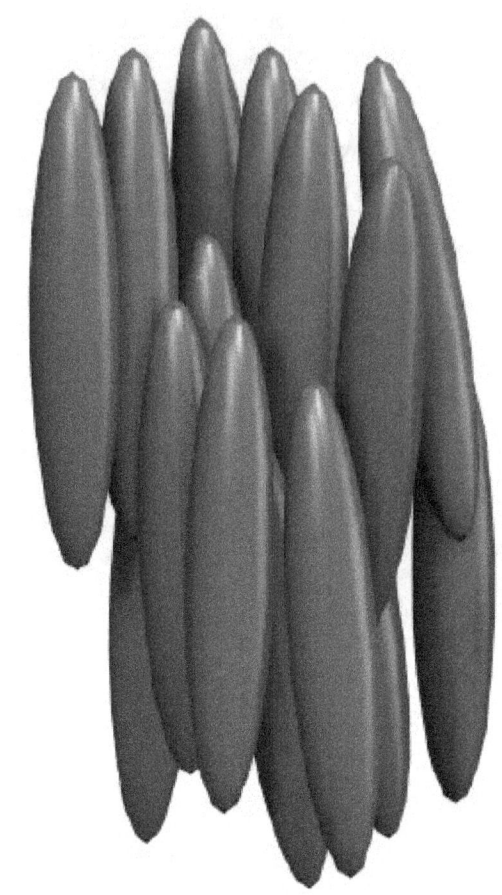

Alignment in a nematic phase.

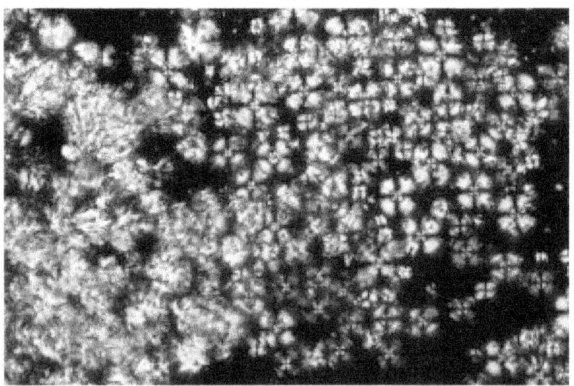

Phase transition between a nematic (left) and smectic A (right) phases observed between crossed polarizers. The black color corresponds to isotropic medium.

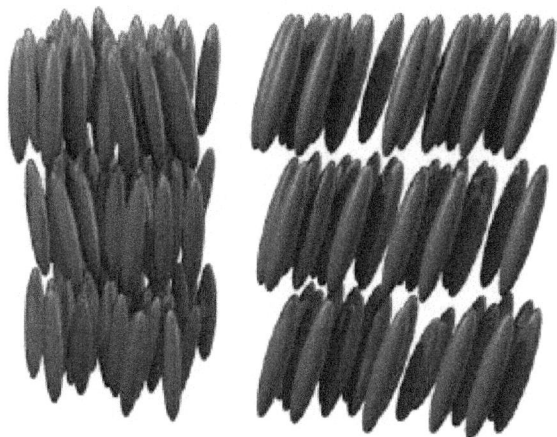

Schematic of alignment in the smectic phases. The smectic A phase (left) has molecules organized into layers. In the smectic C phase (right), the molecules are tilted inside the layers.

Schematic of ordering in chiral liquid crystal phases. The chiral nematic phase (left), also called the cholesteric phase, and the smectic C* phase (right).

Smectic phases

The smectic phases, which are found at lower temperatures than the nematic, form well-defined layers that can slide over one another in a manner similar to that of soap. The word "smectic" originates from the Latin word "smecticus", meaning cleaning, or having soap-like properties.[24] The smectics are thus positionally ordered along one direction. In the Smectic A phase, the molecules are oriented along the layer normal, while in the Smectic C phase they are tilted away from it. These phases are liquid-like within the layers. There are many different smectic phases, all characterized by different types and degrees of positional and orientational order.[1][18]

Chiral phases

The chiral nematic phase exhibits chirality (handedness). This phase is often called the *cholesteric* phase because it was first observed for cholesterol derivatives. Only chiral molecules (i.e., those that have no internal planes of symmetry) can give rise to such a phase. This phase exhibits a twisting of the molecules perpendicular to the director, with the molecular axis parallel to the director. The finite twist angle between adjacent molecules is due to their asymmetric packing, which results in longer-range chiral order. In the smectic C* phase (an asterisk denotes a chiral phase), the molecules have positional ordering in a layered structure (as in the other smectic phases), with the molecules tilted by a finite angle with respect to the layer normal. The chirality induces a finite azimuthal twist from one layer to the next, producing a spiral twisting of the molecular axis along the layer normal.[18][19][20]

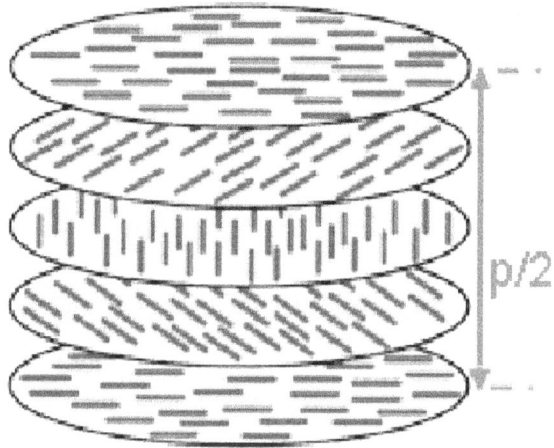

Chiral nematic phase; p refers to the chiral pitch (see text)

The *chiral pitch*, p, refers to the distance over which the LC molecules undergo a full 360° twist (but note that the structure of the chiral nematic phase repeats itself every half-pitch, since in this phase directors at 0° and ±180° are equivalent). The pitch, p, typically changes when the temperature is altered or when other molecules are added to the LC host (an achiral LC host material will form a chiral phase if doped with a chiral material), allowing the pitch of a given material to be tuned accordingly. In some liquid crystal systems, the pitch is of the same order as the wavelength of visible light. This causes these systems to exhibit unique optical properties, such as Bragg reflection and low-threshold laser emission,[25] and these properties are exploited in a number of optical applications.[4][19] For the case of Bragg reflection only the lowest-order reflection is allowed if the light is incident along the helical axis, whereas for oblique incidence higher-order reflections become permitted. Cholesteric liquid crystals also exhibit the unique property that they reflect circularly polarized light when it is incident along the helical axis and elliptically polarized if it comes in obliquely.[26]

Blue phases are liquid crystal phases that appear in the temperature range between a chiral nematic phase and an isotropic liquid phase. Blue phases have a regular three-dimensional cubic structure of defects with lattice periods of several hundred nanometers, and thus they exhibit selective Bragg reflections in the wavelength range of visible light corresponding to the cubic lattice. It was theoretically predicted in 1981 that these phases can possess icosahedral symmetry similar to quasicrystals.[27][28]

Although blue phases are of interest for fast light modulators or tunable photonic crystals, they exist in a very narrow temperature range, usually less than a few kelvin. Recently the stabilization of blue phases over a temperature range of more than 60 K including room temperature (260–326 K) has been demonstrated.[29][30] Blue phases stabilized at room temperature allow electro-optical switching with response times of the order of 10^{-4} s.[31]

In May 2008, the first Blue Phase Mode LCD panel had been developed.[32]

Discotic phases

Disk-shaped LC molecules can orient themselves in a layer-like fashion known as the discotic nematic phase. If the disks pack into stacks, the phase is called a discotic columnar. The columns themselves may be organized into rectangular or hexagonal arrays. Chiral discotic phases, similar to the chiral nematic phase, are also known.

Bowlic phases

Bowl-shaped LC molecules, like in discotics, can form columnar phases. Other phases, such as nonpolar nematic, polar nematic, stringbean, donut and onion phases, have been predicted. Bowlic phases, except nonpolar nematic, are polar phases.

26.3.2 Lyotropic liquid crystals

See also: Lyotropic liquid crystal and Columnar phase

A lyotropic liquid crystal consists of two or more components that exhibit liquid-crystalline properties in certain concentration ranges. In the lyotropic phases, solvent molecules fill the space around the compounds to provide fluidity to the system.[33] In contrast to thermotropic liquid crystals, these lyotropics have another degree of freedom

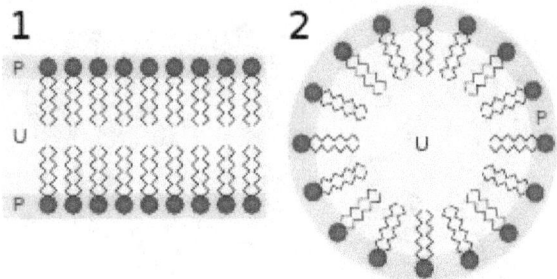

Structure of lyotropic liquid crystal. The red heads of surfactant molecules are in contact with water, whereas the tails are immersed in oil (blue): bilayer (left) and micelle (right).

of concentration that enables them to induce a variety of different phases.

A compound that has two immiscible hydrophilic and hydrophobic parts within the same molecule is called an amphiphilic molecule. Many amphiphilic molecules show lyotropic liquid-crystalline phase sequences depending on the volume balances between the hydrophilic part and hydrophobic part. These structures are formed through the micro-phase segregation of two incompatible components on a nanometer scale. Soap is an everyday example of a lyotropic liquid crystal.

The content of water or other solvent molecules changes the self-assembled structures. At very low amphiphile concentration, the molecules will be dispersed randomly without any ordering. At slightly higher (but still low) concentration, amphiphilic molecules will spontaneously assemble into micelles or vesicles. This is done so as to 'hide' the hydrophobic tail of the amphiphile inside the micelle core, exposing a hydrophilic (water-soluble) surface to aqueous solution. These spherical objects do not order themselves in solution, however. At higher concentration, the assemblies will become ordered. A typical phase is a hexagonal columnar phase, where the amphiphiles form long cylinders (again with a hydrophilic surface) that arrange themselves into a roughly hexagonal lattice. This is called the middle soap phase. At still higher concentration, a lamellar phase (neat soap phase) may form, wherein extended sheets of amphiphiles are separated by thin layers of water. For some systems, a cubic (also called viscous isotropic) phase may exist between the hexagonal and lamellar phases, wherein spheres are formed that create a dense cubic lattice. These spheres may also be connected to one another, forming a bicontinuous cubic phase.

The objects created by amphiphiles are usually spherical (as in the case of micelles), but may also be disc-like (bicelles), rod-like, or biaxial (all three micelle axes are distinct). These anisotropic self-assembled nano-structures can then order themselves in much the same way as thermotropic liquid crystals do, forming large-scale versions of all the thermotropic phases (such as a nematic phase of rod-shaped micelles).

For some systems, at high concentrations, inverse phases are observed. That is, one may generate an inverse hexagonal columnar phase (columns of water encapsulated by amphiphiles) or an inverse micellar phase (a bulk liquid crystal sample with spherical water cavities).

A generic progression of phases, going from low to high amphiphile concentration, is:

- Discontinuous cubic phase (micellar cubic phase)
- Hexagonal phase (hexagonal columnar phase) (middle phase)
- Lamellar phase
- Bicontinuous cubic phase
- Reverse hexagonal columnar phase
- Inverse cubic phase (Inverse micellar phase)

Even within the same phases, their self-assembled structures are tunable by the concentration: for example, in lamellar phases, the layer distances increase with the solvent volume. Since lyotropic liquid crystals rely on a subtle balance of intermolecular interactions, it is more difficult to analyze their structures and properties than those of thermotropic liquid crystals.

Similar phases and characteristics can be observed in immiscible diblock copolymers.

26.3.3 Metallotropic liquid crystals

Liquid crystal phases can also be based on low-melting *inorganic* phases like $ZnCl_2$ that have a structure formed of linked tetrahedra and easily form glasses. The addition of long chain soap-like molecules leads to a series of new phases that show a variety of liquid crystalline behavior both as a function of the inorganic-organic composition ratio and of temperature. This class of materials has been named metallotropic.[34]

26.3.4 Laboratory analysis of mesophases

Thermotropic mesophases are detected and characterized by two major methods, the original method was use of thermal optical microscopy, in which a small sample of the material was placed between two crossed polarizers; the sample was then heated and cooled. As the isotropic phase would not significantly affect the polarization of the light, it would appear very dark, whereas the crystal and liquid crystal phases will both polarize the light in a uniform way, leading to brightness and color gradients. This method allows for the characterization of the particular phase, as the different phases are defined by their particular order, which must be observed. The second method, differential scanning calorimetry (DSC), allows for more precise determination of phase transitions and transition enthalpies. In DSC, a small sample is heated in a way that generates a very precise change in temperature with respect to time. During phase transitions, the heat flow required to maintain this heating or cooling rate will change. These changes can be observed and attributed to various phase transitions, such as key liquid crystal transitions.

Lyotropic mesophases are analyzed in a similar fashion, though these experiments are somewhat more complex, as the concentration of mesogen is a key factor. These experiments are run at various concentrations of mesogen in order to analyze that impact.

26.4 Biological liquid crystals

Lyotropic liquid-crystalline phases are abundant in living systems, the study of which is referred to as lipid polymorphism. Accordingly, lyotropic liquid crystals attract particular attention in the field of biomimetic chemistry. In particular, biological membranes and cell membranes are a form of liquid crystal. Their constituent molecules (e.g. phospholipids) are perpendicular to the membrane surface, yet the membrane is flexible. These lipids vary in shape (see page on lipid polymorphism). The constituent molecules can inter-mingle easily, but tend not to leave the membrane due to the high energy requirement of this process. Lipid molecules can flip from one side of the membrane to the other, this process being catalyzed by flippases and floppases (depending on the direction of movement). These liquid crystal membrane phases can also host important proteins such as receptors freely "floating" inside, or partly outside, the membrane, e.g. CCT.

Many other biological structures exhibit liquid-crystal behavior. For instance, the concentrated protein solution that is extruded by a spider to generate silk is, in fact, a liquid crystal phase. The precise ordering of molecules in silk is critical to its renowned strength. DNA and many polypeptides can also form LC phases and this too forms an important part of current academic research.

26.5 Mineral liquid crystals

Examples of liquid crystals can also be found in the mineral world, most of them being lyotropics. The first discovered was Vanadium(V) oxide, by Zocher in 1925.[35] Since then, few others have been discovered and studied in details.[36] The existence of a true nematic phase in the case of the smectic clays family, was raised by Langmuir in 1938,[37] but remained open for a very long time and was only solved recently.[38][39] With the rapid development of nanosciences and the synthesis of many new anisotropic nanoparticles, the number of such mineral liquid crystals is quickly increasing with for example, carbon nanotubes and graphene. A lamellar phase was even discovered, $H_3Sb_3P_2O_{14}$ that exhibit hyperswelling up to ~250 nm for the interlamellar distance.[40]

26.6 Pattern formation in liquid crystals

See also: Pattern formation

Anisotropy of liquid crystals is a property not observed in other fluids. This anisotropy makes flows of liquid crystals behave more differentially than those of ordinary fluids. For example, injection of a flux of a liquid crystal between two close parallel plates (viscous fingering) causes orientation of the molecules to couple with the flow, with the resulting emergence of dendritic patterns.[41] This anisotropy is also manifested in the interfacial energy (surface tension) between different liquid crystal phases. This anisotropy determines the equilibrium shape at the coexistence temperature, and is so strong that usually facets appear. When temperature is changed one of the phases grows, forming different morphologies depending on the temperature change.[42] Since growth is controlled by heat diffusion, anisotropy in thermal conductivity favors growth in specific directions, which has also an effect on the final shape.[43]

26.7 Theoretical treatment of liquid crystals

Microscopic theoretical treatment of fluid phases can become quite complicated, owing to the high material density, meaning that strong interactions, hard-core repulsions, and many-body correlations cannot be ignored. In the case of liquid crystals, anisotropy in all of these interactions further complicates analysis. There are a number of fairly simple theories, however, that can at least predict the general behavior of the phase transitions in liquid crystal systems.

26.7.1 Director

As we already saw above, the nematic liquid crystals are composed of rod-like molecules with the long axes of neighboring molecules aligned approximately to one another. To describe this anisotropic structure, a dimensionless unit vector n called the *director*, is introduced to represent the direction of preferred orientation of molecules in the neighborhood of any point. Because there is no physical polarity along the director axis, n and $-n$ are fully equivalent.[18]

26.7.2 Order parameter

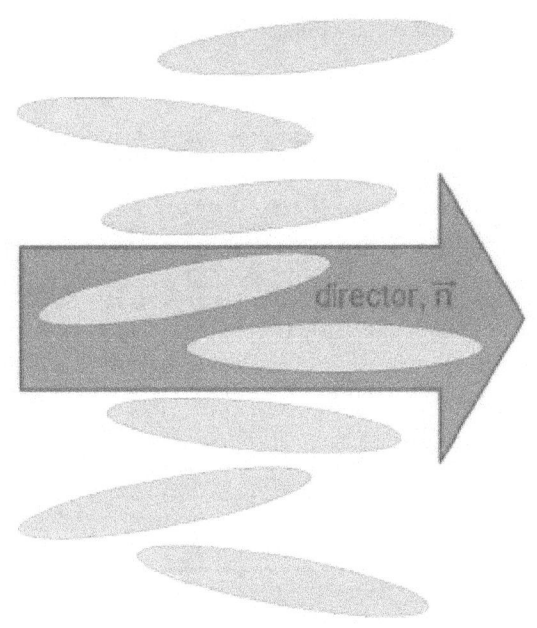

The local nematic director, *which is also the* local optical axis, *is given by the spatial and temporal average of the long molecular axes*

The description of liquid crystals involves an analysis of order. A second rank symmetric traceless tensor order parameter is used to describe the orientational order of a nematic liquid crystal, although a scalar order parameter is usually sufficient to describe uniaxial nematic liquid crystals. To make this quantitative, an orientational order parameter is usually defined based on the average of the second Legendre polynomial:

$$S = \langle P_2(\cos\theta)\rangle = \left\langle \frac{3\cos^2\theta - 1}{2}\right\rangle$$

where θ is the angle between the liquid-crystal molecular axis and the *local director* (which is the 'preferred direction' in a volume element of a liquid crystal sample, also representing its *local optical axis*). The brackets denote both a temporal and spatial average. This definition is convenient, since for a completely random and isotropic sample, S=0, whereas for a perfectly aligned sample S=1. For a typical liquid crystal sample, S is on the order of 0.3 to 0.8, and generally decreases as the temperature is raised. In particular, a sharp drop of the order parameter to 0 is observed when the system undergoes a phase transition from an LC phase into the isotropic phase.[44] The order parameter can be measured experimentally in a number of ways; for instance, diamagnetism, birefringence, Raman scattering, NMR and EPR can be used to determine S.[20]

The order of a liquid crystal could also be characterized by using other even Legendre polynomials (all the odd polynomials average to zero since the director can point in either of two antiparallel directions). These higher-order averages are more difficult to measure, but can yield additional information about molecular ordering.[1]

A positional order parameter is also used to describe the ordering of a liquid crystal. It is characterized by the variation of the density of the center of mass of the liquid crystal molecules along a given vector. In the case of positional variation along the z-axis the density $\rho(z)$ is often given by:

$$\rho(\mathbf{r}) = \rho(z) = \rho_0 + \rho_1 \cos(q_s z - \phi) + \cdots$$

The complex positional order parameter is defined as $\psi(\mathbf{r}) = \rho_1(\mathbf{r})e^{i\phi(\mathbf{r})}$ and ρ_0 the average density. Typically only the first two terms are kept and higher order terms are ignored since most phases can be described adequately using sinusoidal functions. For a perfect nematic $\psi = 0$ and for a smectic phase ψ will take on complex values. The complex nature of this order parameter allows for many parallels between nematic to smectic phase transitions and conductor to superconductor transitions.[18]

26.7.3 Onsager hard-rod model

A simple model which predicts lyotropic phase transitions is the hard-rod model proposed by Lars Onsager. This theory considers the volume excluded from the center-of-mass of one idealized cylinder as it approaches another. Specifically, if the cylinders are oriented parallel to one another, there is very little volume that is excluded from the center-of-mass of the approaching cylinder (it can come quite close to the other cylinder). If, however, the cylinders are at some angle to one another, then there is a large volume surrounding the cylinder which the approaching cylinder's center-of-mass cannot enter (due to the hard-rod repulsion between

the two idealized objects). Thus, this angular arrangement sees a *decrease* in the net positional entropy of the approaching cylinder (there are fewer states available to it).[45][46]

The fundamental insight here is that, whilst parallel arrangements of anisotropic objects lead to a decrease in orientational entropy, there is an increase in positional entropy. Thus in some case greater positional order will be entropically favorable. This theory thus predicts that a solution of rod-shaped objects will undergo a phase transition, at sufficient concentration, into a nematic phase. Although this model is conceptually helpful, its mathematical formulation makes several assumptions that limit its applicability to real systems.[46]

26.7.4 Maier–Saupe mean field theory

This statistical theory, proposed by Alfred Saupe and Wilhelm Maier, includes contributions from an attractive intermolecular potential from an induced dipole moment between adjacent liquid crystal molecules. The anisotropic attraction stabilizes parallel alignment of neighboring molecules, and the theory then considers a mean-field average of the interaction. Solved self-consistently, this theory predicts thermotropic nematic-isotropic phase transitions, consistent with experiment.[47][48][49]

26.7.5 McMillan's model

McMillan's model, proposed by William McMillan,[50] is an extension of the Maier–Saupe mean field theory used to describe the phase transition of a liquid crystal from a nematic to a smectic A phase. It predicts that the phase transition can be either continuous or discontinuous depending on the strength of the short-range interaction between the molecules. As a result, it allows for a triple critical point where the nematic, isotropic, and smectic A phase meet. Although it predicts the existence of a triple critical point, it does not successfully predict its value. The model utilizes two order parameters that describe the orientational and positional order of the liquid crystal. The first is simply the average of the second Legendre polynomial and the second order parameter is given by:

$$\sigma = \left\langle \cos\left(\frac{2\pi z_i}{d}\right) \left(\frac{3}{2}\cos^2\theta_i - \frac{1}{2}\right) \right\rangle$$

The values z_i, θ_i, and d are the position of the molecule, the angle between the molecular axis and director, and the layer spacing. The postulated potential energy of a single molecule is given by:

$$U_i(\theta_i, z_i) = -U_0 \left(S + \alpha\sigma \cos\left(\frac{2\pi z_i}{d}\right) \right) \left(\frac{3}{2}\cos^2\theta_i - \frac{1}{2} \right)$$

Here constant α quantifies the strength of the interaction between adjacent molecules. The potential is then used to derive the thermodynamic properties of the system assuming thermal equilibrium. It results in two self-consistency equations that must be solved numerically, the solutions of which are the three stable phases of the liquid crystal.[20]

26.7.6 Elastic continuum theory

In this formalism, a liquid crystal material is treated as a continuum; molecular details are entirely ignored. Rather, this theory considers perturbations to a presumed oriented sample. The distortions of the liquid crystal are commonly described by the Frank free energy density. One can identify three types of distortions that could occur in an oriented sample: (1) **twists** of the material, where neighboring molecules are forced to be angled with respect to one another, rather than aligned; (2) **splay** of the material, where bending occurs perpendicular to the director; and (3) **bend** of the material, where the distortion is parallel to the director and molecular axis. All three of these types of distortions incur an energy penalty. They are distortions that are induced by the boundary conditions at domain walls or the enclosing container. The response of the material can then be decomposed into terms based on the elastic constants corresponding to the three types of distortions. Elastic continuum theory is a particularly powerful tool for modeling liquid crystal devices [51] and lipid bilayers.[52]

26.8 External influences on liquid crystals

Scientists and engineers are able to use liquid crystals in a variety of applications because external perturbation can cause significant changes in the macroscopic properties of the liquid crystal system. Both electric and magnetic fields can be used to induce these changes. The magnitude of the fields, as well as the speed at which the molecules align are important characteristics industry deals with. Special surface treatments can be used in liquid crystal devices to force specific orientations of the director.

26.8.1 Electric and magnetic field effects

The ability of the director to align along an external field is caused by the electric nature of the molecules. Permanent electric dipoles result when one end of a molecule has a net positive charge while the other end has a net negative charge. When an external electric field is applied to the liquid crystal, the dipole molecules tend to orient themselves along the direction of the field.

Even if a molecule does not form a permanent dipole, it can still be influenced by an electric field. In some cases, the field produces slight re-arrangement of electrons and protons in molecules such that an induced electric dipole results. While not as strong as permanent dipoles, orientation with the external field still occurs. The effects of magnetic fields on liquid crystal molecules are analogous to electric fields. Because magnetic fields are generated by moving electric charges, permanent magnetic dipoles are produced by electrons moving about atoms. When a magnetic field is applied, the molecules will tend to align with or against the field.

26.8.2 Surface preparations

In the absence of an external field, the director of a liquid crystal is free to point in any direction. It is possible, however, to force the director to point in a specific direction by introducing an outside agent to the system. For example, when a thin polymer coating (usually a polyimide) is spread on a glass substrate and rubbed in a single direction with a cloth, it is observed that liquid crystal molecules in contact with that surface align with the rubbing direction. The currently accepted mechanism for this is believed to be an epitaxial growth of the liquid crystal layers on the partially aligned polymer chains in the near surface layers of the polyimide.

26.8.3 Fredericks transition

The competition between orientation produced by surface anchoring and by electric field effects is often exploited in liquid crystal devices. Consider the case in which liquid crystal molecules are aligned parallel to the surface and an electric field is applied perpendicular to the cell. At first, as the electric field increases in magnitude, no change in alignment occurs. However at a threshold magnitude of electric field, deformation occurs. Deformation occurs where the director changes its orientation from one molecule to the next. The occurrence of such a change from an aligned to a deformed state is called a Fredericks transition and can also be produced by the application of a magnetic field of sufficient strength.

The Fredericks transition is fundamental to the operation of many liquid crystal displays because the director orientation (and thus the properties) can be controlled easily by the application of a field.

26.9 Effect of chirality

As already described, chiral liquid-crystal molecules usually give rise to chiral mesophases. This means that the molecule must possess some form of asymmetry, usually a stereogenic center. An additional requirement is that the system not be racemic: a mixture of right- and left-handed molecules will cancel the chiral effect. Due to the cooperative nature of liquid crystal ordering, however, a small amount of chiral dopant in an otherwise achiral mesophase is often enough to select out one domain handedness, making the system overall chiral.

Chiral phases usually have a helical twisting of the molecules. If the pitch of this twist is on the order of the wavelength of visible light, then interesting optical interference effects can be observed. The chiral twisting that occurs in chiral LC phases also makes the system respond differently from right- and left-handed circularly polarized light. These materials can thus be used as polarization filters.[53]

It is possible for chiral LC molecules to produce essentially achiral mesophases. For instance, in certain ranges of concentration and molecular weight, DNA will form an achiral line hexatic phase. An interesting recent observation is of the formation of chiral mesophases from achiral LC molecules. Specifically, bent-core molecules (sometimes called banana liquid crystals) have been shown to form liquid crystal phases that are chiral.[54] In any particular sample, various domains will have opposite handedness, but within any given domain, strong chiral ordering will be present. The appearance mechanism of this macroscopic chirality is not yet entirely clear. It appears that the molecules stack in layers and orient themselves in a tilted fashion inside the layers. These liquid crystals phases may be ferroelectric or anti-ferroelectric, both of which are of interest for applications.[55][56]

Chirality can also be incorporated into a phase by adding a chiral dopant, which may not form LCs itself. Twisted-nematic or super-twisted nematic mixtures often contain a small amount of such dopants.

26.10 Applications of liquid crystals

See also: Liquid crystal display

Liquid crystals find wide use in liquid crystal displays, which rely on the optical properties of certain liquid crystalline substances in the presence or absence of an electric field. In a typical device, a liquid crystal layer (typically 4 μm thick) sits between two polarizers that are crossed (oriented at 90° to one another). The liquid crystal alignment is chosen so that its relaxed phase is a twisted one (see Twisted

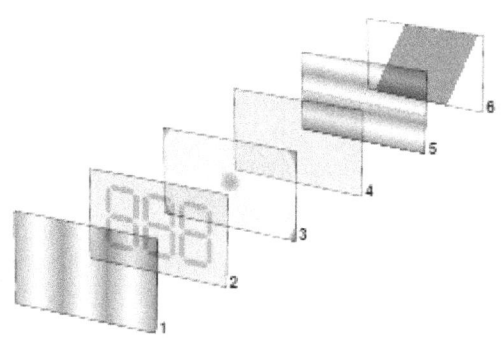

Structure of liquid crystal display: 1 – vertical polarization filter, 2,4 – glass with electrodes, 3 – liquid crystals, 5 – horizontal polarization filter, 6 – reflector

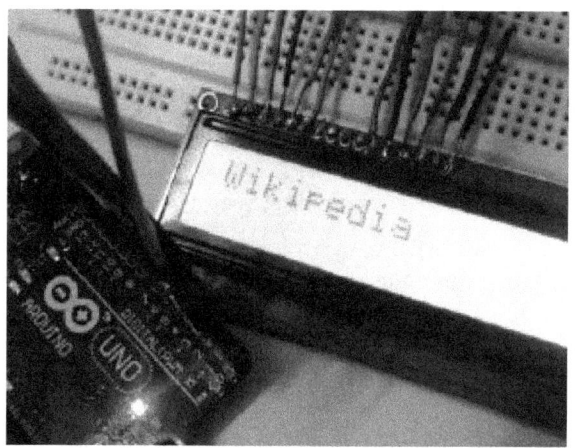

"Wikipedia" displayed on an LCD

nematic field effect).[8] This twisted phase reorients light that has passed through the first polarizer, allowing its transmission through the second polarizer (and reflected back to the observer if a reflector is provided). The device thus appears transparent. When an electric field is applied to the LC layer, the long molecular axes tend to align parallel to the electric field thus gradually untwisting in the center of the liquid crystal layer. In this state, the LC molecules do not reorient light, so the light polarized at the first polarizer is absorbed at the second polarizer, and the device loses transparency with increasing voltage. In this way, the electric field can be used to make a pixel switch between transparent or opaque on command. Color LCD systems use the same technique, with color filters used to generate red, green, and blue pixels.[8] Chiral smectic liquid crystals are used in ferroelectric LCDs which are fast-switching binary light modulators. Similar principles can be used to make other liquid crystal based optical devices.[57]

Liquid crystal tunable filters are used as electrooptical devices, e.g., in hyperspectral imaging.

Thermotropic chiral LCs whose pitch varies strongly with temperature can be used as crude liquid crystal thermometers, since the color of the material will change as the pitch is changed. Liquid crystal color transitions are used on many aquarium and pool thermometers as well as on thermometers for infants or baths.[58] Other liquid crystal materials change color when stretched or stressed. Thus, liquid crystal sheets are often used in industry to look for hot spots, map heat flow, measure stress distribution patterns, and so on. Liquid crystal in fluid form is used to detect electrically generated hot spots for failure analysis in the semiconductor industry.[59]

Liquid crystal lasers use a liquid crystal in the lasing medium as a distributed feedback mechanism instead of external mirrors. Emission at a photonic bandgap created by the periodic dielectric structure of the liquid crystal gives a low-threshold high-output device with stable monochromatic emission.[60][61]

Polymer Dispersed Liquid Crystal (PDLC) sheets and rolls are available as adhesive backed Smart film which can be applied to windows and electrically switched between transparent and opaque to provide privacy.

Many common fluids, such as soapy water, are in fact liquid crystals. Soap forms a variety of LC phases depending on its concentration in water.[62]

Bowlic columns could be used for fast switches.[63]

26.11 See also

- Biaxial nematic
- Columnar phase
- Chromonic
- LCD classification
- Liquid crystal display
- Liquid Crystal on Silicon
- Liquid crystal polymer
- Liquid crystal tunable filter
- Lyotropic liquid crystal
- Pattern formation
- Plastic crystallinity
- Smart glass
- Thermochromics
- Thermotropic crystal
- Twisted nematic field effect
- Nematicon
- Liquid crystal thermometer
- Mood ring

26.12 References

[1] Chandrasekhar, S. (1992). *Liquid Crystals* (2nd ed.). Cambridge: Cambridge University Press. ISBN 0-521-41747-3.

[2] Reinitzer, Friedrich (1888). "Beiträge zur Kenntniss des Cholesterins". *Monatshefte für Chemie (Wien)* **9** (1): 421–441. doi:10.1007/BF01516710.

[3] Lehmann, O. (1889). "Über fliessende Krystalle". *Zeitschrift für Physikalische Chemie* **4**: 462–72.

[4] Sluckin, T. J.; Dunmur, D. A. & Stegemeyer, H. (2004). *Crystals That Flow – classic papers from the history of liquid crystals*. London: Taylor & Francis. ISBN 0-415-25789-1.

[5] Gray, G. W. (1962) *Molecular Structure and the Properties of Liquid Crystals*, Academic Press

[6] Stegemeyer, H (1994). "Professor Horst Sackmann, 1921 – 1993". *Liquid Crystals Today* **4**: 1. doi:10.1080/13583149408628630.

[7] Liquid Crystals. kfupm.edu.sa

[8] Castellano, Joseph A. (2005). *Liquid Gold: The Story of Liquid Crystal Displays and the Creation of an Industry*. World Scientific Publishing. ISBN 978-981-238-956-5.

[9] Goldmacher, Joel E. and Castellano, Joseph A. "Electro-optical Compositions and Devices," U.S. Patent 3,540,796, Issue date: November 17, 1970.

[10] Heilmeier, G. H.; Zanoni, L. A.; Barton, L. A. (1968). "Dynamic Scattering in Nematic Liquid Crystals". *Applied Physics Letters* **13**: 46. Bibcode:1968ApPhL..13...46H. doi:10.1063/1.1652453.

[11] Kelker, H.; Scheurle, B. (1969). "A Liquid-crystalline (Nematic) Phase with a Particularly Low Solidification Point". *Angew. Chem. Int. Ed.* **8** (11): 884. doi:10.1002/anie.196908841.

[12] Gray, G.W.; Harrison, K.J.; Nash, J.A. (1973). "New family of nematic liquid crystals for displays". *Electronics Lett.* **9** (6): 130. doi:10.1049/el:19730096.

[13] Lin, Lei (Lam, Lui) (1982). "Liquid crystal phases and the 'dimensionality' of molecules". *Wuli (Physics)* **11**, 171-178.

[14] "History and Properties of Liquid Crystals". Nobelprize.org. Retrieved June 6, 2009.

[15] Lam, Lui (1994). "Bowlics". *Liquid Crystalline and Mesomorphic Polymers*, eds. Valery P. Shibaev and Lui Lam. New York: Springer.

[16] Lin, Lei (Lam, Lui) (1987). "Bowlic liquid crytals". *Mol. Cryst. Liq. Cryst* **146**: 41-54.

[17]

[18] de Gennes, P.G. and Prost, J (1993). *The Physics of Liquid Crystals*. Oxford: Clarendon Press. ISBN 0-19-852024-7.

[19] Dierking, I. (2003). *Textures of Liquid Crystals*. Weinheim: Wiley-VCH. ISBN 3-527-30725-7.

[20] Collings, P.J. & Hird, M (1997). *Introduction to Liquid Crystals*. Bristol, PA: Taylor & Francis. ISBN 0-7484-0643-3.

[21] Shao, Y.; Zerda, T. W. (1998). "Phase Transitions of Liquid Crystal PAA in Confined Geometries". *Journal of Physical Chemistry B* **102** (18): 3387–3394. doi:10.1021/jp9734437.

[22] Rego, J.A.; Harvey, Jamie A.A.; MacKinnon, Andrew L.; Gatdula, Elysse (January 2010). "Asymmetric synthesis of a highly soluble 'trimeric' analogue of the chiral nematic liquid crystal twist agent Merck S1011" (PDF). *Liquid Crystals* **37** (1): 37–43. doi:10.1080/02678290903359291.

[23] Madsen, L. A.; Dingemans, T. J.; Nakata, M.; Samulski, E. T. (2004). "Thermotropic Biaxial Nematic Liquid Crystals". *Phys. Rev. Lett.* **92** (14): 145505. Bibcode:2004PhRvL..92n5505M. doi:10.1103/PhysRevLett.92.145505. PMID 15089552.

[24] "smectic". Merriam-Webster Dictionary.

[25] Kopp, V. I.; Fan, B.; Vithana, H. K. M.; Genack, A. Z.; Fan; Vithana; Genack (1998). "Low threshold lasing at the edge of a photonic stop band in cholesteric liquid crystals". *Opt. Lett* **23** (21): 1707–1709. Bibcode:1998OptL...23.1707K. doi:10.1364/OL.23.001707. PMID 18091891.

[26] Priestley, E. B.; Wojtowicz, P. J. & Sheng, P. (1974). *Introduction to Liquid Crystals*. Plenum Press. ISBN 0-306-30858-4.

[27] Kleinert H. and Maki K. (1981). "Lattice Textures in Cholesteric Liquid Crystals" (PDF). *Fortschritte der Physik* **29** (5): 219–259. Bibcode:1981ForPh..29..219K. doi:10.1002/prop.19810290503.

[28] Seideman, T (1990). "The liquid-crystalline blue phases" (PDF). *Rep. Prog. Phys.* **53** (6): 659–705. Bibcode:1990RPPh...53..659S. doi:10.1088/0034-4885/53/6/001.

[29] Coles, Harry J.; Pivnenko, Mikhail N. (2005). "Liquid crystal 'blue phases' with a wide temperature range". *Nature* **436** (7053): 997–1000. Bibcode:2005Natur.436..997C. doi:10.1038/nature03932. PMID 16107843.

[30] Yamamoto, Jun; Nishiyama, Isa; Inoue, Miyoshi; Yokoyama, Hiroshi (2005). "Optical isotropy and iridescence in a smectic blue phase". *Nature* **437** (7058): 525. Bibcode:2005Natur.437..525Y. doi:10.1038/nature04034.

[31] Kikuchi H, Yokota M, Hisakado Y, Yang H, Kajiyama T.; Yokota; Hisakado; Yang; Kajiyama (2002). "Polymer-stabilized liquid crystal blue phases". *Nature Materials* **1** (1): 64–8. Bibcode:2002NatMa...1...64K. doi:10.1038/nmat712. PMID 12618852.

[32] "Samsung Develops World's First 'Blue Phase' Technology to Achieve 240 Hz Driving Speed for High-Speed Video". Retrieved April 23, 2009.

[33] Qizhen Liang; Pengtao Liu; Cheng Liu; Xigao Jian; Dingyi Hong; Yang Li. (2005). "Synthesis and Properties of Lyotropic Liquid Crystalline Copolyamides Containing Phthalazinone Moieties and Ether Linkages". *Polymer* **46** (16): 6258–6265. doi:10.1016/j.polymer.2005.05.059.

[34] Martin, James D.; Keary, Cristin L.; Thornton, Todd A.; Novotnak, Mark P.; Knutson, Jeremey W.; Folmer, Jacob C. W. (2006). "Metallotropic liquid crystals formed by surfactant templating of molten metal halides". *Nature Materials* **5** (4): 271–5. Bibcode:2006NatMa...5..271M. doi:10.1038/nmat1610. PMID 16547520.

[35] Zocher, H (1925). "Uber freiwillige Strukturbildung in Solen. (Eine neue Art anisotrop flqssiger Medien)". *Z Anorg Allg Chem* **147**: 91.

[36] Davidson, Patrick; Gabriel, Jean-Christophe P. (2003). "Mineral Liquid Crystals from Self-Assembly of Anisotropic Nanosystems". *Top Curr Chem* **226**: 119. doi:10.1007/b10827.

[37] Langmuir, I (1938). "The role of attractive and repulsive forces in the formation of tactoids, thixotropic gels, protein crystals and coacervates". *J Chem Phys* **6**: 873. Bibcode:1938JChPh...6..873L. doi:10.1063/1.1750183.

[38] Gabriel, jean-Christophe P.; Sanchez, Clément; Davidson, Patrick (1996). "Observation of Nematic Liquid-Crystal Textures in Aqueous Gels of Smectite Clays". *J. Phys. Chem.* **100**: 11139.

[39] Paineau, E; Philippe, A. M.; Antonova, K.; Bihannic, I.; Davidson, P.; Dozov, I; Gabriel, J.C. P.; Impéror-Clerc, M.; Levitz, P.; Meneau, F.; Michot, L. (2013). "Liquid–crystalline properties of aqueous suspensions of natural clay nanosheets". *Liquid Crystals Reviews* **1**: 110. doi:10.1080/21680396.2013.842130.

[40] Gabriel, Jean-Christophe P.; Camerel, Franck; Lemaire, Bruno J.; Desvaux, Hervé; Davidson, Patrick; Batail, Patrick (2001). "Swollen liquid-crystalline lamellar phase based on extended solid-like sheets". *Nature* **413**. 504. Bibcode:2001Natur.413..504G. doi:10.1038/35097046.

[41] Buka, A.; Palffy-Muhoray, P.; Rácz, Z. (1987). "Viscous fingering in liquid crystals". *Phys. Rev.*

A **36** (8): 3984. Bibcode:1987PhRvA..36.3984B. doi:10.1103/PhysRevA.36.3984.

[42] González-Cinca, R.; Ramírez-Piscina, L.; Casademunt, J.; Hernández-Machado, A.; Kramer, L.; Tóth Katona, T.; Börzsönyi, T.; Buka, Á. (1996). "Phase-field simulations and experiments of faceted growth in liquid crystal". *Physica D* **99** (2–3): 359. Bibcode:1996PhyD...99..359G. doi:10.1016/S0167-2789(96)00162-5.

[43] González-Cinca, R; RamíRez-Piscina, L; Casademunt, J; Hernández-Machado, A; Tóth-Katona, T; Börzsönyi, T; Buka, Á (1998). "Heat diffusion anisotropy in dendritic growth: phase field simulations and experiments in liquid crystals". *Journal of Crystal Growth* **193** (4): 712. Bibcode:1998JCrGr.193..712G. doi:10.1016/S0022-0248(98)00505-3.

[44] Ghosh, S. K. (1984). "A model for the orientational order in liquid crystals". *Il Nuovo Cimento D* **4** (3): 229. Bibcode:1984NCimD...4..229G. doi:10.1007/BF02453342.

[45] Onsager, Lars (1949). "The effects of shape on the interaction of colloidal particles". *Annals of the New York Academy of Sciences* **51** (4): 627. Bibcode:1949NYASA..51..627O. doi:10.1111/j.1749-6632.1949.tb27296.x.

[46] Vroege, G J; Lekkerkerker, H N W (1992). "Phase transitions in lyotropic colloidal and polymer liquid crystals". *Rep. Progr. Phys.* **55** (8): 1241. Bibcode:1992RPPh...55.1241V. doi:10.1088/0034-4885/55/8/003.

[47] Maier W. and Saupe A.; Saupe (1958). "Eine einfache molekulare theorie des nematischen kristallinflussigen zustandes". *Z. Naturforsch. A* (in German) **13**: 564. Bibcode:1958ZNatA..13..564M. doi:10.1515/zna-1958-0716.

[48] Maier W. and Saupe A.; Saupe (1959). "Eine einfache molekular-statistische theorie der nematischen kristallinflussigen phase .1". *Z. Naturforsch. A* (in German) **14**: 882. Bibcode:1959ZNatA..14..882M. doi:10.1515/zna-1959-1005.

[49] Maier W. and Saupe A.; Saupe (1960). "Eine einfache molekular-statistische theorie der nematischen kristallinflussigen phase .2". *Z. Naturforsch. A* (in German) **15**: 287. Bibcode:1960ZNatA..15..287M. doi:10.1515/zna-1960-0401.

[50] McMillan, W. (1971). "Simple Molecular Model for the Smectic A Phase of Liquid Crystals". *Phys. Rev. A* **4** (3): 1238. Bibcode:1971PhRvA...4.1238M. doi:10.1103/PhysRevA.4.1238.

[51] Leslie, F. M. (1992). "Continuum theory for nematic liquid crystals". *Continuum Mechanics and Thermodynamics* **4** (3): 167. Bibcode:1992CMT.....4..167L. doi:10.1007/BF01130288.

[52] Watson, M. C.; Brandt, E. G.; Welch, P. M.; Brown, F. L. H. (2012). "Determining Biomembrane Bending Rigidities from Simulations of Modest Size". *Physical Review Letters* **109** (2): 028102. Bibcode:2012PhRvL.109b8102W. doi:10.1103/PhysRevLett.109.028102.

[53] Fujikake, H.; Takizawa, K.; Aida, T.; Negishi, T.; Kobayashi, M. (1998). "Video camera system using liquid-crystal polarizing filter toreduce reflected light". *IEEE Transactions on Broadcasting* **44** (4): 419. doi:10.1109/11.735903.

[54] Achard, M.F.; Bedel, J.Ph.; Marcerou, J.P.; Nguyen, H.T.; Rouillon, J.C. (2003). "Switching of banana liquid crystal mesophases under field". *European Physical Journal E* **10** (2): 129–34. Bibcode:2003EPJE...10..129A. doi:10.1140/epje/e2003-00016-y. PMID 15011066.

[55] Baus, Marc; Colot, Jean-Louis (1989). "Ferroelectric nematic liquid-crystal phases of dipolar hard ellipsoids". *Phys. Rev. A* **40** (9): 5444. Bibcode:1989PhRvA..40.5444B. doi:10.1103/PhysRevA.40.5444.

[56] Uehara, Hiroyuki; Hatano, Jun (2002). "Pressure-Temperature Phase Diagrams of Ferroelectric Liquid Crystals". *J. Phys. Soc. Jpn.* **71** (2): 509. Bibcode:2002JPSJ...71..509U. doi:10.1143/JPSJ.71.509.

[57] Alkeskjold, Thomas Tanggaard; Scolari, Lara; Noordegraaf, Danny; Lægsgaard, Jesper; Weirich, Johannes; Wei, Lei; Tartarini, Giovanni; Bassi, Paolo; Gauza, Sebastian; Wu, Shin-Tson; Bjarklev, Anders (2007). "Integrating liquid crystal based optical devices in photonic crystal". *Optical and Quantum Electronics* **39** (12–13): 1009. doi:10.1007/s11082-007-9139-8.

[58] Plimpton, R. Gregory "Pool thermometer" U.S. Patent 4,738,549 Issued on April 19, 1988

[59] "Hot-spot detection techniques for ICs". *acceleratedanalysis.com*. Retrieved May 5, 2009.

[60] Kopp, V. I.; Fan, B.; Vithana, H. K. M.; Genack, A. Z. (1998). "Low-threshold lasing at the edge of a photonic stop band in cholesteric liquid crystals". *Optics Express* **23** (21): 1707–1709. Bibcode:1998OptL...23.1707K. doi:10.1364/OL.23.001707. PMID 18091891.

[61] Dolgaleva, Ksenia; Simon K.H. Wei; Svetlana G. Lukishova; Shaw H. Chen; Katie Schwertz; Robert W. Boyd (2008). "Enhanced laser performance of cholesteric liquid crystals doped with oligofluorene dye". *Journal of the Optical Society of America* **25** (9): 1496–1504. Bibcode:2008JOSAB..25.1496D. doi:10.1364/JOSAB.25.001496.

[62] Luzzati, V.; Mustacchi, H.; Skoulios, A. (1957). "Structure of the Liquid-Crystal Phases of the Soap–water System: Middle Soap and Neat Soap". *Nature* **180** (4586): 600. Bibcode:1957Natur.180..600L. doi:10.1038/180600a0.

[63] Bock, H.; Helfrich, W.; Heppke, G. (1992). "Switchable columnar liquid crystalline systems". European Patent EP0529439B1 (filing date: 08/14/1992; publication date: 02/14/1996).

26.13 External links

- "History and Properties of Liquid Crystals". Nobelprize.org. Retrieved June 6, 2009.
- Definitions of basic terms relating to low-molar-mass and polymer liquid crystals (IUPAC Recommendations 2001)
- An intelligible introduction to liquid crystals from Case Western Reserve University
- Liquid Crystal Physics tutorial from the Liquid Crystals Group, University of Colorado
- Liquid Crystals & Photonics Group – Ghent University (Belgium), good tutorial
- Simulation of light propagation in liquid crystals, free program
- Liquid Crystals Interactive Online
- Liquid Crystal Institute Kent State University
- Liquid Crystals a journal by Taylor&Francis
- Molecular Crystals and Liquid Crystals a journal by Taylor & Francis
- Hot-spot detection techniques for ICs
- What are liquid crystals? from Chalmers University of Technology, Sweden
- H. Kleinert & K. Maki (1981). "Lattice Textures in Cholesteric Liquid Crystals" (PDF). *Fortschritte der Physik* **29** (5): 219. Bibcode:1981ForPh..29..219K. doi:10.1002/prop.19810290503.
- Progress in liquid crystal chemistry Thematic series in the Open Access Beilstein Journal of Organic Chemistry
- DoITPoMS Teaching and Learning Package- "Liquid Crystals"
- Bowlic liquid crystal from San Jose State University
- Liquid crystals are distributed by Merck Group (DE), and Yancheng Smiling (CN).

Chapter 27

Grid complex

Latticial metal complex or **grid complex** is a Supramolecular complex of several metal atoms and coordinating ligands which form a grid-like structural motif. The structure formation usually occurs while on thermodynamic molecular self-assembly. They have properties that make them interesting for information technology as the future storage materials.[1] Chelate ligands are used as ligands in Tetrahedral or octahedral structures, which mostly use nitrogen atoms in pyridine like ring systems other than donor centers. Suitable metal ions are in accordance with octahedral coordinating transition metal ions such as Mn or rare tetrahedral Coordinating such as Ag used. [1]

27.1 Nomencluture

The nomenclature is based on [n × m] G, n corresponds to the number of ligands above the metal ion level, m the number below ones. In case of using only one ligand type, the homoleptic grid is formed in a square [nxn] structure. When using different ligands arise heteroleptic complexes, however, compete with the homoleptic. The number of metal ions is always n + m.

27.2 Application

The grid complexes exhibit pH-dependent changes in the optical absorption, electronic spin states and reversible redox states. The latticial metal complexes may thus be used theoretically for information storage and processing in the future.[2][3][4]

27.3 References

[1] J.-M. Lehn et al., Angew. Chem., 2004, 116, S. 3728–3747.

[2] Ruben, Lehn, Chem. Commun., 2003, S. 1338–1339.

[3] Ruben et al., Chem. Eur. J., 2003, 9, S. 291–299.

[4] Müller, Lehn et al., Angew. Chem., 2005, 117, S. 8109–8113.

Chapter 28

Colloidal crystal

A **colloidal crystal** is an ordered array of colloid particles, analogous to a standard crystal whose repeating subunits are atoms or molecules.[1] A natural example of this phenomenon can be found in the gem opal, where spheres of silica assume a close-packed locally periodic structure under moderate compression.[2][3] Bulk properties of a colloidal crystal depend on composition, particle size, packing arrangement, and degree of regularity. Applications include photonics, materials processing, and the study of self-assembly and phase transitions.

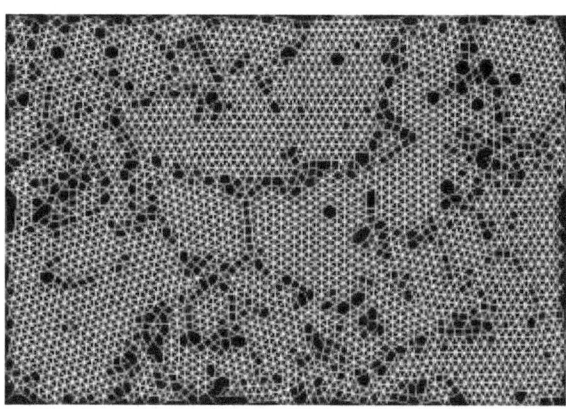

The connectivity of the crystals in the colloidal crystals above. Connections in white indicate that particle has six equally spaced neighbours and therefore forms part of a crystalline domain.

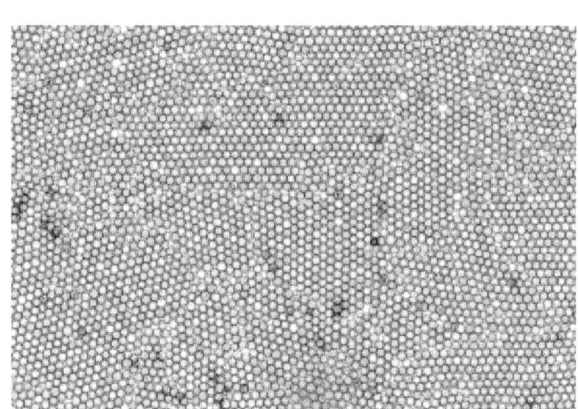

A collection of small 2D colloidal crystals with grain boundaries between them. Spherical glass particles (10 μm diameter) in water.

IUPAC definition

Assembly of colloid particles with a periodic structure that conforms to symmetries familiar from molecular or atomic crystals.

Note: Colloidal crystals may be formed in a liquid medium or during
drying of particle suspension.[4]

28.1 Introduction

A colloidal crystal is a highly ordered array of particles which can be formed over a long range (to about a centimeter). Arrays such as this appear to be analogous to their atomic or molecular counterparts with proper scaling considerations. A good natural example of this phenomenon can be found in precious opal, where brilliant regions of pure spectral color result from close-packed domains of colloidal spheres of amorphous silicon dioxide, SiO_2 (see above illustration). The spherical particles precipitate in highly siliceous pools and form highly ordered arrays after years of sedimentation and compression under hydrostatic and gravitational forces. The periodic arrays of spherical particles make similar arrays of interstitial voids, which act as a natural diffraction grating for light waves in photonic crystals, especially when the interstitial spacing is of the same order of magnitude as the incident lightwave. [5][6]

28.2 Origins

The origins of colloidal crystals go back to the mechanical properties of bentonite sols, and the optical properties of Schiller layers in iron oxide sols. The properties are supposed to be due to the ordering of monodisperse inorganic particles.[7] Monodisperse colloids, capable of forming long-range ordered arrays, existing in nature. The discovery by W.M. Stanley of the crystalline forms of the tobacco and tomato viruses provided examples of this. Using X-ray diffraction methods, it was subsequently determined that when concentrated by centrifuging from dilute water suspensions, these virus particles often organized themselves into highly ordered arrays.

Rod-shaped particles in the tobacco mosaic virus could form a two-dimensional triangular lattice, while a body-centered cubic structure was formed from the almost spherical particles in the tomato Bushy Stunt Virus.[8] In 1957, a letter describing the discovery of "*A Crystallizable Insect Virus*" was published in the journal *Nature*.[9] Known as the Tipula Iridescent Virus, from both square and triangular arrays occurring on crystal faces, the authors deduced the face-centered cubic close-packing of virus particles. This type of ordered array has also been observed in cell suspensions, where the symmetry is well adapted to the mode of reproduction of the organism.[10] The limited content of genetic material places a restriction on the size of the protein to be coded by it. The use of a large number of the same proteins to build a protective shell is consistent with the limited length of RNA or DNA content.[11][12]

It has been known for many years that, due to repulsive Coulombic interactions, electrically charged macromolecules in an aqueous environment can exhibit long-range crystal-like correlations with interparticle separation distances often being considerably greater than the individual particle diameter. In all of the cases in nature, the same iridescence is caused by the diffraction and constructive interference of visible lightwaves which falls under Bragg's law.

Because of the rarity and pathological properties, neither opal nor any of the organic viruses have been very popular in scientific laboratories. The number of experiments exploring the physics and chemistry of these "colloidal crystals" has emerged as a result of the simple methods which have evolved in 20 years for preparing synthetic monodisperse colloids, both polymer and mineral, and, through various mechanisms, implementing and preserving their long-range order formation.

28.3 Trends

Colloidal crystals are receiving increased attention, largely due to their mechanisms of ordering and self-assembly, cooperative motion, structures similar to those observed in condensed matter by both liquids and solids, and structural phase transitions.[13][14] Phase equilibrium has been considered within the context of their physical similarities, with appropriate scaling, to elastic solids. Observations of the interparticle separation distance has shown a decrease on ordering. This led to a re-evaluation of Langmuir's beliefs about the existence of a long-range attractive component in the interparticle potential.[15]

Colloidal crystals have found application in optics as photonic crystals. Photonics is the science of generating, controlling, and detecting photons (packets of light), particularly in the visible and near Infrared, but also extending to the Ultraviolet, Infrared and far IR portions of the electromagnetic spectrum. The science of photonics includes the emission, transmission, amplification, detection, modulation, and switching of lightwaves over a broad range of frequencies and wavelengths. Photonic devices include electro-optic components such as lasers (Light Amplification by Stimulated Emission of Radiation) and optical fiber. Applications include telecommunications, information processing, illumination, spectroscopy, holography, medicine (surgery, vision correction, endoscopy), military (guided missile) technology, agriculture and robotics.

Polycrystalline colloidal structures have been identified as the basic elements of submicrometre colloidal materials science.[16] Molecular self-assembly has been observed in various biological systems and underlies the formation of a wide variety of complex biological structures. This includes an emerging class of mechanically superior biomaterials based on microstructure features and designs found in nature.

The principal mechanical characteristics and structures of biological ceramics, polymer composites, elastomers, and cellular materials are being re-evaluated, with an emphasis on bioinspired materials and structures. Traditional approaches focus on design methods of biological materials using conventional synthetic materials.[17] The uses have been identified in the synthesis of bioinspired materials through processes that are characteristic of biological systems in nature. This includes the nanoscale self-assembly of the components and the development of hierarchical structures.[18]

28.4 Bulk crystals

28.4.1 Aggregation

Aggregation in colloidal dispersions (or stable suspensions) has been characterized by the degree of interparticle attraction.[19] For attractions strong relative to the thermal energy (given by kT), Brownian motion produces irreversibly flocculated structures with growth rates limited by the rate of particle diffusion. This leads to a description using such parameters as the degree of branching, ramification or fractal dimensionality. A reversible growth model has been constructed by modifying the cluster-cluster aggregation model with a finite inter-particle attraction energy.[20][21]

In systems where forces of attraction forces are buffered to some degree, a balance of forces leads to an equilibrium phase separation, that is particles coexist with equal chemical potential in two distinct structural phases. The role of the ordered phase as an elastic colloidal solid has been evidenced by the elastic (or reversible) deformation due to the force of gravity. This deformation can be quantified by the distortion of the lattice parameter, or interparticle spacing.[22]

28.4.2 Viscoelasticity

Periodic ordered lattices behave as linear viscoelastic solids when subjected to small amplitude mechanical deformations. Okano's group experimentally correlated the shear modulus to the frequency of standing shear modes using mechanical resonance techniques in the ultrasonic range (40 to 70 kHz).[23][24] In oscillatory experiments at lower frequencies (< 40 Hz), the fundamental mode of vibration as well as several higher frequency partial overtones (or harmonics) have been observed. Structurally, most systems exhibit a clear instability toward the formation of periodic domains of relatively short-range order Above a critical amplitude of oscillation, plastic deformation is the primary mode of structural rearrangement.[25]

28.4.3 Phase transitions

Equilibrium phase transitions (e.g. order/disorder), an equation of state, and the kinetics of colloidal crystallization have all been actively studied, leading to the development of several methods to control the self-assembly of the colloidal particles.[26] Examples include colloidal epitaxy and space-based reduced-gravity techniques, as well as the use of temperature gradients to define a density gradient.[27] This is somewhat counterintuitive as temperature does not play a role in determining the hard-sphere phase diagram. However, hard-sphere single crystals (size 3 mm) have been obtained from a sample in a concentration regime that would remain in the liquid state in the absence of a temperature gradient.[28]

28.4.4 Phonon dispersion

Using a single colloidal crystal, phonon dispersion of the normal modes of vibration modes were investigated using photon correlation spectroscopy, or dynamic light scattering. This technique relies on the relaxation or decay of concentration (or density) fluctuations. These are often associated with longitudinal modes in the acoustic range. A distinctive increase in the sound wave velocity (and thus the elastic modulus) by a factor of 2.5 has been observed at the structural transition from colloidal liquid to colloidal solid, or point of ordering.[29][30]

28.4.5 Kossel lines

Using a single body-centered cubic colloidal crystal, the occurrence of Kossel lines in diffraction patterns were used to monitor the initial nucleation and subsequent motion caused distortion of the crystal. Continuous or homogeneous deformations occurring beyond the elastic limit produce a 'flowing crystal', where the nucleation site density increases significantly with increasing particle concentration.[31] Lattice dynamics have been investigated for longitudinal as well as transverse modes. The same technique was used to evaluate the crystallization process near the edge of a glass tube. The former might be considered analogous to a homogeneous nucleation event—whereas the latter would clearly be considered a heterogeneous nucleation event, being catalyzed by the surface of the glass tube.

28.4.6 Growth rates

Small-angle laser light scattering has provided information about spatial density fluctuations or the shape of growing crystal grains.[31][32] In addition, confocal laser scanning microscopy has been used to observe crystal growth near a glass surface. Electro-optic shear waves have been induced by an ac pulse, and monitored by reflection spectroscopy as well as light scattering. Kinetics of colloidal crystallization have been measured quantitatively, with nucleation rates being depending on the suspension concentration.[33][34][35] Similarly, crystal growth rates have been shown to decrease linearly with increasing reciprocal concentration.

28.4.7 Microgravity

Experiments performed in microgravity on the Space Shuttle Columbia suggest that the typical face-centered cubic structure may be induced by gravitational stresses. Crystals tend to exhibit the hcp structure alone (random stacking of hexagonally close-packed crystal planes), in contrast with a mixture of (rhcp) and face-centred cubic packing when allowed sufficient time to reach mechanical equilibrium under gravitational forces on Earth.[36] Glassy (disordered or amorphous) colloidal samples have become fully crystallized in microgravity in less than two weeks.

28.5 Thin films

Two-dimensional (thin film) semi-ordered lattices have been studied using an optical microscope, as well as those collected at electrode surfaces. Digital video microscopy has revealed the existence of an equilibrium hexatic phase as well as a strongly first-order liquid-to-hexatic and hexatic-to-solid phase transition.[37] These observations are in agreement with the explanation that melting might proceed via the unbinding of pairs of lattice dislocations.

28.5.1 Long-range order

Long-range order has been observed in thin films of colloidal liquids under oil—with the faceted edge of an emerging single crystal in alignment with the diffuse streaking pattern in the liquid phase. Structural defects have been directly observed in the ordered solid phase as well as at the interface of the solid and liquid phases. Mobile lattice defects have been observed via Bragg reflections, due to the modulation of the light waves in the strain field of the defect and its stored elastic strain energy.[16]

28.5.2 Mobile lattice defects

All of the experiments have led to at least one common conclusion: colloidal crystals may indeed mimic their atomic counterparts on appropriate scales of length (spatial) and time (temporal). Defects have been reported to flash by in the blink of an eye in thin films of colloidal crystals under oil using a simple optical microscope. But quantitatively measuring the rate of its propagation provides an entirely different challenge, which has been measured at somewhere near the speed of sound.

28.6 Non-spherical colloid based crystals

Crystalline thin-films from non-spherical colloids were produced using convective assembly techniques. Colloid shapes included dumbbell, hemisphere, disc, and spherocylinder shapes. Both purely crystalline and plastic crystal phases could be produced, depending on the aspect ratio of the colloidal particle. The particles were crystallized both as 2D (i.e., monolayer) and 3D (i.e., multilayer) structures.[38][39][40][41] The observed lattice and particle orientations experimentally confirmed a body of theoretical work on the condensed phases of non-spherical objects.

28.7 Applications

28.7.1 Photonics

Technologically, colloidal crystals have found application in the world of optics as photonic band gap (PBG) materials (or photonic crystals). Synthetic opals as well as inverse opal configurations are being formed either by natural sedimentation or applied forces, both achieving similar results: long-range ordered structures which provide a natural diffraction grating for lightwaves of wavelength comparable to the particle size.

Novel PBG materials are being formed from opal-semiconductor-polymer composites, typically utilizing the ordered lattice to create an ordered array of holes (or pores) which is left behind after removal or decomposition of the original particles. Residual hollow honeycomb structures provide a relative index of refraction (ratio of matrix to air) sufficient for selective filters. Variable index liquids or liquid crystals injected into the network alter the ratio and band gap.

Such frequency-sensitive devices may be ideal for optical switching and frequency selective filters in the ultraviolet, visible, or infrared portions of the spectrum, as well as higher efficiency antennae at microwave and millimeter wave frequencies.

28.7.2 Self-assembly

Self-assembly is the most common term in use in the modern scientific community to describe the spontaneous aggregation of particles (atoms, molecules, colloids, micelles, etc.) without the influence of any external forces.[18] Large groups of such particles are known to assemble themselves into thermodynamically stable, structurally well-defined arrays, quite reminiscent of one of the 7 crystal systems found in metallurgy and mineralogy (e.g. face-centered cubic, body-centered cubic, etc.). The fundamental difference in equilibrium structure is in the spatial scale of the unit cell (or lattice parameter) in each particular case.

Molecular self-assembly is found widely in biological systems and provides the basis of a wide variety of complex biological structures. This includes an emerging class of mechanically superior biomaterials based on microstructural features and designs found in nature. Thus, self-assembly is also emerging as a new strategy in chemical synthesis and nanotechnology.[17] Molecular crystals, liquid crystals, colloids, micelles, emulsions, phase-separated polymers, thin films and self-assembled monolayers all represent examples of the types of highly ordered structures which are obtained using these techniques. The distinguishing feature of these methods is self-organization.

28.8 See also

- Crystal growth
- Crystal structure
- Ceramic engineering
- Diffusion-limited aggregation
- Nanomaterials
- Nanoparticle
- Nucleation
- Photonic crystal
- Opal
- Sol-gel

28.9 References

[1] Pieranski, Pawel (1983). "Colloidal crystals". *Contemporary Physics* 24: 25. Bibcode:1983ConPh..24...25P. doi:10.1080/00107518308227471.

[2] Jones, J. B.; Sanders, J. V.; Segnit, E. R. (1964). "Structure of Opal". *Nature* 204 (4962): 990. Bibcode:1964Natur.204..990J. doi:10.1038/204990a0.

[3] Darragh, P.J., et al., Opal, Scientific American, Vol. 234, p. 84, (1976)

[4] "Terminology of polymers and polymerization processes in dispersed systems (IUPAC Recommendations 2011)" (PDF). *Pure and Applied Chemistry* 83 (12): 2229–2259. 2011. doi:10.1351/PAC-REC-10-06-03.

[5] Luck, W. (1963). *Ber. Busenges Phys. Chem.* 67: 84. Missing or empty |title= (help)

[6] Hiltner, P. Anne; Krieger, Irvin M. (1969). "Diffraction of light by ordered suspensions". *The Journal of Physical Chemistry* 73 (7): 2386. doi:10.1021/j100727a049.

[7] Langmuir, Irving (1938). "The Role of Attractive and Repulsive Forces in the Formation of Tactoids, Thixotropic Gels, Protein Crystals and Coacervates". *The Journal of Chemical Physics* 6 (12): 873. Bibcode:1938JChPh...6..873L. doi:10.1063/1.1750183.

[8] Bernal, J. D.; Fankuchen, I (1941). "X-Ray and Crystallographic Studies of Plant Virus Preparations: I. Introduction and Preparation of Specimens Ii. Modes of Aggregation of the Virus Particles". *The Journal of General Physiology* 25 (1): 111–46. doi:10.1085/jgp.25.1.111. PMC 2142030. PMID 19873255.

[9] Williams, Robley C.; Smith, Kenneth M. (1957). "A Crystallizable Insect Virus". *Nature* 179 (4551): 119–20. Bibcode:1957Natur.179..119W. doi:10.1038/179119a0. PMID 13400114.

[10] Watson, J.D., Molecular Biology of the Gene, Benjamin, Inc. (1970)

[11] Stanley, W.M. (1937). "Crystalline Form of the Tobacco Mosaic Virus Protein". *American Journal of Botany* 24 (2): 59–68. doi:10.2307/2436720. JSTOR 2436720.

[12] Nobel Lecture: The Isolation and Properties of Crystalline TMV (1946)

[13] Murray, Cherry A.; Grier, David G. (1996). "Video Microscopy of Monodisperse Colloidal Systems". *Annual Review of Physical Chemistry* 47: 421. Bibcode:1996ARPC...47..421M. doi:10.1146/annurev.physchem.47.1.421.

[14] Grier, David G.; Murray, Cherry A. (1994). "The microscopic dynamics of freezing in supercooled colloidal fluids". *The Journal of Chemical Physics* 100 (12): 9088. Bibcode:1994JChPh.100.9088G. doi:10.1063/1.466662.

[15] Russel, W.B., et al., Eds. Colloidal Dispersions (Cambridge Univ. Press, 1989) [see cover]

[16] Ref.14 in Mangels, J.A. and Messing, G.L., Eds., Forming of Ceramics, Microstructural Control Through Colloidal Consolidation, I.A. Aksay, Advances in Ceramics, Vol. 9, p. 94, Proc. Amer. Ceramic Soc. (1984)

[17] Whitesides, G.; Mathias, J.; Seto, C. (1991). "Molecular self-assembly and nanochemistry: A chemical strategy for the synthesis of nanostructures". *Science* 254 (5036): 1312–9. Bibcode:1991Sci...254.1312W. doi:10.1126/science.1962191. PMID 1962191.

[18] Dabbs, Daniel M.; Aksay, Ilhan A. (2000). "Self-Assembledceramicsproduced Bycomplex-Fluidtemplation". *Annual Review of Physical Chemistry* 51 (1): 601–22. Bibcode:2000ARPC...51..601D. doi:10.1146/annurev.physchem.51.1.601. PMID 11031294.

[19] Aubert, Claude; Cannell, David (1986). "Restructuring of colloidal silica aggregates". *Physical Review Letters* 56 (7): 738–741. Bibcode:1986PhRvL..56..738A. doi:10.1103/PhysRevLett.56.738. PMID 10033272.

28.9. REFERENCES

[20] Witten, T.; Sander, L. (1981). "Diffusion-Limited Aggregation, a Kinetic Critical Phenomenon". *Physical Review Letters* **47** (19): 1400. Bibcode:1981PhRvL..47.1400W. doi:10.1103/PhysRevLett.47.1400.

[21] Witten, T.; Sander, L. (1983). "Diffusion-limited aggregation". *Physical Review B* **27** (9): 5686. Bibcode:1983PhRvB..27.5686W. doi:10.1103/PhysRevB.27.5686.

[22] Crandall, R. S.; Williams, R. (1977). "Gravitational Compression of Crystallized Suspensions of Polystyrene Spheres". *Science* **198** (4314): 293–5. Bibcode:1977Sci...198..293C. doi:10.1126/science.198.4314.293. PMID 17770503.

[23] Mitaku, Shigeki; Ohtsuki, Toshiya; Enari, Katsumi; Kishimoto, Akihiko; Okano, Koji (1978). "Studies of Ordered Monodisperse Polystyrene Latexes. I. Shear Ultrasonic Measurements". *Japanese Journal of Applied Physics* **17** (2): 305. Bibcode:1978JaJAP..17..305M. doi:10.1143/JJAP.17.305.

[24] Ohtsuki, Toshiya; Mitaku, Sigeki; Okano, Koji (1978). "Studies of Ordered Monodisperse Latexes. II. Theory of Mechanical Properties". *Japanese Journal of Applied Physics* **17** (4): 627. Bibcode:1978JaJAP..17..627O. doi:10.1143/JJAP.17.627.

[25] Russel, W (1981). "The viscoelastic properties of ordered latices: A self-consistent field theory". *Journal of Colloid and Interface Science* **83**: 163. doi:10.1016/0021-9797(81)90021-7.

[26] Phan, See-Eng; Russel, William; Cheng, Zhengdong; Zhu, Jixiang; Chaikin, Paul; Dunsmuir, John; Ottewill, Ronald (1996). "Phase transition, equation of state, and limiting shear viscosities of hard sphere dispersions". *Physical Review E* **54** (6): 6633. Bibcode:1996PhRvE..54.6633P. doi:10.1103/PhysRevE.54.6633.

[27] Chaikin, P. M.; Cheng, Zhengdong; Russel, William B. (1999). "Controlled growth of hard-sphere colloidal crystals". *Nature* **401** (6756): 893. Bibcode:1999Natur.401..893C. doi:10.1038/44785.

[28] Davis, K. E.; Russel, W. B.; Glantschnig, W. J. (1989). "Disorder-to-Order Transition in Settling Suspensions of Colloidal Silica: X-ray Measurements". *Science* **245** (4917): 507–10. Bibcode:1989Sci...245..507D. doi:10.1126/science.245.4917.507. PMID 17750261.

[29] Cheng, Zhengdong; Zhu, Jixiang; Russel, William; Chaikin, P. (2000). "Phonons in an Entropic Crystal". *Physical Review Letters* **85** (7): 1460–3. Bibcode:2000PhRvL..85.1460C. doi:10.1103/PhysRevLett.85.1460. PMID 10970529.

[30] Penciu, R. S; Kafesaki, M; Fytas, G; Economou, E. N; Steffen, W; Hollingsworth, A; Russel, W. B (2002). "Phonons in colloidal crystals". *Europhysics Letters (EPL)* **58** (5): 699. Bibcode:2002EL.....58..699P. doi:10.1209/epl/i2002-00322-3.

[31] Sogami, I. S.; Yoshiyama, T. (1990). "Kossel line analysis on crystallization in colloidal suspensions". *Phase Transitions* **21** (2–4): 171. doi:10.1080/01411599008206889.

[32] Schätzel, Klaus (1993). "Light scattering – diagnostic methods for colloidal dispersions". *Advances in Colloid and Interface Science* **46**: 309. doi:10.1016/0001-8686(93)80046-E.

[33] Ito, Kensaku; Okumura, Hiroya; Yoshida, Hiroshi; Ise, Norio (1990). "Growth of local structure in colloidal suspensions". *Physical Review B* **41** (8): 5403. Bibcode:1990PhRvB..41.5403I. doi:10.1103/PhysRevB.41.5403.

[34] Yoshida, Hiroshi; Ito, Kensaku; Ise, Norio (1991). "Localized ordered structure in polymer latex suspensions as studied by a confocal laser scanning microscope". *Physical Review B* **44**: 435. Bibcode:1991PhRvB..44..435Y. doi:10.1103/PhysRevB.44.435.

[35] Yoshida, Hiroshi; Ito, Kensaku; Ise, Norio (1991). "Colloidal crystal growth". *Journal of the Chemical Society, Faraday Transactions* **87** (3): 371. doi:10.1039/FT9918700371.

[36] Chaikin, P. M.; Zhu, Jixiang; Li, Min; Rogers, R.; Meyer, W.; Ottewill, R. H.; Sts-73 Space Shuttle Crew; Russel, W. B. (1997). "Crystallization of hard-sphere colloids in microgravity". *Nature* **387** (6636): 883. Bibcode:1997Natur.387..883Z. doi:10.1038/43141.

[37] Armstrong, A J; Mockler, R C; O'Sullivan, W J (1989). "Isothermal-expansion melting of two-dimensional colloidal monolayers on the surface of water". *Journal of Physics: Condensed Matter* **1** (9): 1707. Bibcode:1989JPCM....1.1707A. doi:10.1088/0953-8984/1/9/015.

[38] Hosein, Ian D.; Liddell, Chekesha M. (2007). "Convectively Assembled Asymmetric Dimer-Based Colloidal Crystals". *Langmuir* **23** (21): 10479–85. doi:10.1021/la7007254. PMID 17629310.

[39] Hosein, Ian D.; Liddell, Chekesha M. (2007). "Convectively Assembled nonspherical Mushroom Cap-Based Colloidal Crystals". *Langmuir* **23** (17): 8810–4. doi:10.1021/la700865t. PMID 17630788.

[40] Hosein, Ian D.; John, Bettina S.; Lee, Stephanie H.; Escobedo, Fernando A.; Liddell, Chekesha M. (2009). "Rotator and crystalline films via self-assembly of short-bond-length colloidal dimers". *Journal of Materials Chemistry* **19** (3): 344. doi:10.1039/B818613H.

[41] Hosein, Ian D.; Lee, Stephanie H.; Liddell, Chekesha M. (2010). "Dimer-Based Three-Dimensional Photonic Crystals". *Advanced Functional Materials* **20** (18): 3085. doi:10.1002/adfm.201000134.

28.10 Further reading

- M.W. Barsoum, *Fundamentals of Ceramics*, McGraw-Hill Co., Inc., 1997, ISBN 978-0-07-005521-6.

- W.D. Callister, Jr., *Materials Science and Engineering: An Introduction*, 7th Ed., John Wiley & Sons, Inc., 2006, ISBN 978-0-471-73696-7.

- W.D. Kingery, H.K. Bowen and D.R. Uhlmann, *Introduction to Ceramics*, John Wiley & Sons, Inc., 1976, ISBN 0-471-47860-1.

- M.N. Rahaman, *Ceramic Processing and Sintering*, 2nd Ed., Marcel Dekker Inc., 2003, ISBN 0-8247-0988-8.

- J.S. Reed, *Introduction to the Principles of Ceramic Processing*, John Wiley & Sons, Inc., 1988, ISBN 0-471-84554-X.

- D.W. Richerson, *Modern Ceramic Engineering*, 2nd Ed., Marcel Dekker Inc., 1992, ISBN 0-8247-8634-3.

- W.F. Smith, *Principles of Materials Science and Engineering*, 3rd Ed., McGraw-Hill, Inc., 1996, ISBN 978-0-07-059241-4.

- Wachtman, John B. (1996). *Mechanical Properties of Ceramics*. New York: Wiley-Interscience, John Wiley & Son's. ISBN 0-471-13316-7.

- L.H. VanVlack, *Physical Ceramics for Engineers*, Addison-Wesley Publishing Co., Inc., 1964, ISBN 0-201-08068-0.

- *Colloidal Dispersions*, Russel, W.B., et al., Eds., Cambridge Univ. Press (1989)

- *Sol-Gel Science: The Physics and Chemistry of Sol-Gel Processing* by C. Jeffrey Brinker and George W. Scherer, Academic Press (1990)

- *Sol-Gel Materials: Chemistry and Applications* by John D. Wright, Nico A.J.M. Sommerdijk

- *Sol-Gel Technologies for Glass Producers and Users* by Michel A. Aegerter and M. Mennig

- *Sol-Gel Optics: Processing and Applications*, Lisa Klein, Springer Verlag (1994)

28.11 External links

- University of Utrecht
- Nucleation and Growth

Chapter 29

Self-assembled monolayer

Self-assembled monolayers (SAM) of organic molecules are molecular assemblies formed spontaneously on surfaces by adsorption and are organized into more or less large ordered domains.[1][2] In some cases molecules that form the monolayer do not interact strongly with the substrate. This is the case for instance of the two-dimensional supramolecular networks[3] of e.g. Perylene-tetracarboxylicacid-dianhydride (PTCDA) on gold[4] or of e.g. porphyrins on highly oriented pyrolitic graphite (HOPG).[5] In other cases the molecules possess a head group that has a strong affinity to the substrate and anchors the molecule to it.[1] Such a SAM consisting of a head group, tail and functional end group is depicted in Figure 1. Common head groups include thiols, silanes, phosphonates, etc.

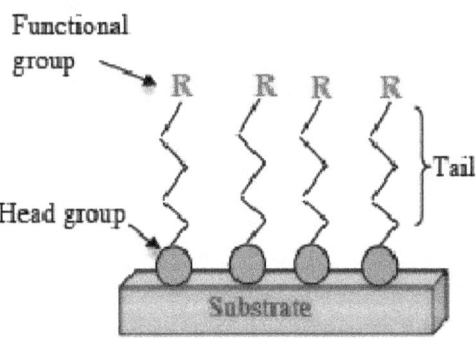

Figure 1. Representation of a SAM structure

SAMs are created by the chemisorption of "head groups" onto a substrate from either the vapor or liquid phase[6][7] followed by a slow organization of "tail groups".[8] Initially, at small molecular density on the surface, adsorbate molecules form either a disordered mass of molecules or form an ordered two-dimensional "lying down phase",[6] and at higher molecular coverage, over a period of minutes to hours, begin to form three-dimensional crystalline or semicrystalline structures on the substrate surface.[9] The "head groups" assemble together on the substrate, while the tail groups assemble far from the substrate. Areas of close-packed molecules nucleate and grow until the surface of the substrate is covered in a single monolayer.

Adsorbate molecules adsorb readily because they lower the surface free-energy of the substrate[1] and are stable due to the strong chemisorption of the "head groups." These bonds create monolayers that are more stable than the physisorbed bonds of Langmuir–Blodgett films.[10][11] A Trichlorosilane based "head group", for example in a FDTS molecule reacts with an hydroxyl group on a substrate, and forms very stable, covalent bond [R-Si-O-substrate] with an energy of 452 kJ/mol. Thiol-metal bonds, that are on the order of 100 kJ/mol, making the bond a fairly stable in a variety of temperature, solvents, and potentials.[9] The monolayer packs tightly due to van der Waals interactions,[1][11] thereby reducing its own free energy.[1] The adsorption can be described by the Langmuir adsorption isotherm if lateral interactions are neglected. If they cannot be neglected, the adsorption is better described by the Frumkin isotherm.[9]

29.1 Types

Selecting the type of head group depends on the application of the SAM.[1] Typically, head groups are connected to a molecular chain in which the terminal end can be functionalized (i.e. adding –OH, –NH2, –COOH, or –SH groups) to vary the wetting and interfacial properties.[10][12] An appropriate substrate is chosen to react with the head group. Substrates can be planar surfaces, such as silicon and metals, or curved surfaces, such as nanoparticles. Alkanethiols are the most commonly used molecules for SAMs. Alkanethiols are molecules with an alkyl chain, (C-C)^n chain, as the back bone, a tail group, and a S-H head group. Other types of interesting molecules include aromatic thiols, of interest in molecular electronics, in which the alkane chain is (partly) replaced by aromatic rings. An example is the dithiol 1,4-Benzenedimethanethiol (SHCH$_2$C$_6$H$_4$CH$_2$SH)). Interest in such dithiols stems from the possibility of linking the two sulfur ends to metallic contacts, which was first used in molecular conduction measurements.[13] Thiols are frequently used on noble metal substrates because of the strong affinity of sulfur for these metals. The sulfur gold interaction is semi-covalent and has a strength of approximately 45kcal/mol. In addition, gold is an inert and biocompatible material that is easy to acquire. It is also easy to pattern via lithography, a useful feature for applications in nanoelectromechanical systems (NEMS).[1] Additionally, it can withstand harsh chemical cleaning treatments.[9] Recently other chalcogenide SAMs: selenides and tellurides have attracted attention[14][15] in a search for different bonding characteristics to substrates affecting the SAM characteristics and which could be of interest in some applications such as molecular electronics. Silanes are generally used on nonmetallic oxide surfaces;[1] however monolayers formed from covalent bonds between silicon and carbon or oxygen cannot be considered self assembled because they do not form reversibly. Self-assembled monolayers of thiolates on noble metals are a special case because the metal-metal bonds become reversible after the formation of the thiolate-metal complex.[16] This reversibility is what gives rise to vacancy islands and it is why SAMs of alkanethiolates can be thermally desorbed and undergo exchange with free thiols.[17]

29.2 Preparation

Metal substrates for use in SAMs can be produced through physical vapor deposition techniques, electrodeposition or electroless deposition.[1] Thiol or selenium SAMs produced by adsorption from solution are typically made by immersing a substrate into a dilute solution of alkane thiol in ethanol, though many different solvents can be used[1] besides use of pure liquids.[15] While SAMs are often allowed to form over 12 to 72 hours at room temperature,[9][18] SAMs of alkanethiolates form within minutes.[19][20] Special attention is essential in some cases, such as that of dithiol SAMs to avoid problems due to oxidation or photoinduced processes, which can affect terminal groups and lead to disorder and multilayer formation.[21][22] In this case appropriate choice of solvents, their degassing by inert gasses and preparation in the absence of light is crucial[21][22] and allows formation of "standing up" SAMs with free –SH groups. Self-assembled monolayers can also be adsorbed from the vapor phase.[7][23] In some cases when obtaining an ordered assembly is difficult or when different density phases need to be obtained substitutional self-assembly is used. Here one first forms the SAM of a given type of molecules, which give rise to ordered assembly and then a second assembly phase is performed (e.g. by immersion into a different solution). This method has also been used to give information on relative binding strengths of SAMs with different head groups and more generally on self-assembly characteristics.[17][24]

29.3 Characterization

The thicknesses of SAMs can be measured using ellipsometry and X-ray photoelectron spectroscopy (XPS), which also give information on interfacial properties.[21][25] The order in the SAM and orientation of molecules can be probed by Near Edge Xray Absorption Fine Structure (NEXAFS) and Fourier Transform Infrared Spectroscopy in Reflection Absorption Infrared Spectroscopy (RAIRS)[19][22] studies. Numerous other spectroscopic techniques are used[7] such as Second-harmonic generation (SHG), Sum-frequency generation (SFG), Surface-enhanced Raman scattering (SERS), as well as[26] High-resolution electron energy loss spectroscopy (HREELS). The structures of SAMs are commonly determined using scanning probe microscopy techniques such as atomic force microscopy (AFM) and scanning tunneling microscopy (STM). STM has been able to help understand the mechanisms of SAM formation as well as determine the important structural features that lend SAMs their integrity as surface-stable entities. In particular STM can image the shape, spatial distribution, terminal groups and their packing structure. AFM offers an equally powerful tool without the requirement of the SAM being conducting or semi-conducting. AFM has been used to determine chemical functionality, conductance, magnetic properties, surface charge, and frictional forces of SAMs.[27] More recently, however, diffractive methods have also been used.[1] The structure can be used to characterize the kinetics and defects found on the monolayer surface. These techniques have also shown physical differences between SAMs with planar substrates and nanoparticle substrates. An alternative characterisation instrument for measuring the self-assembly in real time is dual polarisation interferometry where the refractive index, thickness, mass and birefringence of the self assembled layer are quantified at high resolution. Contact angle measurements can be used to determine the surface free-energy which reflects the average composition of the surface of the SAM and can be used to probe the kinetics and thermodynamics of the formation of SAMs.[19][20] The kinetics of adsorption and temperature induced desorption as well as information on structure can also be obtained in real time by ion scattering techniques such as low energy ion scattering (LEIS) and time of flight direct recoil spectroscopy (TOFDRS).[23]

29.3.1 Defects

Defects due to both external and intrinsic factors may appear. External factors include the cleanliness of the substrate, method of preparation, and purity of the adsorbates.[1][9] SAMs intrinsically form defects due to the thermodynamics of formation, e.g. thiol SAMs on gold typically exhibit etch pits (monatomic vacancy islands) likely due to extraction of adatoms from the substrate and formation of adatom-adsorbate moieties. Recently, a new type of fluorosurfactants have found that can form nearly perfect monolayer on gold substrate due to the increase of mobility of gold surface atoms.[28][29][30]

29.3.2 Nanoparticle properties

The structure of SAMs is also dependent on the curvature of the substrate. SAMs on nanoparticles, including colloids and nanocrystals, "stabilize the reactive surface of the particle and present organic functional groups at the particle-solvent interface".[1] These organic functional groups are useful for applications, such as immunoassays, that are dependent on chemical composition of the surface.[1]

29.4 Kinetics

There is evidence that SAM formation occurs in two steps: an initial fast step of adsorption and a second slower step of monolayer organization. Adsorption occurs at the liquid–liquid, liquid–vapor, and liquid-solid interfaces. The transport of molecules to the surface occurs due to a combination of diffusion and convective transport. According to the Langmuir or Avrami kinetic model the rate of deposition onto the surface is proportional to the free space of the surface.[6]

$$\mathbf{k}(1 - \theta) = \frac{d\theta}{dt}.$$

Where θ is the proportional amount of area deposited and $\mathbf{k}$ is the rate constant. Although this model is robust it is only used for approximations because it fails to take into account intermediate processes.[6] Dual polarisation interferometry being a real time technique with ~10 Hz resolution can measure the kinetics of monolayer self-assembly directly.

Once the molecules are at the surface the self-organization occurs in three phases:[6]

1. A low-density phase with random dispersion of molecules on the surface.

2. An intermediate-density phase with conformational disordered molecules or molecules lying flat on the surface.

3. A high-density phase with close-packed order and molecules standing normal to the substrate's surface.

The phase transitions in which a SAM forms depends on the temperature of the environment relative to the triple point temperature, the temperature in which the tip of the low-density phase intersects with the intermediate-phase region. At temperatures below the triple point the growth goes from phase 1 to phase 2 where many islands form with the final SAM structure, but are surrounded by random molecules. Similar to nucleation in metals, as these islands grow larger they intersect forming boundaries until they end up in phase 3, as seen below.[6]

At temperatures above the triple point the growth is more complex and can take two paths. In the first path the heads of the SAM organize to their near final locations with the tail groups loosely formed on top. Then as they transit to phase 3, the tail groups become ordered and straighten out. In the second path the molecules start in a lying down position along the surface. These then form into islands of ordered SAMs, where they grow into phase 3, as seen below.[6]

The nature in which the tail groups organize themselves into a straight ordered monolayer is dependent on the intermolecular attraction, or Van der Waals forces, between the tail groups. To minimize the free energy of the organic layer the molecules adopt conformations that allow high degree of Van der Waals forces with some hydrogen bonding. The small size of the SAM molecules are important here because Van der Waals forces arise from the dipoles of molecules and are thus much weaker than the surrounding surface forces at larger scales. The assembly process begins with a small group of molecules, usually two, getting close enough that the Van der Waals forces overcome the surrounding force. The forces between the molecules orient them so they are in their straight, optimal, configuration. Then as other molecules come close by they interact with these already organized molecules in the same fashion and become a part of the conformed group. When this occurs across a large area the molecules support each other into forming their SAM shape seen in Figure 1. The orientation of the molecules can be described with two parameters: α and β. α is the angle of tilt of the backbone from the surface normal. In typical applications α varies from 0 to 60 degrees depending on the substrate and type of SAM molecule. β is the angle of rotation along the long axis of tee molecule. β is usually between 30 and 40 degrees.[1] In some cases existence of kinetic traps hindering the final ordered orientation has been pointed out.[7] Thus in case of dithiols formation of a "lying down" phase[7] was considered an impediment to formation of "standing up" phase, however various recent studies indicate this is not the case.[21][22]

Many of the SAM properties, such as thickness, are determined in the first few minutes. However, it may take hours for defects to be eliminated via annealing and for final SAM properties to be determined.[6][9] The exact kinetics of SAM formation depends on the adsorbate, solvent and substrate properties. In general, however, the kinetics are dependent on both preparations conditions and material properties of the solvent, adsorbate and substrate.[6] Specifically, kinetics for adsorption from a liquid solution are dependent on:[1]

- Temperature – room-temperature preparation improves kinetics and reduces defects.

- Concentration of adsorbate in the solution – low concentrations require longer immersion times[1][9] and often create highly crystalline domains.[9]

- Purity of the adsorbate – impurities can affect the final physical properties of the SAM

- Dirt or contamination on the substrate – imperfections can cause defects in the SAM

The final structure of the SAM is also dependent on the chain length and the structure of both the adsorbate and the substrate. Steric hindrance and metal substrate properties, for example, can affect the packing density of the film,[1][9] while chain length affects SAM thickness.[11] Longer chain length also increases the thermodynamic stability.[1]

29.5 Patterning

29.5.1 1. Locally attract

This first strategy involves locally depositing self-assembled monolayers on the surface only where the nanostructure will later be located. This strategy is advantageous because it involves high throughput methods that generally involve fewer steps than the other two strategies. The major techniques that use this strategy are:[31]

- Micro-contact printing

    Micro-contact printing or soft lithography is analogous to printing ink with a rubber stamp. The SAM molecules are inked onto a pre-shaped elastomeric stamp with a solvent and transferred to the substrate surface by stamping. The SAM solution is applied to the entire stamp but only areas that make contact with the surface allow transfer of the SAMs. The transfer of the SAMs is a complex diffusion process that depends on the type of molecule, concentration, duration of contact, and pressure applied. Typical stamps use PDMS because its elastomeric properties, E = 1.8 MPa, allow it to fit the countour of micro surfaces and its low surface energy, $\gamma = 21.6$ dyn/cm^2. This is a parallel process and can thus place nanoscale objects over a large area in a short time.[1]

- Dip-pen nanolithography

    Dip-pen nanolithography is a process that uses an atomic force microscope to transfer molecules on the tip to a substrate. Initially the tip is dipped into a reservoir with an ink. The ink on the tip evaporates and leaves the desired molecules attached to the tip. When the tip is brought into contact with the surface a water meniscus forms between the tip and the surface resulting in the diffusion of molecules from the tip to the surface. These tips can have radii in the tens of nanometers, and thus SAM molecules can be very precisely deposited onto a specific location of the surface. This process was discovered by Chad Mirkin and co-workers at Northwestern University.[32]

29.5.2 2. Locally remove

The locally remove strategy begins with covering the entire surface with a SAM. Then individual SAM molecules are removed from locations where the deposition of nanostructures is not desired. The end result is the same as in the locally attract strategy, the difference being in the way this is achieved. The major techniques that use this strategy are:[31]

- Scanning tunneling microscope

    The scanning tunneling microscope can remove SAM molecules in many different ways. The first is to remove them mechanically by dragging the tip across the substrate surface. This is not the most desired technique as these tips are expensive and dragging them causes a lot of wear and reduction of the tip quality. The second way is to degrade or desorb the SAM molecules by shooting them with an electron beam. The scanning tunneling microscope can also remove SAMs by field desorption and field enhanced surface diffusion.[31]

- Atomic force microscope

The most common use of this technique is to remove the SAM molecules in a process called shaving, where the atomic force microscope tip is dragged along the surface mechanically removing the molecules. An atomic force microscope can also remove SAM molecules by local oxidation nanolithography.[31]

- Ultraviolet irradiation

  In this process, UV light is projected onto the surface with a SAM through a pattern of apperatures in a chromium film. This leads to photo oxidation of the SAM molecules. These can then be washed away in a polar solvent. This process has 100 nm resolutions and requires exposure time of 15–20 minutes.[1]

29.5.3 3. Modify tail groups

The final strategy focuses not on the deposition or removal of SAMS, but the modification of terminal groups. In the first case the terminal group can be modified to remove functionality so that SAM molecule will be inert. In the same regards the terminal group can be modified to add functionality[33] so it can accept different materials or have different properties than the original SAM terminal group. The major techniques that use this strategy are:[31]

- Focused electron beam and ultraviolet irradiation

  Exposure to electron beams and UV light changes the terminal group chemistry. Some of the changes that can occur include the cleavage of bonds, the forming of double carbon bonds, cross-linking of adjacent molecules, fragmentation of molecules, and confromational disorder.[1]

- Atomic force microscope

  A conductive AFM tip can create an electrochemical reaction that can change the terminal group.[31]

29.6 Applications

29.6.1 Thin-film SAMs

SAMs are an inexpensive and versatile surface coating for applications including control of wetting and adhesion, chemical resistance, bio compatibility, sensitization, and molecular recognition for sensors and nano fabrication.[6] Areas of application for SAMs include biology, electrochemistry and electronics, nanoelectromechanical systems (NEMS) and microelectromechanical systems (MEMS), and everyday household goods. SAMs can serve as models for studying membrane properties of cells and organelles and cell attachment on surfaces.[1] SAMs can also be used to modify the surface properties of electrodes for electrochemistry, general electronics, and various NEMS and MEMS.[1] For example, the properties of SAMs can be used to control electron transfer in electrochemistry.[34] They can serve to protect metals from harsh chemicals and etchants. SAMs can also reduce sticking of NEMS and MEMS components in humid environments. In the same way, SAMs can alter the properties of glass. A common household product, Rain-X, utilizes SAMs to create a hydrophobic monolayer on car windshields to keep them clear of rain. Another application is an anti-adhesion coating on nanoimprint lithography (NIL) tools and stamps. One can also coat injection molding tools for polymer replication with a Perfluordecyltrichlorosilane SAM.[35]

Thin film SAMs can also be placed on nanostructures. In this way they functionalize the nanostructure. This is advantageous because the nanostructure can now selectively attach itself to other molecules or SAMs. This technique is useful in biosensors or other MEMS devices that need to separate one type of molecule from its environment. One example is the use of magnetic nanoparticles to remove a fungus from a blood stream. The nanoparticle is coated with a SAM that binds to the fungus. As the contaminated blood is filtered through a MEMS device the magnetic nanoparticles are inserted into the blood where they bind to the fungus and are then magnetically driven out of the blood stream into a nearby laminar waste stream.[36]

29.6.2 Patterned SAMs

SAMs are also useful in depositing nanostructures, because each adsorbate molecule can be tailored to attract two different materials. Current techniques utilize the head to attract to a surface, like a plate of gold. The terminal group is then modified to attract a specific material like a particular nanoparticle, wire, ribbon, or other nanostructure. In this way, wherever the a SAM is patterned to a surface there will be nanostructures attached to the tail groups. One example is the use of two types of SAMs to align single wall carbon nanotubes, SWNTs. Dip pen nanolithography was used to pattern a 16-mercaptohexadecanoic acid (MHA)SAM and the rest of the surface was passivated with 1-octadecanethiol (ODT) SAM. The polar solvent that is carrying the SWNTs is attracted to the hydrophilic MHA; as the solvent evaporates, the SWNTs are close enough to the MHA SAM to attach to it due to Van der Waals forces. The nanotubes thus line up with the MHA-ODT boundary. Using this technique Chad Mirkin, Schatz and their co-workers were able to make complex two-dimensional shapes, a representation of a shape created is shown to the right.[31][37] Another application of patterned SAMs is the functionalization of biosensors. The tail groups can be modified so they have an affinity for cells, proteins, or molecules. The SAM can then be placed onto a biosensor so that binding of these molecules can be detected. The ability to pattern these SAMs allows them to be placed in configurations that increase sensitivity and do not damage or interfere with other components of the biosensor.[27]

29.6.3 Metal organic superlattices

There has been considerable interest in use of SAMs for new materials e.g. via formation of two- or three-dimensional metal organic superlattices by assembly of SAM capped nanoparticles[38] or layer by layer SAM-nanoparticle arrays using dithiols.[39]

29.7 References

[1] Love; et al. (2005). "Self-Assembled Monolayers of Thiolates on Metals as a Form of Nanotechnology". *Chem. Rev.* **105** (4): 1103–1170. doi:10.1021/cr0300789. PMID 15826011.

[2] Barlow, S.M.; Raval R.. (2003). "Complex organic molecules at metal surfaces: bonding, organisation and chirality". *Surface Science reports* **50** (6–8): 201–341. Bibcode:2003SurSR..50..201B. doi:10.1016/S0167-5729(03)00015-3.

[3] Elemans, J.A.A.W.; Lei S., De Feyter S. (2009). "Molecular and Supramolecular Networks on Surfaces: From Two-Dimensional Crystal Engineering to Reactivity". *Angew. Chem. Int. Ed.* **48** (40): 7298–7332. doi:10.1002/anie.200806339.

[4] Witte, G.; Wöll Ch. (2004). "Growth of aromatic molecules on solid substrates for applications in organic electronics". *Journal of Material Research* **19** (7): 1889–1916. Bibcode:2004JMatR..19.1889W. doi:10.1557/JMR.2004.0251.

[5] De Feyter, S.; De Schreyer F.C. (2003). "Two-dimensional supramolecular self-assembly probed by scanning tunneling microscopy". *Chemical Society Reviews* **32** (3): 139–150. doi:10.1039/b206566p. PMID 12792937.

[6] Schwartz, D.K., Mechanisms and Kinetics of Self-Assembled Monolayer Formation (2001). "Mechanisms and kinetics of self-assembled monolayer formation". *Annu. Rev. Phys. Chem.* **52**: 107–37. Bibcode:2001ARPC...52..107S. doi:10.1146/annurev.physchem.52.1.107. PMID 11326061.

[7] Schreiber, F (30 November 2000). "Structure and growth of self-assembling monolayers". *Progress in Surface Science* **65** (5–8): 151–257. Bibcode:2000PrSS...65..151S. doi:10.1016/S0079-6816(00)00024-1.

[8] Wnek, Gary, Gary L. Bowlin (2004). *Encyclopedia of Biomaterials and Biomedical Engineering*. Informa Healthcare. pp. 1331–1333.

[9] Vos, Johannes G., Robert J. Forster, Tia E. Keyes (2003). *Interfacial Supramolecular Assemblies*. Wiley. pp. 88–94.

[10] Madou, Marc (2002). *Fundamentals of Microfabrication: The Science of Miniaturization*. CRC. pp. 62–63.

[11] Kaifer, Angel (2001). *Supramolecular Electrochemistry. Coral Gables*. Wiley VCH. pp. 191–193.

[12] Saliterman, Steven (2006). *Self-assembled monolayers (SAMs). Fundamentals of BioMEMS and Medical Microdevices*. SPIE Press. pp. 94–96.

[13] Andres, R.P.; Bein T.; Dorogi M.; Feng S.; Henderson J.I.; Kubiak C.P.; Mahoney W.; Osifchin R.G.; Reifenberger R. (1996). "Coulomb Staircase at Room Temperature in a Self-Assembled Molecular Nanostructure". *Science* **272** (5266): 1323–1325. Bibcode:1996Sci...272.1323A. doi:10.1126/science.272.5266.1323. PMID 8662464.

[14] Shaporenko, A.; Muller J.; Weidner T.; Terfort A.; Zharnikov M. (2007). "Balance of Structure-Building Forces in Selenium-Based Self-Assembled Monolayers". *Journal of the American Chemical Society* **129** (8): 2232–2233. doi:10.1021/ja068916e.

[15] Subramanian, S.; Sampath S. (2007). "Enhanced stability of short- and long-chain diselenide self-assembled monolayers on gold probed by electrochemistry, spectroscopy, and microscopy". *Journal of Colloid and Interface Science* **312** (2): 413–424. doi:10.1016/j.jcis.2007.03.021. PMID 17451727.

[16] Bucher, Jean-Pierre; Santesson, Lars, Kern, Klaus (31 March 1994). "Thermal Healing of Self-Assembled Organic Monolayers: Hexane- and Octadecanethiol on Au(111) and Ag(111)". *Langmuir* **10** (4): 979–983. doi:10.1021/la00016a001.

[17] Schlenoff, Joseph B.; Li, Ming, Ly, Hiep (30 November 1995). "Stability and Self-Exchange in Alkanethiol Monolayers". *Journal of the American Chemical Society* **117** (50): 12528–12536. doi:10.1021/ja00155a016.

[18] Wysocki. "Self-Assembled Monolayers (SAMs) as Collision Surfaces for Ion Activation" (PDF).

[19] Nuzzo, Ralph G.; Allara, David L. (31 May 1983). "Adsorption of bifunctional organic disulfides on gold surfaces". *Journal of the American Chemical Society* **105** (13): 4481–4483. doi:10.1021/ja00351a063.

29.7. REFERENCES

[20] Bain, Colin D.; Troughton, E. Barry; Tao, Yu Tai; Evall, Joseph; Whitesides, George M.; Nuzzo, Ralph G. (31 December 1988). "Formation of monolayer films by the spontaneous assembly of organic thiols from solution onto gold". *Journal of the American Chemical Society* 111 (1): 321–335. doi:10.1021/ja00183a049.

[21] Hamoudi, H.; Prato M., Dablemont C., Cavalleri O., Canepa M., Esaulov, V. A. (2010). "Self-Assembly of 1,4-Benzenedimethanethiol Self-Assembled Monolayers on Gold". *Langmuir* 26 (10): 7242–7247. doi:10.1021/la904317b. PMID 20199099.

[22] Hamoudi, H.; Guo Z.,Prato M., Dablemont C., Zheng W.Q., Bourguignon B., Canepa M., Esaulov, V. A.; Prato, Mirko; Dablemont, Céline; Zheng, Wan Quan; Bourguignon, Bernard; Canepa, Maurizio; Esaulov, Vladimir A. (2008). "On the self assembly of short chain alkanedithiols". *Physical Chemistry Chemical Physics* 10 (45): 6836–6841. Bibcode:2008PCCP...10.6836H. doi:10.1039/B809760G. PMID 19015788.

[23] Alarcon, L.S.; Chen L., Esaulov, V. A., Gayone J.E., Sanchez E., Grizzi O. (2010). "Thiol Terminated 1,4-Benzenedimethanethiol Self-Assembled Monolayers on Au(111) and InP(110) from Vapor Phase". *Journal of Physical Chemistry C* 114 (47): 19993–19999. doi:10.1021/jp1044157.

[24] Chaudhari, V.; Harish N.M.K.; Sampath S.; Esaulov V.A. (2011). "Substitutional Self-Assembly of Alkanethiol and Selenol SAMs from a Lying-Down Doubly Tethered Butanedithiol SAM on Gold". *Journal of Physical Chemistry C* 115 (33): 16518–16523. doi:10.1021/jp2042922.

[25] Prato, M.; Moroni R.; Bisio F.; Rolandi R.; Mattera L.; Cavalleri O.; Canepa M. (2008). "Optical Characterization of Thiolate Self-Assembled Monolayers on Au(111)". *Journal of Physical Chemistry C* 112 (10): 3899–3906. doi:10.1021/jp711194s.

[26] Kato, H.; Noh J.; Hara M.; Kawai M. (2002). "An HREELS Study of Alkanethiol Self-Assembled Monolayers on Au(111)". *Journal of Physical Chemistry C* 106 (37): 9655–9658. doi:10.1021/jp020968c.

[27] Smith; et al. (2004). "Patterning Self-Assembled Monolayers". *Progress in Surface Science* 75: 1–68. Bibcode:2004PrSS...75....1S. doi:10.1016/j.progsurf.2003.12.001.

[28] Yongan Tang, Jiawei Yan, Xiaoshun Zhou, Yongchun Fu, Bingwei Mao. An STM study on nonionic fluorosurfactant zonyl FSN self-assembly on Au(111): large domains, few defects, and good stability. Langmuir 2008, 24, 13245-13249.

[29] Jiawei Yan, Yongan Tang, Chunfeng Sun, Yuzhuan Su, Bingwei Mao. STM Study on Nonionic Fluorosurfactant Zonyl FSN Self-Assembly on Au(100): (3/1/−1/1) Molecular Lattice, Corrugations, and Adsorbate-Enhanced Mobility. Langmuir 2010, 26, 3829-3834

[30] Yongan Tang, Jiawei Yan, Feng Zhu, Chunfeng Sun, Bingwei Mao. Comparative electrochemical scanning tunneling microscopy study of nonionic fluorosurfactant zonyl FSN self-assembled monolayers on Au(111) and Au(100): a potential-induced structural transition. Langmuir 2011, 27, 943-947

[31] Seong, Jin Koh (2007). "Strategies for Controlled Placement of Nanoscale Building Blocks". *Nanoscale Res Lett* 2 (11): 519–545. Bibcode:2007NRL.....2..519K. doi:10.1007/s11671-007-9091-3. PMC 3246612. PMID 21794185.

[32] Piner, R.D; Zhu, J; Xu, F; Hong, S; Mirkin, C.A (1999). "Dip-Pen Nanolithography". *Science* 283 (5402): 661–663. doi:10.1126/science.283.5402.661. PMID 9924019.

[33] Lud, S.Q; Neppl, S; Xu, F; Feulner, P; Stutzmann, M; Jordan, Rainer; Feulner, Peter; Stutzmann, Martin; Garrido, Jose A. (2010). "Controlling Surface Functionality through Generation of Thiol Groups in a Self-Assembled Monolayer". *Langmuir* 26 (20): 15895–900. doi:10.1021/la102225r.

[34] Lud, S.Q; Steenackers, M; Bruno, P; Gruen, D.M; Feulner, P; Garrido, J.A; Stutzmann, M; Stutzmann, M (2006). "Chemical Grafting of Biphenyl Self-Assembled Monolayers on Ultrananocrystalline Diamond". *Journal of the American Chemical Society* 128 (51): 16884–16891. doi:10.1021/ja0657049. PMID 17177439.

[35] Cech J; Taboryski R (2012). "Stability of FDTS monolayer coating on aluminum injection molding tools". *Applied Surface Science* 259: 538–541. Bibcode:2012ApSS..259..538C. doi:10.1016/j.apsusc.2012.07.078.

[36] Yung "et all"; Fiering, J; Mueller, AJ; Ingber, DE (2009). "Micromagnetic–microfluidic blood cleansing device". *Lab on a Chip* 9 (9): 1171–1177. doi:10.1039/b816986a. PMID 19370233.

[37] Garcia, R.; Martinez, R.V; Martinez, J (2005). "Nano Chemistry and Scanning Probe Nanolithographies". *Chemical Society Reviews* 35 (1): 29–38. doi:10.1039/b501599p. PMID 16365640.

[38] Kiely, C.J.; Fink J., Brust M., Bethell D? Schiffrin D.J. (1999). "Spontaneous ordering of bimodal ensembles of nanoscopic gold clusters". *Nature* 396 (3): 444–446. doi:10.1038/24808.

[39] Vijaya Sarathy, K.; John Thomas P.,Kulkarni G.U., Rao C.N.R. (1999). "Superlattices of Metal and Metal–Semiconductor Quantum Dots Obtained by Layer-by-Layer Deposition of Nanoparticle Arrays". *Journal of Physical Chemistry* 103 (3): 399–401. doi:10.1021/jp983836l.

29.8 Further reading

- Sagiv, J.; Polymeropoulos, E.E. (1978). "ADSORBED MONOLAYERS - MOLECULAR-ORGANIZATION AND ELECTRICAL-PROPERTIES". *BERICHTE DER BUNSEN-GESELLSCHAFT-PHYSICAL CHEMISTRY CHEMICAL PHYSICS* **82** (9): 883–883. doi:10.1002/bbpc.19780820917.

- I. Rubinstein, E. Sabatani, R. Maoz and J. Sagiv, Organized Monolayers on Gold Electrodes, in *Electrochemical Sensors for Biomedical Applications*, C.K.N. Li (Ed.), The Electrochemical Society 1986: 175.

- Faucheux, N.; Schweiss, R.; Lützow, K.; Werner, C.; Groth, T. (2004). "Self-assembled monolayers with different terminating groups as model substrates for cell adhesion studies". *Biomaterials* **25**: 2721–2730. doi:10.1016/j.biomaterials.2003.09.069.

- Wasserman, S. R.; Tao, Y. T.; Whitesides, G. M. (1989). "Structure and Reactivity of Alkylsiloxane Monolayers Formed by Reaction of Alkyltrichlorosilanes on Silicon Substrates". *Langmuir* **5**: 1074–1087. doi:10.1021/la00088a035.

- Hoster, H.E.; Roos, M.; Breitruck, A.; Meier, C.; Tonigold, K.; Waldmann, T.; Ziener, U.; Landfester, K.; Behm, R.J. (2007). "Structure Formation in Bis(terpyridine)Derivative Adlayers – Molecule-Substrate vs. Molecule-Molecule Interactions". *Langmuir* **23**: 11570–11579. doi:10.1021/la701382n.

- Molecular-Self Assembly Webinar

- Sigma-Aldrich "Material Matters", Molecular Self-Assembly

- Structure and Dynamics of Organic Nanostructures

- Metal organic coordination networks of oligopyridines and Cu on graphite

- Surface Alloys

29.9 External links

- Schwartz Research Group, University of Colorado Boulder

Chapter 30

Micelle

IUPAC definition

Micelle: Particle of colloidal dimensions that exists in equilibrium with the
molecules or ions in solution from which it is formed.[1][2]

Micelle (polymers): Organized auto-assembly formed in a liquid and
composed of amphiphilic *macromolecules*, in general amphiphilic di-
or tri-block copolymers made of solvophilic and solvophobic blocks.

Note 1: An amphiphilic behavior can be observed for water and an organic
solvent or between two organic solvents.

Note 2: Polymeric micelles have a much lower critical micellar concentration
(CMC) than soap or surfactant micelles, but are nevertheless at equilibrium
with isolated macromolecules called unimers. Therefore, micelle formation
and stability are concentration-dependent.[3]

A **micelle** (/maɪˈsɛl/) or **micella** (/maɪˈsɛlə/) (plural **micelles** or **micellae**, respectively) is an aggregate (or supramolecular assembly) of surfactant molecules dispersed in a liquid colloid. A typical micelle in aqueous solution forms an aggregate with the hydrophilic "head" regions in contact with surrounding solvent, sequestering the hydrophobic single-tail regions in the micelle centre. This phase is caused by the packing behavior of single-tail lipids in a bilayer. The difficulty filling all the volume of the interior of a bilayer, while accommodating the area per head group forced on the molecule by the hydration of the lipid head group, leads to the formation of the micelle. This type of micelle is known as a normal-phase micelle (oil-in-water micelle). Inverse micelles have the head groups at the centre with the tails extending out (water-in-oil micelle). Micelles are approximately spherical in shape. Other phases, including shapes such as ellipsoids, cylinders, and bilayers, are also possible. The shape and size of a micelle are a function of the molecular geometry of its surfactant molecules and solution conditions such as surfactant concentration, temperature, pH, and ionic strength. The process of forming micelles is known as micellisation and forms part of the phase behaviour of many lipids according to their polymorphism.

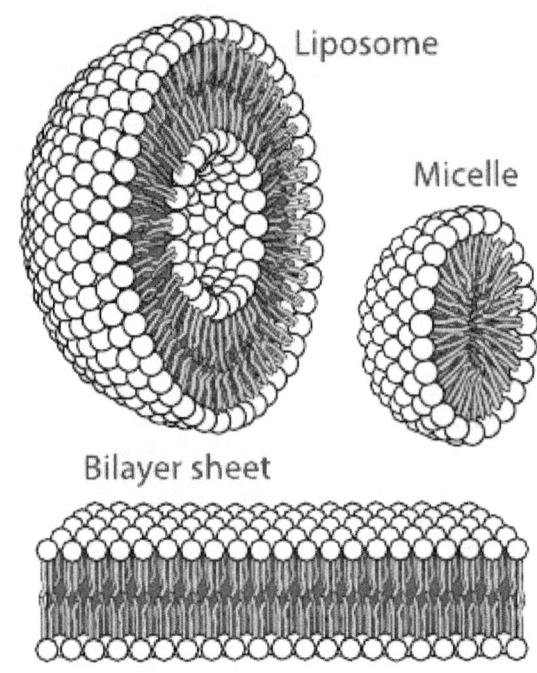

Cross-section view of the structures that can be formed by phospholipids in aqueous solutions (Note that, unlike this illustration, micelles are usually formed by single-chain lipids, since it is tough to fit two chains into this shape)

30.1 History

The ability of a soapy solution to act as a detergent has been recognized for centuries. However, it is only at the

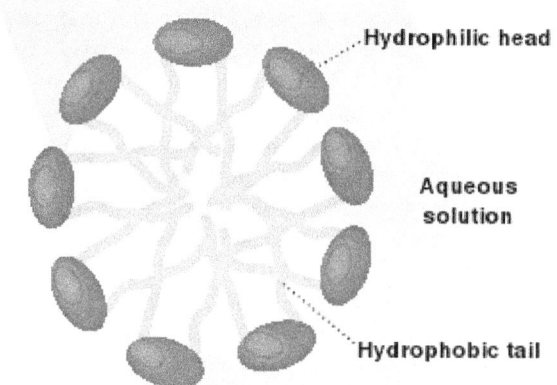

Scheme of a micelle formed by phospholipids in an aqueous solution

beginning of the twentieth century that the constitution of such solutions was scientifically studied. Pioneering work in this area was carried out by James William McBain at the University of Bristol. As early as 1913, he postulated the existence of "colloidal ions" to explain the good electrolytic conductivity of sodium palmitate solutions.[4] These highly mobile, spontaneously formed clusters came to be called micelles, a term borrowed from biology and popularized by G.S. Hartley in his classic book *Paraffin Chain Salts: A Study in Micelle Formation*.[5]

30.2 Solvation

Individual surfactant molecules that are in the system but are not part of a micelle are called "monomers". Micelles represent a molecular assembly, in which the individual components are thermodynamically in equilibrium with monomers of the same species in the surrounding medium. In water, the hydrophilic "heads" of surfactant molecules are always in contact with the solvent, regardless of whether the surfactants exist as monomers or as part of a micelle. However, the lipophilic "tails" of surfactant molecules have less contact with water when they are part of a micelle—this being the basis for the energetic drive for micelle formation. In a micelle, the hydrophobic tails of several surfactant molecules assemble into an oil-like core, the most stable form of which having no contact with water. By contrast, surfactant monomers are surrounded by water molecules that create a "cage" or solvation shell connected by hydrogen bonds. This water cage is similar to a clathrate and has an ice-like crystal structure and can be characterized according to the hydrophobic effect. The extent of lipid solubility is determined by the unfavorable entropy contribution due to the ordering of the water structure according to the hydrophobic effect.

Micelles composed of ionic surfactants have an electrostatic attraction to the ions that surround them in solution, the latter known as counterions. Although the closest counterions partially mask a charged micelle (by up to 90%), the effects of micelle charge affect the structure of the surrounding solvent at appreciable distances from the micelle. Ionic micelles influence many properties of the mixture, including its electrical conductivity. Adding salts to a colloid containing micelles can decrease the strength of electrostatic interactions and lead to the formation of larger ionic micelles.[6] This is more accurately seen from the point of view of an effective charge in hydration of the system.

30.3 Energy of formation

See also: Thermodynamics of micellization

Micelles form only when the concentration of surfactant is greater than the critical micelle concentration (CMC), and the temperature of the system is greater than the critical micelle temperature, or Krafft temperature. The formation of micelles can be understood using thermodynamics: Micelles can form spontaneously because of a balance between entropy and enthalpy. In water, the hydrophobic effect is the driving force for micelle formation, despite the fact that assembling surfactant molecules is unfavorable in terms of both enthalpy and entropy of the system. At very low concentrations of the surfactant, only monomers are present in solution. As the concentration of the surfactant is increased, a point is reached at which the unfavorable entropy contribution, from clustering the hydrophobic tails of the molecules, is overcome by a gain in entropy due to release of the solvation shells around the surfactant tails. At this point, the lipid tails of a part of the surfactants must be segregated from the water. Hence, they start to form micelles. In broad terms, above the CMC, the loss of entropy due to assembly of the surfactant molecules is less than the gain in entropy by setting free the water molecules that were "trapped" in the solvation shells of the surfactant monomers. Also important are enthalpic considerations, such as the electrostatic interactions that occur between the charged parts of surfactants.

30.4 Micelle packing parameter

The micelle packing parameter equation is utilized to help "predict molecular self-assembly in surfactant solutions":[7]

$$\frac{v_o}{a_e \cdot l_o}$$

where v_o is the surfactant tail volume, l_o is the tail length, and a_e is the equilibrium area per molecule at the aggregate surface.

30.5 Block copolymer micelles

The concept of micelles was introduced to describe the core-corona aggregates of small surfactant molecules, however it has also extended to describe aggregates of amphiphilic block copolymers in selective solvents. It is important to know the difference between these two systems. The major difference between these two types of aggregates is in the size of their building blocks. Surfactant molecules have a molecular weight which is generally of a few hundreds of grams per mole while block copolymers are generally one or two orders of magnitude larger. Moreover, thanks to the larger hydrophilic and hydrophobic parts, block copolymers can have a much more pronounced amphiphilic nature when compared to surfactant molecules.

Because of these differences in the building blocks, some block copolymer micelles behave like surfactant ones, while others don't. It is necessary therefore to make a distinction between the two situations. The former ones will belong to the *dynamic micelles* while the latter will be called *kinetically frozen micelles*.

30.5.1 Dynamic micelles

Certain amphiphilic block copolymer micelles display a similar behavior as surfactant micelles. These are generally called dynamic micelles and are characterized by the same relaxation processes assigned to surfactant exchange and micelle scission/recombination. Although the relaxation processes are the same between the two types of micelles, the kinetics of unimer exchange are very different. While in surfactant systems the unimers leave and join the micelles through a diffusion-controlled process, for copolymers the entry rate constant is slower than a diffusion controlled process. The rate of this process was found to be a decreasing power-law of the degree of polymerization of the hydrophobic block to the power 2/3. This difference is due to the coiling of the hydrophobic block of a copolymer exiting the core of a micelle.[8]

Block copolymers which form dynamic micelles are some of the tri-block Poloxamers under the right conditions.

30.5.2 Kinetically frozen micelles

When block copolymer micelles don't display the characteristic relaxation processes of surfactant micelles, these are called *kinetically frozen micelles*. These can be achieved in two ways: when the unimers forming the micelles are not soluble in the solvent of the micelle solution, or if the core forming blocks are glassy at the temperature in which the micelles are found. Kinetically frozen micelles are formed when either of these conditions is met. A special example in which both of these conditions are valid is that of polystyrene-b-poly(ethylene oxide). This block copolymer is characterized by the high hydrophobicity of the core forming block, PS, which causes the unimers to be insoluble in water. Moreover, PS has a high glass transition temperature which is, depending on the molecular weight, higher than room temperature. Thanks to these two characteristics, a water solution of PS-PEO micelles of sufficiently high molecular weight can be considered kinetically frozen. This means that none of the relaxation processes, which would drive the micelle solution towards thermodynamic equilibrium, are possible.[9] Pioneering work on these micelles was done by Adi Eisenberg.[10] It was also shown how the lack of relaxation processes allowed great freedom in the possible morphologies formed.[11][12][12] Moreover, the stability against dilution and vast range of morphologies of kinetically frozen micelles make them particularly interesting, for example, for the development of long circulating drug delivery nanoparticles.[13]

30.6 Inverse/reverse micelles

In a non-polar solvent, it is the exposure of the hydrophilic head groups to the surrounding solvent that is energetically unfavourable, giving rise to a water-in-oil system. In this case, the hydrophilic groups are sequestered in the micelle core and the hydrophobic groups extend away from the centre. These inverse micelles are proportionally less likely to form on increasing headgroup charge, since hydrophilic sequestration would create highly unfavorable electrostatic interactions.

30.7 Supermicelles

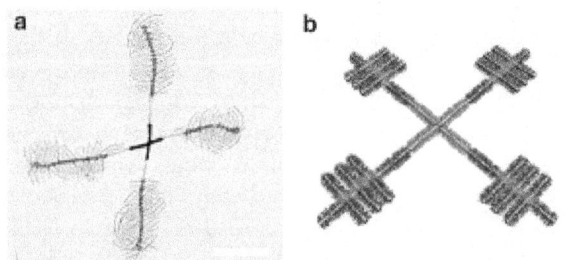

Electron micrograph of a windmill-like supermicelle, scale bar 500 nm.[14]

Supermicelle is a hierarchical micelle structure (supramolecular assembly) where individual components are also micelles. Supermicelles are formed via bottom-up chemical approaches, such as self-assembly of long cylindrical micelles into radial cross-, star- or dandelion-like patterns in a specially selected solvent; solid nanoparticles may be added to the solution to act as nucleation centers and form the central core of the supermicelle. The stems of the primary cylindrical micelles are composed of various block copolymers connected by strong covalent bonds; within the supermicelle structure they are loosely held together by hydrogen bonds, electrostatic or solvophobic interactions.[14][15]

30.8 Uses

When surfactants are present above the critical micelle concentration (CMC), they can act as emulsifiers that will allow a compound that is normally insoluble (in the solvent being used) to dissolve. This occurs because the insoluble species can be incorporated into the micelle core, which is itself solubilized in the bulk solvent by virtue of the head groups' favorable interactions with solvent species. The most common example of this phenomenon is detergents, which clean poorly soluble lipophilic material (such as oils and waxes) that cannot be removed by water alone. Detergents clean also by lowering the surface tension of water, making it easier to remove material from a surface. The emulsifying property of surfactants is also the basis for emulsion polymerization.

Micelle formation is essential for the absorption of fat-soluble vitamins and complicated lipids within the human body. Bile salts formed in the liver and secreted by the gall bladder allow micelles of fatty acids to form. This allows the absorption of complicated lipids (e.g., lecithin) and lipid-soluble vitamins (A, D, E, and K) within the micelle by the small intestine.

During the process of milk-clotting, proteases act on the soluble portion of caseins, κ-casein, thus originating an unstable micellar state that results in clot formation.

Micelles can also be used for targeted drug delivery as gold nanoparticles.[16]

30.9 See also

- Liposome
- Vesicle (biology)
- Lipid bilayer
- Surfactant
- Micellar liquid chromatography

30.10 References

[1] MacNaught, Alan D.; Wilkinson, Andrew R. (eds.). *Compendium of Chemical Terminology: IUPAC Recommendations* (2nd ed.). Oxford: Blackwell Science. ISBN 0865426848.

[2] Slomkowski, Stanislaw; Alemán, José V.; Gilbert, Robert G.; Hess, Michael; Horie, Kazuyuki; Jones, Richard G.; Kubisa, Przemyslaw; Meisel, Ingrid; Mormann, Werner; Penczek, Stanisław; Stepto, Robert F. T. (2011). "Terminology of polymers and polymerization processes in dispersed systems (IUPAC Recommendations 2011)". *Pure and Applied Chemistry* **83** (12): 2229–2259. doi:10.1351/PAC-REC-10-06-03.

[3] Vert, Michel; Doi, Yoshiharu; Hellwich, Karl-Heinz; Hess, Michael; Hodge, Philip; Kubisa, Przemyslaw; Rinaudo, Marguerite; Schué, François (2012). "Terminology for biorelated polymers and applications (IUPAC Recommendations 2012)". *Pure and Applied Chemistry* **84** (2): 377–410. doi:10.1351/PAC-REC-10-12-04.

[4] McBain, J.W., Trans. Faraday Soc. 1913, 9, 99

[5] Hartley, G.S. (1936) *Aqueous Solutions of Paraffin Chain Salts, A Study in Micelle Formation*, Hermann et Cie, Paris

[6] Turro, Nicholas J.; Yekta, Ahmad (1978). "Luminescent probes for detergent solutions. A simple procedure for determination of the mean aggregation number of micelles". *Journal of the American Chemical Society* **100** (18): 5951–5952. doi:10.1021/ja00486a062.

[7] Nagarajan, R. (2002). "Molecular Packing Parameter and Surfactant Self-Assembly: The Neglected Role of the Surfactant Tail†". *Langmuir* **18**: 31–38. doi:10.1021/la010831y.

[8] Zana, Raoul; Marques, Carlos; Johner, Albert (2006-11-16). "Dynamics of micelles of the triblock copolymers poly(ethylene oxide)–poly(propylene oxide)–poly(ethylene oxide) in aqueous solution". *Advances in Colloid and Interface Science*. Special Issue in Honor of Dr. K. L. Mittal. 123-126: 345–351. doi:10.1016/j.cis.2006.05.011. PMID 16854361.

[9] Nicolai, Taco; Colombani, Olivier; Chassenieux, Christophe (2010). "Dynamic polymeric micelles versus frozen nanoparticles formed by block copolymers". *Soft Matter* **6** (14): 3111. doi:10.1039/b925666k.

[10] Prescott, R.J. (1983). "Communications to the editor". *Journal of Psychosomatic Research* **27** (4): 327–329. doi:10.1016/0022-3999(83)90056-9.

[11] Zhang, L; Eisenberg, A (1995). "Multiple Morphologies of "Crew-Cut" Aggregates of Polystyrene-b-poly(acrylic acid) Block Copolymers". *Science* **268** (5218): 1728–31. doi:10.1126/science.268.5218.1728. PMID 17834990.

30.10. REFERENCES

[12] Zhu, Jintao; Hayward, Ryan C. (2008-06-01). "Spontaneous Generation of Amphiphilic Block Copolymer Micelles with Multiple Morphologies through Interfacial Instabilities". *Journal of the American Chemical Society* **130** (23): 7496–7502. doi:10.1021/ja801268e. PMID 18479130.

[13] D'Addio, Suzanne M.; Saad, Walid; Ansell, Steven M.; Squiers, John J.; Adamson, Douglas H.; Herrera-Alonso, Margarita; Wohl, Adam R.; Hoye, Thomas R.; Macosko, Christopher W. (2012-08-20). "Effects of block copolymer properties on nanocarrier protection from in vivo clearance". *Journal of Controlled Release* **162** (1): 208–217. doi:10.1016/j.jconrel.2012.06.020. PMC 3416956. PMID 22732478.

[14] Li, Xiaoyu; Gao, Yang; Boott, Charlotte E.; Winnik, Mitchell A.; Manners, Ian (2015). "Non-covalent synthesis of supermicelles with complex architectures using spatially confined hydrogen-bonding interactions". *Nature Communications* **6**: 8127. doi:10.1038/ncomms9127. PMC 4569713. PMID 26337527.

[15] Gould, Oliver E.C.; Qiu, Huibin; Lunn, David J.; Rowden, John; Harniman, Robert L.; Hudson, Zachary M.; Winnik, Mitchell A.; Miles, Mervyn J.; Manners, Ian (2015). "Transformation and patterning of supermicelles using dynamic holographic assembly". *Nature Communications* **6**: 10009. doi:10.1038/ncomms10009. PMID 26627644.

[16] Chen, Xi; An, Yingli; Zhao, Dongyun; He, Zhenping; Zhang, Yan; Cheng, Jing; Shi, Linqi (August 2008). "Core–Shell–Corona Au–Micelle Composites with a Tunable Smart Hybrid Shell". *Langmuir* **24** (15): 8198–8204. doi:10.1021/la800244g. PMID 18576675.

Chapter 31

Copolymer

IUPAC definition for copolymer

A polymer derived from more than one species of monomer.

Note: Copolymers that are obtained by copolymerization of two monomer species
are sometimes termed bipolymers, those obtained from three monomers terpolymers,
those obtained from four monomers quaterpolymers, etc.[1]

Alternating copolymers: A copolymer consisting of macromolecules comprising
two species of monomeric units in alternating sequence.

Note: An alternating copolymer may be considered as a homopolymer derived from
an implicit or hypothetical monomer.[1]

Block copolymers: A portion of a macromolecule, comprising many constitutional
units, that has at least one feature which is not present in the adjacent portions.[1]

Graft macromolecule: A macromolecule with one or more species of
block connected to the main chain as side-chains, these side-chains having constitutional
or configurational features that differ from those in the main chain.[2]

When two or more different monomers unite together to polymerize, their result is called a **copolymer** and its process is called copolymerization.

Commercially relevant copolymers include acrylonitrile butadiene styrene (ABS), styrene/butadiene co-polymer (SBR), nitrile rubber, styrene-acrylonitrile, styrene-isoprene-styrene (SIS) and ethylene-vinyl acetate.

Vinyl Copolymer Milk

Different types of copolymers

31.1 Types of copolymers

Since a copolymer consists of at least two types of constituent units (also structural units), copolymers can be classified based on how these units are arranged along the chain.[3] These include:

- **Alternating copolymers** with regular alternating A and B units (2)

- **Periodic copolymers** with A and B units arranged in a repeating sequence (e.g. (A-B-A-B-B-A-A-A-A-B-B-B)$_n$)

- **Statistical copolymers** are copolymers in which the sequence of monomer residues follows a statistical rule. If the probability of finding a given type monomer residue at a particular point in the chain is equal to the mole fraction of that monomer residue in the chain, then the polymer may be referred to as a truly **random copolymer**[4] (3).

- **Block copolymers** comprise two or more homopolymer subunits linked by covalent bonds (4). The union of the homopolymer subunits may require an intermediate non-repeating subunit, known as a **junction block**. Block copolymers with two or three distinct blocks are called **diblock copolymers** and **triblock copolymers**, respectively.

Copolymers may also be described in terms of the existence of or arrangement of **branches** in the polymer structure. **Linear copolymers** consist of a single main chain whereas **branched copolymers** consist of a single main chain with one or more polymeric side chains.

Other special types of branched copolymers include **star copolymers**, **brush copolymers**, and **comb copolymers**. In gradient copolymers the monomer composition changes gradually along the chain.

A **terpolymer** is a copolymer consisting of three distinct monomers. The term is derived from *ter* (Latin), meaning thrice, and polymer.

- Stereoblock copolymers

$$\sim(-CH_2-C-)_m \sim (-CH_2-C-)_n \sim$$
$$\overset{|}{H}\,\overset{|}{X}\overset{|}{X}\,\overset{|}{H}$$

A special structure can be formed from one monomer where now the distinguishing feature is the tacticity of each block.

31.1.1 Graft copolymers

Graft copolymers are a special type of branched copolymer in which the side chains are structurally distinct from the main chain. The illustration (5) depicts a special case where the main chain and side chains are composed of distinct homopolymers. However, the individual chains of a graft copolymer may be homopolymers or copolymers. Note that different copolymer sequencing is sufficient to define a structural difference, thus an A-B diblock copolymer with A-B alternating copolymer side chains is properly called a graft copolymer.

For example, suppose we perform a free-radical polymerization of styrene in the presence of polybutadiene, a synthetic rubber, which retains one reactive C=C double bond per residue. We get polystyrene chains growing out in either direction from some of the places where there were double bonds, with a one-carbon rearrangement. Or to look at it the other way around, the result is a polystyrene backbone with polybutadiene chains growing out of it in both directions. This is an interesting copolymer variant in that one of the ingredients was a polymer to begin with.

As with block copolymers, the quasi-composite product has properties of both "components". In the example cited, the rubbery chains absorb energy when the substance is hit, so it is much less brittle than ordinary polystyrene. The product is called high-impact polystyrene, or HIPS.

31.1.2 Block copolymers

One kind of copolymer is called a "block copolymer". Block copolymers are made up of blocks of different polymerized monomers and is usually made by first polymerizing styrene, and then subsequently polymerizing methyl methacrylate (MMA) from the reactive end of the polystyrene chains. This polymer is a "diblock copolymer" because it contains two different chemical blocks. Triblocks, tetrablocks, multiblocks, etc. can also be made. Diblock copolymers are made using living polymerization techniques, such as atom transfer free radical polymerization (ATRP), reversible addition fragmentation chain transfer (RAFT), ring-opening metathesis polymerization (ROMP), and living cationic or living anionic polymerizations.[5] An emerging technique is chain shuttling polymerization.

The "**blockiness**" of a copolymer is a measure of the adjacency of comonomers vs their statistical distribution. Many or even most synthetic polymers are in fact copolymers, containing about 1-20% of a minority monomer. In such cases, blockiness is undesirable.[6]

Phase separation

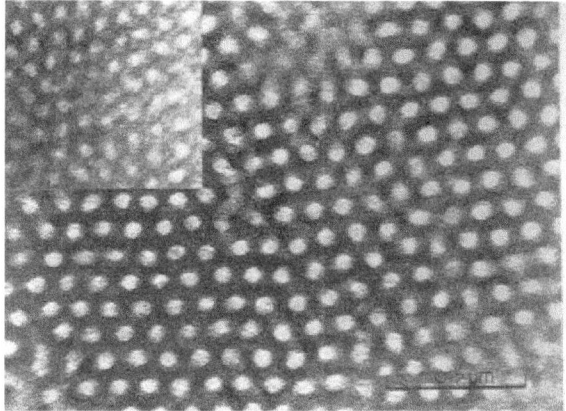

SBS block copolymer in TEM

Block copolymers are interesting because they can "microphase separate" to form periodic nanostructures, as in the styrene-butadiene-styrene block copolymer shown at right. The polymer is known as Kraton and is used for shoe soles and adhesives. Owing to the microfine structure, the transmission electron microscope or TEM was needed to examine the structure. The butadiene matrix was stained with osmium tetroxide to provide contrast in the image. The material was made by living polymerization so that the blocks are almost monodisperse, so helping to create a very regular microstructure. The molecular weight of the polystyrene blocks in the main picture is 102,000; the inset picture has a molecular weight of 91,000, producing slightly smaller domains.

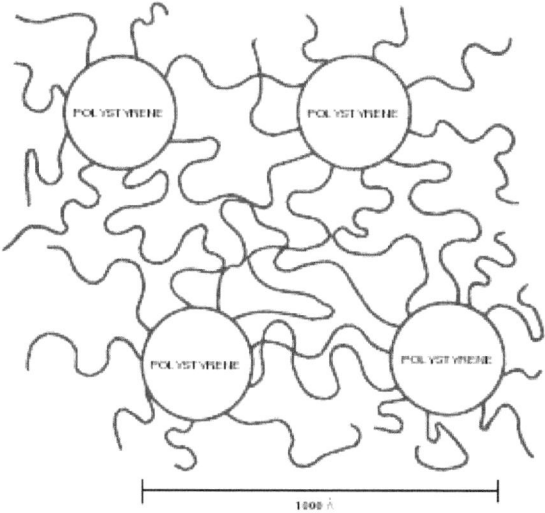

SBS block copolymer schematic microstructure

Microphase separation is a situation similar to that of oil and water. Oil and water are immiscible - they phase separate. Due to incompatibility between the blocks, block copolymers undergo a similar phase separation. Because the blocks are covalently bonded to each other, they cannot demix macroscopically as water and oil. In "microphase separation" the blocks form nanometer-sized structures. Depending on the relative lengths of each block, several morphologies can be obtained. In diblock copolymers, sufficiently different block lengths lead to nanometer-sized spheres of one block in a matrix of the second (for example PMMA in polystyrene). Using less different block lengths, a "hexagonally packed cylinder" geometry can be obtained. Blocks of similar length form layers (often called lamellae in the technical literature). Between the cylindrical and lamellar phase is the gyroid phase. The nanoscale structures created from block copolymers could potentially be used for creating devices for use in computer memory, nanoscale-templating and nanoscale separations.[7]

Polymer scientists use thermodynamics to describe how the different blocks interact.[8][9] The product of the degree of polymerization, n, and the Flory-Huggins interaction parameter, χ, gives an indication of how incompatible the two blocks are and whether or not they will microphase separate. For example, a diblock copolymer of symmetric composition will microphase separate if the product χN is greater than 10.5. If χN is less than 10.5, the blocks will mix and microphase separation is not observed. The incompatibility between the blocks also affects the solution behavior of these copolymers and their adsorption behavior on various surfaces.[10]

31.2 Copolymer equation

An alternating copolymer has the formula: -A-B-A-B-A-B-A-B-A-B-, or -(-A-B-)$_n$-. The molar ratios of the monomer in the polymer is close to one, which happens when the reactivity ratios r_1 & r_2 are close to zero, as given by the Mayo–Lewis equation also called the **copolymerization equation**:[11]

$$\frac{d[M_1]}{d[M_2]} = \frac{[M_1](r_1[M_1]+[M_2])}{[M_2]([M_1]+r_2[M_2])}$$

where $r_1 = k_{11}/k_{12}$ & $r_2 = k_{22}/k_{21}$

31.3 Copolymer engineering

Copolymerization is used to modify the properties of manufactured plastics to meet specific needs, for example to reduce crystallinity, modify glass transition temperature or to improve solubility. It is a way of improving mechanical properties, in a technique known as rubber toughening. Elastomeric phases within a rigid matrix act as crack arrestors, and so increase the energy absorption when the material is impacted for example. Acrylonitrile butadiene styrene is a common example.

31.4 See also

- Copolymers section of Polymer article
- Thermoplastic elastomer
- Tholin

31.5 References

[1] McNaught, A. D.; Wilkinson, A. (1996). "Glossary of basic terms in polymer science (IUPAC Recommendations 1996)". *Pure and Applied Chemistry* **68**: 2287–2311. doi:10.1351/goldbook.C01335.

[2] "Glossary of basic terms in polymer science (IUPAC Recommendations 1996)" (PDF). *Pure and Applied Chemistry* **68** (12): 2287–2311. 1996. doi:10.1351/pac199668122287.

[3] Jenkins, A. D.; Kratochvíl, P.; Stepto, R. F. T.; Suter, U. W. (1996). "Glossary of Basic Terms in Polymer Science". *Pure Appl. Chem.* **68** (12): 2287–2311. doi:10.1351/pac199668122287.

[4] Painter P. C. and Coleman M. M., *Fundamentals of Polymer Science*, CRC Press, 1997, p 14.

[5] Hadjichristidis N., Pispas S., Floudas G. Block copolymers: synthetic strategies, physical properties, and applications – Wiley, 2003.

[6] Chum, P. S.; Swogger, K. W., "Olefin Polymer Technologies-History and Recent Progress at the Dow Chemical Company", Progress in Polymer Science 2008, volume 33, 797-819. doi:10.1016/j.progpolymsci.2008.05.003

[7] Gazit, Oz; Khalfin, Rafail; Cohen, Yachin; Tannenbaum, Rina (2009). "Self-assembled diblock copolymer "nanoreactors" as catalysts for metal nanoparticle synthesis". *Journal of Physical Chemistry C* **113**: 576–583. doi:10.1021/jp807668h.

[8] Bates, Frank S.; Fredrickson, Glenn H. (2014). "Block Copolymer Thermodynamics: Theory and Experiment". *Annual Review of Physical Chemistry* **41**: 525–557. doi:10.1146/annurev.pc.41.100190.002521.

[9] Chremos, Alexandros; Nikoubashman, Arash; Panagiotopoulos, Athanassios (2014). "Flory-Huggins parameter χ, from binary mixtures of Lennard-Jones particles to block copolymer melts". *J. Chem. Phys.* **140**: 054909. doi:10.1063/1.4863331.

[10] Hershkovitz, Eli; Tannenbaum, Allen; Tannenbaum, Rina (2008). "Adsorption of block co-polymers from selective solvents on curved surfaces". *Macromolecules* **41**: 3190–3198. doi:10.1021/ma702706p.

[11] Mayo, Frank R.; Lewis, Frederick M. (1944). "Copolymerization. I. A Basis for Comparing the Behavior of Monomers in Copolymerization; The Copolymerization of Styrene and Methyl Methacrylate". *J. Am. Chem. Soc.* **66** (9): 1594–1601. doi:10.1021/ja01237a052.

31.6 External links

- Introduction to Polymer Chemistry

Chapter 32

Langmuir–Blodgett film

Sarfus image of one Langmuir–Blodgett monolayer of stearic acid (thickness=2.4nm).

A **Langmuir–Blodgett film** contains one or more monolayers of an organic material, deposited from the surface of a liquid onto a solid by immersing (or emersing) the solid substrate into (or from) the liquid. A monolayer is adsorbed homogeneously with each immersion or emersion step, thus films with very accurate thickness can be formed. This thickness is accurate because the thickness of each monolayer is known and can therefore be added to find the total thickness of a Langmuir–Blodgett film. The monolayers are assembled vertically and are usually composed of amphiphilic molecules (see Chemical polarity) with a hydrophilic head and a hydrophobic tail (example: fatty acids). Langmuir–Blodgett films are named after Irving Langmuir and Katharine B. Blodgett, who invented this technique while working in Research and Development for General Electric Co. An alternative technique of creating single monolayers on surfaces is that of self-assembled monolayers.

Langmuir–Blodgett films should not be confused with Langmuir films, which tends to describe an organic monolayer submersed in an aqueous solution.

32.1 Historical background

Advances to the discovery of Langmuir–Blodgett films began with Benjamin Franklin in 1773 when he dropped about a teaspoon of oil onto a pond. Franklin noticed that the waves were calmed almost instantly and that the calming of the waves spread for about half an acre. What Franklin did not realize was that the oil had formed a monolayer on top of the pond surface. Over a century later, Lord Rayleigh quantified what Benjamin Franklin had seen. Knowing that the oil, oleic acid, had spread evenly over the water, Rayleigh calculated that the thickness of the film was 1.6 nm by knowing the volume of oil dropped and the area of coverage. In addition, he used these calculations to prove the existence of the Avogadro number.

With the help of her kitchen sink, Agnes Pockels showed that area of films can be controlled with barriers. She added that surface tension varies with contamination of water. She used different oils to deduce that surface pressure would not change until area was confined to about 0.2 nm^2. This work was originally written as a letter to Lord Rayleigh who then helped Agnes Pockels become published in the journal, *Nature*, in 1891.

Agnes Pockels' work set the stage for Irving Langmuir who continued to work and confirmed Pockels' results. Using Pockels' idea, he developed the Langmuir (or Langmuir–Blodgett) trough. His observations indicated that chain length did not impact the affected area since the organic molecules were arranged vertically.

Langmuir's breakthrough did not occur until he hired Katherine Blodgett as his assistant. Blodgett initially went to seek for a job at General Electric (GE) with Langmuir during her Christmas break of her senior year at Bryn Mawr College, where she received a BA in Physics. Langmuir advised to Blodgett that she should continue her education before working for him. She thereafter attended University of Chicago for her MA in Chemistry. Upon her completion of her Master's, Langmuir hired her as his assistant. However, breakthroughs in surface chemistry happened after she received her PhD degree in 1926 from Cambridge University.

While working for GE, Langmuir and Blodgett discovered that when a solid surface is inserted into an aqueous solution containing organic moities, the organic molecules will deposit a monolayer homogeneously over the surface. This is the Langmuir–Blodgett film deposition process. Through this work in surface chemistry and with the help of Blodgett, Langmuir was awarded the Nobel Prize in 1932. In addition, Blodgett used Langmuir–Blodgett film to create 99% transparent anti-reflective glass by coating glass with fluorinated organic compounds, forming a simple anti-reflective coating.

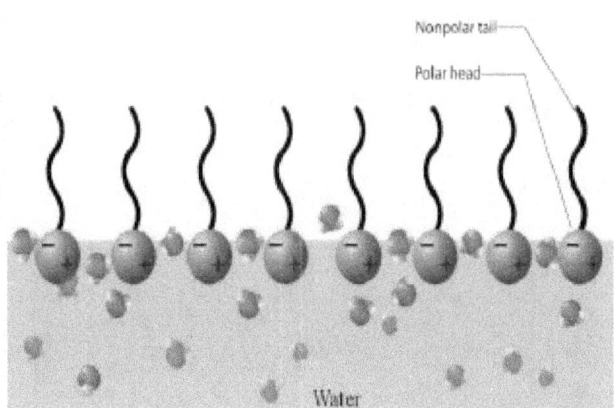

Figure 1: Surfactant molecules arranged on an air–water interface

32.2 Physical insight

LB films are formed when amphiphilic molecules like surfactants interact with air at an air–water interface. Surfactants (or surface-acting agents) are molecules with hydrophobic 'tails' and hydrophilic 'heads'. When surfactant concentration is less than critical micellar concentration (CMC), the surfactant molecules arrange themselves as shown in Figure 1 below. This tendency can be explained by surface-energy considerations. Since the tails are hydrophobic, their exposure to air is favoured over that to water. Similarly, since the heads are hydrophilic, the head–water interaction is more favourable than air–water interaction. The overall effect is reduction in the surface energy (or equivalently, surface tension of water).

For very small concentrations, far less than critical micellar concentration (CMC), the surfactant molecules execute a random motion on the water–air interface. This motion can be thought to be similar to the motion of ideal-gas molecules enclosed in a container. The corresponding thermodynamic variables for the surfactant system are, surface pressure (Π), surface area (A) and number of surfactant molecules (N). This system behaves similar to a gas in a container. The density of surfactant molecules as well as the surface pressure increases upon reducing the surface area A ('compression' of the 'gas'). Further compression of the surfactant molecules on the surface shows behavior similar to phase transitions. The 'gas' gets compressed into 'liquid' and ultimately into a perfectly closed packed array of the surfactant molecules on the surface corresponding to a 'solid' state. Instruments like the Langmuir–Blodgett trough can be used to quantify such phenomena.

32.3 Pressure–area characteristics

Adding a monolayer to the surface reduces the surface tension, and the surface pressure, Π is given by the following equation:

$$\Pi = \gamma_0 - \gamma$$

where, γ_0 is equal to the surface tension of the water and γ is the surface tension due to the monolayer. But the concentration-dependence of surface tension (similar to Langmuir isotherm) is as follows:

$$\gamma_0 - \gamma = \mathbf{RTKHC} = -\mathbf{RT}\,\Gamma$$

Thus,

$$\Pi = RT\Gamma \text{ or,}$$

$$\Pi A = RT.$$

The last equation indicates a relationship similar to ideal gas law. However, it should be noted that the concentration-dependence of surface tension is valid only when the solutions are dilute and concentrations are low. Hence, at very low concentrations of the surfactant, the molecules behave like ideal gas molecules.

Experimentally, the surface pressure is usually measured using the Wilhelmy plate. A pressure sensor/electrobalance arrangement detects the pressure exerted by the monolayer. Also monitored is the area to the side of the barrier which the monolayer resides.

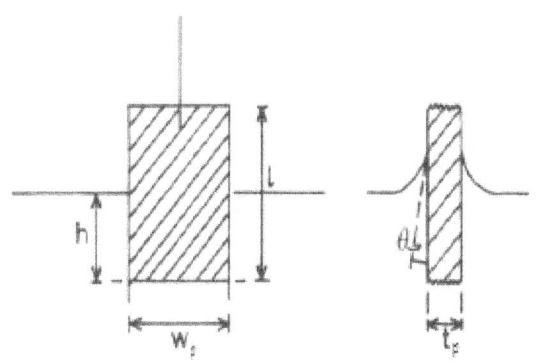

2. A Wilhelmy plate

A simple force balance on the plate leads to the following equation for the surface pressure:

$$\Pi = -\Delta\gamma = -\left[\frac{\Delta F}{2(t_p + w_p)}\right] \approx -\frac{\Delta F}{2w_p},$$

only when $w_p \gg t_p$.

Here, l_p, w_p and t_p are the dimensions of the plate, and ΔF is the difference in forces. The Wilhelmy plate measurements give pressure – area isotherms that show phase transition-like behaviour of the LB films, as mentioned before (see figure below). In the gaseous phase, there is minimal pressure increase for a decrease in area. This continues until the first transition occurs and there is a proportional increase in pressure with decreasing area. Moving into the solid region is accompanied by another sharp transition to a more severe area dependent pressure. This trend continues up to a point where the molecules are relatively close packed and have very little room to move. Applying an increasing pressure at this point causes the monolayer to become unstable and destroy the monolayer.

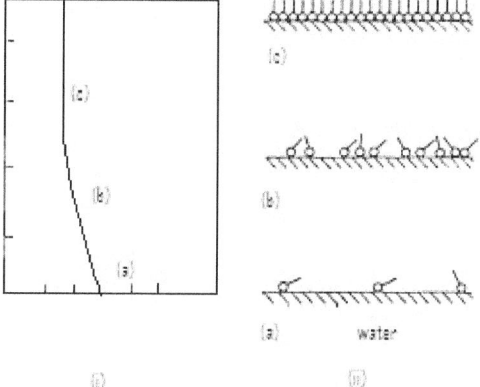

Figure 3. (i) Surface pressure – Area isotherms. (ii) Molecular configuration in the three regions marked in the Π-A curve; (a) gaseous phase, (b) liquid-expanded phase, and (c) condensed phase. (Adapted from Osvaldo N. Oliveira Jr., Brazilian Journal of Physics, vol. 22, no. 2, June 1992)

32.4 Applications

Many possible applications have been suggested over years for Langmuir–Blodgett films. Their characteristics are extremely thin films and high degree of structural order. These films have different optical, electrical and biological properties which are composed of some specific organic compounds. Organic compounds usually have more positive responses than inorganic materials for outside factors (pressure, temperature or gas change).

- LB films can be used as passive layers in MIS (metal-insulator-semiconductor) which have more open structure than silicon oxide, and they allow gases to penetrate to the interface more effectively.

- LB films also can be used as biological membranes. Lipid molecules with the fatty acid moiety of long carbon chains attached to a polar group have received extended attention because of being naturally suited to the Langmuir method of film production. This type of biological membrane can be used to investigate: the modes of drug action, the permeability of biologically active molecules, and the chain reactions of biological systems.

- Also, it is possible to propose field effect devices for observing the immunological response and enzyme-substrate reactions by collecting biological molecules such as antibodies and enzymes in insulating LB films.

- Anti-reflective glass can be produced with successive layers of fluorinated organic film.

- The glucose biosensor can be made of poly(3-hexyl thiopene) as Langmuir–Blodgett film, which entraps glucose-oxide and transfers it to a coated indium-tin-oxide glass plate.

- UV resists can be made of poly(N-alkylmethacrylamides) Langmuir–Blodgett film.

- UV light and conductivity of a Langmuir–Blodgett film.

- Langmuir–Blodgett films are inherently 2D-structures and can be built up layer by layer, by dipping hydrophobic or hydrophilic substrates into a liquid subphase.

- **Langmuir–Blodgett patterning** is a new paradigm for large-area patterning with mesostructured features[1][2]

32.5 See also

- Langmuir–Blodgett trough
- Self-assembled monolayers
- Wilhelmy plate
- Surfactants
- Hydrophobicity

32.6 References

[1] Chen, Xiaodong; Lenhert, Steven; Hirtz, Michael; Lu, Nan; Fuchs, Harald; Chi, Lifeng (2007). "Langmuir–Blodgett Patterning: A Bottom–Up Way to Build Mesostructures over Large Areas". *Accounts of Chemical Research* 40 (6): 393–401. doi:10.1021/ar600019r. PMID 17441679.

[2] Purrucker, Oliver; Förtig, Anton; Lüdtke, Karin; Jordan, Rainer; Tanaka, Motomu (2005). "Confinement of Transmembrane Cell Receptors in Tunable Stripe Micropatterns". *Journal of the American Chemical Society* 127 (4): 1258–64. doi:10.1021/ja045713m. PMID 15669865.

32.7 Bibliography

- R. W. Corkery, Langmuir, 1997, 13 (14), 3591–3594

- Osvaldo N. Oliveira Jr., Brazilian Journal of Physics, vol. 22, no. 2, June 1992

- Roberts G G, Pande K P and Barlow, Phys. Technol., Vol. 12, 1981

- Singhal, Rahul. Poly-3-Hexyl Thiopene Langmuir–Blodgett Films for Application to Glucose Biosensor. National Physics Laboratory: Biotechnology and Bioengineering, p 277-282, February 5, 2004. John and Wiley Sons Inc.

- Guo, Yinzhong. Preparation of poly(N-alkylmethacrylamide) Langmuir–Blodgett films for the application to a novel dry-developed positive deep UV resist. Macromolecules, p1115-1118, February 23, 1999. ACS

- Franklin, Benjamin, Of the stilling of Waves by means of Oil. Letter to William Brownrigg and the Reverend Mr. Farish. London, November 7, 1773.

- Pockels, A., Surface Tension, Nature, 1891, 43, 437.

- Blodgett, Katherine B., Use of Interface to Extinguish Reflection of Light from Glass. Physical Review, 1939, 55,

- A. Ulman, An Introduction to Ultrathin Organic Films From Langmuir-Blodgett to Self-Assembly, Academic Press, Inc.: San Diego (1991).

- I.R. Peterson, "Langmuir Blodgett Films ", J. Phys. D 23, 4, (1990) 379–95.

- I.R. Peterson, "Langmuir Monolayers", in T.H. Richardson, Ed., Functional Organic and Polymeric Materials Wiley: NY (2000).

- L.S. Miller, D.E. Hookes, P.J. Travers and A.P. Murphy, "A New Type of Langmuir-Blodgett Trough", J. Phys. E 21 (1988) 163–167.

- I.R.Peterson, J.D.Earls, I.R.Girling and G.J.Russell, "Disclinations and Annealing in Fatty-Acid Monolayers", Mol. Cryst. Liq. Cryst. 147 (1987) 141–147.

- A.M.Bibo, C.M.Knobler and I.R.Peterson, "A Monolayer Phase Miscibility Comparison of the Long Chain Fatty Acids and Their Ethyl Esters", J. Phys. Chem. 95 (1991) 5591–5599.

32.8 External links

- http://www.apexicindia.com

- http://www.kibron.com

- http://www.ksvinc.com/LB.htm

- http://www.nima.co.uk

- http://www.edisonexploratorium.org/bio/blodgett.htm

- http://www.aist.go.jp/NIMC/overview/v2.html

- KSV Instruments LTD. Helsinki, Finland. http://www.ksvltd.fi/Literature/Application%20notes/LB.pdf

- http://home.frognet.net/~{}ejcov/blodgett2.html

- Sarfus images of Langmuir–Blodgett films: http://www.nano-lane.com/langmuir-blodgett.php

Chapter 33

Biological organisation

"Hierarchy of life" and "Levels of organization" redirect here. For the hierarchical ordering and organization of all organisms, see Biological classification. For the evolutionary hierarchy of organisms and interspecial relationships, see Phylogenetic tree. For the concept of hierarchical organization outside of biology, see Integrative level.

Biological organization is the hierarchy of complex

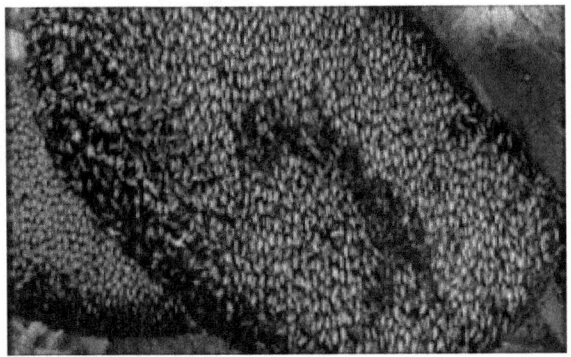

A population of bees shimmers in response to a predator.

biological structures and systems that define life using a reductionistic approach.[1] The traditional hierarchy, as detailed below, extends from atoms to biospheres. The higher levels of this scheme are often referred to as a **ecological organization** concept, or as the field, hierarchical ecology.

Each level in the hierarchy represents an increase in organizational complexity, with each "object" being primarily composed of the previous level's basic unit.[2] The basic principle behind the organization is the concept of *emergence*—the properties and functions found at a hierarchical level are not present and irrelevant at the lower levels.

Organization furthermore is the high degree of order of an organism (in comparison to general objects).[3] Ideally, individual organisms of the same species have the same arrangement of the same structures. For example, the typical human has a torso with two legs at the bottom and two arms on the sides and a head on top. It is extremely rare (and usually impossible, due to physiological and biomechanical factors) to find a human that has all of these structures but in a different arrangement.

The biological organization of life is a fundamental premise for numerous areas of scientific research, particularly in the medical sciences. Without this necessary degree of organization, it would be much more difficult—and likely impossible—to apply the study of the effects of various physical and chemical phenomena to diseases and physiology (body function). For example, fields such as cognitive and behavioral neuroscience could not exist if the brain was not composed of specific types of cells, and the basic concepts of pharmacology could not exist if it was not known that a change at the cellular level can affect an entire organism. These applications extend into the ecological levels as well. For example, DDT's direct inseciticidal effect occurs at the subcellular level, but affects higher levels up to and including multiple ecosystems. Theoretically, a change in one atom could change the entire biosphere.

33.1 Levels

The simple standard biological organization scheme, from the lowest level to the highest level, is as follows:[1]

More complex schemes incorporate many more levels. For example, a molecule can be viewed as a grouping of elements, and an atom can be further divided into subatomic particles (these levels are outside the scope of biological organization). Each level can also be broken down into its own hierarchy, and specific types of these biological objects can have their own hierarchical scheme. For example, genomes can be further subdivided into a hierarchy of genes.[5]

Each level in the hierarchy can be described by its lower levels. For example, the organism may be described at any of its component levels, including the atomic, molecular, cellular, histological (tissue), organ and organ system levels. Furthermore, at every level of the hierarchy, new functions necessary for the control of life appear. These new roles are not functions that the lower level components are capable of and are thus referred to as *emergent properties*.

Every organism is organized, though not necessarily to the same degree.[6] An organism can not be organized at the histological (tissue) level if it is not composed of tissues in the first place.[7]

33.2 Fundamentals

Empirically, a large proportion of the (complex) biological systems we observe in nature exhibit hierarchic structure. On theoretical grounds we could expect complex systems to be hierarchies in a world in which complexity had to evolve from simplicity. System hierarchies analysis performed in the 1950s,[8][9] laid the empirical foundations for a field that would be, from 1980's, **hierarchical ecology**.[10][11][12][13][14]

The theoretical foundations are summarized by Thermodynamics. When biological systems are modeled as physical systems, in its most general abstraction, they are thermodynamic open systems that exhibit self-organized behavior, and the set/subset relations between dissipative structures can be characterized in an hierarchy.

Another way, more simple and direct to explain the fundamentals of the "hierarchical organization of life", was introduced in Ecology by Odum and others as the "Simon's hierarchical principle";[15] Simon[16] emphasized that hierarchy "*emerges almost inevitably through a wide variety of evolutionary processes, for the simple reason that hierarchical structures are stable*".

To motivate this deep idea, he offered his "parable" about imaginary watchmakers.

33.3 See also

- Abiogenesis
- Cell theory
- Cellular differentiation
- Composition of the human body
- Evolutionary biology
- Gaia hypothesis
- Holon hierarchy
- Human ecology
- Living systems
- Noogenesis
- Self-organization
- Spontaneous order

33.4 Notes

[1] Solomon, Berg & Martin 2002, pp. 9–10

[2] Pavé 2006, p. 40

[3] Postlethwait & Hopson 2006, p. 6

[4] Huggett 1999

[5] Pavé 2006, p. 39

[6] Postlethwait & Hopson 2006, p. 7

[7] Witzany, G., (2014) Biological Self-Organization. International Journal of Signs and Semiotic Systems 3(2), 1-11.

[8] Evans 1951

[9] Evans 1956

[10] Margalef 1975

[11] O'Neill 1986

[12] Wicken & Ulanowicz 1988

[13] Pumain 2006

[14] Jordan & Jørgensen 2012

[15] Simon 1969, pp. 192–229

[16] Simon's texts at polaris.gseis.ucla.edu/pagre/simon or johncarlosbaez/2011/08/29 transcriptions

33.4.1 References

- Evans, F. C. (1951), "Ecology and urban areal research", *Scientific Monthly* (73)
- Evans, F. C. (1956), *Ecosystem as basic unit in ecology*
- Griswold, Joseph G.; McDaniel, Nichole (Spring 2006), "Module 1:Overview and Hierarchy of Life", *Progressions* (New York, New York) **7** (3), ISSN 1539-1752
- Huggett, R. J. (1999). "Ecosphere, biosphere, or Gaia? What to call the global ecosystem. ECOLOGICAL SOUNDING". *Global Ecology and Biogeography* **8** (6): 425–431. doi:10.1046/j.1365-2699.1999.00158.x. ISSN 1466-822X.

- Jordan, F.; Jørgensen, S. E. (2012), *Models of the Ecological Hierarchy: From Molecules to the Ecosphere*, ISBN 9780444593962

- Margalef, R. (1975), *External factors and ecosystem stability*, doi:10.1007/BF02505181

- O'Neill, R. V. (1986), *A Hierarchical Concept of Ecosystems*, ISBN 0691084378

- Pavé, Alain (2006), "Biological and Ecological Systems Hierarchical Organization", in Pumain, D., *Hierarchy in Natural and Social Sciences*, New York, New York: Springer-Verlag, ISBN 978-1-4020-4126-6

- Postlethwait, John H.; Hopson, Janet L. (2006), *Modern Biology*, Holt, Rinehart and Winston, ISBN 0-03-065178-6

- Pumain, D. (2006), *Hierarchy in Natural and Social Sciences*, ISBN 978-1-4020-4127-3

- Simon, H. A. (1969), "The architecture of complexity", *The Sciences of the Artificial*, Cambridge, MA: MIT Press

- Solomon, Eldra P.; Berg, Linda R.; Martin, Diana W. (2002), *Biology* (6th ed.), Brooks/Cole, ISBN 0-534-39175-3, LCCN 2001095366

- Wicken, J. S.; Ulanowicz, R. E. (1988), *On quantifying hierarchical connections in ecology*, doi:10.1016/0140-1750(88)90066-8

33.5 External links

- Cell physiology (in *Human Physiology*) at Wikibooks

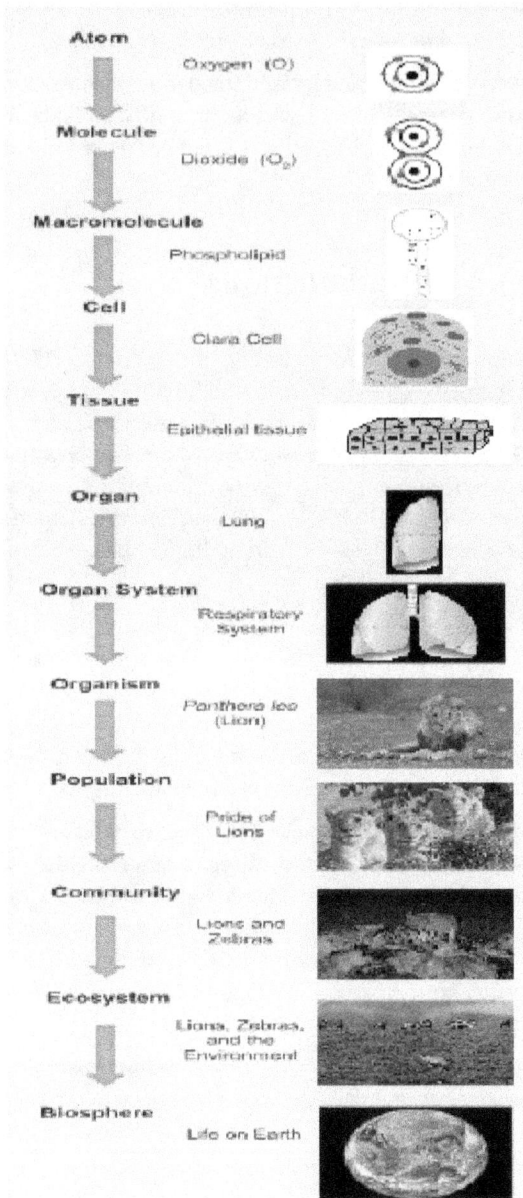

The simplest unit of life is the atom, like oxygen. Two or more atoms is a molecule, like dioxide. Many molecules is a macromolecule, such as a phospholipid. Multiple macromolecules form a cell, like a club cell. A group of cells functioning together is a tissue, for example, Epithelial tissue. Different tissues make up an organ, like a lung. Organs work together to form an organ system, such as the Respiratory System. All of the organ systems make a living organism, like a lion. A group of the same organism living together in an area is a population, such as a pride of lions. Two or more populations interacting with each other form a community, for example, lion and zebra populations interacting with each other. Communities interacting not only with each other but also with the physical environment encompass an ecosystem, such as the Savanna ecosystem. All of the ecosystems make up the biosphere, the area of life on Earth.

Chapter 34

Protein folding

"Protein thermodynamics" redirects here. For the thermodynamics of reactions catalyzed by proteins, see Enzyme.
Protein folding is the physical process by which a protein

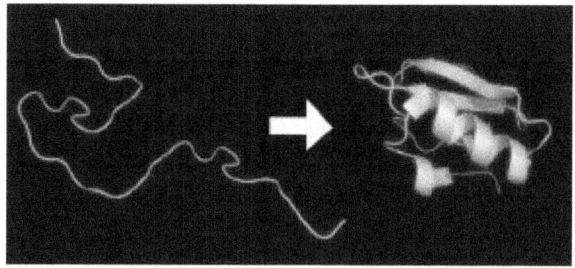

Protein before and after folding.

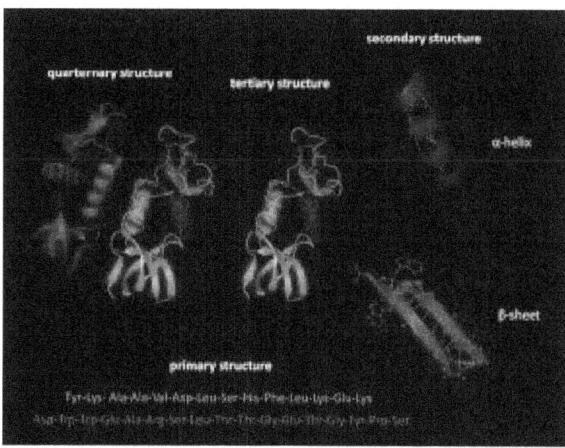

Results of protein folding.

chain acquires its native 3-dimensional structure, a conformation that is usually biologically functional, in an expeditious and reproducible manner. It is the physical process by which a polypeptide folds into its characteristic and functional three-dimensional structure from random coil.[1] Each protein exists as an unfolded polypeptide or random coil when translated from a sequence of mRNA to a linear chain of amino acids. This polypeptide lacks any stable (long-lasting) three-dimensional structure (the left hand side of the first figure). Amino acids interact with each other to produce a well-defined three-dimensional structure, the folded protein (the right hand side of the figure), known as the native state. The resulting three-dimensional structure is determined by the amino acid sequence (Anfinsen's dogma).[2] Experiments [3] beginning in the 1980s indicate the codon for an amino acid can also influence protein structure.

The correct three-dimensional structure is essential to function, although some parts of functional proteins may remain unfolded,[4] so that protein dynamics is important. Failure to fold into native structure generally produces inactive proteins, but in some instances misfolded proteins have modified or toxic functionality. Several neurodegenerative and other diseases are believed to result from the accumulation of amyloid fibrils formed by *misfolded* proteins.[5] Many allergies are caused by incorrect folding of some proteins, because the immune system does not produce antibodies for certain protein structures.[6]

34.1 Known facts

34.1.1 Relationship between folding and amino acid sequence

The amino-acid sequence of a protein determines its native conformation.[7] A protein molecule folds spontaneously during or after biosynthesis. While these macromolecules may be regarded as "folding themselves", the process also depends on the solvent (water or lipid bilayer),[8] the concentration of salts, the pH, the temperature, the possible presence of cofactors and of molecular chaperones.

Minimizing the number of hydrophobic side-chains exposed to water is an important driving force behind the folding process.[9] Formation of intramolecular hydrogen bonds provides another important contribution to protein stability.[10] The strength of hydrogen bonds depends on their environment, thus H-bonds enveloped in a hydropho-

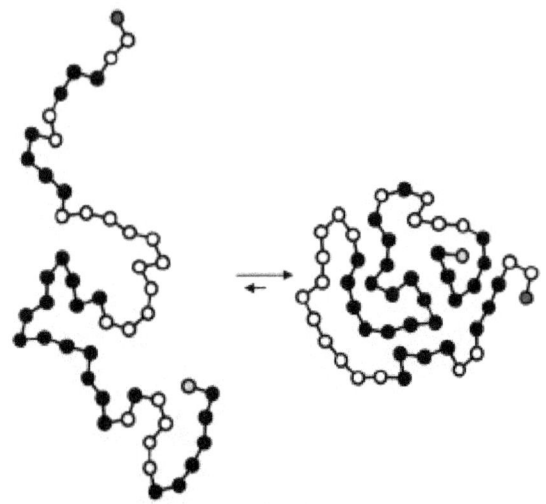

Illustration of the main driving force behind protein structure formation. In the compact fold (to the right), the hydrophobic amino acids (shown as black spheres) are in general shielded from the solvent.

bic core contribute more than H-bonds exposed to the aqueous environment to the stability of the native state.[11]

The process of folding often begins co-translationally, so that the N-terminus of the protein begins to fold while the C-terminal portion of the protein is still being synthesized by the ribosome. Specialized proteins called chaperones assist in the folding of other proteins.[12] A well studied example is the bacterial GroEL system, which assists in the folding of globular proteins. In eukaryotic organisms chaperones are known as heat shock proteins. Although most globular proteins are able to assume their native state unassisted, chaperone-assisted folding is often necessary in the crowded intracellular environment to prevent aggregation; chaperones are also used to prevent misfolding and aggregation that may occur as a consequence of exposure to heat or other changes in the cellular environment.

There are two models of protein folding that are currently being confirmed:

- The diffusion collision model, in which a nucleus is formed, then the secondary structure is formed, and finally these secondary structures are collided together and pack tightly together.

- The nucleation-condensation model, in which the secondary and tertiary structures of the protein are made at the same time.

Recent studies have shown that some proteins show characteristics of both of these folding models.

For the most part, scientists have been able to study many identical molecules folding together *en masse*. At the coarsest level, it appears that in transitioning to the native state, a given amino acid sequence takes on roughly the same route and proceeds through roughly the same intermediates and transition states. Often folding involves first the establishment of regular secondary and supersecondary structures, in particular alpha helices and beta sheets, and afterward tertiary structure. Formation of quaternary structure usually involves the "assembly" or "coassembly" of subunits that have already folded. The regular alpha helix and beta sheet structures fold rapidly because they are stabilized by intramolecular hydrogen bonds, as was first characterized by Linus Pauling. Protein folding may involve covalent bonding in the form of disulfide bridges formed between two cysteine residues or the formation of metal clusters. Shortly before settling into their more energetically favourable native conformation, molecules may pass through an intermediate "molten globule" state.

The essential fact of folding, however, remains that the amino acid sequence of each protein contains the information that specifies both the native structure and the pathway to attain that state. This is not to say that nearly identical amino acid sequences always fold similarly.[13] Conformations differ based on environmental factors as well; similar proteins fold differently based on where they are found. Folding is a spontaneous process independent of energy inputs from nucleoside triphosphates. The passage of the folded state is mainly guided by hydrophobic interactions, formation of intramolecular hydrogen bonds, and van der Waals forces, and it is opposed by conformational entropy.

34.1.2 Disruption of the native state

Under some conditions proteins will not fold into their biochemically functional forms. Temperatures above or below the range that cells tend to live in will cause thermally unstable proteins to unfold or "denature" (this is why boiling makes an egg white turn opaque). High concentrations of solutes, extremes of pH, mechanical forces, and the presence of chemical denaturants can do the same. Protein thermal stability is far from constant, however. For example, hyperthermophilic bacteria have been found that grow at temperatures as high as 122 °C,[14] which of course requires that their full complement of vital proteins and protein assemblies be stable at that temperature or above.

A fully denatured protein lacks both tertiary and secondary structure, and exists as a so-called random coil. Under certain conditions some proteins can refold; however, in many cases, denaturation is irreversible.[15] Cells sometimes protect their proteins against the denaturing influence of heat with enzymes known as chaperones or heat shock proteins, which assist other proteins both in folding and in remaining folded. Some proteins never fold in cells at all except with the assistance of chaperone molecules, which either isolate individual proteins so that their folding is not interrupted by interactions with other proteins or help to unfold misfolded proteins, giving them a second chance to refold properly. This function is crucial to prevent the risk of precipitation into insoluble amorphous aggregates.

34.1.3 Incorrect protein folding and neurodegenerative disease

Main article: Proteopathy

Aggregated proteins are associated with prion-related illnesses such as Creutzfeldt-Jakob disease, bovine spongiform encephalopathy (mad cow disease), amyloid-related illnesses such as Alzheimer's disease and familial amyloid cardiomyopathy or polyneuropathy,[16] as well as intracytoplasmic aggregation diseases such as Huntington's and Parkinson's disease.[5][17] These age onset degenerative diseases are associated with the aggregation of misfolded proteins into insoluble, extracellular aggregates and/or intracellular inclusions including cross-beta sheet amyloid fibrils. It is not completely clear whether the aggregates are the cause or merely a reflection of the loss of protein homeostasis, the balance between synthesis, folding, aggregation and protein turnover. Recently the European Medicines Agency approved the use of Tafamidis or Vyndaqel (a kinetic stabilizer of tetrameric transthyretin) for the treatment of transthyretin amyloid diseases. This suggests that the process of amyloid fibril formation (and not the fibrils themselves) causes the degeneration of post-mitotic tissue in human amyloid diseases.[18] Misfolding and excessive degradation instead of folding and function leads to a number of proteopathy diseases such as antitrypsin-associated emphysema, cystic fibrosis and the lysosomal storage diseases, where loss of function is the origin of the disorder. While protein replacement therapy has historically been used to correct the latter disorders, an emerging approach is to use pharmaceutical chaperones to fold mutated proteins to render them functional.

34.1.4 Effect of external factors on the folding of proteins

Several external factors such as temperature, external fields (electric, magnetic),[19] molecular crowding,[20] and limitation of space could have a big influence on the folding of proteins.[21] Modification of the local minima by external factors can also induce modifications of the folding trajectory.

Protein folding is a very finely tuned process. Hydrogen bonding between different atoms provides the force required.[22] Hydrophobic interactions between hydrophobic amino acids pack the hydrophobic residues.

34.1.5 The Levinthal paradox and kinetics

Levinthal's paradox is a thought experiment, also constituting a self-reference in the theory of protein folding. In 1969, Cyrus Levinthal noted that, because of the very large number of degrees of freedom in an unfolded polypeptide chain, the molecule has an astronomical number of possible conformations. An estimate of 3^{300} or 10^{143} was made in one of his papers.

The Levinthal paradox[23] observes that if a protein were folded by sequentially sampling of all possible conformations, it would take an astronomical amount of time to do so, even if the conformations were sampled at a rapid rate (on the nanosecond or picosecond scale). Based upon the observation that proteins fold much faster than this, Levinthal then proposed that a random conformational search does not occur, and the protein must, therefore, fold through a series of meta-stable intermediate states.

The duration of the folding process varies dramatically depending on the protein of interest. When studied outside the cell, the slowest folding proteins require many minutes or hours to fold primarily due to proline isomerization, and must pass through a number of intermediate states, like checkpoints, before the process is complete.[24] On the other hand, very small single-domain proteins with lengths of up to a hundred amino acids typically fold in a single step.[25] Time scales of milliseconds are the norm and the very fastest known protein folding reactions are complete within a few microseconds.[26]

34.2 Experimental techniques for studying protein folding

While inferences about protein folding can be made through mutation studies; typically, experimental techniques for studying protein folding rely on the gradual unfolding or folding of proteins and observing conformational changes using standard non-crystallographic techniques.

34.2.1 Protein nuclear magnetic resonance spectroscopy

Main article: Protein NMR

Protein folding is routinely studied using NMR spectroscopy, for example by monitoring hydrogen-deuterium exchange of backbone amide protons of proteins in their native state which provides both the residue-specific stability and overall stability of proteins.[27]

34.2.2 Circular dichroism

Main article: Circular dichroism

Circular dichroism is one of the most general and basic tools to study protein folding. Circular dichroism spectroscopy measures the absorption of circularly polarized light. In proteins, structures such as alpha helices and beta sheets are chiral, and thus absorb such light. The absorption of this light acts as a marker of the degree of foldedness of the protein ensemble. This technique has been used to measure equilibrium unfolding of the protein by measuring the change in this absorption as a function of denaturant concentration or temperature. A denaturant melt measures the free energy of unfolding as well as the protein's m value, or denaturant dependence. A temperature melt measures the melting temperature (T_m) of the protein. This type of spectroscopy can also be combined with fast-mixing devices, such as stopped flow, to measure protein folding kinetics and to generate chevron plots.

34.2.3 Dual polarisation interferometry

Main article: Dual polarisation interferometry

Dual polarisation interferometry is a surface based technique for measuring the optical properties of molecular layers. When used to characterise protein folding, it measures the conformation by determining the overall size of a monolayer of the protein and its density in real time at sub-Angstrom resolution.[28] Although real time, measurement of the kinetics of protein folding are limited to processes that occur slower than ~10 Hz. Similar to circular dichroism the stimulus for folding can be a denaturant or temperature.

34.2.4 Vibrational circular dichroism of proteins

The more recent developments of vibrational circular dichroism (VCD) techniques for proteins, currently involving Fourier transform (FFT) instruments, provide powerful means for determining protein conformations in solution even for very large protein molecules. Such VCD studies of proteins are often combined with X-ray diffraction of protein crystals, FT-IR data for protein solutions in heavy water (D_2O), or *ab initio* quantum computations to provide unambiguous structural assignments that are unobtainable from CD.

34.2.5 Studies of folding with high time resolution

The study of protein folding has been greatly advanced in recent years by the development of fast, time-resolved techniques. These are experimental methods for rapidly triggering the folding of a sample of unfolded protein, and then observing the resulting dynamics. Fast techniques in use include neutron scattering,[29] ultrafast mixing of solutions, photochemical methods, and laser temperature jump spectroscopy. Among the many scientists who have contributed to the development of these techniques are Jeremy Cook, Heinrich Roder, Harry Gray, Martin Gruebele, Brian Dyer, William Eaton, Sheena Radford, Chris Dobson, Alan Fersht, Bengt Nölting and Lars Konermann.

34.2.6 Proteolysis

Proteolysis is routinely used to probe the fraction unfolded under a wide range of solution conditions (e.g. Fast parallel proteolysis (FASTpp).<ref name "Minde>Minde DP, Maurice MM, Rüdiger SG (2012). "Determining biophysical protein stability in lysates by a fast proteolysis assay, FASTpp". *PloS One* 7 (10): e46147. Bibcode:2012PLoSO...746147M. doi:10.1371/journal.pone.0046147. PMC 3463568. PMID 23056252.</ref>[30]

34.2.7 Optical tweezers

Single molecule techniques, such as optical tweezers and AFM, have been used to understand protein folding mechanisms of isolated proteins as well as proteins with chaperones.[31] Optical tweezers have been used to stretch single protein molecules from their C- and N-termini and unfold them and study the subsequent refolding.[32] The technique allows one to measure folding rates at single-molecule level. For example, optical tweezers have been recently applied to study folding and unfolding of proteins involved in blood coagulation. von Willebrand factor (vWF) is a protein with an essential role in blood clot formation process. It is discovered -using single molecule optical tweezers measurement - that calcium-bound vWF acts as a shear force sensor in the blood. Shear force leads to unfolding of the A2 domain of vWF whose refolding rate is dramatically enhanced in the presence of calcium.[33] Recently, it was also shown that the simple src SH3 domain accesses multiple unfolding pathways under force.[34]

34.3 Computational methods for studying protein folding

Main article: Protein structure prediction

The study of protein folding includes three main aspects related to the prediction of protein stability, kinetics and structure. A recent review summarizes the available computational methods for protein folding. [35]

34.3.1 Energy landscape of protein folding

The protein folding phenomenon was largely an experimental endeavor until the formulation of an energy landscape theory of proteins by Joseph Bryngelson and Peter Wolynes in the late 1980s and early 1990s. This approach introduced the *principle of minimal frustration*.[36] This principle says that nature has chosen amino acid sequences so that the folded state of the protein is very stable. In addition, the undesired interactions between amino acids along the folding pathway are reduced making the acquisition of the folded state a very fast process. Even though nature has reduced the level of *frustration* in proteins, some degree of it remains up to now as can be observed in the presence of local minima in the energy landscape of proteins. A consequence of these evolutionarily selected sequences is that proteins are generally thought to have globally "funneled energy landscapes" (coined by José Onuchic)[37] that are largely directed toward the native state. This "folding funnel" landscape allows the protein to fold to the native state through any of a large number of pathways and intermediates, rather than being restricted to a single mechanism. The theory is supported by both computational simulations of model proteins and experimental studies,[36] and it has been used to improve methods for protein structure prediction and design.[36] The description of protein folding by the leveling free-energy landscape is also consistent with the 2nd law of thermodynamics.[38] Physically, thinking of landscapes in terms of visualizable potential or total energy surfaces simply with maxima, saddle points, minima, and funnels, rather like geographic landscapes, is perhaps a little misleading. The relevant description is really a high-dimensional phase space in which manifolds might take a variety of more complicated topological forms.[39]

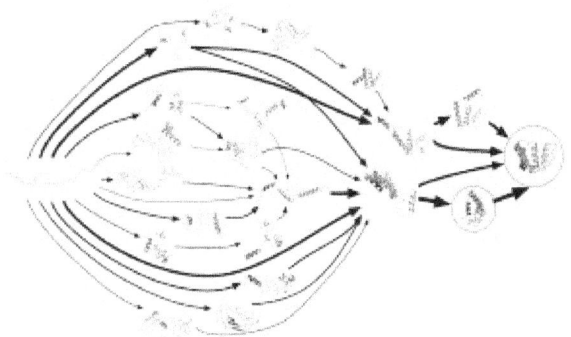

Folding@home uses Markov state models, like the one diagrammed here, to model the possible shapes and folding pathways a protein can take as it condenses from its initial randomly coiled state (left) into its native 3D structure (right).

34.3.2 Modeling of protein folding

De novo or *ab initio* techniques for computational protein structure prediction are related to, but strictly distinct from experimental studies of protein folding. Molecular Dynamics (MD) is an important tool for studying protein folding and dynamics in silico.[40] First equilibrium folding simulations were done using implicit solvent model and umbrella sampling.[41] Because of computational cost, ab initio MD folding simulations with explicit water are limited to peptides and very small proteins.[42][43] MD simulations of larger proteins remain restricted to dynamics of the experimental structure or its high-temperature unfolding. In order to simulate long-time folding processes (beyond about 1 microsecond), like folding of small-size proteins (about 50 residues) or larger, some approximations or simplifications in protein models may be introduced to speed-up the calculation process.[44][45]

The 40-petaFLOP distributed computing project Folding@home created by Vijay Pande's group at Stanford University simulates protein folding using the idle processing time of CPUs and GPUs of personal computers from volunteers. The project aims to understand protein misfolding and accelerate drug design for disease research.

Long continuous-trajectory simulations have been performed on Anton, a massively parallel supercomputer designed and built around custom ASICs and interconnects by D. E. Shaw Research. The longest published result of a simulation performed using Anton is a 1.112 millisecond simulation of NTL9 at 355 K.[46]

34.4 See also

- Anton (computer)

- Chaperone (protein)
- Chevron plot
- Denaturation (biochemistry)
- Denaturation midpoint
- Downhill folding
- Folding (chemistry)
- Folding@Home
- Foldit computer game
- Levinthal paradox
- Potential energy of protein
- Protein design
- Protein dynamics
- Protein Misfolding Cyclic Amplification
- Protein structure prediction
- Protein structure prediction software
- Proteopathy
- Rosetta@home
- Software for molecular mechanics modeling
- Statistical potential
- Time-resolved mass spectrometry

34.5 References

[1] Alberts B, Johnson A, Lewis J, Raff M, Roberts K, Walters P (2002). "The Shape and Structure of Proteins". *Molecular Biology of the Cell; Fourth Edition*. New York and London: Garland Science. ISBN 0-8153-3218-1.

[2] Anfinsen CB (Jul 1972). "The formation and stabilization of protein structure". *The Biochemical Journal* **128** (4): 737–49. PMC 1173893. PMID 4565129.

[3] Saunders R, Deane CM (Oct 2010). "Synonymous codon usage influences the local protein structure observed". *Nucleic Acids Research* (Oxford University Press) **38** (19): 6719–28. doi:10.1093/nar/gkq495. PMC 2965230. PMID 20530529.

[4] Berg JM, Tymoczko JL, Stryer L (2002). "3. Protein Structure and Function". *Biochemistry*. San Francisco: W. H. Freeman. ISBN 0-7167-4684-0.

[5] Selkoe DJ (Dec 2003). "Folding proteins in fatal ways". *Nature* **426** (6968): 900–4. Bibcode:2003Natur.426..900S. doi:10.1038/nature02264. PMID 14685251.

[6] Alberts B, Bray D, Hopkin K, Johnson A, Lewis J, Raff M, Roberts K, Walter P (2010). "Protein Structure and Function". *Essential cell biology* (Third ed.). New York, NY: Garland Science. pp. 120–170. ISBN 978-0-8153-4454-4.

[7] Anfinsen CB (Jul 1973). "Principles that govern the folding of protein chains". *Science* **181** (4096): 223–30. Bibcode:1973Sci...181..223A. doi:10.1126/science.181.4096.223. PMID 4124164.

[8] van den Berg B, Wain R, Dobson CM, Ellis RJ (Aug 2000). "Macromolecular crowding perturbs protein refolding kinetics: implications for folding inside the cell". *The EMBO Journal* **19** (15): 3870–5. doi:10.1093/emboj/19.15.3870. PMC 306593. PMID 10921869.

[9] Pace CN, Shirley BA, McNutt M, Gajiwala K (Jan 1996). "Forces contributing to the conformational stability of proteins". *FASEB Journal* **10** (1): 75–83. PMID 8566551.

[10] Rose GD, Fleming PJ, Banavar JR, Maritan A (Nov 2006). "A backbone-based theory of protein folding". *Proceedings of the National Academy of Sciences of the United States of America* **103** (45): 16623–33. Bibcode:2006PNAS..10316623R. doi:10.1073/pnas.0606843103. PMC 1636505. PMID 17075053.

[11] Deechongkit S, Nguyen H, Powers ET, Dawson PE, Gruebele M, Kelly JW (Jul 2004). "Context-dependent contributions of backbone hydrogen bonding to beta-sheet folding energetics". *Nature* **430** (6995): 101–5. Bibcode:2004Natur.430..101D. doi:10.1038/nature02611. PMID 15229605.

[12] Lee S, Tsai FT (May 2005). "Molecular chaperones in protein quality control". *Journal of Biochemistry and Molecular Biology* **38** (3): 259–65. doi:10.5483/BMBRep.2005.38.3.259. PMID 15943899.

[13] Alexander PA, He Y, Chen Y, Orban J, Bryan PN (Jul 2007). "The design and characterization of two proteins with 88% sequence identity but different structure and function". *Proceedings of the National Academy of Sciences of the United States of America* **104** (29): 11963–8. Bibcode:2007PNAS..10411963A. doi:10.1073/pnas.0700922104. PMC 1906725. PMID 17609385.

[14] Takai K, Nakamura K, Toki T, Tsunogai U, Miyazaki M, Miyazaki J, Hirayama H, Nakagawa S, Nunoura T, Horikoshi K (Aug 2008). "Cell proliferation at 122 degrees C and isotopically heavy CH4 production by a hyperthermophilic methanogen under high-pressure cultivation". *Proceedings of the National Academy of Sciences of the United States of America* **105** (31): 10949–54. Bibcode:2008PNAS..10510949T.

doi:10.1073/pnas.0712334105. PMC 2490668. PMID 18664583.

[15] Shortle D (Jan 1996). "The denatured state (the other half of the folding equation) and its role in protein stability". *FASEB Journal* **10** (1): 27–34. PMID 8566543.

[16] Hammarström P, Wiseman RL, Powers ET, Kelly JW (2003). "Prevention of transthyretin amyloid disease by changing protein misfolding energetics". *Science (New York, N.Y.)* **299** (5607): 713–6. doi:10.1126/science.1079589. PMID 12560553.

[17] Chiti F, Dobson CM (2006). "Protein misfolding, functional amyloid, and human disease". *Annual Review of Biochemistry* **75**: 333–66. doi:10.1146/annurev.biochem.75.101304.123901. PMID 16756495.

[18] Johnson SM, Wiseman RL, Sekijima Y, Green NS, Adamski-Werner SL, Kelly JW (2005). "Native state kinetic stabilization as a strategy to ameliorate protein misfolding diseases: a focus on the transthyretin amyloidoses". *Accounts of Chemical Research* **38** (12): 911–21. doi:10.1021/ar020073i. PMID 16359163.

[19] Ojeda-May P, Garcia ME (Jul 2010). "Electric field-driven disruption of a native beta-sheet protein conformation and generation of a helix-structure". *Biophysical Journal* **99** (2): 595–9. Bibcode:2010BpJ....99..595O. doi:10.1016/j.bpj.2010.04.040. PMC 2905109. PMID 20643079.

[20] van den Berg B, Ellis RJ, Dobson CM (Dec 1999). "Effects of macromolecular crowding on protein folding and aggregation". *The EMBO Journal* **18** (24): 6927–33. doi:10.1093/emboj/18.24.6927. PMC 1171756. PMID 10601015.

[21] Ellis RJ (Jul 2006). "Molecular chaperones: assisting assembly in addition to folding". *Trends in Biochemical Sciences* **31** (7): 395–401. doi:10.1016/j.tibs.2006.05.001. PMID 16716593.

[22] Pace CN (Jul 2009). "Energetics of protein hydrogen bonds". *Nature Structural & Molecular Biology* **16** (7): 681–2. doi:10.1038/nsmb0709-681. PMID 19578376.

[23] Levinthal, Cyrus (1968). "Are there pathways for protein folding?" (PDF). *Journal de Chimie Physique et de Physico-Chimie Biologique* **65**: 44–45. Archived from the original (PDF) on 2009-09-02.

[24] Kim PS, Baldwin RL (1990). "Intermediates in the folding reactions of small proteins". *Annual Review of Biochemistry* **59** (1): 631–60. doi:10.1146/annurev.bi.59.070190.003215. PMID 2197986.

[25] Jackson SE (August 1998). "How do small single-domain proteins fold?". *Folding & Design* **3** (4): R81–91. doi:10.1016/S1359-0278(98)00033-9. PMID 9710577.

[26] Kubelka J, Hofrichter J, Eaton WA (Feb 2004). "The protein folding 'speed limit'". *Current Opinion in Structural Biology* **14** (1): 76–88. doi:10.1016/j.sbi.2004.01.013. PMID 15102453.

[27] Beatrice M.P. Huyghues-Despointes, C. Nick Pace, S. Walter Englander, and J. Martin Scholtz. "Measuring the Conformational Stability of a Protein by Hydrogen Exchange." Methods in Molecular Biology. Kenneth P. Murphy Ed. Humana Press, Totowa, New Jersey, 2001. Pg. 69-92

[28] Cross GH, Freeman NJ, Swann MJ (2008). "Dual Polarization Interferometry: A Real-Time Optical Technique for Measuring (Bio)molecular Orientation, Structure and Function at the Solid/Liquid Interface". doi:10.1002/9780470061565.hbb055.

[29] Bu Z, Cook J, Callaway DJ (Sep 2001). "Dynamic regimes and correlated structural dynamics in native and denatured alpha-lactalbumin". *Journal of Molecular Biology* **312** (4): 865–73. doi:10.1006/jmbi.2001.5006. PMID 11575938.

[30] Park C, Marqusee S (Mar 2005). "Pulse proteolysis: a simple method for quantitative determination of protein stability and ligand binding". *Nature Methods* **2** (3): 207–12. doi:10.1038/nmeth740. PMID 15782190.

[31] Mashaghi A, Kramer G, Lamb DC, Mayer MP, Tans SJ (2014). "Chaperone action at the single-molecule level". *Chemical Reviews* **114** (1): 660–76. doi:10.1021/cr400326k. PMID 24001118.

[32] Jagannathan B, Marqusee S (Nov 2013). "Protein folding and unfolding under force". *Biopolymers* **99** (11): 860–9. doi:10.1002/bip.22321. PMID 23784721.

[33] Jakobi AJ, Mashaghi A, Tans SJ, Huizinga EG. Calcium modulates force sensing by the von Willebrand factor A2 domain. Nature Commun. 2011 Jul 12;2:385.

[34] Jagannathan B, Elms PJ, Bustamante C, Marqusee S (Oct 2012). "Direct observation of a force-induced switch in the anisotropic mechanical unfolding pathway of a protein". *Proceedings of the National Academy of Sciences of the United States of America* **109** (44): 17820–5. doi:10.1073/pnas.1201800109. PMID 22949695.

[35] Compiani M, Capriotti E (Dec 2013). "Computational and theoretical methods for protein folding" (PDF). *Biochemistry* **52** (48): 8601–24. doi:10.1021/bi4001529. PMID 24187909.

[36] Bryngelson JD, Onuchic JN, Socci ND, Wolynes PG (Mar 1995). "Funnels, pathways, and the energy landscape of protein folding: a synthesis" (PDF). *Proteins* **21** (3): 167–95. doi:10.1002/prot.340210302. PMID 7784423.

[37] Leopold PE, Montal M, Onuchic JN (Sep 1992). "Protein folding funnels: a kinetic approach to the sequence-structure relationship" (PDF). *Proceedings of the National Academy of Sciences of the United States of America* **89** (18): 8721–5. Bibcode:1992PNAS...89.8721L.

doi:10.1073/pnas.89.18.8721. PMC 49992. PMID 1528885.

[38] Sharma V, Kaila VR, Annila A (2009). "Protein folding as an evolutionary process". *Physica A* **388** (6): 851–862. Bibcode:2009PhyA..388..851S. doi:10.1016/j.physa.2008.12.004.

[39] Robson B, Vaithilingam A (2008). "Protein folding revisited". *Progress in Molecular Biology and Translational Science* **84**: 161–202. doi:10.1016/S0079-6603(08)00405-4. PMID 19121702.

[40] Rizzuti B, Daggett V (Mar 2013). "Using simulations to provide the framework for experimental protein folding studies". *Archives of Biochemistry and Biophysics* **531** (1-2): 128–35. doi:10.1016/j.abb.2012.12.015. PMID 23266569.

[41] Schaefer M, Bartels C, Karplus M (Dec 1998). "Solution conformations and thermodynamics of structured peptides: molecular dynamics simulation with an implicit solvation model". *Journal of Molecular Biology* **284** (3): 835–48. doi:10.1006/jmbi.1998.2172. PMID 9826519.

[42] "Fragment-based Protein Folding Simulations".

[43] "Protein folding" (by Molecular Dynamics).

[44] Kmiecik S, Kolinski A (Jul 2007). "Characterization of protein-folding pathways by reduced-space modeling". *Proceedings of the National Academy of Sciences of the United States of America* **104** (30): 12330–5. Bibcode:2007PNAS..10412330K. doi:10.1073/pnas.0702265104. PMC 1941469. PMID 17636132.

[45] Adhikari AN, Freed KF, Sosnick TR (Oct 2012). "De novo prediction of protein folding pathways and structure using the principle of sequential stabilization". *Proceedings of the National Academy of Sciences of the United States of America* **109** (43): 17442–7. doi:10.1073/pnas.1209000109. PMID 23045636.

[46] Lindorff-Larsen K, Piana S, Dror RO, Shaw DE (2011). "How fast-folding proteins fold". *Science (New York, N.Y.)* **334** (6055): 517–20. doi:10.1126/science.1208351. PMID 22034434.

34.6 External links

- FoldIt - Folding Protein Game
- Folding@Home
- Rosetta@Home
- Human Proteome Folding Project
- BHAGEERATH-H: Protein tertiary structure prediction server

Chapter 35

Lipid bilayer

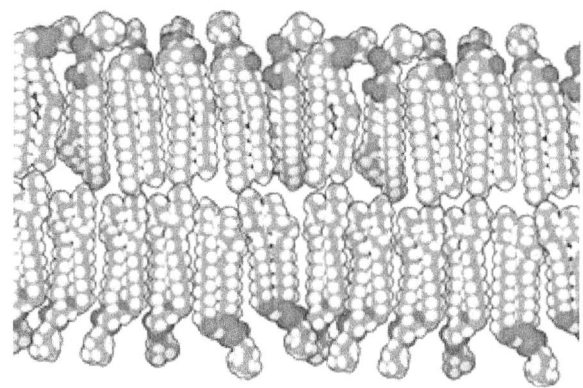

This fluid lipid bilayer cross section is made up entirely of phosphatidylcholine.

The **lipid bilayer** is a thin polar membrane made of two layers of lipid molecules. These membranes are flat sheets that form a continuous barrier around all cells. The cell membranes of almost all living organisms and many viruses are made of a lipid bilayer, as are the membranes surrounding the cell nucleus and other sub-cellular structures. The lipid bilayer is the barrier that keeps ions, proteins and other molecules where they are needed and prevents them from diffusing into areas where they should not be. Lipid bilayers are ideally suited to this role because, even though they are only a few nanometers in width,[1] they are impermeable to most water-soluble (hydrophilic) molecules. Bilayers are particularly impermeable to ions, which allows cells to regulate salt concentrations and pH by transporting ions across their membranes using proteins called ion pumps.

Biological bilayers are usually composed of amphiphilic phospholipids that have a hydrophilic phosphate head and a hydrophobic tail consisting of two fatty acid chains. Phospholipids with certain head groups can alter the surface chemistry of a bilayer and can, for example, serve as signals as well as "anchors" for other molecules in the membranes of cells.[2] Just like the heads, the tails of lipids can also affect membrane properties, for instance by determining the phase of the bilayer. The bilayer can adopt a solid gel phase state at lower temperatures but undergo phase transition to

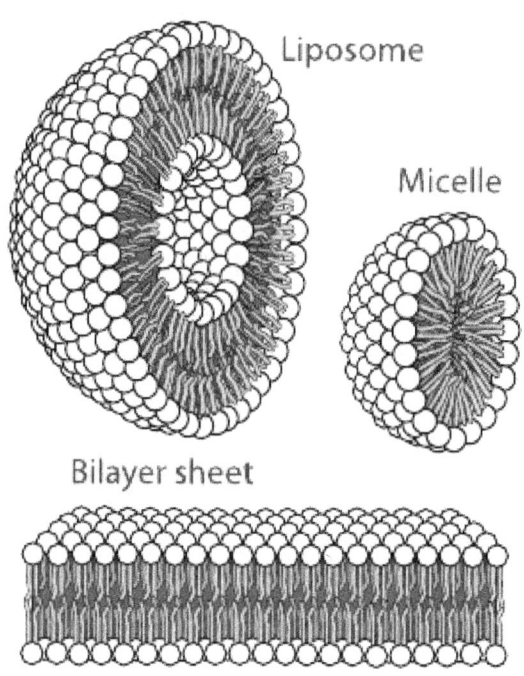

The three main structures phospholipids form in solution; the liposome (a closed bilayer), the micelle and the bilayer.

a fluid state at higher temperatures, and the chemical properties of the lipids' tails influence at which temperature this happens. The packing of lipids within the bilayer also affects its mechanical properties, including its resistance to stretching and bending. Many of these properties have been studied with the use of artificial "model" bilayers produced in a lab. Vesicles made by model bilayers have also been used clinically to deliver drugs.

Biological membranes typically include several types of molecules other than phospholipids. A particularly important example in animal cells is cholesterol, which helps strengthen the bilayer and decrease its permeability. Cholesterol also helps regulate the activity of certain integral membrane proteins. Integral membrane proteins function when incorporated into a lipid bilayer, and they are held tightly to lipid bilayer with the help of an annular lipid shell. Because bilayers define the boundaries of the cell and its compartments, these membrane proteins are involved in many intra- and inter-cellular signaling processes. Certain kinds of membrane proteins are involved in the process of fusing two bilayers together. This fusion allows the joining of two distinct structures as in the fertilization of an egg by sperm or the entry of a virus into a cell. Because lipid bilayers are quite fragile and invisible in a traditional microscope, they are a challenge to study. Experiments on bilayers often require advanced techniques like electron microscopy and atomic force microscopy.

35.1 Structure and organization

When phospholipids are exposed to water, they self-assemble into a two-layered sheet with the hydrophobic tails pointing toward the center of the sheet. This arrangement results in two "leaflets" that are each a single molecular layer. The center of this bilayer contains almost no water and excludes molecules like sugars or salts that dissolve in water. The assembly process is driven by interactions between hydrophobic molecules (also called the hydrophobic effect). An increase in interactions between hydrophobic molecules (causing clustering of hydrophobic regions) allows water molecules to bond more freely with each other, increasing the entropy of the system. This complex process includes noncovalent interactions such as van der Waals, electrostatic and hydrogen bonds.

35.1.1 Cross section analysis

The lipid bilayer is very thin compared to its lateral dimensions. If a typical mammalian cell (diameter ~10 micrometers) were magnified to the size of a watermelon (~1 ft/30 cm), the lipid bilayer making up the plasma membrane would be about as thick as a piece of office paper. Despite being only a few nanometers thick, the bilayer is composed of several distinct chemical regions across its cross-section. These regions and their interactions with the surrounding water have been characterized over the past several decades with x-ray reflectometry,[4] neutron scattering[5] and nuclear magnetic resonance techniques.

The first region on either side of the bilayer is the hydrophilic headgroup. This portion of the membrane is completely hydrated and is typically around 0.8-0.9 nm thick. In phospholipid bilayers the phosphate group is located within this hydrated region, approximately 0.5 nm outside the hydrophobic core.[6] In some cases, the hydrated re-

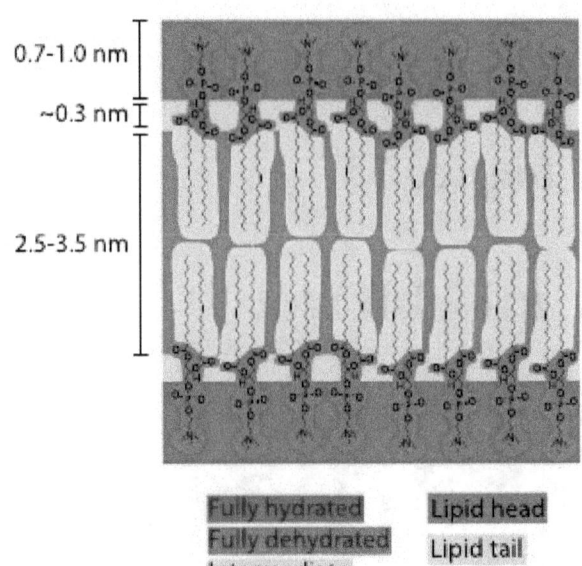

Schematic cross sectional profile of a typical lipid bilayer. There are three distinct regions: the fully hydrated headgroups, the fully dehydrated alkane core and a short intermediate region with partial hydration. Although the head groups are neutral, they have significant dipole moments that influence the molecular arrangement.[3]

gion can extend much further, for instance in lipids with a large protein or long sugar chain grafted to the head. One common example of such a modification in nature is the lipopolysaccharide coat on a bacterial outer membrane,[7] which helps retain a water layer around the bacterium to prevent dehydration.

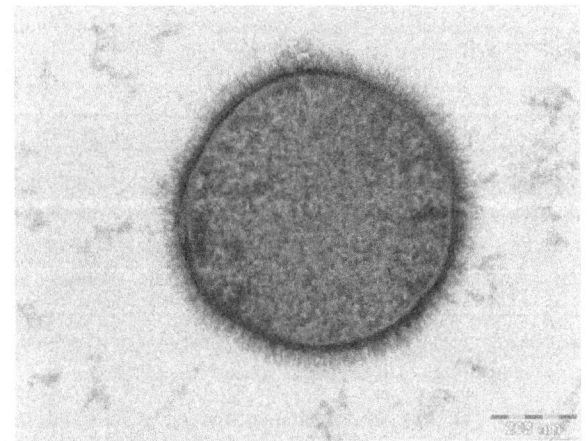

TEM image of a bacterium. The furry appearance on the outside is due to a coat of long-chain sugars attached to the cell membrane. This coating helps trap water to prevent the bacterium from becoming dehydrated.

Next to the hydrated region is an intermediate region that is only partially hydrated. This boundary layer is approximately 0.3 nm thick. Within this short distance, the water concentration drops from 2M on the headgroup side to nearly zero on the tail (core) side.[8][9] The hydrophobic core of the bilayer is typically 3-4 nm thick, but this value varies with chain length and chemistry.[4][10] Core thickness also varies significantly with temperature, in particular near a phase transition.[11]

35.1.2 Asymmetry

In many naturally occurring bilayers, the compositions of the inner and outer membrane leaflets are different. In human red blood cells, the inner (cytoplasmic) leaflet is composed mostly of phosphatidylethanolamine, phosphatidylserine and phosphatidylinositol and its phosphorylated derivatives. By contrast, the outer (extracellular) leaflet is based on phosphatidylcholine, sphingomyelin and a variety of glycolipids.[12][13] In some cases, this asymmetry is based on where the lipids are made in the cell and reflects their initial orientation.[14] The biological functions of lipid asymmetry are imperfectly understood, although it is clear that it is used in several different situations. For example, when a cell undergoes apoptosis, the phosphatidylserine — normally localised to the cytoplasmic leaflet — is transferred to the outer surface: There, it is recognised by a macrophage that then actively scavenges the dying cell.

Lipid asymmetry arises, at least in part, from the fact that most phospholipids are synthesised and initially inserted into the inner monolayer: those that constitute the outer monolayer are then transported from the inner monolayer by a class of enzymes called flippases.[15][16] Other lipids, such as sphingomyelin, appear to be synthesised at the external leaflet. Flippases are members of a larger family of lipid transport molecules that also includes floppases, which transfer lipids in the opposite direction, and scramblases, which randomize lipid distribution across lipid bilayers (as in apoptotic cells). In any case, once lipid asymmetry is established, it does not normally dissipate quickly because spontaneous flip-flop of lipids between leaflets is extremely slow.[17]

It is possible to mimic this asymmetry in the laboratory in model bilayer systems. Certain types of very small artificial vesicle will automatically make themselves slightly asymmetric, although the mechanism by which this asymmetry is generated is very different from that in cells.[18] By utilizing two different monolayers in Langmuir-Blodgett deposition[19] or a combination of Langmuir-Blodgett and vesicle rupture deposition[20] it is also possible to synthesize an asymmetric planar bilayer. This asymmetry may be lost over time as lipids in supported bilayers can be prone to flip-flop.[21]

35.1.3 Phases and phase transitions

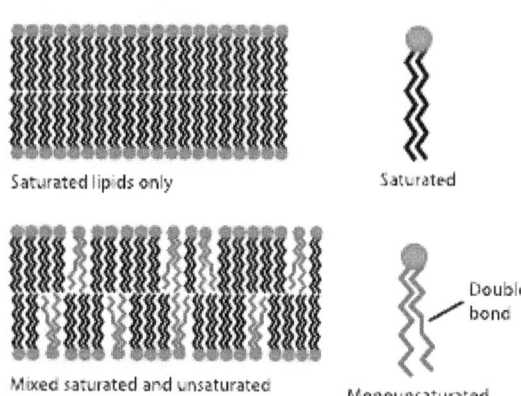

Diagram showing the effect of unsaturated lipids on a bilayer. The lipids with an unsaturated tail (blue) disrupt the packing of those with only saturated tails (black). The resulting bilayer has more free space and is, as a consequence, more permeable to water and other small molecules.

Further information: Lipid bilayer phase behavior

At a given temperature a lipid bilayer can exist in either a liquid or a gel (solid) phase. All lipids have a characteristic temperature at which they transition (melt) from the gel to liquid phase. In both phases the lipid molecules are prevented from flip-flopping across the bilayer, but in liquid phase bilayers a given lipid will exchange locations with its neighbor millions of times a second. This random walk exchange allows lipid to diffuse and thus wander across the surface of the membrane.[22] Unlike liquid phase bilayers, the lipids in a gel phase bilayer are locked in place.

The phase behavior of lipid bilayers is determined largely by the strength of the attractive Van der Waals interactions between adjacent lipid molecules. Longer-tailed lipids have more area over which to interact, increasing the strength of this interaction and, as a consequence, decreasing the lipid mobility. Thus, at a given temperature, a short-tailed lipid will be more fluid than an otherwise identical long-tailed lipid.[10] Transition temperature can also be affected by the degree of unsaturation of the lipid tails. An unsaturated double bond can produce a kink in the alkane chain, disrupting the lipid packing. This disruption creates extra free space within the bilayer that allows additional flexibility in the adjacent chains.[10] An example of this effect can be noted in everyday life as butter, which has a large percentage saturated fats, is solid at room temperature while vegetable oil, which is mostly unsaturated, is liquid.

Most natural membranes are a complex mixture of different lipid molecules. If some of the components are liquid at a given temperature while others are in the gel phase, the two phases can coexist in spatially separated regions, rather like an iceberg floating in the ocean. This phase separation plays a critical role in biochemical phenomena because membrane components such as proteins can partition into one or the other phase[23] and thus be locally concentrated or activated. One particularly important component of many mixed phase systems is cholesterol, which modulates bilayer permeability, mechanical strength, and biochemical interactions.

35.1.4 Surface chemistry

While lipid tails primarily modulate bilayer phase behavior, it is the headgroup that determines the bilayer surface chemistry. Most natural bilayers are composed primarily of phospholipids, although sphingolipids such as sphingomyelin and sterols such as cholesterol are also important components. Of the phospholipids, the most common headgroup is phosphatidylcholine (PC), accounting for about half the phospholipids in most mammalian cells.[24] PC is a zwitterionic headgroup, as it has a negative charge on the phosphate group and a positive charge on the amine but, because these local charges balance, no net charge.

Other headgroups are also present to varying degrees and can include phosphatidylserine (PS) phosphatidylethanolamine (PE) and phosphatidylglycerol (PG). These alternate headgroups often confer specific biological functionality that is highly context-dependent. For instance, PS presence on the extracellular membrane face of erythrocytes is a marker of cell apoptosis,[25] whereas PS in growth plate vesicles is necessary for the nucleation of hydroxyapatite crystals and subsequent bone mineralization.[26][27] Unlike PC, some of the other headgroups carry a net charge, which can alter the electrostatic interactions of small molecules with the bilayer.[28]

35.2 Biological roles

35.2.1 Containment and separation

The primary role of the lipid bilayer in biology is to separate aqueous compartments from their surroundings. Without some form of barrier delineating "self" from "non-self," it is difficult to even define the concept of an organism or of life. This barrier takes the form of a lipid bilayer in all known life forms except for a few species of archaea that utilize a specially adapted lipid monolayer.[7] It has even been proposed that the very first form of life may have been a simple lipid vesicle with virtually its sole biosynthetic capability being the production of more phospholipids.[29] The partitioning ability of the lipid bilayer is based on the fact that hydrophilic molecules cannot easily cross the hydrophobic bilayer core, as discussed in Transport across the bilayer below. The nucleus, mitochondria and chloroplasts have two lipid bilayers, while other sub-cellular structures are surrounded by a single lipid bilayer (such as the plasma membrane, endoplasmic reticula, Golgi apparatus and lysosomes). See Organelle.[30]

Prokaryotes have only one lipid bilayer- the cell membrane (also known as the plasma membrane). Many prokaryotes also have a cell wall, but the cell wall is composed of proteins or long chain carbohydrates, not lipids. In contrast, eukaryotes have a range of organelles including the nucleus, mitochondria, lysosomes and endoplasmic reticulum. All of these sub-cellular compartments are surrounded by one or more lipid bilayers and, together, typically comprise the majority of the bilayer area present in the cell. In liver hepatocytes for example, the plasma membrane accounts for only two percent of the total bilayer area of the cell, whereas the endoplasmic reticulum contains more than fifty percent and the mitochondria a further thirty percent.[31]

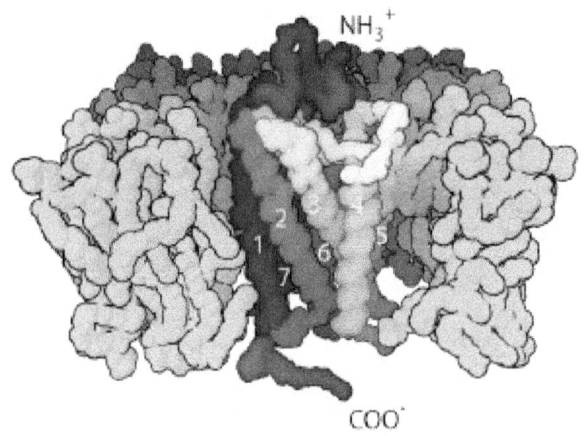

Illustration of a GPCR signaling protein. In response to a molecule such as a hormone binding to the exterior domain (blue) the GPCR changes shape and catalyzes a chemical reaction on the interior domain (red). The gray feature is the surrounding bilayer.

35.2.2 Signaling

See also: Neurotransmission and Lipid raft

Probably the most familiar form of cellular signaling is synaptic transmission, whereby a nerve impulse that has reached the end of one neuron is conveyed to an adjacent neuron via the release of neurotransmitters. This transmission is made possible by the action of synaptic vesicles loaded with the neurotransmitters to be released. These vesicles fuse with the cell membrane at the pre-synaptic terminal and release its contents to the exterior of the cell. The contents then diffuse across the synapse to the post-synaptic terminal.

Lipid bilayers are also involved in signal transduction through their role as the home of integral membrane proteins. This is an extremely broad and important class of biomolecule. It is estimated that up to a third of the human proteome may be membrane proteins.[32] Some of these proteins are linked to the exterior of the cell membrane. An example of this is the CD59 protein, which identifies cells as "self" and thus inhibits their destruction by the immune system. The HIV virus evades the immune system in part by grafting these proteins from the host membrane onto its own surface.[31] Alternatively, some membrane proteins penetrate all the way through the bilayer and serve to relay individual signal events from the outside to the inside of the cell. The most common class of this type of protein is the G protein-coupled receptor (GPCR). GPCRs are responsible for much of the cell's ability to sense its surroundings and, because of this important role, approximately 40% of all modern drugs are targeted at GPCRs.[33]

In addition to protein- and solution-mediated processes, it is also possible for lipid bilayers to participate directly in signaling. A classic example of this is phosphatidylserine-triggered phagocytosis. Normally, phosphatidylserine is asymmetrically distributed in the cell membrane and is present only on the interior side. During programmed cell death a protein called a scramblase equilibrates this distribution, displaying phosphatidylserine on the extracellular bilayer face. The presence of phosphatidylserine then triggers phagocytosis to remove the dead or dying cell.

35.3 Characterization methods

Further information: Lipid bilayer characterization
The lipid bilayer is a very difficult structure to study because it is so thin and fragile. In spite of these limitations dozens of techniques have been developed over the last seventy years to allow investigations of its structure and function.

Electrical measurements are a straightforward way to characterize an important function of a bilayer: its ability to segregate and prevent the flow of ions in solution. By applying a voltage across the bilayer and measuring the resulting current, the resistance of the bilayer is determined. This resistance is typically quite high (10^8 Ohm-cm^2 or more) [34] since the hydrophobic core is impermeable to charged species. The presence of even a few nanometer-scale holes results in a dramatic increase in current.[35] The sensitivity of this system is such that even the activity of single ion channels can be resolved.[36]

Electrical measurements do not provide an actual picture

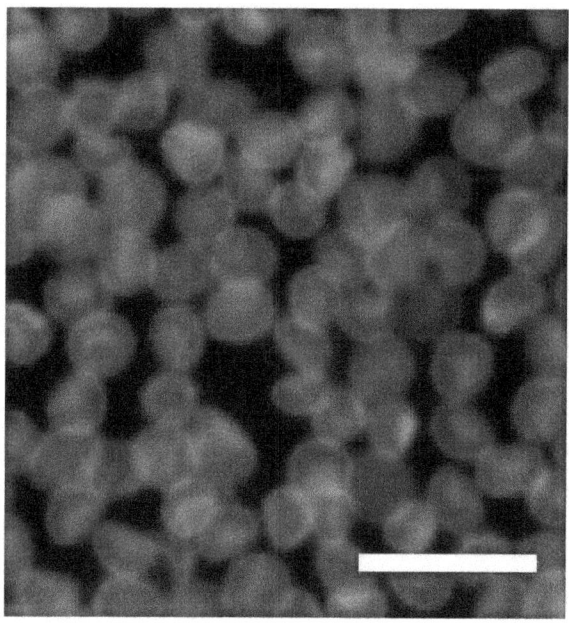

Human red blood cells viewed through a fluorescence microscope. The cell membrane has been stained with a fluorescent dye. Scale bar is 20μm.

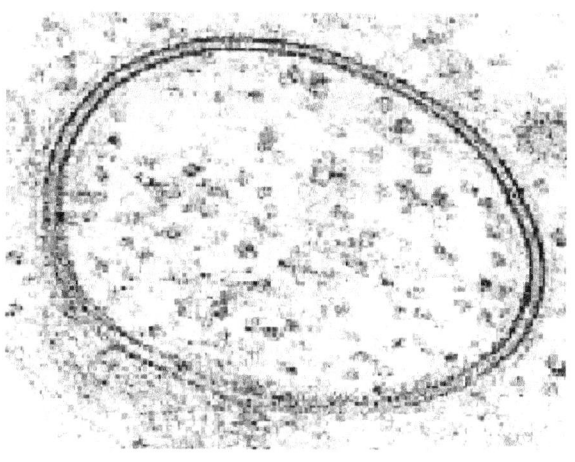

Transmission Electron Microscope (TEM) image of a lipid vesicle. The two dark bands around the edge are the two leaflets of the bilayer. Historically, similar images confirmed that the cell membrane is a bilayer

like imaging with a microscope can. Lipid bilayers cannot be seen in a traditional microscope because they are too thin. In order to see bilayers, researchers often use fluorescence microscopy. A sample is excited with one wavelength of light and observed in a different wavelength, so that only fluorescent molecules with a matching excitation and emission profile will be seen. Natural lipid bilayers are not fluorescent, so a dye is used that attaches to the desired molecules in the bilayer. Resolution is usually limited to a few hundred nanometers, much smaller than a typical cell but much larger than the thickness of a lipid bilayer.

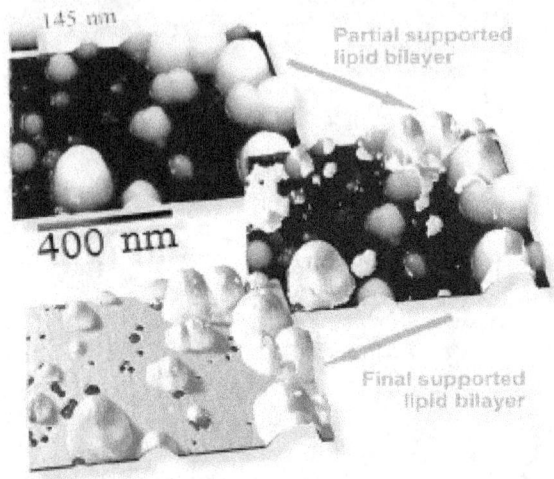

3d-Adapted AFM images showing formation of transmembrane pores (holes) in supported lipid bilayer[37]

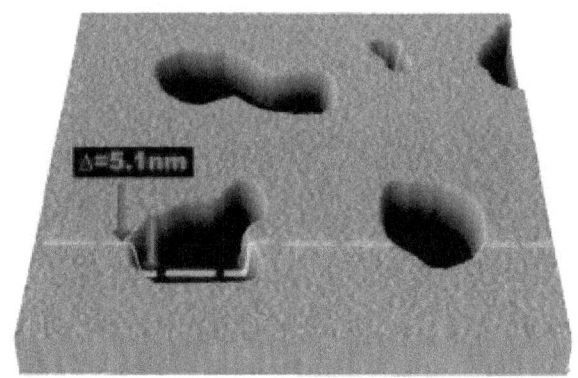

Illustration of a typical AFM scan of a supported lipid bilayer. The pits are defects in the bilayer, exposing the smooth surface of the substrate underneath.

Electron microscopy offers a higher resolution image. In an electron microscope, a beam of focused electrons interacts with the sample rather than a beam of light as in traditional microscopy. In conjunction with rapid freezing techniques, electron microscopy has also been used to study the mechanisms of inter- and intracellular transport, for instance in demonstrating that exocytotic vesicles are the means of chemical release at synapses.[38]

^{31}P-NMR(nuclear magnetic resonance) spectroscopy is widely used for studies of phospholipid bilayers and biological membranes in native conditions. The analysis[39] of ^{31}P-NMR spectra of lipids could provide a wide range of information about lipid bilayer packing, phase transitions (gel phase, physiological liquid crystal phase, ripple phases, non bilayer phases), lipid head group orientation/dynamics, and elastic properties of pure lipid bilayer and as a result of binding of proteins and other biomolecules.

A new method to study lipid bilayers is Atomic force microscopy (AFM). Rather than using a beam of light or particles, a very small sharpened tip scans the surface by making physical contact with the bilayer and moving across it, like a record player needle. AFM is a promising technique because it has the potential to image with nanometer resolution at room temperature and even under water or physiological buffer, conditions necessary for natural bilayer behavior. Utilizing this capability, AFM has been used to examine dynamic bilayer behavior including the formation of transmembrane pores (holes)[37] and phase transitions in supported bilayers.[40] Another advantage is that AFM does not require fluorescent or isotopic labeling of the lipids, since the probe tip interacts mechanically with the bilayer surface. Because of this, the same scan can image both lipids and associated proteins, sometimes even with single-molecule resolution.[37][41] AFM can also probe the mechanical nature of lipid bilayers.[42]

Lipid bilayers exhibit high levels of birefringence where the refractive index in the plane of the bilayer differs from that perpendicular by as much as 0.1 refractive index units. This has been used to characterise the degree of order and disruption in bilayers using dual polarisation interferometry to understand mechanisms of protein interaction.

Lipid bilayers are complicated molecular systems with many degrees of freedom. Thus atomistic simulation of membrane and in particular ab initio calculations of its properties is difficult and computationally expensive. Quantum chemical calculations has recently been successfully performed to estimate dipole and quadrupole moments of lipid membranes.[43]

35.4 Transport across the bilayer

35.4.1 Passive diffusion

Most polar molecules have low solubility in the hydrocarbon core of a lipid bilayer and, as a consequence, have low permeability coefficients across the bilayer. This effect is particularly pronounced for charged species, which have even lower permeability coefficients than neutral polar molecules.[44] Anions typically have a higher rate of diffusion through bilayers than cations.[45][46] Compared to ions, water molecules actually have a relatively large permeability through the bilayer, as evidenced by osmotic swelling. When a cell or vesicle with a high interior salt concentration is placed in a solution with a low salt concentration it will swell and eventually burst. Such a result would not be observed unless water was able to pass through the bilayer with relative ease. The anomalously large permeability of water through bilayers is still not completely understood and continues to be the subject of active debate.[47] Small uncharged apolar molecules diffuse through lipid bilayers many orders of magnitude faster than ions or water. This applies both to fats and organic solvents like chloroform and ether. Regardless of their polar character larger molecules diffuse more slowly across lipid bilayers than small molecules.[48]

lows conduction of an action potential along neurons. All ion pumps have some sort of trigger or "gating" mechanism. In the previous example it was electrical bias, but other channels can be activated by binding a molecular agonist or through a conformational change in another nearby protein.[49]

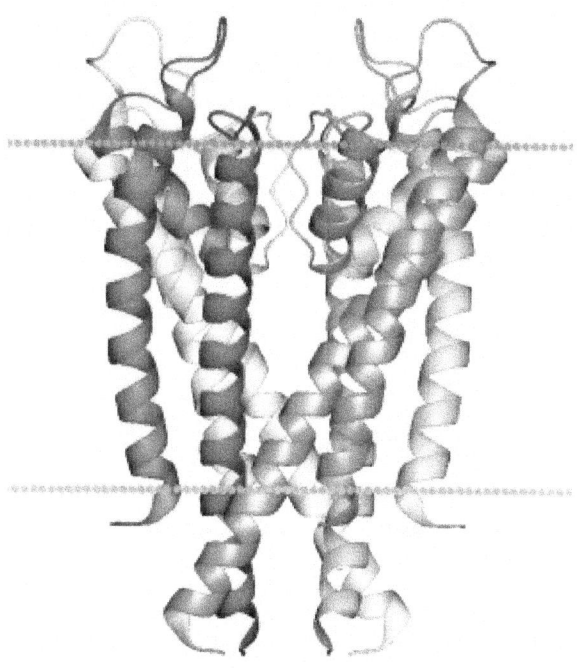

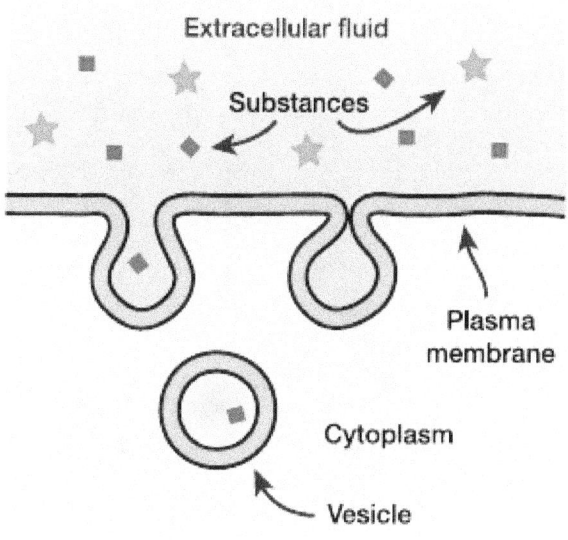

Schematic illustration of pinocytosis, a type of endocytosis

35.4.3 Endocytosis and exocytosis

See also: Endocytosis and Exocytosis

Some molecules or particles are too large or too hydrophilic to pass through a lipid bilayer. Other molecules could pass through the bilayer but must be transported rapidly in such large numbers that channel-type transport is impractical. In both cases, these types of cargo can be moved across the cell membrane through fusion or budding of vesicles. When a vesicle is produced inside the cell and fuses with the plasma membrane to release its contents into the extracellular space, this process is known as exocytosis. In the reverse process, a region of the cell membrane will dimple inwards and eventually pinch off, enclosing a portion of the extracellular fluid to transport it into the cell. Endocytosis and exocytosis rely on very different molecular machinery to function, but the two processes are intimately linked and could not work without each other. The primary mechanism of this interdependence is the sheer volume of lipid material involved.[50] In a typical cell, an area of bilayer equivalent to the entire plasma membrane will travel through the endocytosis/exocytosis cycle in about half an hour.[51] If these two

Structure of a potassium ion channel. The alpha helices penetrate the bilayer (boundaries indicated by red and blue lines), opening a hole through which potassium ions can flow

35.4.2 Ion pumps and channels

Two special classes of protein deal with the ionic gradients found across cellular and sub-cellular membranes in nature-ion channels and ion pumps. Both pumps and channels are integral membrane proteins that pass through the bilayer, but their roles are quite different. Ion pumps are the proteins that build and maintain the chemical gradients by utilizing an external energy source to move ions against the concentration gradient to an area of higher chemical potential. The energy source can be ATP, as is the case for the Na^+-K^+ ATPase. Alternatively, the energy source can be another chemical gradient already in place, as in the Ca^{2+}/Na^+ antiporter. It is through the action of ion pumps that cells are able to regulate pH via the pumping of protons.

In contrast to ion pumps, ion channels do not build chemical gradients but rather dissipate them in order to perform work or send a signal. Probably the most familiar and best studied example is the voltage-gated Na^+ channel, which al-

processes were not balancing each other, the cell would either balloon outward to an unmanageable size or completely deplete its plasma membrane within a matter of minutes.

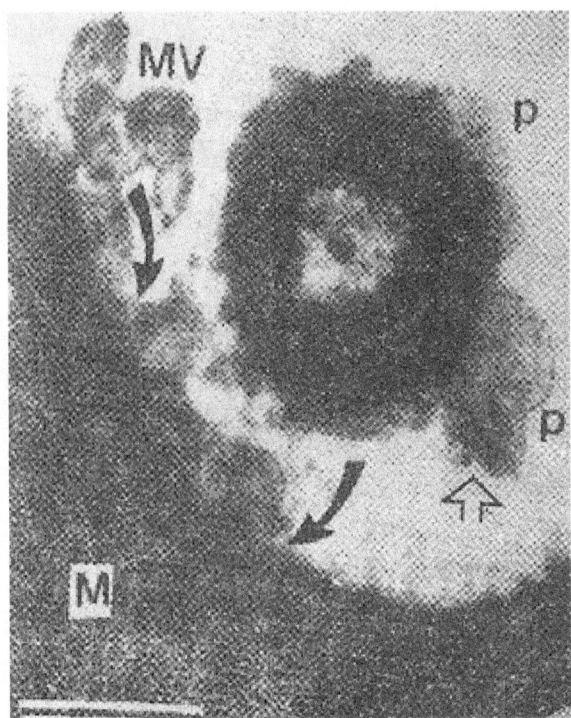

Exocytosis of outer membrane vesicles (MV) liberated from inflated periplasmic pockets (p) on surface of human Salmonella *3,10:r:- pathogens docking on plasma membrane of macrophage cells (M) in chicken ileum, for host-pathogen signaling* in vivo.

Exocytosis in prokaryotes: Membrane vesicular exocytosis, popularly known as membrane vesicle trafficking, a Nobel prize-winning (year, 2013) process, is traditionally regarded as a prerogative of eukaryotic cells.[52] This *myth* was however broken with the revelation that nanovesicles, popularly known as bacterial outer membrane vesicles, released by gram-negative microbes, translocate bacterial signal molecules to host or target cells[53] to carry out multiple processes in favour of the secreting microbe e.g., in *host cell invasion*[54] and microbe-environment interactions, in general.[55]

35.4.4 Electroporation

Further information: Electroporation

Electroporation is the rapid increase in bilayer permeability induced by the application of a large artificial electric field across the membrane. Experimentally, electroporation is used to introduce hydrophilic molecules into cells. It is a particularly useful technique for large highly charged molecules such as DNA, which would never passively diffuse across the hydrophobic bilayer core.[56] Because of this, electroporation is one of the key methods of transfection as well as bacterial transformation. It has even been proposed that electroporation resulting from lightning strikes could be a mechanism of natural horizontal gene transfer.[57]

This increase in permeability primarily affects transport of ions and other hydrated species, indicating that the mechanism is the creation of nm-scale water-filled holes in the membrane. Although electroporation and dielectric breakdown both result from application of an electric field, the mechanisms involved are fundamentally different. In dielectric breakdown the barrier material is ionized, creating a conductive pathway. The material alteration is thus chemical in nature. In contrast, during electroporation the lipid molecules are not chemically altered but simply shift position, opening up a pore that acts as the conductive pathway through the bilayer as it is filled with water.

35.5 Mechanics

Further information: Lipid bilayer mechanics

Lipid bilayers are large enough structures to have some of

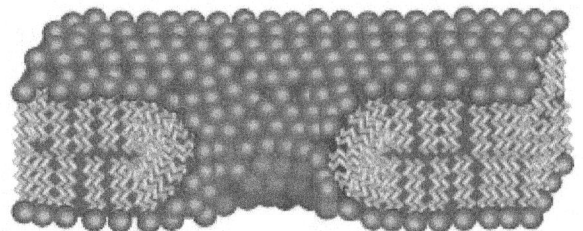

Schematic showing two possible conformations of the lipids at the edge of a pore. In the top image the lipids have not rearranged, so the pore wall is hydrophobic. In the bottom image some of the lipid heads have bent over, so the pore wall is hydrophilic.

the mechanical properties of liquids or solids. The area compression modulus K_a, bending modulus K_b, and edge energy Λ, can be used to describe them. Solid lipid bilayers also have a shear modulus, but like any liquid, the shear modulus is zero for fluid bilayers. These mechanical properties affect how the membrane functions. K_a and K_b affect the ability of proteins and small molecules to insert into the bilayer,[58][59] and bilayer mechanical properties have been shown to alter the function of mechanically activated ion channels.[60] Bilayer mechanical properties also govern what types of stress a cell can withstand without tearing. Although lipid bilayers can easily bend, most cannot stretch more than a few percent before rupturing.[61]

As discussed in the Structure and organization section, the hydrophobic attraction of lipid tails in water is the primary force holding lipid bilayers together. Thus, the elastic modulus of the bilayer is primarily determined by how much extra area is exposed to water when the lipid molecules are stretched apart.[62] It is not surprising given this understanding of the forces involved that studies have shown that K_a varies strongly with osmotic pressure[63] but only weakly with tail length and unsaturation.[10] Because the forces involved are so small, it is difficult to experimentally determine K_a. Most techniques require sophisticated microscopy and very sensitive measurement equipment.[42][64]

In contrast to K_a, which is a measure of how much energy is needed to stretch the bilayer, K_b is a measure of how much energy is needed to bend or flex the bilayer. Formally, bending modulus is defined as the energy required to deform a membrane from its intrinsic curvature to some other curvature. Intrinsic curvature is defined by the ratio of the diameter of the head group to that of the tail group. For two-tailed PC lipids, this ratio is nearly one so the intrinsic curvature is nearly zero. If a particular lipid has too large a deviation from zero intrinsic curvature it will not form a bilayer and will instead form other phases such as micelles or inverted micelles. Addition of *small hydrophilic molecules* like *sucrose* into mixed lipid *lamellar liposomes* made from galactolipid-rich thylakoid membranes destabilises bilayers into **micellar** phase.[65] Typically, K_b is not measured experimentally but rather is calculated from measurements of K_a and bilayer thickness, since the three parameters are related.

Λ is a measure of how much energy it takes to expose a bilayer edge to water by tearing the bilayer or creating a hole in it. The origin of this energy is the fact that creating such an interface exposes some of the lipid tails to water, but the exact orientation of these border lipids is unknown. There is some evidence that both hydrophobic (tails straight) and hydrophilic (heads curved around) pores can coexist.[66]

35.6 Fusion

See also: Lipid bilayer fusion and Interbilayer forces in membrane fusion

Fusion is the process by which two lipid bilayers merge, re-

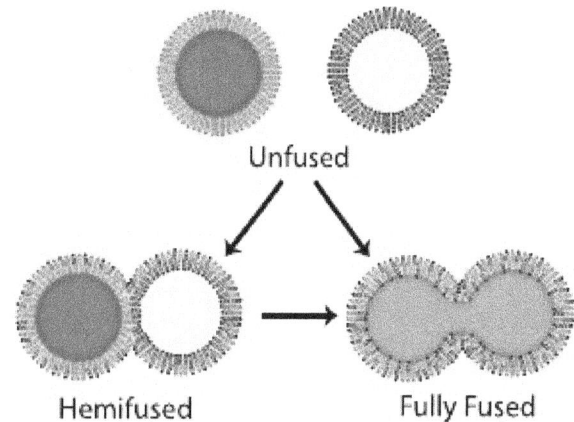

Illustration of lipid vesicles fusing showing two possible outcomes: hemifusion and full fusion. In hemifusion, only the outer bilayer leaflets mix. In full fusion both leaflets as well as the internal contents mix.

sulting in one connected structure. If this fusion proceeds completely through both leaflets of both bilayers, a water-filled bridge is formed and the solutions contained by the bilayers can mix. Alternatively, if only one leaflet from each bilayer is involved in the fusion process, the bilayers are said to be hemifused. Fusion is involved in many cellular processes, in particular in eukaryotes, since the eukaryotic cell is extensively sub-divided by lipid bilayer membranes. Exocytosis, fertilization of an egg by sperm and transport of waste products to the lysozome are a few of the many eukaryotic processes that rely on some form of fusion. Even the entry of pathogens can be governed by fusion, as many bilayer-coated viruses have dedicated fusion proteins to gain entry into the host cell.

There are four fundamental steps in the fusion process.[24] First, the involved membranes must aggregate, approaching each other to within several nanometers. Second, the two bilayers must come into very close contact (within a few angstroms). To achieve this close contact, the two surfaces must become at least partially dehydrated, as the bound surface water normally present causes bilayers to strongly repel. The presence of ions, in particular divalent cations like magnesium and calcium, strongly affects this step.[67][68] One of the critical roles of calcium in the body is regulating membrane fusion. Third, a destabilization must form at one point between the two bilayers, locally distorting their structures. The exact nature of this distortion is not known. One theory is that a highly curved "stalk" must form between the two bilayers.[69] Proponents of this theory believe that it explains why phosphatidylethanolamine, a highly curved lipid, promotes fusion.[70] Finally, in the last step of fusion, this point defect grows and the components of the two bilayers mix and diffuse away from the site of contact.

The situation is further complicated when considering fu-

Schematic illustration of the process of fusion through stalk formation.

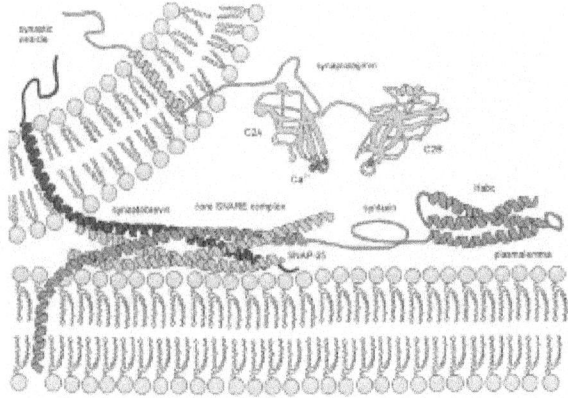

Diagram of the action of SNARE proteins docking a vesicle for exocytosis. Complementary versions of the protein on the vesicle and the target membrane bind and wrap around each other, drawing the two bilayers close together in the process.[71]

sion *in vivo* since biological fusion is almost always regulated by the action of membrane-associated proteins. The first of these proteins to be studied were the viral fusion proteins, which allow an enveloped virus to insert its genetic material into the host cell (enveloped viruses are those surrounded by a lipid bilayer; some others have only a protein coat). Eukaryotic cells also use fusion proteins, the best-studied of which are the SNAREs. SNARE proteins are used to direct all vesicular intracellular trafficking. Despite years of study, much is still unknown about the function of this protein class. In fact, there is still an active debate regarding whether SNAREs are linked to early docking or participate later in the fusion process by facilitating hemifusion.[72]

In studies of molecular and cellular biology it is often desirable to artificially induce fusion. The addition of polyethylene glycol (PEG) causes fusion without significant aggregation or biochemical disruption. This procedure is now used extensively, for example by fusing B-cells with melanoma cells.[73] The resulting "hybridoma" from this combination expresses a desired antibody as determined by the B-cell involved, but is immortalized due to the melanoma component. Fusion can also be artificially induced through electroporation in a process known as electrofusion. It is believed that this phenomenon results from the energetically active edges formed during electroporation, which can act as the local defect point to nucleate stalk growth between two bilayers.[74]

35.7 Model systems

Further information: Model lipid bilayers

Lipid bilayers can be created artificially in the lab to allow researchers to perform experiments that cannot be done with natural bilayers. These synthetic systems are called model lipid bilayers. There are many different types of model bilayers, each having experimental advantages and disadvantages. They can be made with either synthetic or natural lipids. Among the most common model systems are:

- Black lipid membranes (BLM)
- Supported lipid bilayers (SLB)
- Tethered Bilayer Lipid Membranes (t-BLM)
- Vesicles

35.8 Commercial applications

To date, the most successful commercial application of lipid bilayers has been the use of liposomes for drug delivery, especially for cancer treatment. (Note- the term "liposome" is in essence synonymous with "vesicle" except that vesicle is a general term for the structure whereas liposome refers to only artificial not natural vesicles) The basic idea of liposomal drug delivery is that the drug is encapsulated in solution inside the liposome then injected into the patient. These drug-loaded liposomes travel through the system until they bind at the target site and rupture, releasing the drug. In theory, liposomes should make an ideal drug delivery system since they can isolate nearly any hydrophilic drug, can be grafted with molecules to target specific tissues and can be relatively non-toxic since the body possesses biochemical pathways for degrading lipids.[75]

The first generation of drug delivery liposomes had a simple lipid composition and suffered from several limitations. Circulation in the bloodstream was extremely limited due to both renal clearing and phagocytosis. Refinement of the lipid composition to tune fluidity, surface charge density, and surface hydration resulted in vesicles that adsorb fewer proteins from serum and thus are less readily recognized by the immune system.[76] The most significant advance in this area was the grafting of polyethylene glycol (PEG) onto the liposome surface to produce "stealth" vesicles, which circulate over long times without immune or renal clearing.[77]

The first stealth liposomes were passively targeted at tumor tissues. Because tumors induce rapid and uncontrolled angiogenesis they are especially "leaky" and allow liposomes to exit the bloodstream at a much higher rate than normal tissue would.[78] More recently work has been undertaken to graft antibodies or other molecular markers onto the liposome surface in the hope of actively binding them to a specific cell or tissue type.[79] Some examples of this approach are already in clinical trials.[80]

Another potential application of lipid bilayers is the field of biosensors. Since the lipid bilayer is the barrier between the interior and exterior of the cell, it is also the site of extensive signal transduction. Researchers over the years have tried to harness this potential to develop a bilayer-based device for clinical diagnosis or bioterrorism detection. Progress has been slow in this area and, although a few companies have developed automated lipid-based detection systems, they are still targeted at the research community. These include Biacore (now GE Healthcare Life Sciences), which offers a disposable chip for utilizing lipid bilayers in studies of binding kinetics[81] and Nanion Inc., which has developed an automated patch clamping system.[82] Other, more exotic applications are also being pursued such as the use of lipid bilayer membrane pores for DNA sequencing by Oxford Nanolabs. To date, this technology has not proven commercially viable.

A supported lipid bilayer (SLB) as described above has achieved commercial success as a screening technique to measure the permeability of drugs. This parallel artificial membrane permeability assay PAMPA technique measures the permeability across specifically formulated lipid cocktail(s) found to be highly correlated with Caco-2 cultures,[83][84] the gastrointestinal tract,[85] blood–brain barrier[86] and skin.[87]

35.9 History

Further information: History of cell membrane theory

By the early twentieth century scientists had come to believe that cells are surrounded by a thin oil-like barrier,[88] but the structural nature of this membrane was not known. Two experiments in 1925 laid the groundwork to fill in this gap. By measuring the capacitance of erythrocyte solutions, Hugo Fricke determined that the cell membrane was 3.3 nm thick.[89]

Although the results of this experiment were accurate, Fricke misinterpreted the data to mean that the cell membrane is a single molecular layer. Prof. Dr. Evert Gorter[90] (1881–1954) and F. Grendel of Leiden University approached the problem from a different perspective, spreading the erythrocyte lipids as a monolayer on a Langmuir-Blodgett trough. When they compared the area of the monolayer to the surface area of the cells, they found a ratio of two to one.[91] Later analyses showed several errors and incorrect assumptions with this experiment but, serendipitously, these errors canceled out and from this flawed data Gorter and Grendel drew the correct conclusion- that the cell membrane is a lipid bilayer.[24]

This theory was confirmed through the use of electron microscopy in the late 1950s. Although he did not publish the first electron microscopy study of lipid bilayers[92] J. David Robertson was the first to assert that the two dark electron-dense bands were the headgroups and associated proteins of two apposed lipid monolayers.[93][94] In this body of work, Robertson put forward the concept of the "unit membrane." This was the first time the bilayer structure had been universally assigned to all cell membranes as well as organelle membranes.

Around the same time, the development of model membranes confirmed that the lipid bilayer is a stable structure that can exist independent of proteins. By "painting" a solution of lipid in organic solvent across an aperture, Mueller and Rudin were able to create an artificial bilayer and determine that this exhibited lateral fluidity, high electrical resistance and self-healing in response to puncture,[95] all of which are properties of a natural cell membrane. A few years later, Alec Bangham showed that bilayers, in the form of lipid vesicles, could also be formed simply by exposing a dried lipid sample to water.[96] This was an important advance, since it demonstrated that lipid bilayers form spontaneously via self assembly and do not require a patterned support structure.

35.10 See also

- Category: surfactants
- Membrane protein
- Annular lipid shell
- Membrane vesicle trafficking
- Exocytosis
- Bacterial outer membrane vesicles
- Membrane biophysics
- Lipid polymorphism
- Lipid bilayer phase behavior
- Liposome
- Lipidomics

35.11 References

[1] Andersen, Olaf S.; Koeppe, II, Roger E. (June 2007). "Bilayer Thickness and Membrane Protein Function: An Energetic Perspective". *Annual Review of Biophysics and Biomolecular Structure* **36** (1): 107–130. doi:10.1146/annurev.biophys.36.040306.132643. Retrieved 12 December 2014.

[2] Divecha, Nullin; Irvine, Robin F (27 January 1995). "Phospholipid signaling" (PDF, 0.04 MB). *Cell* **80** (2): 269–278. doi:10.1016/0092-8674(95)90409-3. PMID 7834746.

[3] Mashaghi et al. Hydration strongly affects the molecular and electronic structure of membrane phospholipids. 136, 114709 (2012)

[4] Lewis BA, Engelman DM (May 1983). "Lipid bilayer thickness varies linearly with acyl chain length in fluid phosphatidylcholine vesicles". *J. Mol. Biol.* **166** (2): 211–7. doi:10.1016/S0022-2836(83)80007-2. PMID 6854644.

[5] Zaccai G, Blasie JK, Schoenborn BP (January 1975). "Neutron Diffraction Studies on the Location of Water in Lecithin Bilayer Model Membranes". *Proc. Natl. Acad. Sci. U.S.A.* **72** (1): 376–380. Bibcode:1975PNAS...72..376Z. doi:10.1073/pnas.72.1.376. PMC 432308. PMID 16592215.

[6] Nagle JF, Tristram-Nagle S (November 2000). "Structure of lipid bilayers". *Biochim. Biophys. Acta* **1469** (3): 159–95. doi:10.1016/S0304-4157(00)00016-2. PMC 2747654. PMID 11063882.

[7] Parker J, Madigan MT, Brock TD, Martinko JM (2003). *Brock biology of microorganisms* (10th ed.). Englewood Cliffs, N.J: Prentice Hall. ISBN 0-13-049147-0.

[8] Marsh D (July 2001). "Polarity and permeation profiles in lipid membranes". *Proc. Natl. Acad. Sci. U.S.A.* **98** (14): 7777–82. Bibcode:2001PNAS...98.7777M. doi:10.1073/pnas.131023798. PMC 35418. PMID 11438731.

[9] Marsh D (December 2002). "Membrane water-penetration profiles from spin labels". *Eur. Biophys. J.* **31** (7): 559–62. doi:10.1007/s00249-002-0245-z. PMID 12602343.

[10] Rawicz W, Olbrich KC, McIntosh T, Needham D, Evans E (July 2000). "Effect of chain length and unsaturation on elasticity of lipid bilayers". *Biophys. J.* **79** (1): 328–39. Bibcode:2000BpJ....79..328R. doi:10.1016/S0006-3495(00)76295-3. PMC 1300937. PMID 10866959.

[11] Trauble H, Haynes DH (1971). "The volume change in lipid bilayer lamellae at the crystalline-liquid crystalline phase transition". *Chem. Phys. Lipids.* **7** (4): 324–35. doi:10.1016/0009-3084(71)90010-7.

[12] Bretscher MS (1 March 1972). "Asymmetrical Lipid Bilayer Structure for Biological Membranes". *Nature New Biology* **236** (61): 11–12. doi:10.1038/newbio236011a0. PMID 4502419.

[13] Verkleij AJ, Zwaal RF, Roelofsen B, Comfurius P, Kastelijn D, van Deenen LL (October 1973). "The asymmetric distribution of phospholipids in the human red cell membrane. A combined study using phospholipases and freeze-etch electron microscopy". *Biochim. Biophys. Acta* **323** (2): 178–93. doi:10.1016/0005-2736(73)90143-0. PMID 4356540.

[14] Bell RM, Ballas LM, Coleman RA (1 March 1981). "Lipid topogenesis". *J. Lipid Res.* **22** (3): 391–403. PMID 7017050.

[15] Bretscher MS (August 1973). "Membrane structure: some general principles". *Science* **181** (4100): 622–629. Bibcode:1973Sci...181..622B. doi:10.1126/science.181.4100.622. PMID 4724478.

[16] Rothman JE, Kennedy EP (May 1977). "Rapid transmembrane movement of newly synthesized phospholipids during membrane assembly". *Proc. Natl. Acad. Sci. U.S.A.* **74** (5): 1821–5. Bibcode:1977PNAS...74.1821R. doi:10.1073/pnas.74.5.1821. PMC 431015. PMID 405668.

[17] Kornberg RD, McConnell HM (March 1971). "Inside-outside transitions of phospholipids in vesicle membranes". *Biochemistry* **10** (7): 1111–20. doi:10.1021/bi00783a003. PMID 4324203.

[18] Litman BJ (July 1974). "Determination of molecular asymmetry in the phosphatidylethanolamine surface distribution in mixed phospholipid vesicles". *Biochemistry* **13** (14): 2844–8. doi:10.1021/bi00711a010. PMID 4407872.

[19] Crane JM, Kiessling V, Tamm LK (February 2005). "Measuring lipid asymmetry in planar supported bilayers by fluorescence interference contrast microscopy". *Langmuir* **21** (4): 1377–88. doi:10.1021/la047654w. PMID 15697284.

[20] Kalb E, Frey S, Tamm LK (January 1992). "Formation of supported planar bilayers by fusion of vesicles to supported phospholipid monolayers". *Biochim. Biophys. Acta* **1103** (2): 307–16. doi:10.1016/0005-2736(92)90101-Q. PMID 1311950.

[21] Lin WC, Blanchette CD, Ratto TV, Longo ML (January 2006). "Lipid asymmetry in DLPC/DSPC-supported lipid bilayers: a combined AFM and fluorescence microscopy study". *Biophys. J.* **90** (1): 228–37. Bibcode:2006BpJ....90..228L. doi:10.1529/biophysj.105.067066. PMC 1367021. PMID 16214871.

[22] Berg, Howard C. (1993). *Random walks in biology* (Extended Paperback ed.). Princeton, N.J: Princeton University Press. ISBN 0-691-00064-6.

[23] Dietrich C, Volovyk ZN, Levi M, Thompson NL, Jacobson K (September 2001). "Partitioning of Thy-1, GM1, and cross-linked phospholipid analogs into lipid rafts reconstituted in supported model membrane monolayers". *Proc. Natl. Acad. Sci. U.S.A.* **98** (19): 10642–7. Bibcode:2001PNAS...9810642D. doi:10.1073/pnas.191168698. PMC 58519. PMID 11535814.

[24] Yeagle, Philip (1993). *The membranes of cells* (2nd ed.). Boston: Academic Press. ISBN 0-12-769041-7.

[25] Fadok VA, Bratton DL, Frasch SC, Warner ML, Henson PM (July 1998). "The role of phosphatidylserine in recognition of apoptotic cells by phagocytes". *Cell Death Differ.* **5** (7): 551–62. doi:10.1038/sj.cdd.4400404. PMID 10200509.

[26] Anderson HC, Garimella R, Tague SE (January 2005). "The role of matrix vesicles in growth plate development and biomineralization". *Front. Biosci.* **10** (1–3): 822–37. doi:10.2741/1576. PMID 15569622.

[27] Eanes ED, Hailer AW (January 1987). "Calcium phosphate precipitation in aqueous suspensions of phosphatidylserine-containing anionic liposomes". *Calcif. Tissue Int.* **40** (1): 43–8. doi:10.1007/BF02555727. PMID 3103899.

[28] Kim J, Mosior M, Chung LA, Wu H, McLaughlin S (July 1991). "Binding of peptides with basic residues to membranes containing acidic phospholipids". *Biophys. J.* **60** (1): 135–48. Bibcode:1991BpJ....60..135K. doi:10.1016/S0006-3495(91)82037-9. PMC 1260045. PMID 1883932.

[29] Koch AL (1984). "Primeval cells: possible energy-generating and cell-division mechanisms". *J. Mol. Evol.* **21** (3): 270–7. doi:10.1007/BF02102359. PMID 6242168.

[30] 5.1 Cell Membrane Structure | Life Science | University of Tokyo

[31] Alberts, Bruce (2002). *Molecular biology of the cell* (4th ed.). New York: Garland Science. ISBN 0-8153-4072-9.

[32] Martelli PL, Fariselli P, Casadio R (2003). "An ENSEMBLE machine learning approach for the prediction of all-alpha membrane proteins". *Bioinformatics* **19** (Suppl 1): i205–11. doi:10.1093/bioinformatics/btg1027. PMID 12855459.

[33] Filmore D (2004). "It's A GPCR World". *Modern Drug Discovery* **11**: 24–9.

[34] Montal M, Mueller P (December 1972). "Formation of bimolecular membranes from lipid monolayers and a study of their electrical properties". *Proc. Natl. Acad. Sci.* **69** (12): 3561–6. doi:10.1073/pnas.69.12.3561. PMID 4509315.

[35] Melikov KC, Frolov VA, Shcherbakov A, Samsonov AV, Chizmadzhev YA, Chernomordik LV (April 2001). "Voltage-induced nonconductive pre-pores and metastable single pores in unmodified planar lipid bilayer". *Biophys. J.* **80** (4): 1829–36. Bibcode:2001BpJ....80.1829M. doi:10.1016/S0006-3495(01)76153-X. PMC 1301372. PMID 11259296.

[36] Neher E, Sakmann B (April 1976). "Single-channel currents recorded from membrane of denervated frog muscle fibres". *Nature* **260** (5554): 799–802. Bibcode:1976Natur.260..799N. doi:10.1038/260799a0. PMID 1083489.

[37] Y. Roiter, M. Ornatska, A. R. Rammohan, J. Balakrishnan, D. R. Heine, and S. Minko, Interaction of Nanoparticles with Lipid Membrane, Nano Letters, vol. 8, iss. 3, pp. 941–944 (2008).

[38] Heuser JE, Reese TS, Dennis MJ, Jan Y, Jan L, Evans L (May 1979). "Synaptic vesicle exocytosis captured by quick freezing and correlated with quantal transmitter release". *J. Cell Biol.* **81** (2): 275–300. doi:10.1083/jcb.81.2.275. PMC 2110310. PMID 38256.

[39] Dubinnyi MA, Lesovoy DM, Dubovskii PV, Chupin VV, Arseniev AS (Jun 2006). "Modeling of ^{31}P-NMR spectra of magnetically oriented phospholipid liposomes: A new analytical solution". *Solid State Nucl Magn Reson.* **29** (4): 305–311. doi:10.1016/j.ssnmr.2005.10.009. PMID 16298110.

[40] Tokumasu F, Jin AJ, Dvorak JA (2002). "Lipid membrane phase behavior elucidated in real time by controlled environment atomic force microscopy". *J. Electron Micros.* **51** (1): 1–9. doi:10.1093/jmicro/51.1.1. PMID 12003236.

[41] Richter RP, Brisson A (2003). "Characterization of lipid bilayers and protein assemblies supported on rough surfaces by atomic force microscopy". *Langmuir* **19** (5): 1632–40. doi:10.1021/la026427w.

[42] Steltenkamp S, Müller MM, Deserno M, Hennesthal C, Steinem C, Janshoff A (July 2006). "Mechanical properties of pore-spanning lipid bilayers probed by atomic force microscopy". *Biophys. J.* **91** (1): 217–26. Bibcode:2006BpJ....91..217S. doi:10.1529/biophysj.106.081398. PMC 1479081. PMID 16617084.

[43] Alireza Mashaghi et al., Hydration strongly affects the molecular and electronic structure of membrane phospholipids. J. Chem. Phys. 136, 114709 (2012) http://jcp.aip.org/resource/1/jcpsa6/v136/i11/p114709_s1

[44] Chakrabarti AC (1994). "Permeability of membranes to amino acids and modified amino acids: mechanisms involved in translocation". *Amino Acids* **6** (3): 213–29. doi:10.1007/BF00813743. PMID 11543596.

[45] Hauser H, Phillips MC, Stubbs M (October 1972). "Ion permeability of phospholipid bilayers". *Nature* **239** (5371): 342–4. Bibcode:1972Natur.239..342H. doi:10.1038/239342a0. PMID 12635233.

[46] Papahadjopoulos D, Watkins JC (September 1967). "Phospholipid model membranes. II. Permeability properties of hydrated liquid crystals". *Biochim. Biophys. Acta* **135** (4): 639–52. doi:10.1016/0005-2736(67)90095-8. PMID 6048247.

[47] Paula S, Volkov AG, Van Hoek AN, Haines TH, Deamer DW (January 1996). "Permeation of protons, potassium ions, and small polar molecules through phospholipid bilayers as a function of membrane thickness". *Biophys. J.* **70** (1): 339–48. Bibcode:1996BpJ....70..339P. doi:10.1016/S0006-3495(96)79575-9. PMC 1224932. PMID 8770210.

[48] Xiang TX, Anderson BD (June 1994). "The relationship between permeant size and permeability in lipid bilayer membranes". *J. Membr. Biol.* **140** (2): 111–22. doi:10.1007/bf00232899. PMID 7932645.

[49] Gouaux E, Mackinnon R (December 2005). "Principles of selective ion transport in channels and pumps". *Science* **310** (5753): 1461–5. Bibcode:2005Sci...310.1461G. doi:10.1126/science.1113666. PMID 16322449.

[50] Gundelfinger ED, Kessels MM, Qualmann B (February 2003). "Temporal and spatial coordination of exocytosis and endocytosis". *Nat. Rev. Mol. Cell Biol.* **4** (2): 127–39. doi:10.1038/nrm1016. PMID 12563290.

[51] Steinman RM, Brodie SE, Cohn ZA (March 1976). "Membrane flow during pinocytosis. A stereologic analysis". *J. Cell Biol.* **68** (3): 665–87. doi:10.1083/jcb.68.3.665. PMC 2109655. PMID 1030706.

[52] YashRoy R.C. (1999) 'Exocytosis in prokaryotes' and its role in *salmonella* invasion. ICAR NEWS - A Science and Technology Newsletter, (Oct-Dec) vol. 5(4), page 18.https://www.researchgate.net/publication/230822402_'Exocytosis_in_prokaryotes'{}_and_its_role_in_Salmonella_invasion?ev=prf_pub

[53] YashRoy R C (1993) Electron microscope studies of surface pili and vesicles of *Salmonella* 3,10:r:- organisms. Ind Jl of Anim Sci 63, 99-102.https://www.researchgate.net/publication/230817087_Electron_microscope_studies_of_surface_pilli_and_vesicles_of_Salmonella_310r-_organisms?ev=prf_pub

[54] YashRoy R.C. (1998) Discovery of vesicular exocytosis in prokaryotes and its role in *Salmonella* invasion. *Current Science*, vol. 75(10), pp. 1062-1066.https://www.researchgate.net/publication/230793568_Discovery_of_vesicular_exocytosis_in_prokaryotes_and_its_role_in_Salmonella_invasion?ev=prf_pub

[55] YashRoy R.C. (1998) Exocytosis from gram negative bacteria for *Salmonella* invasion of chicken ileal epithelium. *Indian Journal of Poultry Science*, vol. 33(2), pp. 119–123. https://www.researchgate.net/publication/230856738_Exocytosis_from_gram-negative_bacteria_for_Salmonella_invasion_of_chicken_ileal_epithelium?ev=prf_pub

[56] Neumann E, Schaefer-Ridder M, Wang Y, Hofschneider PH (1982). "Gene transfer into mouse lyoma cells by electroporation in high electric fields". *EMBO J.* **1** (7): 841–5. PMC 553119. PMID 6329708.

[57] Demanèche S, Bertolla F, Buret F, et al. (August 2001). "Laboratory-scale evidence for lightning-mediated gene transfer in soil". *Appl. Environ. Microbiol.* **67** (8): 3440–4. doi:10.1128/AEM.67.8.3440-3444.2001. PMC 93040. PMID 11472916.

[58] Garcia ML (July 2004). "Ion channels: gate expectations". *Nature* **430** (6996): 153–5. Bibcode:2004Natur.430..153G. doi:10.1038/430153a. PMID 15241399.

[59] McIntosh TJ, Simon SA (2006). "Roles of Bilayer Material Properties in Function and Distribution of Membrane Proteins". *Annu. Rev. Biophys. Biomol. Struct.* **35** (1): 177–98. doi:10.1146/annurev.biophys.35.040405.102022. PMID 16689633.

[60] Suchyna TM, Tape SE, Koeppe RE, Andersen OS, Sachs F, Gottlieb PA (July 2004). "Bilayer-dependent inhibition of mechanosensitive channels by neuroactive peptide enantiomers". *Nature* **430** (6996): 235–40. Bibcode:2004Natur.430..235S. doi:10.1038/nature02743. PMID 15241420.

[61] Hallett FR, Marsh J, Nickel BG, Wood JM (February 1993). "Mechanical properties of vesicles. II. A model for osmotic swelling and lysis". *Biophys. J.* **64** (2): 435–42. Bibcode:1993BpJ....64..435H. doi:10.1016/S0006-3495(93)81384-5. PMC 1262346. PMID 8457669.

[62] Boal, David H. (2001). *Mechanics of the cell*. Cambridge, UK: Cambridge University Press. ISBN 0-521-79681-4.

[63] Rutkowski CA, Williams LM, Haines TH, Cummins HZ (June 1991). "The elasticity of synthetic phospholipid vesicles obtained by photon correlation spectroscopy". *Biochemistry* **30** (23): 5688–96. doi:10.1021/bi00237a008. PMID 2043611.

[64] Evans E, Heinrich V, Ludwig F, Rawicz W (October 2003). "Dynamic tension spectroscopy and strength of biomembranes". *Biophys. J.* **85** (4): 2342–50. Bibcode:2003BpJ....85.2342E. doi:10.1016/S0006-3495(03)74658-X. PMC 1303459. PMID 14507698.

[65] YashRoy R.C. (1994) Destabilisation of lamellar dispersion of *thylakoid* membrane lipids by sucrose. *Biochimica et Biophysica Acta*, vol. 1212, pp. 129-133.https://www.researchgate.net/publication/15042978_Destabilisation_of_lamellar_dispersion_of_thylakoid_membrane_lipids_by_sucrose?ev=prf_pub

[66] Weaver JC, Chizmadzhev YA (1996). "Theory of electroporation: A review". *Biochemistry and Bioenergetics* **41** (2): 135–60. doi:10.1016/S0302-4598(96)05062-3.

[67] Papahadjopoulos D, Nir S, Düzgünes N (April 1990). "Molecular mechanisms of calcium-induced membrane fusion". *J. Bioenerg. Biomembr.* **22** (2): 157–79. doi:10.1007/BF00762944. PMID 2139437.

[68] Leventis R, Gagné J, Fuller N, Rand RP, Silvius JR (November 1986). "Divalent cation induced fusion and lipid lateral segregation in phosphatidylcholine-phosphatidic acid vesicles". *Biochemistry* **25** (22): 6978–87. doi:10.1021/bi00370a600. PMID 3801406.

[69] Markin VS, Kozlov MM, Borovjagin VL (October 1984). "On the theory of membrane fusion. The stalk mechanism". *Gen. Physiol. Biophys.* **3** (5): 361–77. PMID 6510702.

[70] Chernomordik LV, Kozlov MM (2003). "Protein-lipid interplay in fusion and fission of biological membranes". *Annu. Rev. Biochem.* **72** (1): 175–207. doi:10.1146/annurev.biochem.72.121801.161504. PMID 14527322.

[71] Georgiev, Danko D.; James F. Glazebrook (2007). "Subneuronal processing of information by solitary waves and stochastic processes". In Lyshevski, Sergey Edward. *Nano and Molecular Electronics Handbook*. Nano and Microengineering Series. CRC Press. pp. 17-1–17-41. ISBN 978-0-8493-8528-5.

[72] Chen YA, Scheller RH (February 2001). "SNARE-mediated membrane fusion". *Nat. Rev. Mol. Cell Biol.* **2** (2): 98–106. doi:10.1038/35052017. PMID 11252968.

[73] Köhler G, Milstein C (August 1975). "Continuous cultures of fused cells secreting antibody of predefined specificity". *Nature* **256** (5517): 495–7. Bibcode:1975Natur.256..495K. doi:10.1038/256495a0. PMID 1172191.

[74] Jordan, Carol A.; Neumann, Eberhard; Sowershi mason, Arthur E. (1989). *Electroporation and electrofusion in cell biology*. New York: Plenum Press. ISBN 0-306-43043-6.

[75] Immordino ML, Dosio F, Cattel L (2006). "Stealth liposomes: review of the basic science, rationale, and clinical applications, existing and potential". *Int J Nanomedicine* **1** (3): 297–315. doi:10.2217/17435889.1.3.297. PMC 2426795. PMID 17717971.

[76] Chonn A, Semple SC, Cullis PR (15 September 1992). "Association of blood proteins with large unilamellar liposomes in vivo. Relation to circulation lifetimes". *J. Biol. Chem.* **267** (26): 18759–65. PMID 1527006.

[77] Boris EH, Winterhalter M, Frederik PM, Vallner JJ, Lasic DD (1997). "Stealth liposomes: from theory to product". *Advanced Drug Delivery Reviews* **24** (2-3): 165–77. doi:10.1016/S0169-409X(96)00456-5.

[78] Maeda H, Sawa T, Konno T (July 2001). "Mechanism of tumor-targeted delivery of macromolecular drugs, including the EPR effect in solid tumor and clinical overview of the prototype polymeric drug SMANCS". *J Control Release* **74** (1-3): 47–61. doi:10.1016/S0168-3659(01)00309-1. PMID 11489482.

[79] Lopes DE, Menezes DE, Kirchmeier MJ, Gagne JF (1999). "Cellular trafficking and cytotoxicity of anti-CD19-targeted liposomal doxorubicin in B lymphoma cells". *Journal of Liposome Research* **9** (2): 199–228. doi:10.3109/08982109909024786.

[80] Matsumura Y, Gotoh M, Muro K, et al. (March 2004). "Phase I and pharmacokinetic study of MCC-465, a doxorubicin (DXR) encapsulated in PEG immunoliposome, in patients with metastatic stomach cancer". *Ann. Oncol.* **15** (3): 517–25. doi:10.1093/annonc/mdh092. PMID 14998859.

[81] . Biacore Inc. Retrieved Feb 12, 2009.

[82] Nanion Technologies. Automated Patch Clamp. Retrieved Feb 28, 2010. (PDF)

[83] Bermejo, M. et al. (2004). PAMPA – a drug absorption in vitro model 7. Comparing rat in situ, Caco-2, and PAMPA permeability of fluoroquinolones. *Pharm. Sci.*, **21**: 429-441.

[84] Avdeef, A. et al. (2005). Caco-2 permeability of weakly basic drugs predicted with the Double-Sink PAMPA pKaflux method. *Pharm. Sci.*, **24**: 333-349.

[85] Avdeef, A. et al. (2004). PAMPA – a drug absorption in vitro model 11. Matching the in vivo unstirred water layer thickness by individual-well stirring in microtitre plates. *Pharm. Sci.*, **22**: 365-374.

[86] Dagenais, C. et al. (2009). P-glycoprotein deficient mouse in situ blood–brain barrier permeability and its prediction using an in combo PAMPA model. *Eur. J. Phar. Sci.*, **38**(2): 121-137.

[87] Sinkó, B. et al. (2009). A PAMPA Study of the Permeability-Enhancing Effect of New Ceramide Analogues. *Chemistry & Biodiversity*, **6**: 1867-1874.

[88] Loeb J (December 1904). "The recent development of Biology". *Science* **20** (519): 777–786. Bibcode:1904Sci....20..777L. doi:10.1126/science.20.519.777. PMID 17730464.

[89] Fricke H (1925). "The electrical capacity of suspensions with special reference to blood". *Journal of General Physiology* **9** (2): 137–52. doi:10.1085/jgp.9.2.137. PMC 2140799. PMID 19872238.

[90] Dooren LJ, Wiedemann LR (1986). "On bimolecular layers of lipids on the chromocytes of the blood". *Journal of European Journal of Pediatrics* **145** (5): 329. doi:10.1007/BF00439232.

[91] Gorter E, Grendel F (1925). "On bimolecular layers of lipids on the chromocytes of the blood". *Journal of Experimental Medicine* **41** (4): 439–43. doi:10.1084/jem.41.4.439. PMC 2130960. PMID 19868999.

[92] Sjöstrand FS, Andersson-Cedergren E, Dewey MM (April 1958). "The ultrastructure of the intercalated discs of frog, mouse and guinea pig cardiac muscle". *J. Ultrastruct. Res.* **1**

(3): 271–87. doi:10.1016/S0022-5320(58)80008-8. PMID 13550367.

[93] Robertson JD (1960). "The molecular structure and contact relationships of cell membranes". *Prog. Biophys. Mol. Biol.* **10**: 343–418. PMID 13742209.

[94] Robertson JD (1959). "The ultrastructure of cell membranes and their derivatives". *Biochem. Soc. Symp.* **16**: 3–43. PMID 13651159.

[95] Mueller P, Rudin DO, Tien HT, Wescott WC (June 1962). "Reconstitution of cell membrane structure in vitro and its transformation into an excitable system". *Nature* **194** (4832): 979–80. Bibcode:1962Natur.194..979M. doi:10.1038/194979a0. PMID 14476933.

[96] Bangham, A. D.; Horne, R. W. (1964). "Negative Staining of Phospholipids and Their Structural Modification by Surface-Active Agents As Observed in the Electron Microscope". *Journal of Molecular Biology* **8** (5): 660–668. doi:10.1016/S0022-2836(64)80115-7. PMID 14187392.

35.12 External links

- Avanti Lipids One of the largest commercial suppliers of lipids. Technical information on lipid properties and handling and lipid bilayer preparation techniques.

- LIPIDAT An extensive database of lipid physical properties

- Structure of Fluid Lipid Bilayers Simulations and publication links related to the cross sectional structure of lipid bilayers.

- Lipid Bilayers and the Gramicidin Channel (requires Java plugin) Pictures and movies showing the results of molecular dynamics simulations of lipid bilayers.

- Structure of Fluid Lipid Bilayers, from the Stephen White laboratory at University of California, Irvine

- Animations of lipid bilayer dynamics (requires Flash plugin)

Chapter 36

Homeostasis

Not to be confused with hemostasis.

Homeostasis or **homoeostasis**[nb 1] is the property of a system in which a variable (for example, the concentration of a substance in solution, or its temperature) is actively regulated to remain very nearly constant. This regulation occurs inside a defined environment (mostly within a living organism's body). Examples of homeostasis include the regulation of the body temperature of an animal, the pH of its extracellular fluids, or the concentrations of sodium (Na^+) and calcium (Ca^{2+}) ions or glucose in the blood plasma, despite changes in the animal's environment, or what it has eaten, or what it is doing (for example, resting or exercising). Each of these variables (for example, body temperature, the pH, or the Na^+, Ca^{2+} and glucose concentrations) is controlled by a separate "homeostat" (or regulator), which, together, maintain life. Homeostats are energy-consuming physiological mechanisms.[3]

The concept was described by French physiologist Claude Bernard in 1865 and the word was coined by Walter Bradford Cannon in 1926.[3][4]

Although the term was originally used to refer to processes within living organisms, it is frequently applied to technological control systems such as thermostats. A homeostat has an absolute requirement for a sensor to detect changes in the controlled entity's value, as well as an effector mechanism that reverses any detected deviation from the desired value (or "setpoint") of the regulated entity. Since the correction of any error detected by the sensor is always in the opposite direction to the error, a homeostat relies on what is known as a negative feedback connection between the sensor and effector.[5][6] The effector's corrective effects are monitored by the sensor, which turns the corrective measures off when setpoint conditions have been restored.[7] Negative feedback systems are therefore referred to as "closed loop", or "negative feedback loops", to distinguish them from "open loop" systems where a stimulus (acting on a sensor) results in an, often, all-or-none response that is not subject to modification once it has been set in motion.[5][8][9]

36.1 Biological

The metabolic processes of all living organisms can only take place in very specific physical and chemical environments. The conditions vary with each organism, and with whether the chemical processes take place inside the cell or in the fluids bathing the cells in multicellular creatures. The best known homeostats in human and other mammalian bodies are regulators that keep the composition of the extracellular fluids (or the "internal environment") constant, especially with regard to the temperature, pH, osmolality, and the concentrations of Na^+, K^+, Ca^{2+}, glucose and CO_2 and O_2. However, a great many other homeostats, encompassing many aspects of human physiology, control other entities in the body. On the other hand, it should be noted that not everything in the body is homeostatically controlled. For instance the signal (be it via neurons or hormones) from the sensor to the effector is, of necessity, highly variable in order to convey information about the direction and magnitude of the error detected by the sensor.[10][11][12] Similarly the effector's response needs to be highly adjustable to reverse the error – in fact it should be very nearly in proportion (but in the opposite direction) to the error that is threatening the internal environment.[8][9] For instance, the arterial blood pressure in mammals is homeostatically controlled, and measured by sensors in the aorta and carotid arteries. The sensors send messages via sensory nerves to the medulla oblongata of the brain indicating whether the blood pressure has fallen or risen, and by how much. The medulla oblongata then distributes messages along motor or efferent nerves belonging to the autonomic nervous system to a wide variety of effector organs, whose activity is consequently changed to reverse the error in the blood pressure. One of the effector organs is the heart whose rate is stimulated to rise (tachycardia) when the arterial blood pressure falls, or to slow down (bradycardia) when the pressure rises above set point. Thus the heart rate (for which there is no sensor in the body) is not homeostatically controlled, but is one of effector responses to errors in the arterial blood pressure. Another example is the rate of sweating. This is one of the effectors in the homeostatic control of body temperature, and therefore highly variable in rough proportion to the heat load that threatens to destabilize the body's core temperature, for which there is a sensor in the hypothalamus of the brain.

Apart from the entities that are homeostatically controlled in the internal environment of the body, and the mechanisms that are responsible for this regulation, there are variables that are neither homeostatically controlled or involved in the operation of homeostats. The blood urea concentration is an example. Mammals do not have "urea sensors". Instead the concentration of urea is determined by a dynamic equilibrium, in much the same way that the water level in a river at any particular point along its course is determined. The level of a river is simply dependent on the rate at which water flows into a particular section and how fast it flows away from there. It therefore varies with the rainfall in the catchment area and obstructions or otherwise to the flow down stream – there is no energy consuming "regulation". The blood urea concentration is comparable to the water level in a natural river. It is manufactured by the liver from the amino groups of the amino acids of proteins that are being degraded in this organ. It is then excreted by the kidneys which simply pass most of the urea in the glomerular filtrate on into the urine without active resorption or excretion by the renal tubules (a relatively small proportion of the urea in the tubules diffuses passively back into the blood as its concentration in the tubules rises when water, without urea, is removed from the tubular fluid). A high protein diet therefore produces high blood urea concentrations, and a protein-poor diet produced low blood plasma urea concentrations, without any physiological attempt to correct or mitigate these fluctuations in the level of urea in the extracellular fluids.

36.2 Examples of some of the better understood physiological homeostats

36.2.1 The core body temperature homeostat

Mammals regulate their core temperatures, using hypothalamic temperature sensors in their brains, but also elsewhere in their bodies. When core body temperature falls behavioral changes are set in motion, which, in humans, include the donning of warmer clothes, the seeking out of wind-free, warmer environments, and, eventually, the curling up in the "fetal position" to reduce the surface area (skin) exposed to the cold.[13] The blood flow to the skin is reduced via sympathetic nerves which constrict the cutaneous blood vessels. The metabolic rate

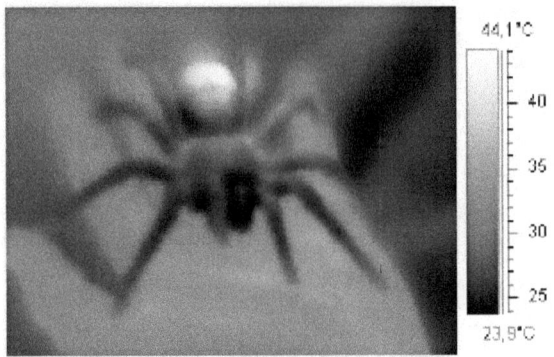

Thermal image of a cold-blooded tarantula (ectothermic) on a warm-blooded human hand (endothermic).

is increased, initially by non-shivering thermogenesis,[14] followed by shivering thermogenesis if the earlier reactions are insufficient to correct the hypothermia.

When body temperature rises, or skin heat sensors detect a threatening rise in body temperature, behavioral changes cause the animal to seek shade, and, in humans, the sweat glands in the skin are stimulated via cholinergic sympathetic nerves to secrete a dilute watery fluid called sweat onto the skin, which, when it evaporates, cools the skin and the blood flowing through it. Panting is an alternative effector in many vertebrates, which cools the body also by the evaporation of water, but this time from the mucous membranes of the throat and mouth.

36.2.2 The blood glucose homeostat

All animals regulate the glucose concentration in their extracellular fluids.[15] In mammals the primary sensor is situated in the beta cells of the pancreatic islets.[16][17] The beta cells respond to a rise in the blood sugar level by secreting insulin into the blood, and simultaneously inhibiting their neighboring alpha cells from secreting glucagon into the blood.[16] This combination (high blood insulin levels and low glucagon levels) act on effector tissues, chief of which are the liver, fat cells and muscle cells. The liver is inhibited from producing glucose, taking it up instead, and converting it to glycogen and triglycerides. The glycogen is stored in the liver, but the triglycerides are secreted into the blood as very low-density lipoprotein (VLDL) particles which are taken up by adipose tissue, there to be stored as fats. The fat cells take up glucose through special glucose transporters (GLUT4), whose numbers in the cell wall are increased as a direct effect of insulin acting on these cells. The glucose that enters the fat cells in this manner is converted into triglycerides (via the same metabolic pathways as are used by the liver) and then stored in those fat cells together with the VLDL-derived triglycerides that were made in the liver.

Muscle cells also take glucose up through insulin-sensitive GLUT4 glucose channels, and convert it into muscle glycogen.

When the beta cells in the pancreatic islets detect lower than normal blood glucose levels, insulin secretion into the blood ceases and the alpha cells are stimulated to secrete glucagon into the blood. This inhibits the uptake of glucose from the blood by the liver, fats cells and muscle. Instead the liver is strongly stimulated to manufacture glucose from glycogen (through glycogenolysis) and from non-carbohydrate sources (such as lactate and de-aminated amino acids) using a process known as gluconeogenesis. The glucose thus produced is discharged into the blood correcting the detected error (hypoglycemia). The glycogen stored in muscles remains in the muscles, and is only broken down, during exercise, to glucose-6-phosphate and thence to pyruvate to be fed into the citric acid cycle or turned into lactate. It is only the lactate and the waste products of the citric acid cycle that are returned to the blood. The liver can take up only the lactate, and by the process of energy consuming gluconeogenesis convert it back to glucose.

36.2.3 The plasma ionized calcium homeostat

The plasma ionized calcium (Ca^{2+}) concentration is very tightly controlled by a pair of homeostats.[18] The sensor for the one is situated in the parathyroid glands, where the chief cells sense the Ca^{2+} level by means of specialized calcium receptors in their membranes. The sensors for the second homeostat are the parafollicular cells in the thyroid gland. The parathyroid chief cells secrete parathyroid hormone (PTH) in response to a fall in the plasma ionized calcium level; the parafollicular cells of the thyroid gland secrete calcitonin in response to a rise in the plasma ionized calcium level.

The effector organs of the first homeostat are the skeleton, the kidney, and, via a hormone released into the blood by the kidney in response to high PTH levels in the blood, the duodenum and jejunum. Parathyroid hormone (in high concentrations in the blood) causes bone resorption, releasing calcium into the plasma. This is a very rapid action which can correct a threatening hypocalcemia within minutes. High PTH concentrations cause the excretion of phosphate ions via the urine. Since phosphates combine with calcium ions to form insoluble salts, a decrease in the level of phosphates in the blood, releases free calcium ions into the plasma ionized calcium pool. PTH has a second action on the kidneys. It stimulates the manufacture and release, by the kidneys, of calcitriol (or 1,25 dihydroxycholecalciferol, or 1,25 dihydroxyvitamin D_3) into the blood. This steroid hormone acts on the epithelial cells of the upper small intestine, increasing their capacity to absorb calcium from the gut contents into the blood.[19]

The second homeostat, with its sensors in the thyroid gland, releases calcitonin into the blood when the blood ionized calcium rises. This hormone acts primarily on bone, causing the rapid removal of calcium from the blood and depositing it, in insoluble form, in the skeleton.

The two homeostats working through PTH on the one hand, and calcitonin on the other, can very rapidly correct any impending error in the plasma ionized calcium level by either removing calcium from the blood and depositing it in the skeleton, or by removing calcium from it. The skeleton acts as an extremely large calcium store (about 1 kg) compared with the plasma calcium store (about 180 mg). Longer term regulation occurs through calcium absorption or loss from the gut (see Regulation of calcium metabolism in the Calcium metabolism article).

36.2.4 The blood partial pressure of oxygen and carbon dioxide homeostats

The partial pressure of oxygen (P_{O_2}) in the arterial blood is measured in the aortic and carotid bodies, near the splitting of the common carotid artery into the internal and external carotid arteries. The partial pressure of carbon dioxide (P_{CO_2}) is measured on the surface of medulla oblongata of the brain. Information from these sets of sensors is sent to the respiratory center in the medulla oblongata of the brain which activates the effector organs, which, in this case, are the skeletal muscles of respiration (particularly the diaphragm). An increase in the P_{CO_2} of the blood, or a decrease in the P_{O_2}, causes deeper and more rapid breathing thus increasing the ventilation rate of the lung alveoli, which blows CO_2 off, out of the blood, and into the outside air, while increasing the uptake of O_2 from the alveolar air into the blood.

Too little CO_2, and, to a lesser extent, too much O_2, in the blood can temporarily halt breathing, which breath-holding divers use to prolong the time they can stay underwater.

36.2.5 The blood oxygen content homeostat

The kidneys measure the oxygen content (rather than the P_{O_2}) of the arterial blood. When the oxygen content of the blood is chronically low, these oxygen-sensitive cells secrete erythropoietin (EPO) into the blood. The effector tissue in this case is the red bone marrow which produces red blood cells (or erythrocytes). This tissue is stimulated by high levels of erythropoietin to increase the rate of red cell production, which leads to an increase in the hematocrit of the blood, and a consequent increase in its oxygen carrying capacity (due to the now high hemoglobin content of the blood). This is the mechanism whereby high altitude dwellers have higher hematocrits than sea-level residents, and also why persons with pulmonary insufficiency or right-to-left shunts in the heart (through which venous blood bypasses the lungs and goes directly into the systemic circulation) have similarly high hematocrits.[20][21]

The distinction between the P_{O_2} of the arterial blood and its oxygen content (or oxygen concentration) is important. The P_{O_2} is the pressure with which the oxygen has been forced into the blood in the alveoli of the lungs. The amount of oxygen that is consequently carried in the blood (at a given P_{O_2}) depends on the hemoglobin concentration in the blood. The greater the hemoglobin concentration the greater the amount of oxygen that can be carried per liter of blood at that P_{O_2}. Thus, in anemia the P_{O_2} of the arterial blood is normal but the oxygen content is below normal. The oxygen content sensors in the kidneys detect this lower than normal oxygen concentration in the arterial blood, and increase their secretion of erythropoietin into the blood. This stimulates a greater rate of red blood cell production in the red bone marrow. This will correct the anemia, and therefore the oxygen concentration in the blood, if there are enough raw materials and co-factors (e.g. iron, vitamin B_{12} and folic acid) to manufacture the extra red cells.[20][22]

36.2.6 The arterial blood pressure homeostat

Stretch receptors in the walls of the aortic arch and carotid sinus (at the beginning of the internal carotid artery) act as arterial blood pressure sensors. As the pressure rises the arteries balloon out, stretching their walls. This information is then conveyed, via sensory nerves, to the medulla oblongata of the brain stem. From here motor nerves belonging to the autonomic nervous system are stimulated to influence the activity of chiefly the heart and the smallest diameter arteries, called arterioles. The arterioles are the main resistance vessels in the arterial tree, and small changes in diameter cause large changes in the resistance to flow through them. When the arterial blood pressure rises the arterioles are stimulated to dilate making it easier for blood to leave the arteries, thus deflating them, and bringing the blood pressure down, back to normal. At the same time the heart is stimulated via cholinergic parasympathetic nerves to beat more slowly (called bradycardia), ensuring that the inflow of blood into the arteries is reduced, thus adding to the reduction in pressure, and correction of the original error.

If the pressure in the arteries falls, the opposite reflex is elicited: constriction of the arterioles, and a speeding up of the heart rate (called tachycardia). If the drop in blood pressure is very rapid or excessive, the medulla oblongata stimulates the adrenal medulla, via "preganglionic" sympathetic nerves, to secrete epinephrine (adrenaline) into the blood. This hormone enhances the tachycardia and causes severe vasoconstriction of the arterioles to all but the essential organ in the body (especially the heart, lungs and brain). These reactions usually correct the low arterial blood pressure (hypotension) very effectively.

36.2.7 The extracellular sodium concentration homeostat

The sodium concentration homeostat is rather more complex than most of the other homeostats described on this page.

The sensor is situated in the juxtaglomerular apparatus of kidneys, which senses the plasma sodium concentration in a surprisingly indirect manner. Instead of measuring it directly in the blood flowing past the juxtaglomerular cells, these cells respond to the sodium concentration in the renal tubular fluid after it has already undergone a certain amount of modification in the proximal convoluted tubule and loop of Henle.[23] These cells also respond to rate of blood flow through the juxtaglomerular apparatus, which, under normal circumstances, is directly proportional to the arterial blood pressure, making this tissue an ancillary arterial blood pressure sensor.

In response to a lowering of the plasma sodium concentration, or to a fall in the arterial blood pressure, the juxtaglomerular cells release renin into the blood.[23][24][25] Renin is an enzyme which cleaves a decapeptide (a short protein chain, 10 amino acids long) from a plasma α−2-globulin called angiotensinogen. This decapeptide is known as angiotensin I.[23] It has no known biological activity. However, when the blood circulates through the lungs a pulmonary capillary endothelial enzyme called angiotensin-converting enzyme (ACE) cleaves a further two amino acids from angiotensin I to form an octapeptide known as angiotensin II. Angiotensin II is a hormone which acts on the adrenal cortex, causing the release into the blood of the steroid hormone, aldosterone. Angiotensin II also acts on the smooth muscle in the walls of the arterioles causing these small diameter vessels to constrict, thereby restricting the outflow of blood from the arterial tree, causing the arterial blood pressure to rise. This therefore reinforces the measures described above (under the heading of *The arterial blood pressure homeostat*), which defend the arterial blood pressure against changes, especially hypotension.

The angiotensin II-stimulated aldosterone released from the zona glomerulosa of the adrenal glands has an effect on particularly the epithelial cells of the distal convoluted tubules and collecting ducts of the kidneys. Here it causes the reabsorption of sodium ions from the renal tubular fluid, in exchange for potassium ions which are secreted from the blood plasma into the tubular fluid to exit the body via the urine.[23][26] The reabsorption of sodium ions from the renal tubular fluid halts further sodium ion losses from the body, and therefore preventing the worsening of hyponatremia. The hyponatremia can only be *corrected* by the consumption of salt in the diet. However, it is not certain whether a "salt hunger" can be initiated by hyponatremia, or by what mechanism this might come about.

When the plasma sodium ion concentration is higher than normal (hypernatremia), the release of renin from the juxtaglomerular apparatus is halted, ceasing the production of angiotensin II, and its consequent aldosterone-release into the blood. The kidneys respond by excreting sodium ions into the urine, thereby normalizing the plasma sodium ion concentration. The low angiotensin II levels in the blood lower the arterial blood pressure as an inevitable concomitant response.

36.2.8 The extracellular potassium concentration homeostat

The extracellular potassium ion (K^+) concentration is sensed by the zona glomerulosa cells of the outer layer of the adrenal cortex, as well as, probably, by sensors in the carotid arteries.[27][28][29] High potassium concentrations in the plasma cause depolarization of the zona glomerulosa cells' membranes.[27] This causes the release of aldosterone into the blood.

Aldosterone acts primarily on the distal convoluted tubules and collecting ducts of the kidneys, stimulating them to excrete potassium ions into the tubular fluid, and thus into the urine.[23] It does so, however, by activating the basolateral Na^+/K^+ pumps of the tubular epithelial cells. These sodium/potassium exchangers pump three sodium ions out of the cell, into the interstitial fluid and two potassium ions into the cell from the interstitial fluid. This creates concentration gradients which result in the reabsorption of sodium (Na^+) ions from the tubular fluid into the blood, and secreting potassium (K^+) ions from the blood into the urine (lumen of collecting duct).[30][31]

This obviously implies that excess potassium in the plasma can only be excreted at the expense of sodium retention by the body. How these two conflicting homeostats (using the same effector) are disentangled to allow the plasma sodium and potassium levels to be regulated independently is currently not clear.

36.2.9 The volume of body water homeostat

The volume of water in the body is measured by stretch receptors in the heart atria, and, somewhat indirectly, by the measurement of the osmolality of the plasma by the hypothalamus. Measurement of the plasma osmolality to give an indication of the water content of the body, relies on the fact that water losses from the body, through sweat, gut fluids (normal fecal water losses, and through vomiting and diarrhea), and the exhaled air, are all hypotonic, meaning that they are less salty than the body fluids (compare, for instance, the taste of saliva with that of tears. The latter have almost the same salt content as the extracellular fluids, whereas the former is hypotonic with respect to plasma. Saliva does not taste salty, whereas tears are decidedly salty). Nearly all normal and abnormal losses of body water therefore cause the extracellular fluids to become hyperosmolar. Conversely excessive water intake (in the form of most regular beverages) dilutes the extracellular fluids causing the hypothalamus to register hypo-osmolar conditions.

When the hypothalamus detects a hyperosmolar extracellular environment, it causes the secretion from the posterior pituitary gland of a peptide hormone called antidiuretic hormone (ADH), which acts on the effector organ, which in this case is the kidney. The effect of ADH on the kidney tubules is to reabsorb water from the distal convoluted tubules and collecting ducts, thus preventing aggravation of the water loss via the urine. The hypothalamus simultaneously stimulates the nearby thirst center causing an almost irresistible (if the hyperosmolarity is severe enough) urge to drink water. The cessation of urine flow prevents the hypovolemia and hypertonicity from getting worse; the drinking of water corrects the defect.

Hypo-osmolality results in very low plasma ADH levels. This results in the inhibition of water reabsorption from the kidney tubules, causing high volumes of very dilute urine to be excreted, thus getting rid of the excess water in the body.

Note that urinary water loss, when the body water homeostat is intact, is a *compensatory* water and solute loss. The volume and osmolality of the urine *corrects* (rather than contributes to) the errors caused by all the other losses and intakes of water and electrolytes. However, since the kidneys cannot generate water, the thirst reflex is the all important second effector mechanism of the body water homeostat.

Stretching of the right atrium of the heart, usually a sign of an excessive blood volume, causes stretch receptors to secrete a hormone known as atrial natriuretic peptide (ANP) into the blood. This also acts on the kidneys causing sodium, and accompanying water loss into the urine, thereby reducing the volume of circulating blood.

36.2.10 The extracellular fluid pH homeostat

The pH of the extracellular fluids (which includes the blood plasma) is regulated by adjusting the ratio of the concentration of carbonic acid (H_2CO_3) to that of the bicarbonate ions (HCO_3^-) to equal 1:20. This ratio and its relationship to the pH is described by the Henderson–Hasselbalch equation, which, when applied to the bicarbonate buffering system in the extracellular fluids, states that:[32]

$$pH = pK_{a\, H_2CO_3} + \log_{10}\left(\frac{[HCO_3^-]}{[H_2CO_3]}\right),$$

where:

- $pK_a\, H_2CO_3$ is the cologarithm of the acid dissociation constant of carbonic acid. It is equal to 6.1.
- $[HCO_3^-]$ is the concentration of bicarbonate in the blood plasma
- $[H_2CO_3]$ is the concentration of carbonic acid in the blood plasma

However, since the carbonic acid concentration is directly proportional to the P_{CO_2} in the extracellular fluid, the Henderson–Hasselbalch equation can be rewritten as follows:[32]

$$pH = 6.1 + \log_{10}\left(\frac{[HCO_3^-]}{0.0307 \times P_{CO_2}}\right),$$

where:

- pH is the acidity in the plasma
- $[HCO_3^-]$ is the concentration of bicarbonate in the plasma
- PCO_2 is the partial pressure of carbon dioxide in the arterial blood plasma

There are therefore at least two homeostats responsible for the regulation of the plasma pH. The first is the P_{CO_2} homeostat described above. The sensor is on the surface of the medulla oblongata of the brain stem, which is also sensitive to the pH of the cerebrospinal fluid.[33] The effector organs are the muscles of respiration, which are stimulated via motor nerves to breathe faster and more deeply when the P_{CO_2} rises and the plasma pH falls, or more slowly and less deeply when the P_{CO_2} falls and the pH rises. Changes in the rate and depth of breathing can change the pH of the arterial plasma within a few seconds.

The sensor for the plasma HCO_3^- concentration is not known for certain. It is very probable that the renal tubular cells of the distal convoluted tubules are themselves sensitive to the pH of the plasma. The metabolism of these cells produces CO_2, which is rapidly converted to H^+ and HCO_3^- through the action of carbonic anhydrase.[17][33] When the extracellular fluids tend towards acidity, the renal tubular cells secrete the H^+ ions into the tubular fluid from where they exit the body via the urine. The HCO_3^- ions are simultaneously secreted into the blood plasma, thus raising the bicarbonate ion concentration in the plasma, increasing the $[HCO_3^-]$: P_{CO_2} ratio, and consequently the pH of the plasma.[33] The converse happens when the plasma pH rises above normal: bicarbonate ions are excreted into the urine, and hydrogen ions into the plasma.[nb 2]

36.3 Homeostatic breakdown

Many diseases are the result the failure of one or other homeostat. Almost any functional component of any homeostat can malfunction, either as a result of an inherited defect, or an acquired disease. Some of the homeostats have inbuilt redundancies, which insures that life is not immediately threatened if a component malfunctions; but in other cases malfunction of a homeostat causes severe disease, which can be fatal if not treated. Here only a few well known examples of homeostat dysfunction are described.

Type 1 diabetes mellitus is probably the best known example. Here the blood glucose homeostat ceases to function because the beta cells of the pancreatic islets are destroyed. This means that the glucose sensor is absent, and its effector pathway (the insulin level in the blood) remains unchanged at zero. The blood glucose concentration therefore rises to very high levels, while the body's proteins are degraded into amino acids which are turned at a very high rate into glucose, via gluconeogenesis, by the liver. The condition is fatal if not treated.

The plasma ionized calcium homeostat can be disrupted by the constant, unchanging, over-production of parathyroid hormone by a parathyroid adenoma resulting in the typically features of hyperparathyroidism, namely high plasma ionized Ca^{2+} levels and the resorption of bone, which can lead to spontaneous fractures. The abnormally high plasma ionized calcium concentrations cause conformational changes in many cell-surface proteins (especially ion channels and hormone or neurotransmitter receptors)[34] giving rise to lethargy, muscle weakness, anorexia, constipation and la-bile emotions.[35]

The body water homeostat can be compromised by the inability to secrete ADH in response to even the normal daily water losses via the exhaled air, the feces, and insensible sweating. On receiving a zero blood ADH signal, the kidneys produce huge unchanging volumes of very dilute urine, causing dehydration and death if not treated.

As organisms age, the efficiency of their control systems becomes reduced. The inefficiencies gradually result in an unstable internal environment that increases the risk of illness, and leads to the physical changes associated with aging.[7]

36.4 Examples from technology

The following are all examples of familiar technological homeostatic mechanisms:

- A thermostat operates by switching heaters or air-conditioners on and off in response to the output of a temperature sensor.

- Cruise control adjusts a car's throttle in response to changes in speed.

- An autopilot operates the steering controls of an aircraft or ship in response to deviation from a pre-set compass bearing or route.

- Process control systems in a chemical plant or oil refinery maintain fluid levels, pressures, temperature, chemical composition, etc. by controlling heaters, pumps and valves.

- The centrifugal governor of a steam engine, as designed by James Watt in 1788, reduces the throttle valve in response to increases in the engine speed, or opens the valve if the speed falls below the pre-set rate.

36.5 Biosphere

In the Gaia hypothesis, James Lovelock[36] stated that the entire mass of living matter on Earth (or any planet with life) functions as a vast homeostatic superorganism that actively modifies its planetary environment to produce the environmental conditions necessary for its own survival. In this view, the entire planet maintains several homeostats (the primary one being temperature homeostasis). Whether this sort of system is present on Earth is open to debate. However, some relatively simple homeostatic mechanisms are generally accepted. For example, it is sometimes claimed that when atmospheric carbon dioxide levels rise, certain plants may be able to grow better and thus act to remove more carbon dioxide from the atmosphere. However, warming has exacerbated droughts, making water the actual limiting factor on land. When sunlight is plentiful and atmospheric temperature climbs, it has been claimed that the phytoplankton of the ocean surface waters, acting as global sunshine, and therefore heat sensors, may thrive and produce more dimethyl sulfide (DMS). The DMS molecules act as cloud condensation nuclei, which produce more clouds, and thus increase the atmospheric albedo, and this feeds back to lower the temperature of the atmosphere. However, rising sea temperature has stratified the oceans, separating warm, sunlit waters from cool, nutrient-rich waters. Thus, nutrients have become the limiting factor, and plankton levels have actually fallen over the past 50 years, not risen. As scientists discover more about Earth, vast numbers of positive and negative feedback loops are being discovered, that, together, maintain a metastable condition, sometimes within very broad range of environmental conditions.

36.6 Predictive

Main article: Predictive homeostasis

Predictive homeostasis is an anticipatory response to an expected challenge in the future, such as the stimulation of insulin secretion by gut hormones which enter the blood in response to a meal.[16] This insulin secretion occurs before the blood sugar level rises, lowering the blood sugar level in anticipation of a large influx into the blood of glucose resulting from the digestion of carbohydrates in the gut. Such anticipatory reactions are open loop systems which are based, essentially, on "guess work", and are not self-correcting.[37] Anticipatory responses always require a closed loop negative feedback system to correct the over- and undershoots to which the anticipatory systems are prone.

36.7 Other fields

The term has come to be used in other fields, for example:

36.7.1 Risk

Main article: Risk homeostasis

An actuary may refer to *risk homeostasis*, where (for example) people *who* have anti-lock brakes have no better safety record than those without anti-lock brakes, because the former unconsciously compensate for the safer vehicle via less-safe driving habits. Previous to the innovation of anti-lock brakes, certain maneuvers involved minor skids, evoking fear and avoidance: Now the anti-lock system moves the boundary for such feedback, and behavior patterns expand into the no-longer punitive area. It has also been suggested that ecological crises are an instance of risk homeostasis in which a particular behavior continues until proven dangerous or dramatic consequences actually occur.

36.7.2 Stress

Sociologists and psychologists may refer to *stress homeostasis*, the tendency of a population or an individual to stay at a certain level of stress, often generating artificial stresses if the "natural" level of stress is not enough.

Jean-François Lyotard, a postmodern theorist, has applied this term to societal 'power centers' that he describes as being 'governed by a principle of homeostasis,' for example, the scientific hierarchy, which will sometimes ignore a radical new discovery for years because it destabilises previously accepted norms. (See *The Postmodern Condition: A Report on Knowledge* by Jean-François Lyotard)

36.8 History of discovery

The conceptual origins of homeostasis reach back to Greek concepts such as balance, harmony, equilibrium, and steady-state; all believed to be fundamental attributes of life and health.[38] Thus, the philosopher Empedocles (495-435 BC) postulated that all matter consisted of elements and qualities that were in dynamic opposition or alliance to one another, and that balance or harmony was a necessary condition for the survival of living organisms. Following these hypotheses, Hippocrates (460-375 BC) compared health to the harmonious balance of the elements, and illness and disease to the systematic disharmony of these elements.[38][39]

Nearly 150 years ago, Claude Bernard published his seminal work, stating that the maintenance of the internal environment, or milieu intérieur, surrounding the body's cells, was essential for the life of the organism.[40] In 1929, Walter B. Cannon published an extrapolation from Bernard's 1865 work naming his theory "homeostasis".[38][40][41] Cannon postulated that homeostasis was a process of synchronized adjustments in the internal environment resulting in the maintenance of specific physiological variables within defined parameters; and that these precise parameters included blood pressure, temperature, pH, and others; all with clearly defined "normal" ranges. Cannon further posited that threats to homeostasis might originate from the external environment (e.g., temperature extremes, traumatic injury) or the internal environment (e.g., pain, infection), and could be physical or psychological, as in emotional distress.[40] Cannon's work outlined that maintenance of this internal physical and psychological balance, homeostasis, demands an internal network of communication, with sensors capable of identifying deviations from the acceptable ranges and effectors to return those deviations back within acceptable limits. Cannon identified these negative feedback systems and emphasized that, regardless of the nature of the threat to homeostasis, the response he mapped within the body would be the same.

36.9 See also

- Acclimatization
- Allostasis
- Apoptosis
- Biological rhythm
- Cybernetics
- Enantiostasis
- Geophysiology
- Homeorhesis
- Le Chatelier's principle
- Lenz's law
- Milieu interieur
- Osmosis
- Proteostasis
- Senescence
- Steady state

36.10 Foot note

[1] The word *homeostasis* (/ˌhoʊmioʊˈsteɪsɪs/[1][2]) uses combining forms of *homeo-* and *-stasis*, New Latin from Greek: ὅμοιος *homoios*, "similar" and στάσις *stasis*, "standing still", yielding the idea of "staying the same".

[2] When H^+ ions are excreted into the urine, and HCO_3^- into the blood, the latter combine with the excess H^+ ions in the plasma that stimulated the kidneys to perform this operation. The resulting reaction in the plasma [$HCO_3^- + H^+ = H_2CO_3$] leads to the formation of carbonic acid, which is in equilibrium with the plasma P_{CO_2}. The latter is tightly regulated by the P_{CO_2} homeostat, ensuring that there is no build up (above normal) of carbonic acid or bicarbonate ions in the

blood plasma. The overall effect is therefore that H$^+$ ions are lost in the urine when the pH of the plasma falls. The concomitant rise in the plasma HCO$_3^-$ ion concentration mops up the excess of H$^+$ ions in the plasma (which caused the fall in plasma pH). The resulting carbonic acid excess is then rapidly disposed of in the lungs as CO$_2$, restoring the normal plasma [HCO$_3^-$] : P_{CO_2} ratio, and therefore the plasma pH. The converse happens when a high plasma pH stimulates the kidneys to excrete HCO$_3^-$ into the urine, and H$^+$ ions into the blood. The H$^+$ ions combine with the excess HCO$_3^-$ ions in the plasma, once again forming an excess of carbonic acid which can be blown off, as CO$_2$, in the lungs, keeping the the plasma bicarbonate ion concentration, the P_{CO_2} and, therefore, the plasma pH, constant.

36.11 References

[1] "Homeostasis". *Merriam-Webster Dictionary*.

[2] "Homeostasis". *Dictionary.com Unabridged*. Random House.

[3] Cannon, W.B. (1932). *The Wisdom of the Body*. New York: W. W. Norton & Company. pp. 177–201.

[4] Cannon, W. B. (1926). "Physiological regulation of normal states: some tentative postulates concerning biological homeostatics". In A. Pettit(ed.). *A Charles Richet : ses amis, ses collègues, ses élèves* (in French). Paris: Les Éditions Médicales. p. 91.

[5] Riggs, D.S. (1963). *The mathematical Approach to Physiological problems*. Baltimore: Williams & Wilkins.

[6] Riggs, D.S. (1970). *Control theory and physiological feedback mechanisms*. Baltimore: Williams & Wilkins.

[7] Marieb, Elaine N., Hoehn, Katja N. (2009). *Essentials of Human Anatomy & Physiology* (9th ed.). San Francisco, CA: Pearson/Benjamin Cummings. ISBN 0321513428.

[8] Guyton, A.C.; Hall, J.E. (1996). *Textbook of medical physiology*. Philadelphia: W.B. Saunders.

[9] Milsum, J.H. (1966). *Biological control systems analysis*. New York: McGraw-Hill.

[10] Shannon, C.E.; Weaver, W. (1949). *The mathematical theory of communication*. Urbana: University of Illinois Press.

[11] Rucker, R. (1987). *Mind tools: the mathematics of information*. Harmondsworth: Penguin Books. pp. 25–30.

[12] Koeslag, Johan H.; Saunders, Peter T.; Wessels, Jabus A. (1999). "The chromogranins and counter-regulatory hormones: do they make homeostatic sense?". *Journal of Physiology* **517**: 643–649.

[13] Mayer, Emeran A. (2011). "Gut feelings: the emerging biology of gut-brain communication". *Nature Reviews Neuroscience* **12** (8): 453–466. doi:10.1038/nrn3071.

[14] Stuart, I.R. (2011). *Human physiology*. (Twelfth ed.). New York: McGraw-Hill. p. 667.

[15] Bhagavan, N. V. (2002). *Medical biochemistry* (4th ed.). Academic Press. p. 499. ISBN 978-0-12-095440-7.

[16] Koeslag, Johan H.; Saunders, Peter T.; Terblanche, Elmarie (2003). "Topical Review: A reappraisal of the blood glucose homeostat which comprehensively explains the type 2 diabetes-syndrome X complex". *Journal of Physiology*. 549.2: 333–346. doi:10.1113/jphysiol.2002.037895.

[17] Stryer, Lubert (1995). *Biochemistry*. (Fourth ed.). New York: W.H. Freeman and Company. pp. 164, 773–774. ISBN 0 7167 2009 4.

[18] Brini M, Ottolini D, Calì T, Carafoli E (2013). "Chapter 4. Calcium in Health and Disease". In Sigel A, Helmut RK. *Interrelations between Essential Metal Ions and Human Diseases*. Metal Ions in Life Sciences **13**. Springer. pp. 81–137. doi:10.1007/978-94-007-7500-8_4.

[19] Stryer, Lubert (1995). "Vitamin D is derived from cholesterol by the ring-splitting action of light.". *In: Biochemistry*. (Fourth ed.). New York: W.H. Freeman and Company. p. 707. ISBN 0 7167 2009 4.

[20] Tortora, Gerard J.; Anagnostakos, Nicholas P. (1987). *Principles of anatomy and physiology* (Fifth ed.). New York: Harper & Row, Publishers. pp. 444–445. ISBN 0-06-350729-3.

[21] Fisher JW, Koury S, Ducey T, Mendel S (1996). "Erythropoietin production by interstitial cells of hypoxic monkey kidneys". *British Journal of Haematology* **95** (1): 27–32. doi:10.1046/j.1365-2141.1996.d01-1864.x. PMID 8857934.

[22] Jelkmann W (2007). "Erythropoietin after a century of research: younger than ever". *European Journal of Haematology* **78** (3): 183–205. doi:10.1111/j.1600-0609.2007.00818.x. PMID 17253966.

[23] Tortora, Gerard J.; Anagnostakos, Nicholas P. (1987). *Principles of anatomy and physiology* (Fifth ed.). New York: Harper & Row, Publishers. pp. 420–421. ISBN 0-06-350729-3.

[24] "JAMA Article Jan 2012".

[25] Williams GH, Dluhy RG (2008). "Chapter 336: Disorders of the Adrenal Cortex". In Loscalzo J, Fauci AS, Braunwald E, Kasper DL, Hauser SL, Longo DL. *Harrison's principles of internal medicine*. New York: McGraw-Hill Medical. ISBN 0-07-146633-9.

[26] Bauer JH, Gauntner WC (March 1979). "Effect of potassium chloride on plasma renin activity and plasma aldosterone during sodium restriction in normal man". *Kidney Int.* **15** (3): 286–93. doi:10.1038/ki.1979.37. PMID 513492.

[27] Hu C, Rusin CG, Tan Z, Guagliardo NA, Barrett PQ (June 2012). "Zona glomerulosa cells of the mouse adrenal cortex are intrinsic electrical oscillators.". *J Clin Invest.* **122** (6): 2046–2053. doi:10.1172/JCI61996. PMID 22546854.

[28] Gann DS, Cruz JF, Casper AG, Bartter FC (May 1962). "Mechanism by which potassium increases aldosterone secretion in the dog". *Am J Physiol.* **202**: 991–6. PMID 13896654.

[29] Bauer JH, Gauntner WC (March 1979). "Effect of potassium chloride on plasma renin activity and plasma aldosterone during sodium restriction in normal man". *Kidney Int.* **15** (3): 286–93. doi:10.1038/ki.1979.37. PMID 513492.

[30] Palmer, LG; Frindt, G (2000). "Aldosterone and potassium secretion by the cortical collecting duct". *Kidney International* **57** (4): 1324–8. doi:10.1046/j.1523-1755.2000.00970.x. PMID 10760062.

[31] Linas SL, Peterson LN, Anderson RJ, Aisenbrey GA, Simon FR, Berl T (June 1979). "Mechanism of renal potassium conservation in the rat". *Kidney International* **15** (6): 601–11. doi:10.1038/ki.1979.79. PMID 222934.

[32] Bray, John J. (1999). *Lecture notes on human physiology.* Malden, Mass.: Blackwell Science. p. 556. ISBN 978-0-86542-775-4.

[33] Tortora, Gerard J.; Anagnostakos, Nicholas P. (1987). *Principles of anatomy and physiology* (Fifth ed.). New York: Harper & Row, Publishers. pp. 581–582, 675–676. ISBN 0-06-350729-3.

[34] Armstrong CM, Cota G (Mar 1999). "Calcium block of Na$^+$ channels and its effect on closing rate". *Proceedings of the National Academy of Sciences of the United States of America* **96** (7): 4154–7. Bibcode:1999PNAS...96.4154A. doi:10.1073/pnas.96.7.4154. PMC 22436. PMID 10097179.

[35] Harrison, T.R. *Principles of Internal Medicine* (third ed.). New York: McGraw-Hill Book Company. pp. 170, 571–579.

[36] Lovelock, James (1991). *Healing Gaia: Practical medicine for the Planet.* New York: Harmony Books. ISBN 0-517-57848-4.

[37] Koeslag, J.H.; Saunders, P.T.; Wessels, J.A. (1997). "Glucose homeostasis with infinite gain: further lessons from the Daisyworld parable?". *Journal of Endocrinology* **134**: 187–192.

[38] Moal, ML (2007). "Historical approach and evolution of the stress concept: a personal account". *Psychoneuroendocrinology* **32**: S3–S9. doi:10.1016/j.psyneuen.2007.03.019. PMID 17659843.

[39] Clendening, L (1942). *Sourcebook of Medical History.* Dover Publications.

[40] Goldstein, DS; Kopin IJ (2007). "Evolution of concepts of stress". *Stress* **10** (2): 109–120. doi:10.1080/10253890701288935. PMID 17514579.

[41] Buchman, TG (2002). "The community of the self". *Nature* **420** (6912): 246–251. doi:10.1038/nature01260. PMID 12432410.

36.12 Further reading

- Banci, Lucia (Ed.), ed. (2013). "Chapter 3 Sodium/Potassium homeostasis, Chapter 5 Calcium homeostasis, Chapter 6 Manganese homeostasis". *Metallomics and the Cell.* Metal Ions in Life Sciences **12**. Springer. doi:10.1007/978-94-007-5561-1_3. ISBN 978-94-007-5560-4. electronic-book ISBN 978-94-007-5561-1 ISSN 1559-0836 electronic-ISSN 1868-0402

36.13 External links

- Homeostasis
- Walter Bradford Cannon, Homeostasis (1932)

Chapter 37

Pattern formation

Pattern formation in a computational model of dendrite growth.

The science of **pattern formation** deals with the visible, (statistically) orderly outcomes of self-organization and the common principles behind similar patterns in nature.

In developmental biology, pattern formation refers to the generation of complex organizations of cell fates in space and time. Pattern formation is controlled by genes. The role of genes in pattern formation is an aspect of morphogenesis, the creation of diverse anatomies from similar genes, now being explored in the science of evolutionary developmental biology or evo-devo. The mechanisms involved are well seen in the anterior-posterior patterning of embryos from the model organism *Drosophila melanogaster* (a fruit fly), one of the first organisms to have its morphogenesis studied, and in the eyespots of butterflies, whose development is a variant of the standard (fruit fly) mechanism.

37.1 Examples

Further information: Patterns in nature

Examples of pattern formation can be found in Biology, Chemistry, Physics and Mathematics,[1] and can readily be simulated with Computer graphics, as described in turn below.

37.1.1 Biology

Further information: Regional specification and Morphogenetic field

Animal markings, segmentation of animals, phyllotaxis,[2] neuronal activation patterns like tonotopy, and predator-prey equations' trajectories are all examples of how natural patterns are formed.

In developmental biology, pattern formation describes the mechanism by which initially equivalent cells in a developing tissue in an embryo assume complex forms and functions.[3] The process of embryogenesis involves coordinated cell fate control.[4][5][6] Pattern formation is genetically controlled, and often involves each cell in a field sensing and responding to its position along a morphogen gradient, followed by short distance cell-to-cell communication through cell signaling pathways to refine the initial pattern. In this context, a field of cells is the group of cells whose fates are affected by responding to the same set positional information cues. This conceptual model was first described as the French flag model in the 1960s.

Anterior-posterior axis patterning in Drosophila

One of the best understood examples of pattern formation is the patterning along the future head to tail (antero-posterior) axis of the fruit fly *Drosophila melanogaster*. The development of this fly is particularly well studied, and it is representative of a major class of animals, the insects. Other multicellular organisms sometimes use similar mechanisms for axis formation, although signal transfer between the earliest cells of many developing organisms is often more important than in *Drosophila*.

See Drosophila embryogenesis

Growth of Colonies

Bacterial colonies show a large variety of beautiful patterns formed during colony growth. The resulting shapes depend on the growth conditions. In particular, stresses (hardness of the culture medium, lack of nutrients, etc.) enhance the complexity of the resulting patterns.[7]

Other organisms such as slime moulds display remarkable patterns caused by the dynamics of chemical signalling.[8]

Vegetation patterns

Main article: patterned vegetation
Vegetation patterns such as tiger bush[9] and fir waves[10]

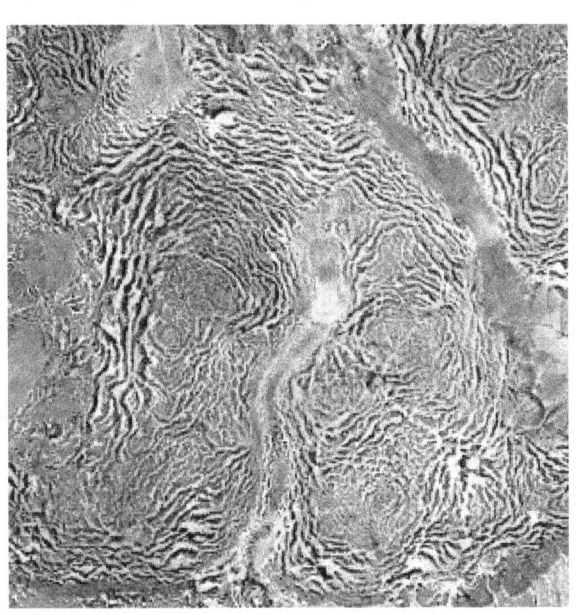

Tiger bush is a vegetation pattern that forms in arid conditions.

form for different reasons. Tiger bush consists of stripes of bushes on arid slopes in countries such as Niger where plant growth is limited by rainfall. Each roughly horizontal stripe of vegetation absorbs rainwater from the bare zone immediately above it.[9] In contrast, fir waves occur in forests on mountain slopes after wind disturbance, during regeneration. When trees fall, the trees that they had sheltered become exposed and are in turn more likely to be damaged, so gaps tend to expand downwind. Meanwhile, on the windward side, young trees grow, protected by the wind shadow of the remaining tall trees.[10]

37.1.2 Chemistry

Further information: reaction–diffusion system and Turing Patterns

- Belousov-Zhabotinsky reaction
- Liesegang rings

37.1.3 Physics

Bénard cells, Laser, cloud formations in stripes or rolls. Ripples in icicles. Washboard patterns on dirtroads. Dendrites in solidification, liquid crystals. Solitons.

37.1.4 Mathematics

Sphere packings and coverings. Mathematics underlies the other pattern formation mechanisms listed.

Further information: Gradient Pattern Analysis

37.1.5 Computer graphics

Pattern resembling a Reaction-diffusion model, produced using sharpen and blur

Further information: Cellular automaton

Some types of automata have been used to generate organic-looking textures for more realistic shading of 3d objects.[11][12]

A popular photoshop plugin, KPT 6, included a filter called 'KPT reaction'. Reaction produced reaction-diffusion style patterns based on the supplied seed image.

A similar effect to the 'KPT reaction' can be achieved with convolution functions in digital image processing, with a little patience, by repeatedly sharpening and blurring an image in a graphics editor. If other filters are used, such as emboss or edge detection, different types of effects can be achieved.

Computers are often used to simulate the biological, physical or chemical processes that lead to pattern formation, and they can display the results in a realistic way. Calculations using models like Reaction-diffusion or MClone are based on the actual mathematical equations designed by the scientists to model the studied phenomena.

37.2 References

[1] Ball, 2009.

[2] Ball, 2009. Shapes, pp. 231–252.

[3] Ball, 2009. Shapes, pp. 261–290.

[4] Eric C. Lai (March 2004). "Notch signaling: control of cell communication and cell fate" (PDF) 131 (5): 965–73. doi:10.1242/dev.01074. PMID 14973298.

[5] Melinda J. Tyler, David A. Cameron (2007). "Cellular pattern formation during retinal regeneration: A role for homotypic control of cell fate acquisition". *Vision Research* 47 (4): 501–511. doi:10.1016/j.visres.2006.08.025.

[6] Hans Meinhard (2001-10-26). "Biological pattern formation".

[7] Ball, 2009. Branches, pp. 52–59.

[8] Ball, 2009. Shapes, pp. 149–151.

[9] Tongway, D.J., Valentin, C. & Seghieri, J. (2001). *Banded vegetation patterning in arid and semiarid environments*. New York: Springer-Verlag. ISBN 978-1461265597.

[10] D'Avanzo, C. (22 February 2004). "Fir Waves: Regeneration in New England Conifer Forests". TIEE. Retrieved 26 May 2012.

[11] Greg Turk, Reaction-Diffusion

[12] Andrew Witkin; Michael Kassy (1991). "Reaction-Diffusion Textures" (PDF). *Proceedings of the 18th annual conference on Computer graphics and interactive techniques*. pp. 299–308. doi:10.1145/122718.122750.

37.3 Bibliography

- Ball, Philip (2009). *Nature's Patterns: a tapestry in three parts. 1:Shapes. 2:Flow. 3:Branches*. Oxford. ISBN 978-0199604869.

37.4 External links

- *SpiralZoom.com*, an educational website about the science of pattern formation, spirals in nature, and spirals in the mythic imagination.

- '15-line Matlab code', A simple 15-line Matlab program to simulate 2D pattern formation for reaction-diffusion model.

Chapter 38

Morphogenesis

This article is about the biological process. For the band, see Morphogenesis (band).

Morphogenesis (from the Greek *morphê* shape and *genesis* creation, literally, "beginning of the shape") is the biological process that causes an organism to develop its shape. It is one of three fundamental aspects of developmental biology along with the control of cell growth and cellular differentiation, unified in evolutionary developmental biology (evo-devo).

The process controls the organized spatial distribution of cells during the embryonic development of an organism. Morphogenesis can take place also in a mature organism, in cell culture or inside tumor cell masses. Morphogenesis also describes the development of unicellular life forms that do not have an embryonic stage in their life cycle, or describes the evolution of a body structure within a taxonomic group.

Morphogenetic responses may be induced in organisms by hormones, by environmental chemicals ranging from substances produced by other organisms to toxic chemicals or radionuclides released as pollutants, and other plants, or by mechanical stresses induced by spatial patterning of the cells.

38.1 History

Some of the earliest ideas and mathematical descriptions on how physical processes and constraints affect biological growth, and hence natural patterns such as the spirals of phyllotaxis, were written by D'Arcy Wentworth Thompson in his 1917 book *On Growth and Form*[1][2][a] and Alan Turing in his *The Chemical Basis of Morphogenesis* (1952).[3] Where Thompson explained animal body shapes as being created by varying rates of growth in different directions, for instance to create the spiral shell of a snail, Turing correctly predicted the diffusion of two different chemical signals, one activating and one deactivating growth, to set up patterns of development. The fuller understanding of the mechanisms involved in actual organisms required the discovery of DNA and the development of molecular biology and biochemistry.

38.2 Molecular basis

Further information: morphogen and transcription factor

Several types of molecules are particularly important during morphogenesis. Morphogens are soluble molecules that can diffuse and carry signals that control cell differentiation decisions in a concentration-dependent fashion. Morphogens typically act through binding to specific protein receptors. An important class of molecules involved in morphogenesis are transcription factor proteins that determine the fate of cells by interacting with DNA. These can be coded for by master regulatory genes and either activate or deactivate the transcription of other genes; in turn, these secondary gene products can regulate the expression of still other genes in a regulatory cascade. At the end of this cascade, another class of molecules involved in morphogenesis are molecules that control cellular behaviors (for example cell migration) or, more generally, their properties, such as cell adhesion or cell contractility. For example, during gastrulation, clumps of stem cells switch off their cell-to-cell adhesion, become migratory, and take up new positions within an embryo where they again activate specific cell adhesion proteins and form new tissues and organs. A number of developmental signaling pathways have been implicated in morphogenesis, including Wnt, Hedgehog, and ephrins.[4] Examples that illustrate the roles of morphogens, transcription factors and cell adhesion molecules in morphogenesis are discussed below.

38.3 Cellular basis

Further information: morphogenetic field

Morphogenesis arises because of changes in the cellular

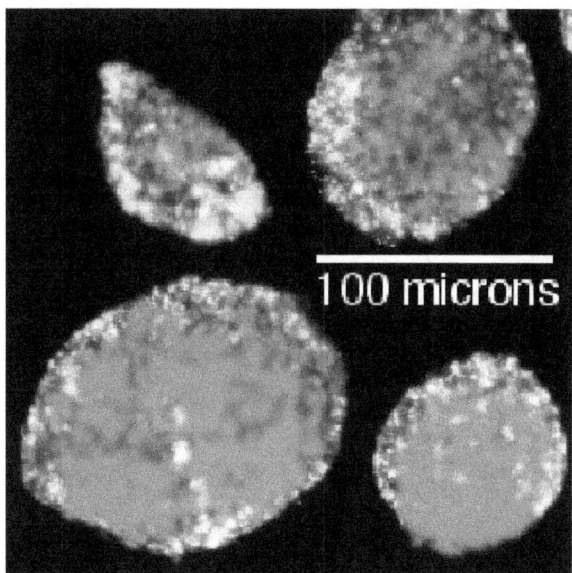

Example of cell sorting out with cultured P19 embryonal carcinoma cells. Live cells were stained with either DiI (red) or DiO (green). The red cells were genetically altered and express higher levels of E-cadherin than the green cells. After labeling, the two populations of cells were mixed and cultured together allowing the cells to form large multi-cellular mixed aggregates. Individual cells are less than 10 micrometres in diameter. The image was captured by scanning confocal microscopy.

structure or how cells interact in tissues.[5] These changes can result in tissue elongation, thinning, folding or separation of one tissue into distinct layers. The latter case is often referred as cell sorting. Cell "sorting out" consists of cells moving so as to sort into clusters that maximize contact between cells of the same type. The ability of cells to do this has been proposed to arise from differential cell adhesion by Malcolm Steinberg through his Differential Adhesion Hypothesis. Tissue separation can also occur via more dramatic cellular differentiation events during which epithelial cells become mesenchymal (see Epithelial-mesenchymal transition). Mesenchymal cells typically leave the epithelial tissue as a consequence of changes in cell adhesive and contractile properties. Following epithelial-mesenchymal transition, cells can migrate away from an epithelium and then associate with other similar cells in a new location.

38.3.1 Cell-cell adhesion

During embryonic development, cells are restricted to different layers due to differential affinities. One of the ways this can occur is when cells share the same cell-to-cell adhesion molecules. For instance, homotypic cell adhesion can maintain boundaries between groups of cells that have different adhesion molecules. Furthermore, cells can sort based upon differences in adhesion between the cells, so even two populations of cells with different levels of the same adhesion molecule can sort out. In cell culture cells that have the strongest adhesion move to the center of a mixed aggregates of cells. Moreover, cell-cell adhesion is often modulated by cell contractility, which can exert forces on the cell-cell contacts so that two cell populations with equal levels of the same adhesion molecule can sort out.

The molecules responsible for adhesion are called cell adhesion molecules (CAMs). Several types of cell adhesion molecules are known and one major class of these molecules are cadherins. There are dozens of different cadherins that are expressed on different cell types. Cadherins bind to other cadherins in a like-to-like manner: E-cadherin (found on many epithelial cells) binds preferentially to other E-cadherin molecules. Mesenchymal cells usually express other cadherin types such as N-cadherin.

38.3.2 Extracellular matrix

The extracellular matrix (ECM) is involved in keeping tissues separated, providing structural support or providing a structure for cells to migrate on. Collagen, laminin, and fibronectin are major ECM molecules that are secreted and assembled into sheets, fibers, and gels. Multisubunit transmembrane receptors called integrins are used to bind to the ECM. Integrins bind extracellularly to fibronectin, laminin, or other ECM components, and intracellularly to microfilament-binding proteins α-actinin and talin to link the cytoskeleton with the outside. Integrins also serve as receptors to trigger signal transduction cascades when binding to the ECM. A well-studied example of morphogenesis that involves ECM is mammary gland ductal branching.[6][7]

38.3.3 Cell contractility

Tissues can change their shape and separate into distinct layers via cell contractility. Just like in muscle cells, myosin can contract different parts of the tissue to change its shape or structure. Typical examples of myosin-driven contractility in tissue morphogenesis occur during the separation of Caenorhabditis elegans, drosophila and zebrafish germ layers. Often, during embryonic morphogenesis, cell contractility occurs via periodic pulses of contraction.

38.4 See also

- Embryogenesis
- Pattern formation
- French flag model
- Reaction-diffusion
- Neurulation
- Gastrulation
- Axon guidance
- Eye development
- Polycystic kidney disease 2
- *Drosophila* embryogenesis
- Cytoplasmic determinant

38.5 Notes

a. ^ Thompson's book is often cited. An abridged version, comprising 349 pages, remains in print and readily obtainable.[8] An unabridged version, comprising 1116 pages, has also been published.[9]

38.6 References

[1] Thompson, D'Arcy Wentworth (1917). *On Growth and Form*. Cambridge University Press.

[2] Montell, Denise J (5 December 2008), "Morphogenetic Cell Movements: Diversity from Modular Mechanical Properties" (PDF), *Science* 322: 1502–1505, Bibcode:2008Sci...322.1502M, doi:10.1126/science.1164073

[3] Turing, A. M. (1952). "The Chemical Basis of Morphogenesis". *Philosophical Transactions of the Royal Society B* 237 (641): 37–72. Bibcode:1952RSPTB.237...37T. doi:10.1098/rstb.1952.0012.

[4] Kouros-Mehr, H.; Werb, Z. (2006). "Candidate regulators of mammary branching morphogenesis identified by genome-wide transcript analysis". *Dev Dyn.* 235 (12): 3404–12. doi:10.1002/dvdy.20978. PMC 2730892. PMID 17039550.

[5] Gilbert, Scott F. (2000). "Morphogenesis and Cell Adhesion". *Developmental biology* (6th ed.). Sunderland, Mass: Sinauer Associates. ISBN 0-87893-243-7.

[6] Fata JE, Werb Z, Bissell MJ (2004). "Regulation of mammary gland branching morphogenesis by the extracellular matrix and its remodeling enzymes". *Breast Cancer Res.* 6 (1): 1–11. doi:10.1186/bcr634. PMC 314442. PMID 14680479.

[7] Sternlicht MD (2006). "Key stages in mammary gland development: the cues that regulate ductal branching morphogenesis". *Breast Cancer Res.* 8 (1): 201. doi:10.1186/bcr1368. PMC 1413974. PMID 16524451.

[8] Thompson, D'Arcy; John Tyler Bonner (editor) (2004 printing. Abridged ed. 1961 (first published 1917)), *On Growth and Form*, Cambridge, U.K., & New York: Cambridge University Press, ISBN 0-521-43776-8, retrieved 11 December 2012 Check date values in: |date= (help)

[9] Thompson, D'Arcy Wentworth (1992), *On Growth and Form: The Complete Revised Edition*, New York: Dover, ISBN 0-486-67135-6

38.7 Sources

- Bard, J.B.L. (1990). *Morphogenesis: The Cellular and Molecular Processes of Developmental Anatomy*. Cambridge, England: Cambridge University Press.
- Slack, J.M.W. (2013). *Essential Developmental Biology*. Oxford: Wiley-Blackwell.

38.8 External links

- Artificial Life model of multicellular morphogenesis with autonomously generated gradients for positional information
- Turing's theory of morphogenesis validated

Chapter 39

Abiogenesis

"Origin of life" redirects here. For non-scientific views on the origins of life, see Creation myth.

Abiogenesis (Brit.: /ˌeɪˌbaɪoʊˈdʒɛnɪsɪs, -ˌbaɪə-, -ˌbiːoʊ-, -ˌbiːə-/[1][2][3][4] *AY-by-oh-JEN-ə-siss* or *AY-bee-oh-JEN-ə-siss*) or **biopoiesis**[5] or **OoL** (**Origins of Life**),[6] is the natural process of life arising from non-living matter, such as simple organic compounds.[7][8][9][10] It is thought to have occurred on Earth between 3.8 and 4.1[11] billion years ago, and is studied through a combination of laboratory experiments and extrapolation from the genetic information of modern organisms in order to make reasonable conjectures about what pre-life chemical reactions may have given rise to a living system.[12]

Precambrian stromatolites in the Siyeh Formation, Glacier National Park. In 2002, a paper in the scientific journal Nature *suggested that these 3.5 Ga (billion years) old geological formations contain fossilized cyanobacteria microbes. This suggests they are evidence of one of the earliest known life forms on Earth.*

The study of abiogenesis involves three main types of considerations: the geophysical, the chemical, and the biological,[13] with more recent approaches attempting a synthesis of all three.[14] Many approaches investigate how self-replicating molecules, or their components, came into existence. It is generally accepted that current life on Earth descended from an RNA world,[15] although RNA-based life may not have been the first life to have existed.[16][17] The Miller–Urey experiment and similar experiments demonstrated that most amino acids, basic chemicals of life, can be synthesized from inorganic compounds in conditions intended to be similar to early Earth. Several mechanisms of organic molecule synthesis have been investigated, including lightning and radiation. Other approaches ("metabolism first" hypotheses) focus on understanding how catalysis in chemical systems on the early Earth might have provided the precursor molecules necessary for self-replication.[18] Complex organic molecules have been found in the Solar System and in interstellar space, and these molecules may have provided starting material for the development of life on Earth.[19][20][21][22]

The panspermia hypothesis suggests that microscopic life was distributed by meteoroids, asteroids and other small Solar System bodies and that life may exist throughout the Universe.[23] It is speculated that the biochemistry of life may have begun shortly after the Big Bang, 13.8 billion years ago, during a habitable epoch when the age of the universe was only 10–17 million years.[24][25] The panspermia hypothesis answers the question of whence, not how life came to be; it only postulates the origin of life to a locale outside the Earth.

Nonetheless, Earth is the only place in the Universe known to harbor life.[26][27] The age of the Earth is about 4.54 billion years.[28][29][30] The earliest undisputed evidence of life on Earth dates at least from 3.5 billion years ago,[31][32][33] during the Eoarchean Era after a geological crust started to solidify following the earlier molten Hadean Eon. There are microbial mat fossils found in 3.48 billion-year-old sandstone discovered in Western Australia.[34][35][36] Other early physical evidence of a biogenic substance is graphite in 3.7 billion-year-old metasedimentary rocks discovered in southwestern Greenland[37] as well as "remains of biotic life" found in 4.1 billion-year-old rocks in Western Australia.[38][39] According to one of the researchers, "If life arose relatively quickly on Earth ... then it could be common in the universe."[38]

39.1 Early geophysical conditions

Earliest humans

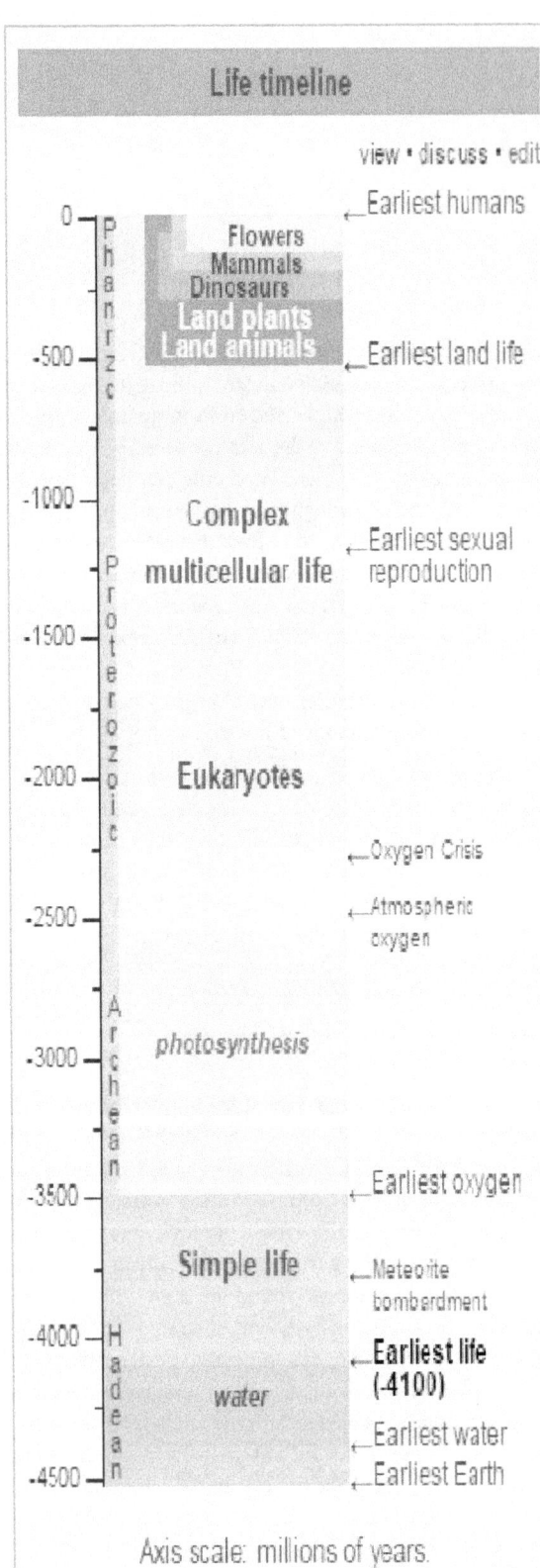

Axis scale: millions of years.
also see {{Human timeline}} and {{Nature timeline}}
Main article: Timeline of the evolutionary history of life

Based on recent computer model studies, the complex organic molecules necessary for life may have formed in the protoplanetary disk of dust grains surrounding the Sun before the formation of the Earth.[40] According to the computer studies, this same process may also occur around other stars that acquire planets.[40] (Also see Extraterrestrial organic molecules).

The Hadean Earth is thought to have had a secondary atmosphere, formed through degassing of the rocks that accumulated from planetesimal impactors. At first, it was thought that the Earth's atmosphere consisted of hydrides—methane, ammonia and water vapour—and that life began under such reducing conditions, which are conducive to the formation of organic molecules. During its formation, the Earth lost a significant part of its initial mass, with a nucleus of the heavier rocky elements of the protoplanetary disk remaining.[41] According to later models, suggested by study of ancient minerals, the atmosphere in the late Hadean period consisted largely of nitrogen and carbon dioxide, with smaller amounts of carbon monoxide, hydrogen, and sulfur compounds.[42] As Earth lacked the gravity to hold any molecular hydrogen, this component of the atmosphere would have been rapidly lost during the Hadean period, along with the bulk of the original inert gases. The solution of carbon dioxide in water is thought to have made the seas slightly acidic, giving it a pH of about 5.5. The atmosphere at the time has been characterized as a "gigantic, productive outdoor chemical laboratory."[43] It may have been similar to the mixture of gases released today by volcanoes, which still support some abiotic chemistry.[43]

Oceans may have appeared first in the Hadean Eon, as soon as two hundred million years (200 Ma) after the Earth was formed, in a hot 100 °C (212 °F) reducing environment, and the pH of about 5.8 rose rapidly towards neutral.[44] This has been supported by the dating of 4.404 Ga-old zircon crystals from metamorphosed quartzite of Mount Narryer in Western Australia, which are evidence that oceans and continental crust existed within 150 Ma of Earth's formation.[45] Despite the likely increased vulcanism and existence of many smaller tectonic "platelets," it has been suggested that between 4.4 and 4.3 Ga (billion year), the Earth was a water world, with little if any continental crust, an extremely turbulent atmosphere and a hydrosphere subject to intense ultraviolet (UV) light, from a T Tauri stage Sun, cosmic radiation and continued bolide impacts.[46]

The Hadean environment would have been highly hazardous to modern life. Frequent collisions with large objects, up to 500 kilometres (310 mi) in diameter, would have been sufficient to sterilise the planet and vaporise the ocean within a few months of impact, with hot steam mixed with rock vapour becoming high altitude clouds that would completely cover the planet. After a few months, the height of these clouds would have begun to decrease but the cloud base would still have been elevated for about the next thousand years. After that, it would have begun to rain at low altitude. For another two thousand years, rains would slowly have drawn down the height of the clouds, returning the oceans to their original depth only 3,000 years after the impact event.[47]

39.1.1 The earliest biological evidence for life on Earth

The earliest life on Earth existed before 3.5 billion years ago,[31][32][33] during the Eoarchean Era when sufficient crust had solidified following the molten Hadean Eon. Physical evidence has been found in biogenic graphite in 3.7 billion-year-old metasedimentary rocks from southwestern Greenland[37] and microbial mat fossils found in 3.48 billion-year-old sandstone from Western Australia.[34][36] Evidence of early life in rocks from Akilia Island, near the Isua supracrustal belt in southwestern Greenland, dating to 3.7 billion years ago have shown biogenic carbon isotopes.[48] At Strelley Pool, in the Pilbarra region of Western Australia, compelling evidence of early life has been found in pyrite-bearing sandstone in a fossilized beach, that showed rounded tubular cells that oxidised sulfur by photosynthesis in the absence of oxygen.[49] More recently, geochemists have found evidence that life likely existed on Earth at least 4.1 billion years ago — 300 million years earlier than previous research suggested.[38][39][50]

In the earlier period between 3.8 and 4.1 Ga, changes in the orbits of the giant planets may have caused a heavy bombardment by asteroids and comets[51] that pockmarked the Moon and the other inner planets (Mercury, Mars, and presumably Earth and Venus). This would likely have repeatedly sterilized the planet, had life appeared before that time.[43] Geologically, the Hadean Earth would have been far more active than at any other time in its history. Studies of meteorites suggests that radioactive isotopes such as aluminium-26 with a half-life of 7.17×10^5 years, and potassium-40 with a half-life of 1.250×10^9 years, isotopes mainly produced in supernovae, were much more common.[52] Coupled with internal heating as a result of gravitational sorting between the core and the mantle, there would have been a great deal of mantle convection, with the probable result of many more smaller and much more active tectonic plates than now exist.

The time periods between such devastating environmental events give time windows for the possible origin of life in the early environments. A study by Kevin A. Maher and David J. Stevenson shows that if the deep marine hydrothermal setting provides a suitable site for the origin of life, then abiogenesis could have happened as early as 4.0 to 4.2 Ga, whereas if it occurred at the surface of the Earth, abiogenesis could only have occurred between 3.7 and 4.0 Ga.[53]

39.2 Conceptual history

39.2.1 Spontaneous generation

Main article: Spontaneous generation

Belief in spontaneous generation of certain forms of life from non-living matter goes back to Aristotle and ancient Greek philosophy and continued to have support in Western scholarship until the 19th century.[54] This belief was paired with a belief in heterogenesis, i.e., that one form of life derived from a different form (e.g., bees from flowers).[55] Classical notions of spontaneous generation held that certain complex, living organisms are generated by decaying organic substances. According to Aristotle, it was a readily observable truth that aphids arise from the dew that falls on plants, flies from putrid matter, mice from dirty hay, crocodiles from rotting logs at the bottom of bodies of water, and so on.[56] In the 17th century, people began to question such assumptions. In 1646, Sir Thomas Browne published his *Pseudodoxia Epidemica* (subtitled *Enquiries into Very many Received Tenets, and commonly Presumed Truths*), which was an attack on false beliefs and "vulgar errors." His contemporary, Alexander Ross, erroneously refuted him, stating: "To question this [Ed.: i.e., spontaneous generation], is to question Reason, Sense, and Experience: If he doubts of this, let him go to *Ægypt*, and there he will finde the fields swarming with mice begot of the mud of *Nylus*, to the great calamity of the Inhabitants."[57][58]

In 1665, Robert Hooke published the first drawings of a microorganism. Hooke was followed in 1676 by Antonie van Leeuwenhoek, who drew and described microorganisms that are now thought to have been protozoa and bacteria.[59] Many felt the existence of microorganisms was evidence in support of spontaneous generation, since microorganisms seemed too simplistic for sexual reproduction, and asexual reproduction through cell division had not yet been observed. Van Leeuwenhoek took issue with the ideas common at the time that fleas and lice could spontaneously result from putrefaction, and that frogs could likewise arise from slime. Using a broad range of experiments ranging from sealed and open meat incubation and the close study of insect reproduction he became, by the 1680s, convinced that spontaneous generation was incorrect.[60]

The first experimental evidence against spontaneous generation came in 1668 when Francesco Redi showed that no maggots appeared in meat when flies were prevented from laying eggs. It was gradually shown that, at least in the case of all the higher and readily visible organisms, the previous sentiment regarding spontaneous generation was false. The alternative seemed to be biogenesis: that every living thing came from a pre-existing living thing (*omne vivum ex ovo*, Latin for "every living thing from an egg").

In 1768, Lazzaro Spallanzani demonstrated that microbes were present in the air, and could be killed by boiling. In 1861, Louis Pasteur performed a series of experiments that demonstrated that organisms such as bacteria and fungi do not spontaneously appear in sterile, nutrient-rich media, but could only appear by invasion from without.

The belief that self-ordering by spontaneous generation was impossible begged for an alternative. By the middle of the 19th century, the theory of biogenesis had accumulated so much evidential support, due to the work of Pasteur and others, that the alternative theory of spontaneous generation had been effectively disproven. John Desmond Bernal, a pioneer in X-ray crystallography, suggested that earlier theories such as spontaneous generation were based upon an explanation that life was continuously created as a result of chance events.[61]

39.2.2 The origin of the terms *biogenesis* and *abiogenesis*

Main article: Biogenesis

The term biogenesis is usually credited to either Henry Charlton Bastian or to Thomas Henry Huxley.[62] Bastian used the term (around 1869) in an unpublished exchange with John Tyndall to mean *life-origination or commencement*. In 1870, Huxley, as new president of the British Association for the Advancement of Science, delivered an address entitled *Biogenesis and Abiogenesis*.[63] In it he introduced the term *biogenesis* (with an opposite meaning to Bastian) and also introduced the term *abiogenesis*:

> And thus the hypothesis that living matter always arises by the agency of pre-existing living matter, took definite shape; and had, henceforward, a right to be considered and a claim to be refuted, in each particular case, before the production of living matter in any other way could be admitted by careful reasoners. It will be necessary for me to refer to this hypothesis so frequently, that, to save circumlocution, I shall call it the hypothesis of *Biogenesis*; and I shall term the contrary doctrine–that living matter may be produced by not living matter–the hypothesis of *Abiogenesis*.[63]

Subsequently, in the preface to Bastian's 1871 book, *The Modes of Origin of Lowest Organisms*,[64] the author refers to the possible confusion with Huxley's usage and he explicitly renounced his own meaning:

> A word of explanation seems necessary with regard to the introduction of the new term

Archebiosis. I had originally, in unpublished writings, adopted the word *Biogenesis* to express the same meaning—viz., life-origination or commencement. But in the mean time the word *Biogenesis* has been made use of, quite independently, by a distinguished biologist [Huxley], who wished to make it bear a totally different meaning. He also introduced the word *Abiogenesis*. I have been informed, however, on the best authority, that neither of these words can— with any regard to the language from which they are derived—be supposed to bear the meanings which have of late been publicly assigned to them. Wishing to avoid all needless confusion, I therefore renounced the use of the word *Biogenesis*, and being, for the reason just given, unable to adopt the other term, I was compelled to introduce a new word, in order to designate the process by which living matter is supposed to come into being, independently of pre-existing living matter.[65]

39.2.3 Louis Pasteur and Charles Darwin

Charles Darwin in 1879

Louis Pasteur remarked, about a finding of his in 1864 which he considered definitive, "Never will the doctrine of spontaneous generation recover from the mortal blow struck by this simple experiment."[66][67] One alternative was that life's origins on Earth had come from somewhere else in the Universe. Periodically resurrected (see Panspermia, above) Bernal said that this approach "is equivalent in the last resort to asserting the operation of metaphysical, spiritual entities... it turns on the argument of creation by design by a creator or demiurge."[68] Such a theory, Bernal said was unscientific and a number of scientists defined life as a result of an inner *life force*, which in the late 19th century was championed by Henri Bergson.

The concept of evolution proposed by Charles Darwin put an end to these metaphysical theologies. In a letter to Joseph Dalton Hooker on 1 February 1871,[69] Darwin discussed the suggestion that the original spark of life may have begun in a "warm little pond, with all sorts of ammonia and phosphoric salts, light, heat, electricity, &c., present, that a proteine compound was chemically formed ready to undergo still more complex changes." He went on to explain that "at the present day such matter would be instantly devoured or absorbed, which would not have been the case before living creatures were formed." He had written to Hooker in 1863 stating that "It is mere rubbish, thinking at present of the origin of life; one might as well think of the origin of matter.". In *On the Origin of Species* he had referred to life having been "created", by which he "really meant 'appeared' by some wholly unknown process", but had soon regretted using the old-testament term "creation".[70]

39.2.4 "Primordial soup" hypothesis

Main article: primordial soup
Further information: Miller–Urey experiment

No new notable research or theory on the subject appeared until 1924, when Alexander Oparin reasoned that atmospheric oxygen prevents the synthesis of certain organic compounds that are necessary building blocks for the evolution of life. In his book *The Origin of Life*,[71][72] Oparin proposed that the "spontaneous generation of life" that had been attacked by Louis Pasteur did in fact occur once, but was now impossible because the conditions found on the early Earth had changed, and preexisting organisms would immediately consume any spontaneously generated organism. Oparin argued that a "primeval soup" of organic molecules could be created in an oxygenless atmosphere through the action of sunlight. These would combine in ever more complex ways until they formed coacervate droplets. These droplets would "grow" by fusion with other droplets, and "reproduce" through fission into daughter droplets, and

39.2. CONCEPTUAL HISTORY

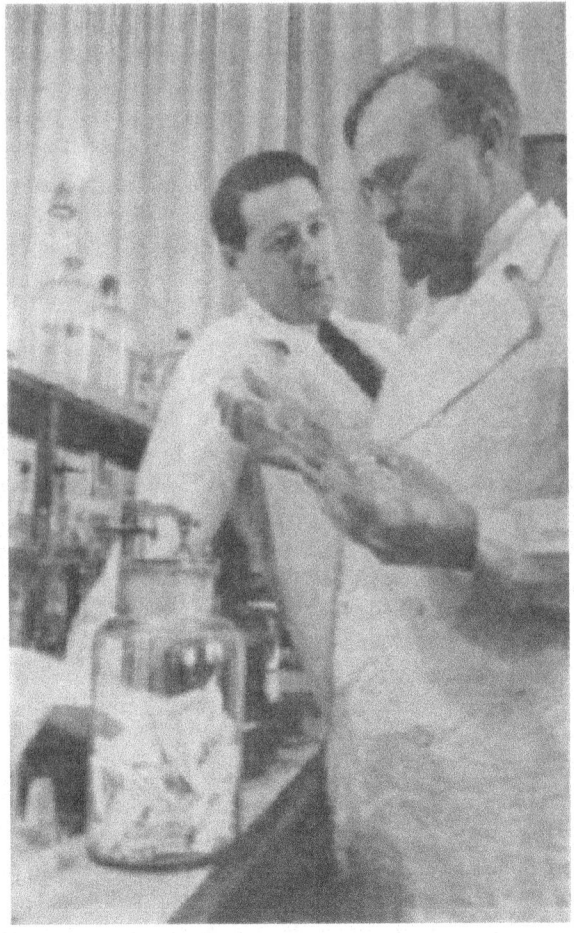

Alexander Oparin (right) at the laboratory

so have a primitive metabolism in which factors that promote "cell integrity" survive, and those that do not become extinct. Many modern theories of the origin of life still take Oparin's ideas as a starting point.

Robert Shapiro has summarized the "primordial soup" theory of Oparin and J. B. S. Haldane in its "mature form" as follows:[73]

1. The early Earth had a chemically reducing atmosphere.

2. This atmosphere, exposed to energy in various forms, produced simple organic compounds ("monomers").

3. These compounds accumulated in a "soup" that may have concentrated at various locations (shorelines, oceanic vents etc.).

4. By further transformation, more complex organic polymers—and ultimately life—developed in the soup.

About this time, Haldane suggested that the Earth's prebiotic oceans—different from their modern counterparts—would have formed a "hot dilute soup" in which organic compounds could have formed. Bernal called this idea *biopoiesis* or *biopoesis*, the process of living matter evolving from self-replicating but nonliving molecules,[61][74] and proposed that biopoiesis passes through a number of intermediate stages.

One of the most important pieces of experimental support for the "soup" theory came in 1952. Stanley L. Miller and Harold C. Urey performed an experiment that demonstrated how organic molecules could have spontaneously formed from inorganic precursors under conditions like those posited by the Oparin-Haldane hypothesis. The now-famous Miller–Urey experiment used a highly reducing mixture of gases—methane, ammonia and hydrogen—to form basic organic monomers, such as amino acids.[75] In the Miller–Urey experiment, a mixture of water, hydrogen, methane, and ammonia was cycled through an apparatus that delivered electrical sparks to the mixture. After one week, it was found that about 10% to 15% of the carbon in the system was now in the form of a racemic mixture of organic compounds, including amino acids, which are the building blocks of proteins. This provided direct experimental support for the second point of the "soup" theory, and it is around the remaining two points of the theory that much of the debate now centers.

Bernal shows that based upon this and subsequent work there is no difficulty in principle in forming most of the molecules we recognise as the basic molecules of life from their inorganic precursors. The underlying hypothesis held by Oparin, Haldane, Bernal, Miller and Urey, for instance, was that multiple conditions on the primeval Earth favored chemical reactions that synthesized the same set of complex organic compounds from such simple precursors. A 2011 reanalysis of the saved vials containing the original extracts that resulted from the Miller and Urey experiments, using current and more advanced analytical equipment and technology, has uncovered more biochemicals than originally discovered in the 1950s. One of the more important findings was 23 amino acids, far more than the five originally found.[76] However, Bernal said that "it is not enough to explain the formation of such molecules, what is necessary," he says, "is a physical-chemical explanation of the origins of these molecules that suggests the presence of suitable sources and sinks for free energy."[77]

39.2.5 Proteinoid microspheres

Main article: Proteinoid

In trying to uncover the intermediate stages of abiogenesis mentioned by Bernal, Sidney W. Fox in the 1950s and 1960s studied the spontaneous formation of peptide structures under conditions that might plausibly have existed early in Earth's history. He demonstrated that amino acids could spontaneously form small chains called peptides. In one of his experiments, he allowed amino acids to dry out as if puddled in a warm, dry spot in prebiotic conditions. He found that, as they dried, the amino acids formed long, often cross-linked, thread-like, submicroscopic polypeptide molecules now named "proteinoid microspheres."[78]

In another experiment using a similar method to set suitable conditions for life to form, Fox collected volcanic material from a cinder cone in Hawaii. He discovered that the temperature was over 100 °C (212 °F) just 4 inches (100 mm) beneath the surface of the cinder cone, and suggested that this might have been the environment in which life was created—molecules could have formed and then been washed through the loose volcanic ash and into the sea. He placed lumps of lava over amino acids derived from methane, ammonia and water, sterilized all materials, and baked the lava over the amino acids for a few hours in a glass oven. A brown, sticky substance formed over the surface and when the lava was drenched in sterilized water a thick, brown liquid leached out. It turned out that the amino acids had combined to form proteinoids, and the proteinoids had combined to form small globules that Fox called "microspheres." His proteinoids were not cells, although they formed clumps and chains reminiscent of cyanobacteria, but they contained no functional nucleic acids or any encoded information. Based upon such experiments, Colin S. Pittendrigh stated in December 1967 that "laboratories will be creating a living cell within ten years," a remark that reflected the typical contemporary levels of innocence of the complexity of cell structures.[79]

39.3 Current models

There is still no "standard model" of the origin of life. Most currently accepted models draw at least some elements from the framework laid out by Alexander Oparin (in 1924) and J. B. S. Haldane (in 1925), who postulated the molecular or chemical evolution theory of life.[80] According to them, the first molecules constituting the earliest cells "were synthesized under natural conditions by a slow process of molecular evolution, and these molecules then organized into the first molecular system with properties with biological order."[80] Oparin and Haldane suggested that the atmosphere of the early Earth may have been chemically reducing in nature, composed primarily of methane (CH_4), ammonia (NH_3), water (H_2O), hydrogen sulfide (H_2S), carbon dioxide (CO_2) or carbon monoxide (CO), and phosphate (PO_4^{3-}), with molecular oxygen (O_2) and ozone (O_3) either rare or absent. According to later models, the atmosphere in the late Hadean period consisted largely of nitrogen (N_2) and carbon dioxide, with smaller amounts of carbon monoxide, hydrogen (H_2), and sulfur compounds;[81] while it did lack molecular oxygen and ozone,[82] it wasn't as chemically reducing as Oparin and Haldane supposed. In the atmosphere proposed by Oparin and Haldane, electrical activity can produce certain basic small molecules (monomers) of life, such as amino acids. This was demonstrated in the Miller–Urey experiment reported in 1953.

Bernal coined the term *biopoiesis* in 1949 to refer to the origin of life.[83] In 1967, he suggested that it occurred in three "stages": 1) the origin of biological monomers; 2) the origin of biological polymers; and 3) the evolution from molecules to cells. He suggested that evolution commenced between stage 1 and 2. The first stage is now fairly well understood, and the discovery of alkaline vents and the similarity with the "proton pump" found as the basis of biological life has begun to provide evidence about the second stage. Bernal considered the third, the discovery of methods by which biological reactions were incorporated behind cell walls, to be the most difficult. Modern work on the self organising capacities by which cell membranes self-assemble, and the work on micropores in various substrates is seen as a halfway house towards the development of independent free-living cells, and research into this is an ongoing effort.[84][85]

The chemical processes that took place on the early Earth are called *chemical evolution*. Both Manfred Eigen and Sol Spiegelman demonstrated that evolution, including replication, variation, and natural selection, can occur in populations of molecules as well as in organisms.[43] Spiegelman took advantage of natural selection to synthesize the Spiegelman Monster, which had a genome with just 218 nucleotide bases, having deconstructively evolved from a 4500 base bacterial RNA. Eigen built on Spiegelman's work and produced a similar system further degraded to just 48 or 54 nucleotides, which was the minimum required for the binding of the replication enzyme.[86]

Chemical evolution was followed by the initiation of biological evolution, which led to the first cells.[43] No one has yet synthesized a "protocell" using basic components with the necessary properties of life (the so-called "bottom-up-approach"). Without such a proof-of-principle, explanations have tended to focus on chemosynthesis.[87] However, some researchers are working in this field, notably Steen Rasmussen and Jack W. Szostak. Others have argued that a "top-down approach" is more feasible. One such approach, successfully attempted by Craig Venter and others at J. Craig Venter Institute, involves engineering existing prokaryotic cells with progressively fewer genes, attempting to discern at which point the most minimal requirements for life were reached.[88][89][90]

39.4 Chemical origin of organic molecules

The elements, except for hydrogen, ultimately derive from stellar nucleosynthesis. Complex molecules, including organic molecules, form naturally both in space and on planets.[19] There are two possible sources of organic molecules on the early Earth:

1. Terrestrial origins – organic molecule synthesis driven by impact shocks or by other energy sources (such as UV light, redox coupling, or electrical discharges) (e.g., Miller's experiments)

2. Extraterrestrial origins – formation of organic molecules in interstellar dust clouds, which rain down on planets.[91][92] (See pseudo-panspermia)

Estimates of the production of organics from these sources suggest that the Late Heavy Bombardment before 3.5 Ga within the early atmosphere made available quantities of organics comparable to those produced by terrestrial sources.[93][94]

Phylogenetic Tree of Life

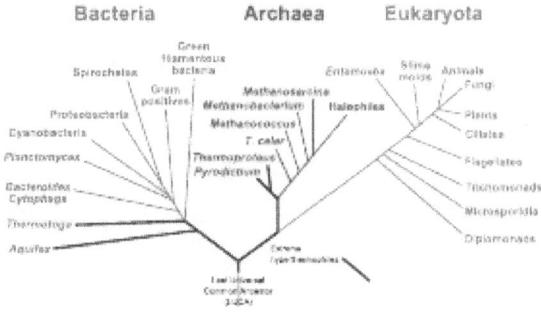

A cladogram demonstrating extreme hyperthermophiles at the base of the phylogenetic tree of life.

It has been estimated that the Late Heavy Bombardment may also have effectively sterilised the Earth's surface to a depth of tens of metres. If life evolved deeper than this, it would have also been shielded from the early high levels of ultraviolet radiation from the T Tauri stage of the Sun's evolution. Simulations of geothermically heated oceanic crust yield far more organics than those found in the Miller-Urey experiments (see below). In the deep hydrothermal vents, Everett Shock has found "there is an enormous thermodynamic drive to form organic compounds, as seawater and hydrothermal fluids, which are far from equilibrium, mix and move towards a more stable state."[95] Shock has found that the available energy is maximised at around 100 – 150 degrees Celsius, precisely the temperatures at which the hyperthermophilic bacteria and thermoacidophilic archaea have been found, at the base of the phylogenetic tree of life closest to the Last Universal Common Ancestor (LUCA).[96]

39.4.1 Chemical synthesis

While features of self-organization and self-replication are often considered the hallmark of living systems, there are many instances of abiotic molecules exhibiting such characteristics under proper conditions. Stan Palasek showed that self-assembly of ribonucleic acid (RNA) molecules can occur spontaneously due to physical factors in hydrothermal vents.[97] Virus self-assembly within host cells has implications for the study of the origin of life,[98] as it lends further credence to the hypothesis that life could have started as self-assembling organic molecules.[99][100]

Multiple sources of energy were available for chemical reactions on the early Earth. For example, heat (such as from geothermal processes) is a standard energy source for chemistry. Other examples include sunlight and electrical discharges (lightning), among others.[43] Unfavorable reactions can also be driven by highly favorable ones, as in the case of iron-sulfur chemistry. For example, this was probably important for carbon fixation (the conversion of carbon from its inorganic form to an organic one).[note 1] Carbon fixation via iron-sulfur chemistry is highly favorable, and occurs at neutral pH and 100 °C (212 °F). Iron-sulfur surfaces, which are abundant near hydrothermal vents, are also capable of producing small amounts of amino acids and other biological metabolites.[43]

Formamide produces all four ribonucleotides and other biological molecules when warmed in the presence of various terrestrial minerals. Formamide is ubiquitous in the Universe, produced by the reaction of water and hydrogen cyanide (HCN). It has several advantages as a biotic precursor, including the ability to easily become concentrated through the evaporation of water.[101][102] Although HCN is poisonous, it only affects aerobic organisms (eukaryotes and aerobic bacteria), which did not yet exist. It can play roles in other chemical processes as well, such as the synthesis of the amino acid glycine.[43]

In 1961, it was shown that the nucleic acid purine base adenine can be formed by heating aqueous ammonium cyanide solutions.[103] Other pathways for synthesizing bases from inorganic materials were also reported.[104] Leslie E. Orgel and colleagues have shown that freezing temperatures are advantageous for the synthesis of purines, due to the concentrating effect for key precursors such as hydrogen cyanide.[105] Research by Stanley L. Miller and colleagues suggested that while adenine and guanine require freezing conditions for synthesis, cytosine and uracil may require boiling temperatures.[106] Research by the Miller group notes the formation of seven different amino acids and 11 types of nucleobases in ice when ammonia and cyanide were left in a freezer from 1972 to 1997.[107][108] Other work demonstrated the formation of s-triazines (alternative nucleobases), pyrimidines (including cytosine and uracil), and adenine from urea solutions subjected to freeze-thaw cycles under a reductive atmosphere (with spark discharges as an energy source).[109] The explanation given for the unusual speed of these reactions at such a low temperature is eutectic freezing. As an ice crystal forms, it stays pure: only molecules of water join the growing crystal, while impurities like salt or cyanide are excluded. These impurities become crowded in microscopic pockets of liquid within the ice, and this crowding causes the molecules to collide more often. Mechanistic exploration using quantum chemical methods provide a more detailed understanding of some of the chemical processes involved in chemical evolution, and a partial answer to the fundamental question of molecular biogenesis.[110]

At the time of the Miller–Urey experiment, scientific consensus was that the early Earth had a reducing atmosphere with compounds relatively rich in hydrogen and poor in oxygen (e.g., CH_4 and NH_3 as opposed to CO_2 and nitrogen dioxide (NO_2)). However, current scientific consensus describes the primitive atmosphere as either weakly reducing or neutral[111][112] (see also Oxygen Catastrophe). Such an atmosphere would diminish both the amount and variety of amino acids that could be produced, although studies that include iron and carbonate minerals (thought present in early oceans) in the experimental conditions have again produced a diverse array of amino acids.[111] Other scientific research has focused on two other potential reducing environments: outer space and deep-sea thermal vents.[113][114][115]

The spontaneous formation of complex polymers from abiotically generated monomers under the conditions posited by the "soup" theory is not at all a straightforward process. Besides the necessary basic organic monomers, compounds that would have prohibited the formation of polymers were also formed in high concentration during the Miller–Urey and Joan Oró experiments.[116] The Miller–Urey experiment, for example, produces many substances that would react with the amino acids or terminate their coupling into peptide chains.[117]

A research project completed in March 2015 by John D. Sutherland and others found that a network of reactions beginning with hydrogen cyanide and hydrogen sulfide, in streams of water irradiated by UV light, could produce the chemical components of proteins and lipids, as well as those of RNA,[118][119] while not producing a wide range of other compounds.[120] The researchers used the term "cyanosulfidic" to describe this network of reactions.[119]

39.4.2 Autocatalysis

Main article: Autocatalysis

Autocatalysts are substances that catalyze the production of themselves and therefore are "molecular replicators." The simplest self-replicating chemical systems are autocatalytic, and typically contain three components: a product molecule and two precursor molecules. The product molecule joins together the precursor molecules, which in turn produce more product molecules from more precursor molecules. The product molecule catalyzes the reaction by providing a complementary template that binds to the precursors, thus bringing them together. Such systems have been demonstrated both in biological macromolecules and in small organic molecules.[121][122] Systems that do not proceed by template mechanisms, such as the self-reproduction of micelles and vesicles, have also been observed.[122]

It has been proposed that life initially arose as autocatalytic chemical networks.[123] British ethologist Richard Dawkins wrote about autocatalysis as a potential explanation for the origin of life in his 2004 book *The Ancestor's Tale*.[124] In his book, Dawkins cites experiments performed by Julius Rebek, Jr. and his colleagues in which they combined amino adenosine and pentafluorophenyl esters with the autocatalyst amino adenosine triacid ester (AATE). One product was a variant of AATE, which catalysed the synthesis of themselves. This experiment demonstrated the possibility that autocatalysts could exhibit competition within a population of entities with heredity, which could be interpreted as a rudimentary form of natural selection.[125][126]

In the early 1970s, Manfred Eigen and Peter Schuster examined the transient stages between the molecular chaos and a self-replicating hypercycle in a prebiotic soup.[127] In a hypercycle, the information storing system (possibly RNA) produces an enzyme, which catalyzes the formation of another information system, in sequence until the product of the last aids in the formation of the first information system. Mathematically treated, hypercycles could create quasispecies, which through natural selection entered into a form of Darwinian evolution. A boost to hypercycle theory was the discovery of ribozymes capable of catalyzing their own chemical reactions. The hypercycle theory requires the existence of complex biochemicals, such as nucleotides, which do not form under the conditions proposed by the Miller–Urey experiment.

It has been shown that early error prone translation machinery can be stable against an error catastrophe of the type that had been envisaged as problematical known as

"Orgel's paradox" caused by catalytic activities that would be disruptive.[128][129][130]

39.4.3 Information theory

A theory that speaks to the origin of life on Earth and other rocky planets posits life as an information system in which information content grows because of selection. Life must start with minimum possible information, or minimum possible departure from thermodynamic equilibrium, and it requires thermodynamically free energy accessible by means of its information content. The most benign circumstances, minimum entropy variations with abundant free energy, suggest the pore space in the first few kilometers of the surface. Free energy is derived from the condensed products of the chemical reactions taking place in the cooling nebula.[131]

39.4.4 Homochirality

Main article: Homochirality

Homochirality refers to the geometric property of some materials that are composed of chiral units. Chiral refers to nonsuperimposable 3D forms that are mirror images of one another, as are left and right hands. Living organisms use molecules that have the same chirality ("handedness"): with almost no exceptions,[132] amino acids are left-handed while nucleotides and sugars are right-handed. Chiral molecules can be synthesized, but in the absence of a chiral source or a chiral catalyst, they are formed in a 50/50 mixture of both enantiomers (called a racemic mixture). Known mechanisms for the production of non-racemic mixtures from racemic starting materials include: asymmetric physical laws, such as the electroweak interaction; asymmetric environments, such as those caused by circularly polarized light, quartz crystals, or the Earth's rotation; and statistical fluctuations during racemic synthesis.[133]

Once established, chirality would be selected for.[134] A small bias (enantiomeric excess) in the population can be amplified into a large one by asymmetric autocatalysis, such as in the Soai reaction.[135] In asymmetric autocatalysis, the catalyst is a chiral molecule, which means that a chiral molecule is catalysing its own production. An initial enantiomeric excess, such as can be produced by polarized light, then allows the more abundant enantiomer to outcompete the other.[136]

Clark has suggested that homochirality may have started in outer space, as the studies of the amino acids on the Murchison meteorite showed that L-alanine is more than twice as frequent as its D form, and L-glutamic acid was more than three times prevalent than its D counterpart. Various chiral crystal surfaces can also act as sites for possible concentration and assembly of chiral monomer units into macromolecules.[137] Compounds found on meteorites suggest that the chirality of life derives from abiogenic synthesis, since amino acids from meteorites show a left-handed bias, whereas sugars show a predominantly right-handed bias, the same as found in living organisms.[138]

39.5 Self-enclosement, reproduction, duplication and the RNA world

39.5.1 Protocells

Main article: Protocell

A protocell is a self-organized, self-ordered, spherical col-

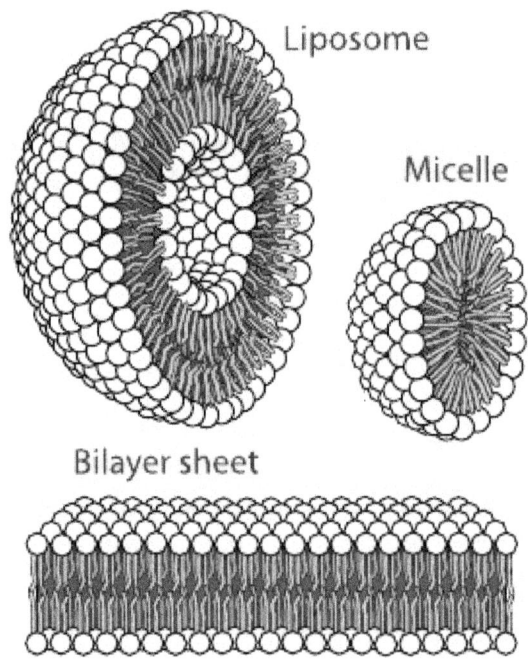

The three main structures phospholipids form spontaneously in solution: the liposome (a closed bilayer), the micelle and the bilayer.

lection of lipids proposed as a stepping-stone to the origin of life.[139] A central question in evolution is how simple protocells first arose and differed in reproductive contribution to the following generation driving the evolution of life. Although a functional protocell has not yet been achieved in a laboratory setting, there are scientists who think the goal is well within reach.[140][141][142]

Self-assembled vesicles are essential components of primitive cells.[139] The second law of thermodynamics requires that the Universe move in a direction in which entropy increases, yet life is distinguished by its great degree of organization. Therefore, a boundary is needed to separate life processes from non-living matter.[143] Researchers Irene A. Chen and Jack W. Szostak amongst others, suggest that simple physicochemical properties of elementary protocells can give rise to essential cellular behaviors, including primitive forms of differential reproduction competition and energy storage. Such cooperative interactions between the membrane and its encapsulated contents could greatly simplify the transition from simple replicating molecules to true cells.[141] Furthermore, competition for membrane molecules would favor stabilized membranes, suggesting a selective advantage for the evolution of cross-linked fatty acids and even the phospholipids of today.[141] Such microencapsulation would allow for metabolism within the membrane, the exchange of small molecules but the prevention of passage of large substances across it.[144] The main advantages of encapsulation include the increased solubility of the contained cargo within the capsule and the storage of energy in the form of a electrochemical gradient.

A 2012 study led by Armen Y. Mulkidjanian of Germany's University of Osnabrück, suggests that inland pools of condensed and cooled geothermal vapour have the ideal characteristics for the origin of life.[145] Scientists confirmed in 2002 that by adding a montmorillonite clay to a solution of fatty acid micelles (lipid spheres), the clay sped up the rate of vesicles formation 100-fold.[142]

Another protocell model is the Jeewanu. First synthesized in 1963 from simple minerals and basic organics while exposed to sunlight, it is still reported to have some metabolic capabilities, the presence of semipermeable membrane, amino acids, phospholipids, carbohydrates and RNA-like molecules.[146][147] However, the nature and properties of the Jeewanu remains to be clarified.

Electrostatic interactions induced by short, positively charged, hydrophobic peptides containing 7 amino acids in length or fewer, can attach RNA to a vesicle membrane, the basic cell membrane.[148]

39.5.2 RNA world

Main article: RNA world

The RNA world hypothesis describes an early Earth with self-replicating and catalytic RNA but no DNA or proteins.[150] It is generally accepted that current life on Earth descends from an RNA world,[15][151] although RNA-based life may not have been the first life to exist.[16][17] This conclusion is drawn from many independent lines of evidence, such as the observations that RNA is central to the

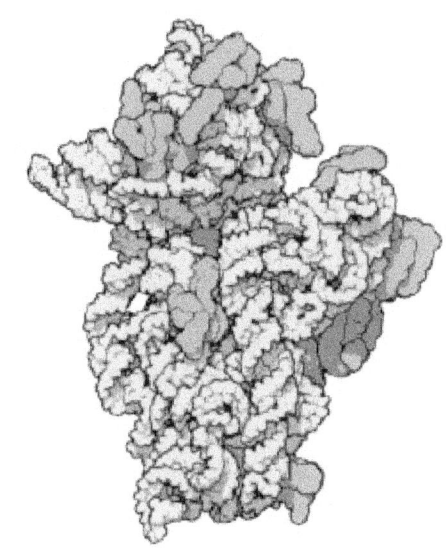

Molecular structure of the ribosome 30S subunit from Thermus thermophilus.[149] *Proteins are shown in blue and the single RNA chain in orange.*

translation process and that small RNAs can catalyze all of the chemical groups and information transfers required for life.[17][152] The structure of the ribosome has been called the "smoking gun," as it showed that the ribosome is a ribozyme, with a central core of RNA and no amino acid side chains within 18 angstroms of the active site where peptide bond formation is catalyzed.[16] The concept of the RNA world was first proposed in 1962 by Alexander Rich,[153] and the term was coined by Walter Gilbert in 1986.[17][154]

Possible precursors for the evolution of protein synthesis include a mechanism to synthesize short peptide cofactors or form a mechanism for the duplication of RNA. It is likely that the ancestral ribosome was composed entirely of RNA, although some roles have since been taken over by proteins. Major remaining questions on this topic include identifying the selective force for the evolution of the ribosome and determining how the genetic code arose.[155]

Eugene Koonin said, "Despite considerable experimental and theoretical effort, no compelling scenarios currently exist for the origin of replication and translation, the key processes that together comprise the core of biological systems and the apparent pre-requisite of biological evolution. The RNA World concept might offer the best chance for the resolution of this conundrum but so far cannot adequately account for the emergence of an efficient RNA replicase or the translation system. The MWO [Ed.: "many worlds in one"] version of the cosmological model of eternal inflation could suggest a way out of this conundrum because, in an infinite multiverse with a finite number of distinct macroscopic histories (each repeated an infinite number of times), emergence of even highly complex systems by chance is not just possible but inevitable."[156]

Viral origins and the RNA World

Recent evidence for a "virus first" hypothesis, which may support theories of the RNA world have been suggested in new research.[157] One of the difficulties for the study viral origins and evolution is their high rate of mutation; this is particularly the case in RNA retroviruses like HIV.[158] A 2015 study compared protein fold structures across different branches of the tree of life, where researchers can reconstruct the evolutionary histories of the folds and of the organisms whose genomes code for those folds. They argue that protein folds are better markers of ancient events as their three-dimensional structures can be maintained even as the sequences that code for those begin to change.[157] Thus, the viral protein repertoire retain traces of ancient evolutionary history that can be recovered using advanced bioinformatics approaches. Those researchers have concluded that, "the prolonged pressure of genome and particle size reduction eventually reduced virocells into modern viruses (identified by the complete loss of cellular makeup), meanwhile other coexisting cellular lineages diversified into modern cells.[159] The data suggest that viruses originated from ancient cells that co-existed with the ancestors of modern cells.[157] These ancient cells likely contained segmented RNA genomes.[157][160]

39.5.3 RNA synthesis and replication

The RNA world hypothesis has spurred scientists to determine if RNA molecules could have spontaneously formed able to catalyze their own replication.[161][162][163] Evidence suggests that the chemical conditions, including the presence of boron, molybdenum and oxygen needed for the initial production of RNA molecules, may have been better on the planet Mars than on the planet Earth.[161][162] If so, life-suitable molecules originating on Mars, may have later migrated to Earth via meteor ejections.[161][162]

A number of hypotheses of formation of RNA have been put forward. As of 1994, there are difficulties in the explanation of the abiotic synthesis of the nucleotides cytosine and uracil.[164] Subsequent research has shown possible routes of synthesis; for example, formamide produces all four ribonucleotides and other biological molecules when warmed in the presence of various terrestrial minerals.[101][102] Early cell membranes could have formed spontaneously from proteinoids, which are protein-like molecules produced when amino acid solutions are heated while in the correct concentration of aqueous solution. These are seen to form micro-spheres which are observed to behave similarly to membrane-enclosed compartments. Other possible means of producing more complicated organic molecules include chemical reactions that take place on clay substrates or on the surface of the mineral pyrite.

Factors supportive of an important role for RNA in early life include its ability to act both to store information and to catalyze chemical reactions (as a ribozyme); its many important roles as an intermediate in the expression of and maintenance of the genetic information (in the form of DNA) in modern organisms; and the ease of chemical synthesis of at least the components of the RNA molecule under the conditions that approximated the early Earth. Relatively short RNA molecules have been artificially produced in labs, which are capable of replication.[165] Such replicase RNA, which functions as both code and catalyst provides its own template upon which copying can occur. Jack W. Szostak has shown that certain catalytic RNAs can join smaller RNA sequences together, creating the potential for self-replication. If these conditions were present, Darwinian natural selection would favour the proliferation of such autocatalytic sets, to which further functionalities could be added.[166] Such autocatalytic systems of RNA capable of self-sustained replication have been identified.[167] The RNA replication systems, which include two ribozymes that catalyze each other's synthesis, showed a doubling time of the product of about one hour, and were subject to natural selection under the conditions that existed in the experiment.[168] In evolutionary competition experiments, this led to the emergence of new systems which replicated more efficiently.[16] This was the first demonstration of evolutionary adaptation occurring in a molecular genetic system.[168]

Depending on the specific definition used, life can be considered to have emerged when RNA chains began to express the basic conditions necessary for natural selection to operate as conceived by Darwin: heritability, variation of type, and differential reproductive output. The fitness of an RNA replicator (its per capita rate of increase) would likely be a function of its adaptive capacities that are intrinsic (in the sense that they were determined by the nucleotide sequence) and the availability of its resources.[169][170] The three primary adaptive capacities may have been (1) the capacity to replicate with moderate fidelity, giving rise to both heritability while allowing variation of type, (2) the capacity to avoid decay, and (3) the capacity to acquire and process resources.[169][170] These capacities would have been determined initially by the folded configurations of the RNA replicators that, in turn, would be encoded in their individual nucleotide sequences. Relative reproductive success, competition, between different replicators would have depended on the relative values of their adaptive capacities.

39.5.4 Pre-RNA world

It is possible that a different type of nucleic acid, such as PNA, TNA or GNA, was the first to emerge as a self-reproducing molecule, only later replaced by RNA.[171][172] Larralde et al., say that "the generally accepted prebiotic synthesis of ribose, the formose reaction, yields numerous sugars without any selectivity."[173] and they conclude that their "results suggest that the backbone of the first genetic material could not have contained ribose or other sugars because of their instability." The ester linkage of ribose and phosphoric acid in RNA is known to be prone to hydrolysis.[174]

Pyrimidine ribonucleosides and their respective nucleotides have been prebiotically synthesised by a sequence of reactions which by-pass the free sugars, and are assembled in a stepwise fashion by using nitrogenous or oxygenous chemistries. Sutherland has demonstrated high yielding routes to cytidine and uridine ribonucleotides built from small 2 and 3 carbon fragments such as glycolaldehyde, glyceraldehyde or glyceraldehyde-3-phosphate, cyanamide and cyanoacetylene. One of the steps in this sequence allows the isolation of enantiopure ribose aminooxazoline if the enantiomeric excess of glyceraldehyde is 60% or greater.[175] This can be viewed as a prebiotic purification step, where the said compound spontaneously crystallised out from a mixture of the other pentose aminooxazolines. Ribose aminooxazoline can then react with cyanoacetylene in a mild and highly efficient manner to give the alpha cytidine ribonucleotide. Photoanomerization with UV light allows for inversion about the 1' anomeric centre to give the correct beta stereochemistry.[176] In 2009 they showed that the same simple building blocks allow access, via phosphate controlled nucleobase elaboration, to 2',3'-cyclic pyrimidine nucleotides directly, which are known to be able to polymerise into RNA. This paper also highlights the possibility for the photo-sanitization of the pyrimidine-2',3'-cyclic phosphates.[177]

39.6 Origin of biological metabolism

Research suggests that metabolism-like reactions could have occurred naturally in early oceans, before the first organisms evolved.[18][178] The findings suggests that metabolism predates the origin of life and evolved through the chemical conditions that prevailed in the world's earliest oceans. Reconstructions in laboratories show that some of these reactions can produce RNA, and some others resemble two essential reaction cascades of metabolism: glycolysis and the pentose phosphate pathway, that provide essential precursors for nucleic acids, amino acids and lipids.[178] Following are some observed discoveries and related hypotheses.

39.6.1 Iron–sulfur world

Main article: Iron–sulfur world theory

In the 1980s, Günter Wächtershäuser, encouraged and supported by Karl R. Popper,[179][180][181] postulated in his iron–sulfur world, a theory of the evolution of pre-biotic chemical pathways as the starting point in the evolution of life. It presents a consistent system of tracing today's biochemistry back to ancestral reactions that provide alternative pathways to the synthesis of organic building blocks from simple gaseous compounds.

In contrast to the classical Miller experiments, which depend on external sources of energy (such as simulated lightning or ultraviolet irradiation), "Wächtershäuser systems" come with a built-in source of energy, sulfides of iron (iron pyrite) and other minerals . The energy released from redox reactions of these metal sulfides is available for the synthesis of organic molecules. It is therefore hypothesized that such systems may be able to evolve into autocatalytic sets of self-replicating, metabolically active entities that predate the life forms known today.[18][178] Experiments with such sulfides in an aqueous environment at 100 °C produced a relatively small yield of dipeptides (0.4% to 12.4%) and a smaller yield of tripeptides (0.10%) although under the same conditions, dipeptides were quickly broken down.[182]

Several models reject the idea of the self-replication of a "naked-gene" but postulate the emergence of a primitive metabolism which could provide a safe environment for the later emergence of RNA replication. The centrality of the Krebs cycle (citric acid cycle) to energy production in aerobic organisms, and in drawing in carbon dioxide and hydrogen ions in biosynthesis of complex organic chemicals, suggests that it was one of the first parts of the metabolism to evolve.[183] Somewhat in agreement with these notions, geochemist Michael Russell has proposed that "the purpose of life is to hydrogenate carbon dioxide" (as part of a "metabolism-first," rather than a "genetics-first," scenario).[184][185] Physicist Jeremy England of MIT has proposed that thermodynamically, life was bound to eventually arrive, as based on established physics, he mathematically indicates "...that when a group of atoms is driven by an external source of energy (like the sun or chemical fuel) and surrounded by a heat bath (like the ocean or atmosphere), it will often gradually restructure itself in order to dissipate increasingly more energy. This could mean that under certain conditions, matter inexorably acquires the key physical attribute associated with life."[186][187]

One of the earliest incarnations of this idea was put forward in 1924 with Oparin's notion of primitive self-replicating vesicles which predated the discovery of the structure of DNA. Variants in the 1980s and 1990s include Wächtershäuser's iron–sulfur world theory and models introduced by Christian de Duve based on the chemistry of thioesters. More abstract and theoretical arguments for the plausibility of the emergence of metabolism without the presence of genes include a mathematical model introduced by Freeman Dyson in the early 1980s and Stuart Kauffman's notion of collectively autocatalytic sets, discussed later in that decade.

Orgel summarized his analysis of the proposal by stating, "There is at present no reason to expect that multistep cycles such as the reductive citric acid cycle will self-organize on the surface of FeS/FeS$_2$ or some other mineral."[188] It is possible that another type of metabolic pathway was used at the beginning of life. For example, instead of the reductive citric acid cycle, the "open" acetyl-CoA pathway (another one of the five recognised ways of carbon dioxide fixation in nature today) would be compatible with the idea of self-organisation on a metal sulfide surface. The key enzyme of this pathway, carbon monoxide dehydrogenase/acetyl-CoA synthase harbours mixed nickel-iron-sulfur clusters in its reaction centers and catalyses the formation of acetyl-CoA (which may be regarded as a modern form of acetyl-thiol) in a single step.

39.6.2 Zn-world hypothesis

The Zn-world (zinc world) theory of Armen Y. Mulkidjanian[189] is an extension of Wächtershäuser's pyrite hypothesis. Wächtershäuser based his theory of the initial chemical processes leading to informational molecules (i.e., RNA, peptides) on a regular mesh of electric charges at the surface of pyrite that may have made the primeval polymerization thermodynamically more favourable by attracting reactants and arranging them appropriately relative to each other.[190] The Zn-world theory specifies and differentiates further.[189][191] Hydrothermal fluids rich in H$_2$S interacting with cold primordial ocean (or Darwin's "warm little pond") water leads to the precipitation of metal sulfide particles. Oceanic vent systems and other hydrothermal systems have a zonal structure reflected in ancient volcanogenic massive sulfide deposits (VMS) of hydrothermal origin. They reach many kilometers in diameter and date back to the Archean Eon. Most abundant are pyrite (FeS$_2$), chalcopyrite (CuFeS$_2$), and sphalerite (ZnS), with additions of galena (PbS) and alabandite (MnS). ZnS and MnS have a unique ability to store radiation energy, e.g., provided by UV light. Since during the relevant time window of the origins of replicating molecules the primordial atmospheric pressure was high enough (>100 bar, about 100 atmospheres) to precipitate near the Earth's surface and UV irradiation was 10 to 100 times more intense than now, the unique photosynthetic properties mediated by ZnS provided just the right energy conditions to energize the synthesis of informational and metabolic molecules and the selection of photostable nucleobases.

The Zn-world theory has been further filled out with experimental and theoretical evidence for the ionic constitution of the interior of the first proto-cells before archaea, bacteria and proto-eukaryotes evolved. Archibald Macallum noted the resemblance of organism fluids such as blood, and lymph to seawater;[192] however, the inorganic composition of all cells differ from that of modern seawater, which led Mulkidjanian and colleagues to reconstruct the "hatcheries" of the first cells combining geochemical analysis with phylogenomic scrutiny of the inorganic ion requirements of universal components of modern cells. The authors conclude that ubiquitous, and by inference primordial, proteins and functional systems show affinity to and functional requirement for K$^+$, Zn^{2+}, Mn^{2+}, and phosphate. Geochemical reconstruction shows that the ionic composition conducive to the origin of cells could not have existed in what we today call marine settings but is compatible with emissions of vapor-dominated zones of what we today call inland geothermal systems. Under the oxygen depleted, CO$_2$-dominated primordial atmosphere, the chemistry of water condensates and exhalations near geothermal fields would resemble the internal milieu of modern cells. Therefore, the precellular stages of evolution may have taken place in shallow "Darwin ponds" lined with porous silicate minerals mixed with metal sulfides and enriched in K$^+$, Zn^{2+}, and phosphorus compounds.[193][194]

39.6.3 Deep sea vent hypothesis

The deep sea vent, or alkaline hydrothermal vent, theory for the origin of life on Earth posits that life may have begun at submarine hydrothermal vents,[195] William Martin and Michael Russell have suggested "that life evolved in structured iron monosulphide precipitates in a seepage site hydrothermal mound at a redox, pH and temperature gradient between sulphide-rich hydrothermal fluid and iron(II)-containing waters of the Hadean ocean floor. The naturally arising, three-dimensional compartmentation observed within fossilized seepage-site metal sulphide precipitates indicates that these inorganic compartments were the precursors of cell walls and membranes found in free-living prokaryotes. The known capability of FeS and NiS to catalyse the synthesis of the acetyl-methylsulphide from carbon monoxide and methylsulphide, constituents of hydrothermal fluid, indicates that pre-biotic syntheses occurred at the inner surfaces of these metal-sulphide-walled compartments,..."[196] These form where hydrogen-

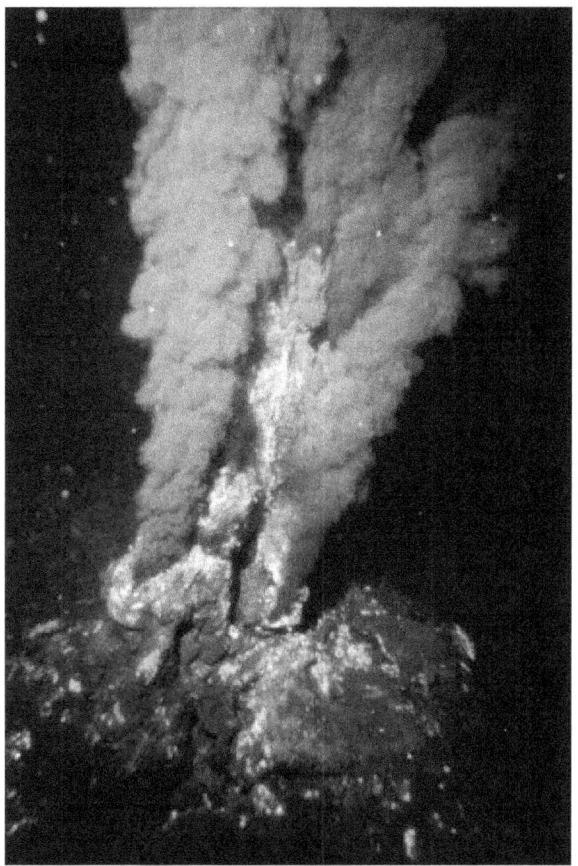

Deep-sea hydrothermal vent or 'black smoker'

gradient—cations like protons H^+ tend to diffuse down the electrical potential, anions in the opposite direction.

These two gradients taken together can be expressed as an electrochemical gradient, providing energy for abiogenic synthesis. The proton motive force can be described as the measure of the potential energy stored as a combination of proton and voltage gradients across a membrane (differences in proton concentration and electrical potential).

White smokers emitting liquid carbon dioxide (CO_2) at the Champagne vent, Marianas Trench Marine National Monument

rich fluids emerge from below the sea floor, as a result of serpentinization of ultra-mafic olivine with seawater and a pH interface with carbon dioxide-rich ocean water. The vents form a sustained chemical energy source derived from redox reactions, in which electron donors, such as molecular hydrogen, react with electron acceptors, such as carbon dioxide (see Iron–sulfur world theory). These are highly exothermic reactions.[note 2]

Michael Russell demonstrated that alkaline vents created an abiogenic proton motive force (PMF) chemiosmotic gradient,[196] in which conditions are ideal for an abiogenic hatchery for life. Their microscopic compartments "provide a natural means of concentrating organic molecules," composed of iron-sulfur minerals such as mackinawite, endowed these mineral cells with the catalytic properties envisaged by Wächtershäuser.[183] This movement of ions across the membrane depends on a combination of two factors:

1. Diffusion force caused by concentration gradient—all particles including ions tend to diffuse from higher concentration to lower.

2. Electrostatic force caused by electrical potential

Jack W. Szostak suggested that geothermal activity provides greater opportunities for the origination of life in open lakes where there is a buildup of minerals. In 2010, based on spectral analysis of sea and hot mineral water, Ignat Ignatov and Oleg Mosin demonstrated that life may have predominantly originated in hot mineral water. The hot mineral water that contains bicarbonate and calcium ions has the most optimal range.[197] This case is similar to the origin of life in hydrothermal vents, but with bicarbonate and calcium ions in hot water. This water has a pH of 9–11 and is possible to have the reactions in seawater. According to Melvin Calvin, certain reactions of condensation-dehydration of amino acids and nucleotides in individual blocks of peptides and nucleic acids can take place in the primary hydrosphere with pH 9-11 at a later evolutionary stage.[198] Some of these compounds like hydrocyanic acid (HCN) have been proven in the experiments of Miller. This is the environment in which the stromatolites have been created. David Ward of Montana State University described the formation of stromatolites in hot mineral water at the Yellowstone National Park. Stromatolites survive in hot mineral water and in proximity to areas with volcanic activity.[199] Processes have evolved in the sea near geysers of hot mineral water. In 2011, Tadashi Sugawara from the University of Tokyo created a protocell in hot water.[200]

Experimental research and computer modeling suggest that the surfaces of mineral particles inside hydrothermal vents have catalytic properties similar to those of enzymes and are able to create simple organic molecules, such as methanol (CH_3OH) and formic, acetic and pyruvic acid out of the dissolved CO_2 in the water.[201][202]

39.6.4 Thermosynthesis

Today's bioenergetic process of fermentation is carried out by either the aforementioned citric acid cycle or the Acetyl-CoA pathway, both of which have been connected to the primordial Iron–sulfur world. In a different approach, the thermosynthesis hypothesis considers the bioenergetic process of chemiosmosis, which plays an essential role in cellular respiration and photosynthesis, more basal than fermentation: the ATP synthase enzyme, which sustains chemiosmosis, is proposed as the currently extant enzyme most closely related to the first metabolic process.[203][204]

First, life needed an energy source to bring about the condensation reaction that yielded the peptide bonds of proteins and the phosphodiester bonds of RNA. In a generalization and thermal variation of the binding change mechanism of today's ATP synthase, the "first protein" would have bound substrates (peptides, phosphate, nucleosides, RNA 'monomers') and condensed them to a reaction product that remained bound until after a temperature change it was released by thermal unfolding.

The energy source under the thermosynthesis hypothesis was thermal cycling, the result of suspension of protocells in a convection current, as is plausible in a volcanic hot spring; the convection accounts for the self-organization and dissipative structure required in any origin of life model. The still ubiquitous role of thermal cycling in germination and cell division is considered a relic of primordial thermosynthesis.

By phosphorylating cell membrane lipids, this "first protein" gave a selective advantage to the lipid protocell that contained the protein. This protein also synthesized a library of many proteins, of which only a minute fraction had thermosynthesis capabilities. As proposed by Dyson,[13] it propagated functionally: it made daughters with similar capabilities, but it did not copy itself. Functioning daughters consisted of different amino acid sequences.

Whereas the Iron–sulfur world identifies a circular pathway as the most simple, the thermosynthesis hypothesis does not even invoke a pathway: ATP synthase's binding change mechanism resembles a physical adsorption process that yields free energy,[205] rather than a regular enzyme's mechanism, which decreases the free energy. It has been claimed that the emergence of cyclic systems of protein catalysts is implausible.[206]

39.7 Other models of abiogenesis

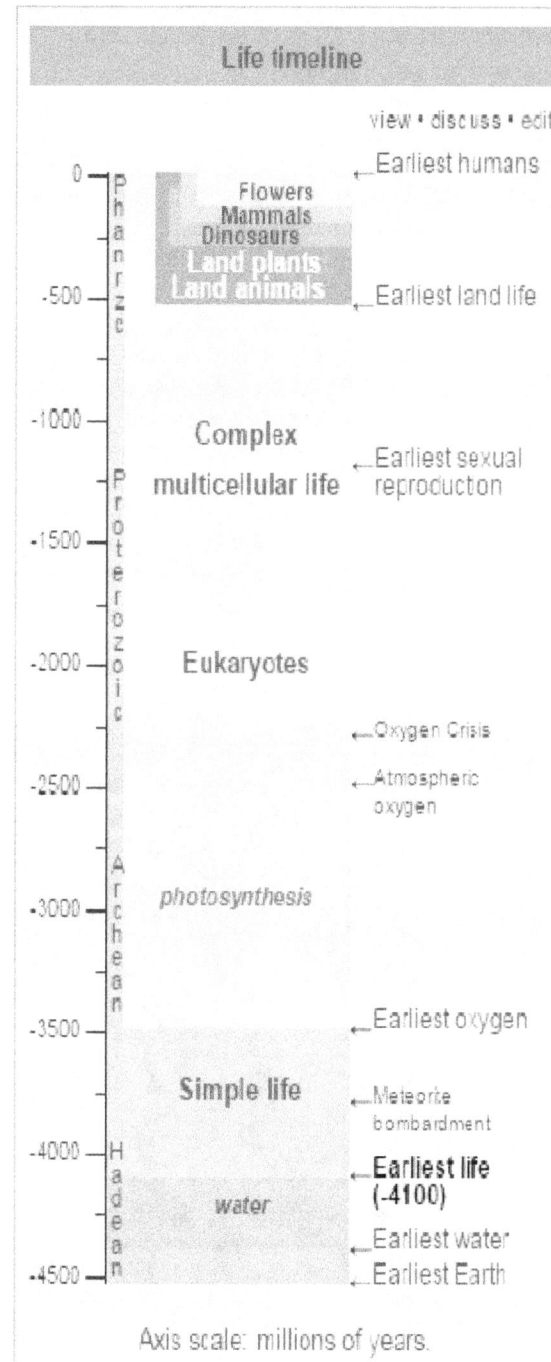

Axis scale: billions of years.

also see {{Human timeline}} and {{Life timeline}}

39.7.1 Clay hypothesis

Montmorillonite, an abundant clay, is a catalyst for the polymerization of RNA and for the formation of membranes from lipids.[207] A model for the origin of life using clay was forwarded by Alexander Graham Cairns-Smith in 1985 and explored as a plausible mechanism by several scientists.[208] The clay hypothesis postulates that complex organic molecules arose gradually on a pre-existing, non-organic replication surfaces of silicate crystals in solution.

At the Rensselaer Polytechnic Institute, James P. Ferris' studies have also confirmed that clay minerals of montmorillonite catalyze the formation of RNA in aqueous solution, by joining nucleotides to form longer chains.[209]

In 2007, Bart Kahr from the University of Washington and colleagues reported their experiments that tested the idea that crystals can act as a source of transferable information, using crystals of potassium hydrogen phthalate. "Mother" crystals with imperfections were cleaved and used as seeds to grow "daughter" crystals from solution. They then examined the distribution of imperfections in the new crystals and found that the imperfections in the mother crystals were reproduced in the daughters, but the daughter crystals also had many additional imperfections. For gene-like behavior to be observed, the quantity of inheritance of these imperfections should have exceeded that of the mutations in the successive generations, but it did not. Thus Kahr concluded that the crystals "were not faithful enough to store and transfer information from one generation to the next."[210]

39.7.2 Gold's "deep-hot biosphere" model

In the 1970s, Thomas Gold proposed the theory that life first developed not on the surface of the Earth, but several kilometers below the surface. It is claimed that discovery of microbial life below the surface of another body in our Solar System would lend significant credence to this theory. Thomas Gold also asserted that a trickle of food from a deep, unreachable, source is needed for survival because life arising in a puddle of organic material is likely to consume all of its food and become extinct. Gold's theory is that the flow of such food is due to out-gassing of primordial methane from the Earth's mantle; more conventional explanations of the food supply of deep microbes (away from sedimentary carbon compounds) is that the organisms subsist on hydrogen released by an interaction between water and (reduced) iron compounds in rocks.

39.7.3 Panspermia

Main article: Panspermia

Panspermia is the hypothesis that life exists throughout the Universe, distributed by meteoroids, asteroids, comets,[211][212] planetoids,[213] and, also, by spacecraft in the form of unintended contamination by microorganisms.[214][215]

Panspermia hypothesis does not attempt to explain how life first originated, but merely shifts it to another planet or a comet. The advantage of an extraterrestrial origin of primitive life is that life is not required to have formed on each planet it occurs on, but rather in a single location, and then spread about the galaxy to other star systems via cometary and/or meteorite impact. Evidence to support the hypothesis is scant, but it finds support in studies of Martian meteorites found in Antarctica and in studies of extremophile microbes' survival in outer space tests.[216][217][218][219] (See also: List of microorganisms tested in outer space.)

39.7.4 Extraterrestrial organic molecules

See also: List of interstellar and circumstellar molecules and Panspermia § Pseudo-panspermia

An organic compound is any member of a large class

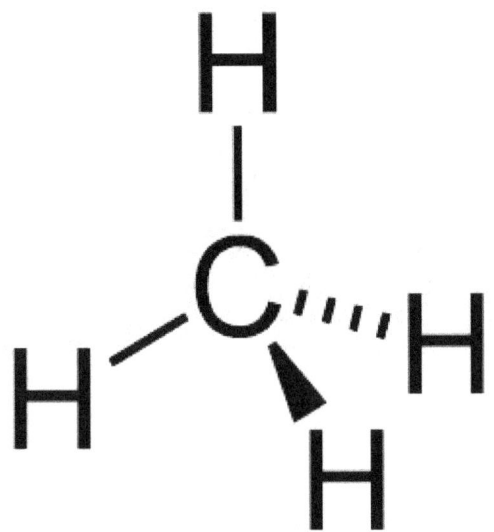

Methane is one of the simplest organic compounds

of gaseous, liquid, or solid chemicals whose molecules contain carbon. Carbon is the fourth most abundant element in the Universe by mass after hydrogen, helium, and oxygen.[220] Carbon is abundant in the Sun, stars, comets, and in the atmospheres of most planets.[221] Organic compounds are relatively common in space, formed by "factories of complex molecular synthesis" which occur in molecular clouds and circumstellar envelopes, and chemically evolve after reactions are initiated mostly by ionizing radiation.[19][222][223][224] Based on computer model studies, the complex organic molecules necessary for life may have formed on dust grains in the protoplanetary disk surrounding the Sun before the formation of the Earth.[40] According to the computer studies, this same process may also occur around other stars that acquire planets.[40]

Observations suggest that the majority of organic compounds introduced on Earth by interstellar dust particles are considered principal agents in the formation of complex molecules, thanks to their peculiar surface-catalytic activities.[225][226] Studies reported in 2008, based on $^{12}C/^{13}C$ isotopic ratios of organic compounds found in the Murchison meteorite, suggested that the RNA component uracil and related molecules, including xanthine, were formed extraterrestrially.[227][228] On 8 August 2011, a report based on NASA studies of meteorites found on Earth was published suggesting DNA components (adenine, guanine and related organic molecules) were made in outer space.[225][229][230] Scientists also found that the cosmic dust permeating the Universe contains complex organics ("amorphous organic solids with a mixed aromatic–aliphatic structure") that could be created naturally, and rapidly, by stars.[231][232][233] Sun Kwok of The University of Hong Kong suggested that these compounds may have been related to the development of life on Earth said that "If this is the case, life on Earth may have had an easier time getting started as these organics can serve as basic ingredients for life."[231]

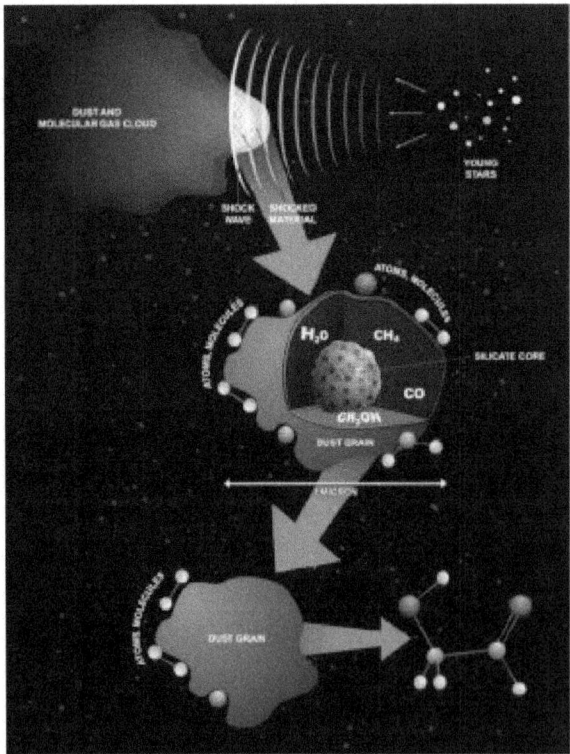

Formation of glycolaldehyde in stardust

Glycolaldehyde, the first example of an interstellar sugar molecule, was detected in the star-forming region near the center of our galaxy. It was discovered in 2000 by Jes Jørgensen and Jan M. Hollis.[234] In 2012, Jørgensen's team reported the detection of glycolaldehyde in a distant star system. The molecule was found around the protostellar binary IRAS 16293-2422 400 light years from Earth.[235][236][237] Glycolaldehyde is needed to form RNA, which is similar in function to DNA. These findings suggest that complex organic molecules may form in stellar systems prior to the formation of planets, eventually arriving on young planets early in their formation.[238] Because sugars are associated with both metabolism and the genetic code, two of the most basic aspects of life, it is thought the discovery of extraterrestrial sugar increases the likelihood that life may exist elsewhere in our galaxy.[234]

NASA announced in 2009 that scientists had identified another fundamental chemical building block of life in a comet for the first time, glycine, an amino acid, which was detected in material ejected from comet Wild 2 in 2004 and grabbed by NASA's *Stardust* probe. Glycine has been detected in meteorites before. Carl Pilcher, who leads the NASA Astrobiology Institute commented that "The discovery of glycine in a comet supports the idea that the fundamental building blocks of life are prevalent in space, and strengthens the argument that life in the Universe may be common rather than rare."[239] Comets are encrusted with outer layers of dark material, thought to be a tar-like substance composed of complex organic material formed from simple carbon compounds after reactions initiated mostly by ionizing radiation. It is possible that a rain of material from comets could have brought significant quantities of such complex organic molecules to Earth.[240][241][242] Amino acids which were formed extraterrestrially may also have arrived on Earth via comets.[43] It is estimated that during the Late Heavy Bombardment, meteorites may have delivered up to five million tons of organic prebiotic elements to Earth per year.[43]

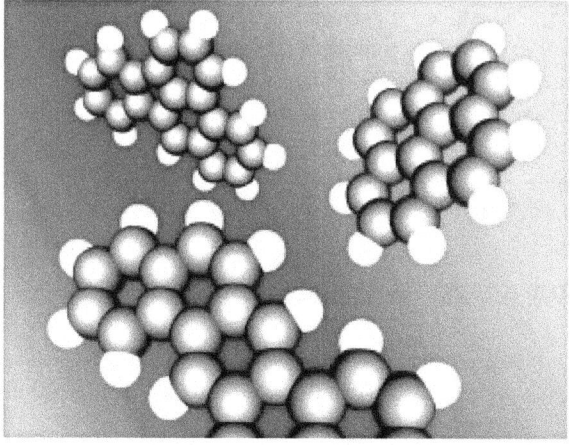

An illustration of typical polycyclic aromatic hydrocarbons. Clockwise from top left. benz(e)acephenanthrylene, pyrene and dibenz(ah)anthracene.

Polycyclic aromatic hydrocarbons (PAH) are the most common and abundant of the known polyatomic molecules in the observable universe, and are considered a likely constituent of the primordial sea.[243][244][245] In 2010, PAHs, along with fullerenes (or "buckyballs"), have been detected in nebulae.[246][247]

In March 2015, NASA scientists reported that, for the first time, complex DNA and RNA organic compounds of life, including uracil, cytosine and thymine, have been formed in the laboratory under outer space conditions, using starting chemicals, such as pyrimidine, found in meteorites. Pyrimidine, like PAHs, the most carbon-rich chemical found in the Universe, may have been formed in red giant stars or in interstellar dust and gas clouds.[248]

39.7.5 Lipid world

Main article: Gard model

The lipid world theory postulates that the first self-replicating object was lipid-like.[249][250] It is known that phospholipids form lipid bilayers in water while under agitation—the same structure as in cell membranes. These molecules were not present on early Earth, but other amphiphilic long-chain molecules also form membranes. Furthermore, these bodies may expand (by insertion of additional lipids), and under excessive expansion may undergo spontaneous splitting which preserves the same size and composition of lipids in the two progenies. The main idea in this theory is that the molecular composition of the lipid bodies is the preliminary way for information storage, and evolution led to the appearance of polymer entities such as RNA or DNA that may store information favorably. Studies on vesicles from potentially prebiotic amphiphiles have so far been limited to systems containing one or two types of amphiphiles. This in contrast to the output of simulated prebiotic chemical reactions, which typically produce very heterogeneous mixtures of compounds.[139] Within the hypothesis of a lipid bilayer membrane composed of a mixture of various distinct amphiphilic compounds there is the opportunity of a huge number of theoretically possible combinations in the arrangements of these amphiphiles in the membrane. Among all these potential combinations, a specific local arrangement of the membrane would have favored the constitution of an hypercycle,[251][252] actually a positive feedback composed of two mutual catalysts represented by a membrane site and a specific compound trapped in the vesicle. Such site/compound pairs are transmissible to the daughter vesicles leading to the emergence of distinct lineages of vesicles which would have allowed Darwinian natural selection.[253]

39.7.6 Polyphosphates

A problem in most scenarios of abiogenesis is that the thermodynamic equilibrium of amino acid versus peptides is in the direction of separate amino acids. What has been missing is some force that drives polymerization. The resolution of this problem may well be in the properties of polyphosphates.[254][255] Polyphosphates are formed by polymerization of ordinary monophosphate ions PO_4^{-3}. Several mechanisms for such polymerization have been suggested. Polyphosphates cause polymerization of amino acids into peptides. They are also logical precursors in the synthesis of such key biochemical compounds as adenosine triphosphate (ATP). A key issue seems to be that calcium reacts with soluble phosphate to form insoluble calcium phosphate (apatite), so some plausible mechanism must be found to keep calcium ions from causing precipitation of phosphate. There has been much work on this topic over the years, but an interesting new idea is that meteorites may have introduced reactive phosphorus species on the early Earth.[256]

39.7.7 PAH world hypothesis

Main article: PAH world hypothesis

Polycyclic aromatic hydrocarbons (PAH) are known to be abundant in the Universe,[243][244][245] including in the interstellar medium, in comets, and in meteorites, and are some of the most complex molecules so far found in space.[221]

Other sources of complex molecules have been postulated, including extraterrestrial stellar or interstellar origin. For example, from spectral analyses, organic molecules are known to be present in comets and meteorites. In 2004, a team detected traces of PAHs in a nebula.[257] In 2010, another team also detected PAHs, along with fullerenes, in nebulae.[246] The use of PAHs has also been proposed as a precursor to the RNA world in the PAH world hypothesis. The Spitzer Space Telescope has detected a star, HH 46-IR, which is forming by a process similar to that by which the Sun formed. In the disk of material surrounding the star, there is a very large range of molecules, including cyanide compounds, hydrocarbons, and carbon monoxide. In September 2012, NASA scientists reported that PAHs, subjected to interstellar medium conditions, are transformed, through hydrogenation, oxygenation and hydroxylation, to more complex organics—"a step along the path toward amino acids and nucleotides, the raw materials of proteins and DNA, respectively."[258][259] Further, as a result of these transformations, the PAHs lose their spectroscopic signature which could be one of the reasons "for the lack of PAH detection in interstellar ice grains, particularly the outer regions of cold, dense clouds or the upper molecular layers of protoplanetary disks."[258][259]

NASA maintains a database for tracking PAHs in the Universe.[221][260] More than 20% of the carbon in the Universe may be associated with PAHs,[221][221] possible starting materials for the formation of life. PAHs seem to have been formed shortly after the Big Bang, are widespread throughout the Universe,[243][244][245] and are associated with new stars and exoplanets.[221]

39.7.8 Radioactive beach hypothesis

Zachary Adam claims that tidal processes that occurred during a time when the Moon was much closer may have concentrated grains of uranium and other radioactive elements at the high-water mark on primordial beaches, where they may have been responsible for generating life's building blocks.[261] According to computer models reported in *Astrobiology*,[262] a deposit of such radioactive materials could show the same self-sustaining nuclear reaction as that found in the Oklo uranium ore seam in Gabon. Such radioactive beach sand might have provided sufficient energy to generate organic molecules, such as amino acids and sugars from acetonitrile in water. Radioactive monazite material also has released soluble phosphate into the regions between sand-grains, making it biologically "accessible." Thus amino acids, sugars, and soluble phosphates might have been produced simultaneously, according to Adam. Radioactive actinides, left behind in some concentration by the reaction, might have formed part of organometallic complexes. These complexes could have been important early catalysts to living processes.

John Parnell has suggested that such a process could provide part of the "crucible of life" in the early stages of any early wet rocky planet, so long as the planet is large enough to have generated a system of plate tectonics which brings radioactive minerals to the surface. As the early Earth is thought to have had many smaller plates, it might have provided a suitable environment for such processes.[263]

39.7.9 Thermodynamic dissipation

Karo Michaelian from the National Autonomous University of Mexico (UNAM) points out that any model for the origin of life must take into account the fact that life is an irreversible thermodynamic process and, like all irreversible processes, its origin and persistence as a "self-organized" system is due to its dissipation an imposed generalized chemical potential, i.e., the production of entropy. That is, entropy production is not incidental to the process of life, but rather the fundamental reason for its existence.

Present day life augments the entropy production of Earth in its solar environment by dissipating ultraviolet and visible photons into heat through organic pigments in water. This heat then catalyzes a host of secondary dissipative processes such as the water cycle, ocean and wind currents, hurricanes, etc.[264][265] Michaelian argues that if the thermodynamic function of life today is to produce entropy through photon dissipation, then this probably was its function at its very beginnings.[266] It turns out that both RNA and DNA when in water solution are very strong absorbers and extremely rapid dissipaters of UV light within the 230–290 nm wavelength region, which is a part of the Sun's spectrum that could have penetrated the prebiotic atmosphere.[267] The amount of ultraviolet (UV-C) light reaching the Earth's surface within this spectral range in the Archean could have been on the order of 4 W/m^2,[268] or some 31 orders of magnitude greater than it is today at 260 nm where RNA and DNA absorb most strongly.[267] In fact, not only RNA and DNA, but many fundamental molecules of life (those common to all three domains of life, archea, bacteria, and eucaryote) are also pigments that absorb in the UV-C, and many of these also have a chemical affinity to RNA and DNA.[269][270] Nucleic acids may thus have acted as acceptor molecules to the UV-C photon excited antenna pigment donor molecules by providing an ultrafast channel for dissipation. Michaelian has shown that there would have existed a non-linear, non-equilibrium thermodynamic imperative to the abiogenic UV-C photochemical synthesis [177] and proliferation of these pigments over the entire Earth surface if they augmented the solar photon dissipation rate.[271]

A simple mechanism to explain enzyme-less replication of RNA and DNA can be given within the same dissipative thermodynamic framework by assuming that life arose when the temperature of the primitive seas had cooled to somewhat below the denaturing temperature of RNA or DNA. The ratio of $^{18}O/^{16}O$ found in cherts of the Barberton greenstone belt of South Africa indicates that the Earth's surface temperature was around 80 °C at 3.8 Ga,[272][273] falling to 70±15 °C about 3.5 to 3.2 Ga,[274] suggestively close to RNA or DNA denaturing (uncoiling and separation) temperatures. During the night, the surface water temperature would drop below the denaturing temperature and single strand RNA/DNA could act as extension template for the formation of double strand RNA/DNA. During the daylight hours, RNA and DNA would absorb UV-C light and convert this directly into heat at the ocean surface, thereby raising the local temperature enough to allow for denaturing of RNA and DNA. Direct experimental evidence for the denaturing of DNA through UV-C light dissipation has now been obtained.[275]

The copying process would have been repeated with each diurnal cycle.[266] Such an ultraviolet and temperature assisted RNA/DNA reproduction (UVTAR) bears similarity to polymerase chain reaction (PCR), a routine laboratory procedure employed to multiply DNA segments. Since denaturation would be most probable in the late afternoon when the Archean sea surface temperature would be highest, and since late afternoon submarine sunlight is somewhat circularly polarized, the homochirality of the organic molecules of life can also be explained within the proposed thermodynamic framework.[266]

The fact that the aromatic amino acids have been shown to have chemical affinity to their codons, or anti-codons, and that they also absorb strongly in the UV-C, suggests that they might have originally acted as antenna pigments to increase dissipation and to provide more local heat for UVTAR replication of RNA and DNA as the sea surface temperature cooled. The accumulation of information, e.g., coding for the aromatic amino acids, in RNA or DNA would thus be related to reproductive success under this mechanism. Michaelian suggests that the traditional origin of life research, that expects to describe the emergence of life without overwhelming reference to entropy production through dissipation, is erroneous and that imposed environmental potentials, such as the solar photon flux, and the dissipation of this flux, must be considered to understand the emergence, proliferation, and evolution of life.

A similar hypothesis has been advanced by Jeremy England, an American physicist at the Massachusetts Institute of Technology, with his statistical physics arguments to explain the spontaneous emergence of life, and consequently, Darwinian evolution.[276][277] England terms this process 'dissipative-driven adaptation', and proposes it as an explanation of the origin of life.[278]

39.7.10 Multiple genesis

Different forms of life with variable origin processes may have appeared quasi-simultaneously in the early history of Earth.[279] The other forms may be extinct (having left distinctive fossils through their different biochemistry—e.g., hypothetical types of biochemistry). It has been proposed that:

> The first organisms were self-replicating iron-rich clays which fixed carbon dioxide into oxalic and other dicarboxylic acids. This system of replicating clays and their metabolic phenotype then evolved into the sulfide rich region of the hotspring acquiring the ability to fix nitrogen. Finally phosphate was incorporated into the evolving system which allowed the synthesis of nucleotides and phospholipids. If biosynthesis recapitulates biopoiesis, then the synthesis of amino acids preceded the synthesis of the

purine and pyrimidine bases. Furthermore the polymerization of the amino acid thioesters into polypeptides preceded the directed polymerization of amino acid esters by polynucleotides.[280]

39.7.11 Fluctuating hydrothermal pools on volcanic islands

Bruce Damer and David Deamer have come to the conclusion that cell membranes cannot be formed in salty seawater, and must therefore have originated in freshwater. Before the continents formed, the only dry land on Earth would be volcanic islands, where rainwater would form ponds where lipids could form the first stages towards cell membranes. These predecessors of true cells are assumed to have behaved more like a superorganism rather than individual structures, where the porous membranes would house molecules which would leak out and enter other protocells. Only when true cells had evolved would they gradually adapt to saltier environments and enter the ocean.[281]

39.8 See also

- Anthropic principle
- Artificial cell
- Astrochemistry
- Biological immortality
- Common descent
- Emergence
- Entropy and life
- GADV protein world
- Mediocrity principle
- Mycoplasma laboratorium
- Nexus for Exoplanet System Science
- Planetary habitability
- Rare Earth hypothesis
- Shadow biosphere
- Stromatolite

39.9 Notes

[1] The reactions are:

$$FeS + H_2S \rightarrow FeS_2 + 2H^+ + 2e^-$$
$$FeS + H_2S + CO_2 \rightarrow FeS_2 + HCOOH$$

[2] The reactions are:
Reaction 1: *Fayalite + water → magnetite + aqueous silica + hydrogen*

$$3Fe_2SiO_4 + 2H_2O \rightarrow 2Fe_3O_4 + 3SiO_2 + 2H_2$$

Reaction 2: *Forsterite + aqueous silica → serpentine*

$$3Mg_2SiO_4 + SiO_2 + 4H_2O \rightarrow 2Mg_3Si_2O_5(OH)_4$$

Reaction 3: *Forsterite + water → serpentine + brucite*

$$2Mg_2SiO_4 + 3H_2O \rightarrow Mg_3Si_2O_5(OH)_4 + Mg(OH)_2$$

Reaction 3 describes the hydration of olivine with water only to yield serpentine and $Mg(OH)_2$ (brucite). Serpentine is stable at high pH in the presence of brucite like calcium silicate hydrate, (C-S-H) phases formed along with portlandite ($Ca(OH)_2$) in hardened Portland cement paste after the hydration of belite (Ca_2SiO_4), the artificial calcium equivalent of forsterite. Analogy of reaction 3 with belite hydration in ordinary Portland cement: *Belite + water → C-S-H phase + portlandite*

$$2\,Ca_2SiO_4 + 4\,H_2O \rightarrow 3\,CaO \cdot 2\,SiO_2 \cdot 3\,H_2O + Ca(OH)_2$$

39.10 References

[1] Pronunciation: "/ˌeɪbaɪə(ʊ)'dʒɛnɪsɪs/". Pearsall, Judy; Hanks, Patrick, eds. (1998). "abiogenesis". *The New Oxford Dictionary of English* (1st ed.). Oxford, UK: Oxford University Press. p. 3. ISBN 0-19-861263-X.

[2] OED On-line (2003)

[3] "Abiogenesis". *Dictionary.com Unabridged*. Random House.

[4] "Abiogenesis". *Merriam-Webster Dictionary*.

[5] Bernal 1960, p. 30

[6] Scharf, Caleb; et al. (18 December 2015). "A Strategy for Origins of Life Research". *Astrobiology (journal)* **15** (12): 1031–1042. doi:10.1089/ast.2015.1113. Retrieved 20 December 2015.

[7] Oparin 1953, p. vi

[8] Warmflash, David; Warmflash, Benjamin (November 2005). "Did Life Come from Another World?". *Scientific American* (Stuttgart: Georg von Holtzbrinck Publishing Group) **293** (5): 64–71. doi:10.1038/scientificamerican1105-64. ISSN 0036-8733.

[9] Yarus 2010, p. 47

[10] Peretó, Juli (2005). "Controversies on the origin of life" (PDF). *International Microbiology* (Barcelona: Spanish Society for Microbiology) **8** (1): 23–31. ISSN 1139-6709. PMID 15906258. Retrieved 2015-06-01.

[11] Elizabeth A. Bell. "Potentially biogenic carbon preserved in a 4.1 billion-year-old zircon".

[12] Voet & Voet 2004, p. 29

[13] Dyson 1999

[14] Davies, Paul (1998) "The Fifth Miracle, Search for the origin and meaning of life" 9Penguin)

[15]
- Copley, Shelley D.; Smith, Eric; Morowitz, Harold J. (December 2007). "The origin of the RNA world: Co-evolution of genes and metabolism" (PDF). *Bioorganic Chemistry* (Amsterdam, the Netherlands: Elsevier) **35** (6): 430–443. doi:10.1016/j.bioorg.2007.08.001. ISSN 0045-2068. PMID 17897696. Retrieved 2015-06-08. The proposal that life on Earth arose from an RNA world is widely accepted.
- Orgel, Leslie E. (April 2003). "Some consequences of the RNA world hypothesis". *Origins of Life and Evolution of the Biosphere* (Kluwer Academic Publishers) **33** (2): 211–218. doi:10.1023/A:1024616317965. ISSN 0169-6149. PMID 12967268. It now seems very likely that our familiar DNA/RNA/protein world was preceded by an RNA world...
- Robertson & Joyce 2012: "There is now strong evidence indicating that an RNA World did indeed exist before DNA- and protein-based life."
- Neveu, Kim & Benner 2013: "[The RNA world's existence] has broad support within the community today."

[16] Robertson, Michael P.; Joyce, Gerald F. (May 2012). "The origins of the RNA world". *Cold Spring Harbor Perspectives in Biology* (Cold Spring Harbor, NY: Cold Spring Harbor Laboratory Press) **4** (5): a003608. doi:10.1101/cshperspect.a003608. ISSN 1943-0264. PMC 3331698. PMID 20739415.

[17] Cech, Thomas R. (July 2012). "The RNA Worlds in Context". *Cold Spring Harbor Perspectives in Biology* (Cold Spring Harbor, NY: Cold Spring Harbor Laboratory Press) **4** (7): a006742. doi:10.1101/cshperspect.a006742. ISSN 1943-0264. PMC 3385955. PMID 21441585.

[18] Keller, Markus A.; Turchyn, Alexandra V.; Ralser, Markus (25 March 2014). "Non-enzymatic glycolysis and pentose phosphate pathway-like reactions in a plausible Archean ocean". *Molecular Systems Biology* (Heidelberg, Germany: EMBO Press on behalf of the European Molecular Biology Organization) **10** (725). doi:10.1002/msb.20145228. ISSN 1744-4292. PMC 4023395. PMID 24771084.

[19] Ehrenfreund, Pascale; Cami, Jan (December 2010). "Cosmic carbon chemistry: from the interstellar medium to the early Earth.". *Cold Spring Harbor Perspectives in Biology* (Cold Spring Harbor, NY: Cold Spring Harbor Laboratory Press) **2** (12): a002097. doi:10.1101/cshperspect.a002097. ISSN 1943-0264. PMC 2982172. PMID 20554702.

[20] Perkins, Sid (8 April 2015). "Organic molecules found circling nearby star". *Science* (News) (Washington, D.C.: American Association for the Advancement of Science). ISSN 1095-9203. Retrieved 2015-06-02.

[21] King, Anthony (14 April 2015). "Chemicals formed on meteorites may have started life on Earth". *Chemistry World* (News) (London: Royal Society of Chemistry). ISSN 1473-7604. Retrieved 2015-04-17.

[22] Saladino, Raffaele; Carota, Eleonora; Botta, Giorgia; et al. (13 April 2015). "Meteorite-catalyzed syntheses of nucleosides and of other prebiotic compounds from formamide under proton irradiation". *Proc. Natl. Acad. Sci. U.S.A.* (Washington, D.C.: National Academy of Sciences) **112** (21): E2746–E2755. doi:10.1073/pnas.1422225112. ISSN 1091-6490. PMID 25870268.

[23] Rampelotto, Pabulo Henrique (26 April 2010). *Panspermia: A Promising Field Of Research* (PDF). Astrobiology Science Conference 2010. Houston, TX: Lunar and Planetary Institute. p. 5224. Bibcode:2010LPICo1538.5224R. Retrieved 2014-12-03. Conference held at League City, TX

[24] Loeb, Abraham (October 2014). "The habitable epoch of the early Universe". *International Journal of Astrobiology* (Cambridge, UK: Cambridge University Press) **13** (4): 337–339. arXiv:1312.0613. Bibcode:2014IJAsB..13..337L. doi:10.1017/S1473550414000196. ISSN 1473-5504.

- Loeb, Abraham (3 June 2014). "The Habitable Epoch of the Early Universe". arXiv:1312.0613v3 [astro-ph.CO].

[25] Dreifus, Claudia (2 December 2014). "Much-Discussed Views That Go Way Back". *The New York Times* (New York: The New York Times Company). p. D2. ISSN 0362-4331. Retrieved 2014-12-03.

[26] Graham, Robert W. (February 1990). "Extraterrestrial Life in the Universe" (PDF) (NASA Technical Memorandum 102363). Lewis Research Center, Cleveland, Ohio: NASA. Retrieved 2015-06-02.

[27] Altermann 2009, p. xvii

[28] "Age of the Earth". United States Geological Survey. 9 July 2007. Retrieved 2006-01-10.

39.10. REFERENCES

[29] Dalrymple 2001, pp. 205–221

[30] Manhesa, Gérard; Allègre, Claude J.; Dupréa, Bernard; Hamelin, Bruno (May 1980). "Lead isotope study of basic-ultrabasic layered complexes: Speculations about the age of the earth and primitive mantle characteristics". *Earth and Planetary Science Letters* (Amsterdam, the Netherlands: Elsevier) **47** (3): 370–382. Bibcode:1980E&PSL..47..370M. doi:10.1016/0012-821X(80)90024-2. ISSN 0012-821X.

[31] Schopf, J. William; Kudryavtsev, Anatoliy B.; Czaja, Andrew D.; Tripathi, Abhishek B. (5 October 2007). "Evidence of Archean life: Stromatolites and microfossils". *Precambrian Research* (Amsterdam, the Netherlands: Elsevier) **158** (3–4): 141–155. doi:10.1016/j.precamres.2007.04.009. ISSN 0301-9268.

[32] Schopf, J. William (29 June 2006). "Fossil evidence of Archaean life". *Philosophical Transactions of the Royal Society B* (London: Royal Society) **361** (1470): 869–885. doi:10.1098/rstb.2006.1834. ISSN 0962-8436. PMC 1578735. PMID 16754604.

[33] Raven & Johnson 2002, p. 68

[34] Borenstein, Seth (13 November 2013). "Oldest fossil found: Meet your microbial mom". *Excite* (Yonkers, NY: Mindspark Interactive Network). Associated Press. Retrieved 2015-06-02.

[35] Pearlman, Jonathan (13 November 2013). "'Oldest signs of life on Earth found'". *The Daily Telegraph* (London: Telegraph Media Group). Retrieved 2014-12-15.

[36] Noffke, Nora; Christian, Daniel; Wacey, David; Hazen, Robert M. (16 November 2013). "Microbially Induced Sedimentary Structures Recording an Ancient Ecosystem in the ca. 3.48 Billion-Year-Old Dresser Formation, Pilbara, Western Australia". *Astrobiology* (New Rochelle, NY: Mary Ann Liebert, Inc.) **13** (12): 1103–1124. Bibcode:2013AsBio..13.1103N. doi:10.1089/ast.2013.1030. ISSN 1531-1074. PMC 3870916. PMID 24205812.

[37] Ohtomo, Yoko; Kakegawa, Takeshi; Ishida, Akizumi; et al. (January 2014). "Evidence for biogenic graphite in early Archaean Isua metasedimentary rocks". *Nature Geoscience* (London: Nature Publishing Group) **7** (1): 25–28. Bibcode:2014NatGe...7...25O. doi:10.1038/ngeo2025. ISSN 1752-0894.

[38] Borenstein, Seth (19 October 2015). "Hints of life on what was thought to be desolate early Earth". *Excite* (Yonkers, NY: Mindspark Interactive Network). Associated Press. Retrieved 2015-10-20.

[39] Bell, Elizabeth A.; Boehnike, Patrick; Harrison, T. Mark; et al. (19 October 2015). "Potentially biogenic carbon preserved in a 4.1 billion-year-old zircon" (PDF). *Proc. Natl. Acad. Sci. U.S.A.* (Washington, D.C.: National Academy of Sciences) **112**: 201517557. doi:10.1073/pnas.1517557112. ISSN 1091-6490. PMC 4664351. PMID 26483481. Retrieved 2015-10-20. Early edition, published online before print.

[40] Moskowitz, Clara (29 March 2012). "Life's Building Blocks May Have Formed in Dust Around Young Sun". *Space.com* (Salt Lake City, UT: Purch). Retrieved 2012-03-30.

[41] Fesenkov 1959, p. 9

[42] Kasting, James F. (12 February 1993). "Earth's Early Atmosphere" (PDF). *Science* (Washington, D.C.: American Association for the Advancement of Science) **259** (5097): 922. doi:10.1126/science.11536547. ISSN 0036-8075. PMID 11536547. Retrieved 2015-07-28.

[43] Follmann, Hartmut; Brownson, Carol (November 2009). "Darwin's warm little pond revisited: from molecules to the origin of life". *Naturwissenschaften* (Berlin: Springer-Verlag) **96** (11): 1265–1292. Bibcode:2009NW.....96.1265F. doi:10.1007/s00114-009-0602-1. ISSN 0028-1042. PMID 19760276.

[44] Morse, John W.; MacKenzie, Fred T. (1998). "Hadean Ocean Carbonate Geochemistry". *Aquatic Geochemistry* (Kluwer Academic Publishers) **4** (3–4): 301–319. doi:10.1023/A:1009632230875. ISSN 1380-6165.

[45] Wilde, Simon A.; Valley, John W.; Peck, William H.; Graham, Colin M. (11 January 2001). "Evidence from detrital zircons for the existence of continental crust and oceans on the Earth 4.4 Gyr ago" (PDF). *Nature* (London: Nature Publishing Group) **409** (6817): 175–178. doi:10.1038/35051550. ISSN 0028-0836. PMID 11196637. Retrieved 2015-06-03.

[46] Rosing, Minik T.; Bird, Dennis K.; Sleep, Norman H.; et al. (22 March 2006). "The rise of continents—An essay on the geologic consequences of photosynthesis" (PDF). *Palaeogeography, Palaeoclimatology, Palaeoecology* (Amsterdam, the Netherlands: Elsevier) **232** (2–4): 99–113. doi:10.1016/j.palaeo.2006.01.007. ISSN 0031-0182. Retrieved 2015-06-08.

[47] Sleep, Norman H.; Zahnle, Kevin J.; Kasting, James F.; et al. (9 November 1989). "Annihilation of ecosystems by large asteroid impacts on early Earth". *Nature* (London: Nature Publishing Group) **342** (6246): 139–142. Bibcode:1989Natur.342..139S. doi:10.1038/342139a0. ISSN 0028-0836. PMID 11536616.

[48] Davies 1999

[49] O'Donoghue, James (21 August 2011). "Oldest reliable fossils show early life was a beach". *New Scientist* (London: Reed Business Information). ISSN 0262-4079. Retrieved 2014-10-13.

- Wacey, David; Kilburn, Matt R.; Saunders, Martin; et al. (October 2011). "Microfossils of sulphur-metabolizing cells in 3.4-billion-year-old rocks of Western Australia". *Nature Geoscience* (London: Nature Publishing Group) **4**

(10): 698–702. Bibcode:2011NatGe...4..698W. doi:10.1038/ngeo1238. ISSN 1752-0894.

[50] Wolpert, Stuart (19 October 2015). "Life on Earth likely started at least 4.1 billion years ago — much earlier than scientists had thought". ULCA. Retrieved 20 October 2015.

[51] Gomes, Rodney; Levison, Hal F.; Tsiganis, Kleomenis; Morbidelli, Alessandro (26 May 2005). "Origin of the cataclysmic Late Heavy Bombardment period of the terrestrial planets". *Nature* (London: Nature Publishing Group) **435** (7041): 466–469. Bibcode:2005Natur.435..466G. doi:10.1038/nature03676. ISSN 0028-0836. PMID 15917802.

[52] Davies 2007, pp. 61–73

[53] Maher, Kevin A.; Stevenson, David J. (18 February 1988). "Impact frustration of the origin of life". *Nature* (London: Nature Publishing Group) **331** (6157): 612–614. Bibcode:1988Natur.331..612M. doi:10.1038/331612a0. ISSN 0028-0836. PMID 11536595.

[54] Sheldon 2005

[55] Vartanian 1973, pp. 307–312

[56] Lennox 2001, pp. 229–258

[57] Balme, D. M. (1962). "Development of Biology in Aristotle and Theophrastus: Theory of Spontaneous Generation". *Phronesis* (Leiden, the Netherlands: Brill Publishers) 7 (1–2): 91–104. doi:10.1163/156852862X00052. ISSN 0031-8868.

[58] Ross 1652

[59] Dobell 1960

[60] Bondeson 1999

[61] Bernal 1967

[62] "Biogenesis". *Hmolpedia*. Ancaster, Ontario, Canada: WikiFoundry, Inc. Retrieved 2014-05-19.

[63] Huxley 1968

[64] Bastian 1871

[65] Bastian 1871, p. xi–xii

[66] Oparin 1953, p. 196

[67] Tyndall 1905, IV, XII (1876), XIII (1878)

[68] Bernal 1967, p. 139

[69] Priscu, John C. "Origin and Evolution of Life on a Frozen Earth". Arlington County, VA: National Science Foundation. Retrieved 2014-03-01.

[70] Darwin 1887, p. 18: "It is often said that all the conditions for the first production of a living organism are now present, which could ever have been present. But if (and oh! what a big if!) we could conceive in some warm little pond, with all sorts of ammonia and phosphoric salts, light, heat, electricity, &c., present, that a proteine compound was chemically formed ready to undergo still more complex changes, at the present day such matter would be instantly devoured or absorbed, which would not have been the case before living creatures were formed." — Charles Darwin, 1 February 1871

[71] Bernal 1967, *The Origin of Life* (A. I. Oparin, 1924), pp. 199–234

[72] Oparin 1953

[73] Shapiro 1987, p. 110

[74] Bryson 2004, pp. 300–302

[75] Miller, Stanley L. (15 May 1953). "A Production of Amino Acids Under Possible Primitive Earth Conditions". *Science* (Washington, D.C.: American Association for the Advancement of Science) **117** (3046): 528–529. Bibcode:1953Sci...117..528M. doi:10.1126/science.117.3046.528. ISSN 0036-8075. PMID 13056598.

[76] Parker, Eric T.; Cleaves, Henderson J.; Dworkin, Jason P.; et al. (5 April 2011). "Primordial synthesis of amines and amino acids in a 1958 Miller H_2S-rich spark discharge experiment" (PDF). *Proc. Natl. Acad. Sci. U.S.A.* (Washington, D.C.: National Academy of Sciences) **108** (14): 5526–5531. Bibcode:2011PNAS..108.5526P. doi:10.1073/pnas.1019191108. ISSN 0027-8424. PMC 3078417. PMID 21422282. Retrieved 2015-06-08.

[77] Bernal 1967, p. 143

[78] Walsh, J. Bruce (1995). "Part 4: Experimental studies of the origins of life". *Origins of life* (Lecture notes). Tucson, AZ: University Of Arizona. Archived from the original on 2008-01-13. Retrieved 2015-06-08.

[79] Woodward 1969, p. 287

[80] Bahadur, Krishna (1973). "Photochemical Formation of Self-sustaining Coacervates" (PDF). *Proceedings of the Indian National Science Academy* (New Delhi: Indian National Science Academy) **39B** (4): 455–467. ISSN 0370-0046.

- Bahadur, Krishna (1975). "Photochemical Formation of Self-Sustaining Coacervates". *Zentralblatt für Bakteriologie, Parasitenkunde, Infektionskrankheiten und Hygiene* (Jena, Germany: Gustav Fischer Verlag) **130** (3): 211–218. doi:10.1016/S0044-4057(75)80076-1. OCLC 641018092. PMID 1242552.

[81] Kasting 1993, p. 922

[82] Kasting 1993, p. 920

[83] Bernal 1951

[84] Bernal, John Desmond (September 1949). "The Physical Basis of Life". *Proceedings of the Physical Society. Section A* (Bristol, UK: Physical Society) **62** (9): 537–558. Bibcode:1949PPSA...62..537B. doi:10.1088/0370-1298/62/9/301. ISSN 0370-1298.

[85] Kauffman 1995

[86] Oehlenschläger, Frank; Eigen, Manfred (December 1997). "30 Years Later – a New Approach to Sol Spiegelman's and Leslie Orgel's in vitro EVOLUTIONARY STUDIES Dedicated to Leslie Orgel on the occasion of his 70th birthday". *Origins of Life and Evolution of Biospheres* (Kluwer Academic Publishers) **27** (5-6): 437–457. doi:10.1023/A:1006501326129. ISSN 0169-6149. PMID 9394469.

[87] McCollom, Thomas; Mayhew, Lisa; Scott, Jim (7 October 2014). "NASA awards CU-Boulder-led team $7 million to study origins, evolution of life in universe" (Press release). Boulder, CO: University of Colorado Boulder. Retrieved 2015-06-08.

[88] Gibson, Daniel G.; Glass, John I.; Lartigue, Carole; et al. (2 July 2010). "Creation of a Bacterial Cell Controlled by a Chemically Synthesized Genome". *Science* (Washington, D.C.: American Association for the Advancement of Science) **329** (5987): 52–56. Bibcode:2010Sci...329...52G. doi:10.1126/science.1190719. ISSN 0036-8075. PMID 20488990.

[89] Swaby, Rachel (20 May 2010). "Scientists Create First Self-Replicating Synthetic Life". *Wired* (New York: Condé Nast). Retrieved 2015-06-08.

[90] Coughlan, Andy (2016) "Smallest ever genome comes to life: Humans built it but we don't know what a third of its genes actually do" (New Scientist 2nd April 2016 No 3067)p.6

[91] Gawlowicz, Susan (6 November 2011). "Carbon-based organic 'carriers' in interstellar dust clouds? Newly discovered diffuse interstellar bands". *Science Daily* (Rockville, MD: ScienceDaily, LLC). Retrieved 2015-06-08. Post is reprinted from materials provided by the Rochester Institute of Technology.

- Geballe, Thomas R.; Najarro, Francisco; Figer, Donald F.; et al. (10 November 2011). "Infrared diffuse interstellar bands in the Galactic Centre region". *Nature* (London: Nature Publishing Group) **479** (7372): 200–202. arXiv:1111.0613. Bibcode:2011Natur.479..200G. doi:10.1038/nature10527. ISSN 0028-0836. PMID 22048316.

[92] Klyce 2001

[93] Chyba, Christopher; Sagan, Carl (9 January 1992). "Endogenous production, exogenous delivery and impact-shock synthesis of organic molecules: an inventory for the origins of life". *Nature* (London: Nature Publishing Group) **355** (6356): 125–132. Bibcode:1992Natur.355..125C. doi:10.1038/355125a0. ISSN 0028-0836. PMID 11538392.

[94] Furukawa, Yoshihiro; Sekine, Toshimori; Oba, Masahiro; et al. (January 2009). "Biomolecule formation by oceanic impacts on early Earth". *Nature Geoscience* (London: Nature Publishing Group) **2** (1): 62–66. Bibcode:2009NatGe...2...62F. doi:10.1038/NGEO383. ISSN 1752-0894.

[95] Davies 1999, p. 155

[96] Bock & Goode 1996

[97] Palasek, Stan (23 May 2013). "Primordial RNA Replication and Applications in PCR Technology". arXiv:1305.5581v1 [q-bio.BM].

[98] Koonin, Eugene V.; Senkevich, Tatiana G.; Dolja, Valerian V. (19 September 2006). "The ancient Virus World and evolution of cells". *Biology Direct* (London: BioMed Central) **1**: 29. doi:10.1186/1745-6150-1-29. ISSN 1745-6150. PMC 1594570. PMID 16984643.

[99] Vlassov, Alexander V.; Kazakov, Sergei A.; Johnston, Brian H.; et al. (August 2005). "The RNA World on Ice: A New Scenario for the Emergence of RNA Information". *Journal of Molecular Evolution* (Berlin: Springer-Verlag) **61** (2): 264–273. doi:10.1007/s00239-004-0362-7. ISSN 0022-2844. PMID 16044244.

[100] Nussinov, Mark D.; Otroshchenko, Vladimir A.; Santoli, Salvatore (1997). "The emergence of the non-cellular phase of life on the fine-grained clayish particles of the early Earth's regolith". *BioSystems* (Amsterdam, the Netherlands: Elsevier) **42** (2–3): 111–118. doi:10.1016/S0303-2647(96)01699-1. ISSN 0303-2647. PMID 9184757.

[101] Saladino, Raffaele; Crestini, Claudia; Pino, Samanta; et al. (March 2012). "Formamide and the origin of life.". *Physics of Life Reviews* (Amsterdam, the Netherlands: Elsevier) **9** (1): 84–104. Bibcode:2012PhLRv...9...84S. doi:10.1016/j.plrev.2011.12.002. ISSN 1571-0645. PMID 22196896.

[102] Saladino, Raffaele; Botta, Giorgia; Pino, Samanta; et al. (July 2012). "From the one-carbon amide formamide to RNA all the steps are prebiotically possible". *Biochimie* (Amsterdam, the Netherlands: Elsevier) **94** (7): 1451–1456. doi:10.1016/j.biochi.2012.02.018. ISSN 0300-9084. PMID 22738728.

[103] Oró, Joan (16 September 1961). "Mechanism of Synthesis of Adenine from Hydrogen Cyanide under Possible Primitive Earth Conditions". *Nature* (London: Nature Publishing Group) **191** (4794): 1193–1194. Bibcode:1961Natur.191.1193O. doi:10.1038/1911193a0. ISSN 0028-0836. PMID 13731264.

[104] Basile, Brenda; Lazcano, Antonio; Oró, Joan (1984). "Prebiotic syntheses of purines and pyrimidines". *Advances in Space Research* (Amsterdam, the Netherlands: Elsevier) 4 (12): 125–131. Bibcode:1984AdSpR...4..125B. doi:10.1016/0273-1177(84)90554-4. ISSN 0273-1177. PMID 11537766.

[105] Orgel, Leslie E. (August 2004). "Prebiotic Adenine Revisited: Eutectics and Photochemistry". *Origins of Life and Evolution of Biospheres* (Kluwer Academic Publishers) 34 (4): 361–369. Bibcode:2004OLEB...34..361O. doi:10.1023/B:ORIG.0000029882.52156.c2. ISSN 0169-6149. PMID 15279171.

[106] Robertson, Michael P.; Miller, Stanley L. (29 June 1995). "An efficient prebiotic synthesis of cytosine and uracil". *Nature* (London: Nature Publishing Group) 375 (6534): 772–774. Bibcode:1995Natur.375..772R. doi:10.1038/375772a0. ISSN 0028-0836. PMID 7596408.

[107] Fox, Douglas (February 2008). "Did Life Evolve in Ice?". *Discover* (Waukesha, WI: Kalmbach Publishing). ISSN 0274-7529. Retrieved 2008-07-03.

[108] Levy, Matthew; Miller, Stanley L.; Brinton, Karen; Bada, Jeffrey L. (June 2000). "Prebiotic Synthesis of Adenine and Amino Acids Under Europa-like Conditions". *Icarus* (Amsterdam, the Netherlands: Elsevier) 145 (2): 609–613. Bibcode:2000Icar..145..609L. doi:10.1006/icar.2000.6365. ISSN 0019-1035. PMID 11543508.

[109] Menor-Salván, César; Ruiz-Bermejo, Marta; Guzmán, Marcelo I.; Osuna-Esteban, Susana; Veintemillas-Verdaguer, Sabino (20 April 2009). "Synthesis of Pyrimidines and Triazines in Ice: Implications for the Prebiotic Chemistry of Nucleobases". *Chemistry: A European Journal* (Weinheim, Germany: Wiley-VCH on behalf of ChemPubSoc Europe) 15 (17): 4411–4418. doi:10.1002/chem.200802656. ISSN 0947-6539. PMID 19288488.

[110] Roy, Debjani; Najafian, Katayoun; von Ragué Schleyer, Paul (30 October 2007). "Chemical evolution: The mechanism of the formation of adenine under prebiotic conditions". *Proc. Natl. Acad. Sci. U.S.A.* (Washington, D.C.: National Academy of Sciences) 104 (44): 17272–17277. Bibcode:2007PNAS..10417272R. doi:10.1073/pnas.0708434104. ISSN 0027-8424. PMC 2077245. PMID 17951429.

[111] Cleaves, H. James; Chalmers, John H.; Lazcano, Antonio; et al. (April 2008). "A Reassessment of Prebiotic Organic Synthesis in Neutral Planetary Atmospheres". *Origins of Life and Evolution of Biospheres* (Dordrecht, the Netherlands: Springer) 38 (2): 105–115. Bibcode:2008OLEB...38..105C. doi:10.1007/s11084-007-9120-3. ISSN 0169-6149. PMID 18204914.

[112] Chyba, Christopher F. (13 May 2005). "Rethinking Earth's Early Atmosphere". *Science* (Washington, D.C.: American Association for the Advancement of Science) 308 (5724): 962–963. doi:10.1126/science.1113157. ISSN 0036-8075. PMID 15890865.

[113] Barton et al. 2007, pp. 93–95

[114] Bada & Lazcano 2009, pp. 56–57

[115] Bada, Jeffrey L.; Lazcano, Antonio (2 May 2003). "Prebiotic Soup--Revisiting the Miller Experiment" (PDF). *Science* (Washington, D.C.: American Association for the Advancement of Science) 300 (5620): 745–746. doi:10.1126/science.1085145. ISSN 0036-8075. PMID 12730584. Retrieved 2015-06-13.

[116] Oró, Joan; Kimball, Aubrey P. (February 1962). "Synthesis of purines under possible primitive earth conditions: II. Purine intermediates from hydrogen cyanide". *Archives of Biochemistry and Biophysics* (Amsterdam, the Netherlands: Elsevier) 96 (2): 293–313. doi:10.1016/0003-9861(62)90412-5. ISSN 0003-9861. PMID 14482339.

[117] Ahuja, Mukesh, ed. (2006). "Origin of Life". *Life Science* 1. Delhi: Isha Books. p. 11. ISBN 81-8205-386-2. OCLC 297208106.

[118] Service, Robert F. (16 March 2015). "Researchers may have solved origin-of-life conundrum". *Science* (News) (Washington, D.C.: American Association for the Advancement of Science). ISSN 1095-9203. Retrieved 2015-07-26.

[119] Patel, Bhavesh H.; Percivalle, Claudia; Ritson, Dougal J.; Duffy, Colm D.; Sutherland, John D. (April 2015). "Common origins of RNA, protein and lipid precursors in a cyanosulfidic protometabolism". *Nature Chemistry* (London: Nature Publishing Group) 7 (4): 301–307. Bibcode:2015NatCh...7..301P. doi:10.1038/nchem.2202. ISSN 1755-4330. PMC 4568310. PMID 25803468. Retrieved 2015-07-22.

[120] Patel et al. 2015, p. 302

[121] Paul, Natasha; Joyce, Gerald F. (December 2004). "Minimal self-replicating systems". *Current Opinion in Chemical Biology* (Amsterdam, the Netherlands: Elsevier) 8 (6): 634–639. doi:10.1016/j.cbpa.2004.09.005. ISSN 1367-5931. PMID 15556408.

[122] Bissette, Andrew J.; Fletcher, Stephen P. (2 December 2013). "Mechanisms of Autocatalysis". *Angewandte Chemie International Edition* (Weinheim, Germany: Wiley-VCH on behalf of the German Chemical Society) 52 (49): 12800–12826. doi:10.1002/anie.201303822. ISSN 1433-7851. PMID 24127341.

[123] Kauffman 1993, chpt. 7

[124] Dawkins 2004

[125] Tjivikua, T.; Ballester, Pablo; Rebek, Julius, Jr. (January 1990). "Self-replicating system". *Journal of the American Chemical Society* (Washington, D.C.: American Chemical Society) 112 (3): 1249–1250. doi:10.1021/ja00159a057. ISSN 0002-7863.

[126] Browne, Malcolm W. (30 October 1990). "Chemists Make Molecule With Hint of Life". *The New York Times* (New York: The New York Times Company). ISSN 0362-4331. Retrieved 2015-07-14.

[127] Eigen & Schuster 1979

[128] Hoffmann, Geoffrey W. (25 June 1974). "On the origin of the genetic code and the stability of the translation apparatus". *Journal of Molecular Biology* (Amsterdam, the Netherlands: Elsevier) 86 (2): 349–362. doi:10.1016/0022-2836(74)90024-2. ISSN 0022-2836. PMID 4414916.

[129] Orgel, Leslie E. (April 1963). "The Maintenance of the Accuracy of Protein Synthesis and its Relevance to Ageing". *Proc. Natl. Acad. Sci. U.S.A.* (Washington, D.C.: National Academy of Sciences) 49 (4): 517–521. Bibcode:1963PNAS...49..517O. doi:10.1073/pnas.49.4.517. ISSN 0027-8424. PMC 299893. PMID 13940312.

[130] Hoffmann, Geoffrey W. (October 1975). "The Stochastic Theory of the Origin of the Genetic Code". *Annual Review of Physical Chemistry* (Palo Alto, CA: Annal Reviews) 26: 123–144. Bibcode:1975ARPC...26..123H. doi:10.1146/annurev.pc.26.100175.001011. ISSN 0066-426X.

[131] Colgate, S. A.; Rasmussen, S.; Solem, J. C.; Lackner, K. (2003). "An astrophysical basis for a universal origin of life". *Advances in Complex Systems* 6 (4): 487–505. doi:10.1142/s0219525903001079.

[132] Chaichian, Rojas & Tureanu 2014, pp. 353–364

[133] Plasson, Raphaël; Kondepudi, Dilip K.; Bersini, Hugues; et al. (August 2007). "Emergence of homochirality in far-from-equilibrium systems: Mechanisms and role in prebiotic chemistry". *Chirality* (Hoboken, NJ: John Wiley & Sons) 19 (8): 589–600. doi:10.1002/chir.20440. ISSN 0899-0042. PMID 17559107. "Special Issue: Proceedings from the Eighteenth International Symposium on Chirality (ISCD-18), Busan, Korea, 2006"

[134] Clark, Stuart (July–August 1999). "Polarized Starlight and the Handedness of Life". *American Scientist* (Research Triangle Park, NC: Sigma Xi) 87 (4): 336. Bibcode:1999AmSci..87..336C. doi:10.1511/1999.4.336. ISSN 0003-0996.

[135] Shibata, Takanori; Morioka, Hiroshi; Hayase, Tadakatsu; et al. (17 January 1996). "Highly Enantioselective Catalytic Asymmetric Automultiplication of Chiral Pyrimidyl Alcohol". *Journal of the American Chemical Society* (Washington, D.C.: American Chemical Society) 118 (2): 471–472. doi:10.1021/ja953066g. ISSN 0002-7863.

[136] Soai, Kenso; Sato, Itaru; Shibata, Takanori (2001). "Asymmetric autocatalysis and the origin of chiral homogeneity in organic compounds". *The Chemical Record* (Hoboken, NJ: John Wiley & Sons on behalf of The Japan Chemical Journal Forum) 1 (4): 321–332. doi:10.1002/tcr.1017. ISSN 1528-0691. PMID 11893072.

[137] Hazen 2005

[138] Mullen, Leslie (5 September 2005). "Building Life from Star-Stuff". *Astrobiology Magazine* (New York: NASA). Retrieved 2015-06-15.

[139] Chen, Irene A.; Walde, Peter (July 2010). "From Self-Assembled Vesicles to Protocells" (PDF). *Cold Spring Harbor Perspectives in Biology* (Cold Spring Harbor, NY: Cold Spring Harbor Laboratory Press) 2 (7): a002170. doi:10.1101/cshperspect.a002170. ISSN 1943-0264. PMC 2890201. PMID 20519344. Retrieved 2015-06-15.

[140] "Exploring Life's Origins: Protocells". *Exploring Life's Origins: A Virtual Exhibit*. Arlington County, VA: National Science Foundation. Retrieved 2014-03-18.

[141] Chen, Irene A. (8 December 2006). "The Emergence of Cells During the Origin of Life". *Science* (Washington, D.C.: American Association for the Advancement of Science) 314 (5805): 1558–1559. doi:10.1126/science.1137541. ISSN 0036-8075. PMID 17158315. Retrieved 2015-06-15.

[142] Zimmer, Carl (26 June 2004). "What Came Before DNA?". *Discover* (Waukesha, WI: Kalmbach Publishing). ISSN 0274-7529.

[143] Shapiro, Robert (June 2007). "A Simpler Origin for Life". *Scientific American* (Stuttgart: Georg von Holtzbrinck Publishing Group) 296 (6): 46–53. doi:10.1038/scientificamerican0607-46. ISSN 0036-8733. PMID 17663224. Retrieved 2015-06-15.

[144] Chang 2007

[145] Switek, Brian (13 February 2012). "Debate bubbles over the origin of life". *Nature* (London: Nature Publishing Group). doi:10.1038/nature.2012.10024. ISSN 0028-0836.

[146] Grote, Mathias (September 2011). "*Jeewanu*, or the 'particles of life'" (PDF). *Journal of Biosciences* (Bangalore, India: Indian Academy of Sciences; Springer) 36 (4): 563–570. doi:10.1007/s12038-011-9087-0. ISSN 0250-5991. PMID 21857103. Retrieved 2015-06-15.

[147] Gupta, V. K.; Rai, R. K. (August 2013). "Histochemical localisation of RNA-like material in photochemically formed self-sustaining, abiogenic supramolecular assemblies 'Jeewanu'". *International Research Journal of Science & Engineering* (Amravati, India) 1 (1): 1–4. ISSN 2322-0015. Retrieved 2015-06-15.

[148] Welter, Kira (10 August 2015). "Peptide glue may have held first protocell components together". *Chemistry World* (News) (London: Royal Society of Chemistry). ISSN 1473-7604. Retrieved 2015-08-29.

- Kamat, Neha P.; Tobé, Sylvia; Hill, Ian T.; Szostak, Jack W. (29 July 2015). "Electrostatic Localization of RNA to Protocell Membranes by Cationic

Hydrophobic Peptides". *Angewandte Chemie International Edition* (Weinheim, Germany: Wiley-VCH on behalf of the German Chemical Society). doi:10.1002/anie.201505742. ISSN 1433-7851. "Early View (Online Version of Record published before inclusion in an issue)"

[149] Wimberly, Brian T.; Brodersen, Ditlev E.; Clemons, William M., Jr.; et al. (21 September 2000). "Structure of the 30S ribosomal subunit". *Nature* (London: Nature Publishing Group) 407 (6802): 327–339. doi:10.1038/35030006. ISSN 0028-0836. PMID 11014182.

[150] Zimmer, Carl (25 September 2014). "A Tiny Emissary From the Ancient Past". *The New York Times* (New York: The New York Times Company). ISSN 0362-4331. Retrieved 2014-09-26.

[151] Wade, Nicholas (4 May 2015). "Making Sense of the Chemistry That Led to Life on Earth". *The New York Times* (New York: The New York Times Company). ISSN 0362-4331. Retrieved 2015-05-10.

[152] Yarus, Michael (April 2011). "Getting Past the RNA World: The Initial Darwinian Ancestor". *Cold Spring Harbor Perspectives in Biology* (Cold Spring Harbor, NY: Cold Spring Harbor Laboratory Press) 3 (4): a003590. doi:10.1101/cshperspect.a003590. ISSN 1943-0264. PMC 3062219. PMID 20719875.

[153] Neveu, Marc; Kim, Hyo-Joong; Benner, Steven A. (22 April 2013). "The 'Strong' RNA World Hypothesis: Fifty Years Old". *Astrobiology* (New Rochelle, NY: Mary Ann Liebert, Inc.) 13 (4): 391–403. Bibcode:2013AsBio..13..391N. doi:10.1089/ast.2012.0868. ISSN 1531-1074. PMID 23551238.

[154] Gilbert, Walter (20 February 1986). "Origin of life: The RNA world". *Nature* (London: Nature Publishing Group) 319 (6055): 618. Bibcode:1986Natur.319..618G. doi:10.1038/319618a0. ISSN 0028-0836.

[155] Noller, Harry F. (April 2012). "Evolution of protein synthesis from an RNA world.". *Cold Spring Harbor Perspectives in Biology* (Cold Spring Harbor, NY: Cold Spring Harbor Laboratory Press) 4 (4): a003681. doi:10.1101/cshperspect.a003681. ISSN 1943-0264. PMC 3312679. PMID 20610545.

[156] Koonin, Eugene V. (31 May 2007). "The cosmological model of eternal inflation and the transition from chance to biological evolution in the history of life". *Biology Direct* (London: BioMed Central) 2: 15. doi:10.1186/1745-6150-2-15. ISSN 1745-6150. PMC 1892545. PMID 17540027.

[157] Yates, Diana (25 September 2015). "Study adds to evidence that viruses are alive" (Press release). Champaign, IL: University of Illinois at Urbana–Champaign. Retrieved 2015-10-20.

[158] Katzourakis, Aris (2013)"Paleovirology: inferring viral evolution from host genome sequence data" (Philosophical Transactions of the Royal Society Published 12 August 2013.DOI: 10.1098/rstb.2012.0493)

[159] Arshan, Nasir; Caetano-Anollés, Gustavo (25 September 2015). "A phylogenomic data-driven exploration of viral origins and evolution". *Science Advances* (Washington, D.C.: American Association for the Advancement of Science) 1 (8): e1500527. doi:10.1126/sciadv.1500527. ISSN 2375-2548.

[160] Nasir, Arshan; Naeem, Aisha; Jawad Khan, Muhammad; et al. (December 2011). "Annotation of Protein Domains Reveals Remarkable Conservation in the Functional Make up of Proteomes Across Superkingdoms". *Genes* (Basel, Switzerland: MDPI) 2 (4): 869–911. doi:10.3390/genes2040869. ISSN 2073-4425. PMC 3927607. PMID 24710297.

[161] Zimmer, Carl (12 September 2013). "A Far-Flung Possibility for the Origin of Life". *The New York Times* (New York: The New York Times Company). ISSN 0362-4331. Retrieved 2015-06-15.

[162] Webb, Richard (29 August 2013). "Primordial broth of life was a dry Martian cup-a-soup". *New Scientist* (London: Reed Business Information). ISSN 0262-4079. Retrieved 2015-06-16.

[163] Wentao Ma; Chunwu Yu; Wentao Zhang; et al. (November 2007). "Nucleotide synthetase ribozymes may have emerged first in the RNA world". *RNA* (Cold Spring Harbor, NY: Cold Spring Harbor Laboratory Press on behalf of the RNA Society) 13 (11): 2012–2019. doi:10.1261/rna.658507. ISSN 1355-8382. PMC 2040096. PMID 17878321.

[164] Orgel, Leslie E. (October 1994). "The origin of life on Earth". *Scientific American* (Stuttgart: Georg von Holtzbrinck Publishing Group) 271 (4): 76–83. doi:10.1038/scientificamerican1094-76. ISSN 0036-8733. PMID 7524147.

[165] Johnston, Wendy K.; Unrau, Peter J.; Lawrence, Michael S.; et al. (18 May 2001). "RNA-Catalyzed RNA Polymerization: Accurate and General RNA-Templated Primer Extension". *Science* (Washington, D.C.: American Association for the Advancement of Science) 292 (5520): 1319–1325. Bibcode:2001Sci...292.1319J. doi:10.1126/science.1060786. ISSN 0036-8075. PMID 11358999.

[166] Szostak, Jack W. (5 February 2015). "The Origins of Function in Biological Nucleic Acids, Proteins, and Membranes". Chevy Chase (CDP), MD: Howard Hughes Medical Institute. Retrieved 2015-06-16.

[167] Lincoln, Tracey A.; Joyce, Gerald F. (27 February 2009). "Self-Sustained Replication of an RNA Enzyme". *Science* (Washington, D.C.: American Association for the Advancement of Science) 323

(5918): 1229–1232. Bibcode:2009Sci...323.1229L. doi:10.1126/science.1167856. ISSN 0036-8075. PMC 2652413. PMID 19131595.

[168] Joyce, Gerald F. (2009). "Evolution in an RNA world" (PDF). *Cold Spring Harbor Perspectives in Biology* (Cold Spring Harbor, NY: Cold Spring Harbor Laboratory Press) **74** (Evolution: The Molecular Landscape): 17–23. doi:10.1101/sqb.2009.74.004. ISSN 1943-0264. PMC 2891321. PMID 19667013. Retrieved 2015-06-16.

[169] Bernstein, Harris; Byerly, Henry C.; Hopf, Frederick A.; et al. (June 1983). "The Darwinian Dynamic". *The Quarterly Review of Biology* (Chicago, IL: University of Chicago Press) **58** (2): 185–207. doi:10.1086/413216. ISSN 0033-5770. JSTOR 2828805.

[170] Michod 1999

[171] Orgel, Leslie E. (17 November 2000). "A Simpler Nucleic Acid". *Science* (Washington, D.C.: American Association for the Advancement of Science) **290** (5495): 1306–1307. doi:10.1126/science.290.5495.1306. ISSN 0036-8075. PMID 11185405.

[172] Nelson, Kevin E.; Levy, Matthew; Miller, Stanley L. (11 April 2000). "Peptide nucleic acids rather than RNA may have been the first genetic molecule". *Proc. Natl. Acad. Sci. U.S.A.* (Washington, D.C.: National Academy of Sciences) **97** (8): 3868–3871. Bibcode:2000PNAS...97.3868N. doi:10.1073/pnas.97.8.3868. ISSN 0027-8424. PMC 18108. PMID 10760258.

[173] Larralde, Rosa; Robertson, Michael P.; Miller, Stanley L. (29 August 1995). "Rates of Decomposition of Ribose and Other Sugars: Implications for Chemical Evolution" (PDF). *Proc. Natl. Acad. Sci. U.S.A.* (Washington, D.C.: National Academy of Sciences) **92** (18): 8158–8160. Bibcode:1995PNAS...92.8158L. doi:10.1073/pnas.92.18.8158. ISSN 0027-8424. PMC 41115. PMID 7667262.

[174] Lindahl, Tomas (22 April 1993). "Instability and decay of the primary structure of DNA". *Nature* (London: Nature Publishing Group) **362** (6422): 709–715. Bibcode:1993Natur.362..709L. doi:10.1038/362709a0. ISSN 0028-0836. PMID 8469282.

[175] Anastasi, Carole; Crowe, Michael A.; Powner, Matthew W.; Sutherland, John D. (18 September 2006). "Direct Assembly of Nucleoside Precursors from Two- and Three-Carbon Units". *Angewandte Chemie International Edition* (Weinheim, Germany: Wiley-VCH on behalf of the German Chemical Society) **45** (37): 6176–6179. doi:10.1002/anie.200601267. ISSN 1433-7851. PMID 16917794.

[176] Powner, Matthew W.; Sutherland, John D. (13 October 2008). "Potentially Prebiotic Synthesis of Pyrimidine β-D-Ribonucleotides by Photoanomerization/Hydrolysis of α-D-Cytidine-2′-Phosphate". *ChemBioChem* (Weinheim, Germany: Wiley-VCH) **9** (15): 2386–2387. doi:10.1002/cbic.200800391. ISSN 1439-4227. PMID 18798212.

[177] Powner, Matthew W.; Gerland, Béatrice; Sutherland, John D. (14 May 2009). "Synthesis of activated pyrimidine ribonucleotides in prebiotically plausible conditions". *Nature* (London: Nature Publishing Group) **459** (7244): 239–242. Bibcode:2009Natur.459..239P. doi:10.1038/nature08013. ISSN 0028-0836. PMID 19444213.

[178] Senthilingam, Meera (25 April 2014). "Metabolism May Have Started in Early Oceans Before the Origin of Life" (Press release). Wellcome Trust. EurekAlert!. Retrieved 2015-06-16.

[179] Yue-Ching Ho, Eugene (July–September 1990). "Evolutionary Epistemology and Sir Karl Popper's Latest Intellectual Interest: A First-Hand Report". *Intellectus* (Hong Kong: Hong Kong Institute of Economic Science) **15**: 1–3. OCLC 26878740. Retrieved 2012-08-13.

[180] Wade, Nicholas (22 April 1997). "Amateur Shakes Up Ideas on Recipe for Life". *The New York Times* (New York: The New York Times Company). ISSN 0362-4331. Retrieved 2015-06-16.

[181] Popper, Karl R. (29 March 1990). "Pyrite and the origin of life". *Nature* (London: Nature Publishing Group) **344** (6265): 387. Bibcode:1990Natur.344..387P. doi:10.1038/344387a0. ISSN 0028-0836.

[182] Huber, Claudia; Wächtershäuser, Günter (31 July 1998). "Peptides by Activation of Amino Acids with CO on (Ni,Fe)S Surfaces: Implications for the Origin of Life". *Science* (Washington, D.C.: American Association for the Advancement of Science) **281** (5377): 670–672. Bibcode:1998Sci...281..670H. doi:10.1126/science.281.5377.670. ISSN 0036-8075. PMID 9685253.

[183] Lane 2009

[184] Musser, George (23 September 2011). "How Life Arose on Earth, and How a Singularity Might Bring It Down". *Observations* (Blog). Scientific American. ISSN 0036-8733. Retrieved 2015-06-17.

[185] Carroll, Sean (10 March 2010). "Free Energy and the Meaning of Life". *Cosmic Variance* (Blog). Discover. ISSN 0274-7529. Retrieved 2015-06-17.

[186] Wolchover, Natalie (22 January 2014). "A New Physics Theory of Life". *Quanta Magazine* (New York: Simons Foundation). Retrieved 2015-06-17.

[187] England, Jeremy L. (28 September 2013). "Statistical physics of self-replication" (PDF). *Journal of Chemical Physics* (College Park, MD: American Institute of Physics) **139**: 121923. arXiv:1209.1179. Bibcode:2013JChPh.139l1923E. doi:10.1063/1.4818538. ISSN 0021-9606. Retrieved 2015-06-18.

[188] Orgel, Leslie E. (7 November 2000). "Self-organizing biochemical cycles". *Proc. Natl. Acad. Sci. U.S.A.* (Washington, D.C.: National Academy of Sciences) 97 (23): 12503–12507. Bibcode:2000PNAS...9712503O. doi:10.1073/pnas.220406697. ISSN 0027-8424. PMC 18793. PMID 11058157.

[189] Mulkidjanian, Armen Y. (24 August 2009). "On the origin of life in the zinc world: 1. Photosynthesizing, porous edifices built of hydrothermally precipitated zinc sulfide as cradles of life on Earth". *Biology Direct* (London: BioMed Central) 4: 26. doi:10.1186/1745-6150-4-26. ISSN 1745-6150.

[190] Wächtershäuser, Günter (December 1988). "Before Enzymes and Templates: Theory of Surface Metabolism" (PDF). *Microbiological Reviews* (Washington, D.C.: American Society for Microbiology) 52 (4): 452–484. ISSN 0146-0749. PMC 373159. PMID 3070320.

[191] Mulkidjanian, Armen Y.; Galperin, Michael Y. (24 August 2009). "On the origin of life in the zinc world. 2. Validation of the hypothesis on the photosynthesizing zinc sulfide edifices as cradles of life on Earth". *Biology Direct* (London: BioMed Central) 4: 27. doi:10.1186/1745-6150-4-27. ISSN 1745-6150.

[192] Macallum, A. B. (1 April 1926). "The Paleochemistry of the body fluids and tissues". *Physiological Reviews* (Bethesda, MD: American Physiological Society) 6 (2): 316–357. ISSN 0031-9333. Retrieved 2015-06-18.

[193] Mulkidjanian, Armen Y.; Bychkov, Andrew Yu.; Dibrova, Daria V.; et al. (3 April 2012). "Origin of first cells at terrestrial, anoxic geothermal fields". *Proc. Natl. Acad. Sci. U.S.A.* (Washington, D.C.: National Academy of Sciences) 109 (14): E821–E830. Bibcode:2012PNAS..109E.821M. doi:10.1073/pnas.1117774109. ISSN 1091-6490. PMC 3325685. PMID 22331915.

[194] For a deeper integrative version of this hypothesis, see in particular Lankenau 2011, pp. 225–286, interconnecting the "Two RNA worlds" concept and other detailed aspects; and Davidovich, Chen; Belousoff, Matthew; Bashan, Anat; Yonath, Ada (September 2009). "The evolving ribosome: from non-coded peptide bond formation to sophisticated translation machinery". *Research in Microbiology* (Amsterdam, the Netherlands: Elsevier) 160 (7): 487–492. doi:10.1016/j.resmic.2009.07.004. ISSN 1769-7123. PMID 19619641.

[195] Schirber, Michael (24 June 2014). "Hydrothermal Vents Could Explain Chemical Precursors to Life". *NASA Astrobiology: Life in the Universe*. NASA. Retrieved 2015-06-19.

[196] Martin, William; Russell, Michael J. (29 January 2003). "On the origins of cells: a hypothesis for the evolutionary transitions from abiotic geochemistry to chemoautotrophic prokaryotes, and from prokaryotes to nucleated cells". *Philosophical Transactions of the Royal Society B* (London: Royal Society) 358 (1429): 59–83; discussion 83–85. doi:10.1098/rstb.2002.1183. ISSN 0962-8436. PMC 1693102. PMID 12594918.

[197] Ignatov, Ignat; Mosin, Oleg V. (2013). "Possible Processes for Origin of Life and Living Matter with modeling of Physiological Processes of Bacterium *Bacillus Subtilis* in Heavy Water as Model System". *Journal of Natural Sciences Research* (New York: International Institute for Science, Technology and Education) 3 (9): 65–76. ISSN 2225-0921.

[198] Calvin 1969

[199] Schirber, Michael (1 March 2010). "First Fossil-Makers in Hot Water". *Astrobiology Magazine* (New York: NASA). Retrieved 2015-06-19.

[200] Kurihara, Kensuke; Tamura, Mieko; Shohda, Koh-ichiroh; et al. (October 2011). "Self-Reproduction of supramolecular giant vesicles combined with the amplification of encapsulated DNA". *Nature Chemistry* (London: Nature Publishing Group) 3 (10): 775–781. Bibcode:2011NatCh...3..775K. doi:10.1038/nchem.1127. ISSN 1755-4330. PMID 21941249.

[201] Usher, Oli (27 April 2015). "Chemistry of seabed's hot vents could explain emergence of life" (Press release). University College London. Retrieved 2015-06-19.

[202] Roldan, Alberto; Hollingsworth, Nathan; Roffey, Anna; Islam, Husn-Ubayda; et al. (May 2015). "Bio-inspired CO2 conversion by iron sulfide catalysts under sustainable conditions" (PDF). *Chemical Communications* (London: Royal Society of Chemistry) 51 (35): 7501–7504. doi:10.1039/C5CC02078F. ISSN 1359-7345. PMID 25835242. Retrieved 2015-06-19.

[203] Muller, Anthonie W. J. (7 August 1985). "Thermosynthesis by biomembranes: Energy gain from cyclic temperature changes". *Journal of Theoretical Biology* (Amsterdam, the Netherlands: Elsevier) 115 (3): 429–453. doi:10.1016/S0022-5193(85)80202-2. ISSN 0022-5193. PMID 3162066.

[204] Muller, Anthonie W. J. (1995). "Were the first organisms heat engines? A new model for biogenesis and the early evolution of biological energy conversion". *Progress in Biophysics and Molecular Biology* (Oxford, UK; New York: Pergamon Press) 63 (2): 193–231. doi:10.1016/0079-6107(95)00004-7. ISSN 0079-6107. PMID 7542789.

[205] Muller, Anthonie W. J.; Schulze-Makuch, Dirk (1 April 2006). "Sorption heat engines: Simple inanimate negative entropy generators". *Physica A: Statistical Mechanics and its Applications* (Utrecht, the Netherlands: Elsevier) 362 (2): 369–381. arXiv:physics/0507173. Bibcode:2006PhyA..362..369M. doi:10.1016/j.physa.2005.12.003. ISSN 0378-4371.

[206] Orgel 1987, pp. 9–16

39.10. REFERENCES

[207] Perry, Caroline (7 February 2011). "Clay-armored bubbles may have formed first protocells" (Press release). Cambridge, MA: Harvard University. EurekAlert!. Retrieved 2015-06-20.

[208] Dawkins 1996, pp. 148–161

[209] Wenhua Huang; Ferris, James P. (12 July 2006). "One-Step, Regioselective Synthesis of up to 50-mers of RNA Oligomers by Montmorillonite Catalysis". *Journal of the American Chemical Society* (Washington, D.C.: American Chemical Society) **128** (27): 8914–8919. doi:10.1021/ja061782k. ISSN 0002-7863. PMID 16819887.

[210] Moore, Caroline (16 July 2007). "Crystals as genes?". *Highlights in Chemical Science* (London: Royal Society of Chemistry). ISSN 2041-5818. Retrieved 2015-06-21.

- Bullard, Theresa; Freudenthal, John; Avagyan, Serine; et al. (2007). "Test of Cairns-Smith's 'crystals-as-genes' hypothesis". *Faraday Discussions* **136**: 231–245. Bibcode:2007FaDi..136..231B. doi:10.1039/b616612c. ISSN 1359-6640.

[211] Wickramasinghe, Chandra (2011). "Bacterial morphologies supporting cometary panspermia: a reappraisal". *International Journal of Astrobiology* **10** (1): 25–30. Bibcode:2011IJAsB..10...25W. doi:10.1017/S1473550410000157.

[212] Napier, William (October 2011). "Exchange of Biomaterial Between Planetary Systems" (PDF) **16**: 6616–6642.

[213] Rampelotto, P. H. (2010). Panspermia: A promising field of research. In: Astrobiology Science Conference. Abs 5224.

[214] Forward planetary contamination like *Tersicoccus phoenicis*, that has shown resistance to methods usually used in spacecraft assembly clean rooms: Madhusoodanan, Jyoti (May 19, 2014). "Microbial stowaways to Mars identified". *Nature*. doi:10.1038/nature.2014.15249. Retrieved May 23, 2014.

[215] Webster, Guy (November 6, 2013). "Rare New Microbe Found in Two Distant Clean Rooms". *NASA.gov*. Retrieved November 6, 2013.

[216] Clark, Stuart (25 September 2002). "Tough Earth bug may be from Mars". *New Scientist* (London: Reed Business Information). ISSN 0262-4079. Retrieved 2015-06-21.

[217] Horneck, Gerda; Klaus, David M.; Mancinelli, Rocco L. (March 2010). "Space Microbiology". *Microbiology and Molecular Biology Reviews* (Washington, D.C.: American Society for Microbiology) **74** (1): 121–156. doi:10.1128/MMBR.00016-09. ISSN 1092-2172. PMC 2832349. PMID 20197502.

[218] Rabbow, Elke; Horneck, Gerda; Rettberg, Petra; et al. (December 2009). "EXPOSE, an Astrobiological Exposure Facility on the International Space Station – from Proposal to Flight". *Origins of Life and Evolution of Biospheres* (Dordrecht, the Netherlands: Springer) **39** (6): 581–598. Bibcode:2009OLEB...39..581R. doi:10.1007/s11084-009-9173-6. ISSN 0169-6149. PMID 19629743.

[219] Onofri, Silvano; de la Torre, Rosa; de Vera, Jean-Pierre; et al. (May 2012). "Survival of Rock-Colonizing Organisms After 1.5 Years in Outer Space". *Astrobiology* (New Rochelle, NY: Mary Ann Liebert, Inc.) **12** (5): 508–516. Bibcode:2012AsBio..12..508O. doi:10.1089/ast.2011.0736. ISSN 1531-1074. PMID 22680696.

[220] "biological abundance of elements". *Encyclopedia of Science*. Dundee, Scotland: David Darling Enterprises. Retrieved 2008-10-09.

[221] Hoover, Rachel (21 February 2014). "Need to Track Organic Nano-Particles Across the Universe? NASA's Got an App for That". *Ames Research Center*. Mountain View, CA: NASA. Retrieved 2015-06-22.

[222] Chang, Kenneth (18 August 2009). "From a Distant Comet, a Clue to Life". *The New York Times* (New York: The New York Times Company). p. A18. ISSN 0362-4331. Retrieved 2015-06-22.

[223] Goncharuk, Vladislav V.; Zui, O. V. (February 2015). "Water and carbon dioxide as the main precursors of organic matter on Earth and in space". *Journal of Water Chemistry and Technology* (Dordrecht, the Netherlands: Springer on behalf of Allerton Press) **37** (1): 2–3. doi:10.3103/S1063455X15010026. ISSN 1063-455X.

[224] Abou Mrad, Ninette; Vinogradoff, Vassilissa; Duvernay, Fabrice; et al. (2015). "Laboratory experimental simulations: Chemical evolution of the organic matter from interstellar and cometary ice analogs" (PDF). *Bulletin de la Société Royale des Sciences de Liège* (Liège, Belgium: Société royale des sciences de Liège) **84**: 21–32. Bibcode:2015BSRSL..84...21A. ISSN 0037-9565. Retrieved 2015-04-06.

[225] Gallori, Enzo (June 2011). "Astrochemistry and the origin of genetic material". *Rendiconti Lincei* (Milan, Italy: Springer) **22** (2): 113–118. doi:10.1007/s12210-011-0118-4. ISSN 2037-4631. "Paper presented at the Symposium 'Astrochemistry: molecules in space and time' (Rome, 4–5 November 2010), sponsored by Fondazione 'Guido Donegani', Accademia Nazionale dei Lincei."

[226] Martins, Zita (February 2011). "Organic Chemistry of Carbonaceous Meteorites". *Elements* (Chantilly, VA: Mineralogical Society of America et al.) **7** (1): 35–40. doi:10.2113/gselements.7.1.35. ISSN 1811-5209.

[227] Martins, Zita; Botta, Oliver; Fogel, Marilyn L.; et al. (15 June 2008). "Extraterrestrial nucleobases in the Murchison meteorite". *Earth and Planetary Science Letters* (Am-

sterdam, the Netherlands: Elsevier) **270** (1–2): 130–136. arXiv:0806.2286. Bibcode:2008E&PSL.270..130M. doi:10.1016/j.epsl.2008.03.026. ISSN 0012-821X.

[228] "We may all be space aliens: study". *ABC News* (Sydney: Australian Broadcasting Corporation). AFP. 14 June 2008. Retrieved 2015-06-22.

[229] Callahan, Michael P.; Smith, Karen E.; Cleaves, H. James, II; et al. (23 August 2011). "Carbonaceous meteorites contain a wide range of extraterrestrial nucleobases". *Proc. Natl. Acad. Sci. U.S.A.* (Washington, D.C.: National Academy of Sciences) **108** (34): 13995–13998. Bibcode:2011PNAS..10813995C. doi:10.1073/pnas.1106493108. ISSN 0027-8424. PMC 3161613. PMID 21836052.

[230] Steigerwald, John (8 August 2011). "NASA Researchers: DNA Building Blocks Can Be Made in Space". *Goddard Space Flight Center*. Greenbelt, MD: NASA. Retrieved 2015-06-23.

[231] Chow, Denise (26 October 2011). "Discovery: Cosmic Dust Contains Organic Matter from Stars". *Space.com* (Ogden, UT: Purch). Retrieved 2015-06-23.

[232] "Astronomers Discover Complex Organic Matter Exists Throughout the Universe". Rockville, MD: ScienceDaily, LLC. 26 October 2011. Retrieved 2015-06-23. Post is reprinted from materials provided by The University of Hong Kong.

[233] Sun Kwok; Yong Zhang (3 November 2011). "Mixed aromatic–aliphatic organic nanoparticles as carriers of unidentified infrared emission features". *Nature* (London: Nature Publishing Group) **479** (7371): 80–83. Bibcode:2011Natur.479...80K. doi:10.1038/nature10542. ISSN 0028-0836. PMID 22031328.

[234] Clemence, Lara; Cohen, Jarrett (7 February 2005). "Space Sugar's a Sweet Find". *Goddard Space Flight Center*. Greenbelt, MD: NASA. Retrieved 2015-06-23.

[235] Than, Ker (30 August 2012). "Sugar Found In Space: A Sign of Life?". *National Geographic News* (Washington, D.C.: National Geographic Society). Retrieved 2015-06-23.

[236] "Sweet! Astronomers spot sugar molecule near star". *Excite* (Yonkers, NY: Mindspark Interactive Network). Associated Press. 29 August 2012. Retrieved 2015-06-23.

[237] "Building blocks of life found around young star". *News & Events*. Leiden, the Netherlands: Leiden University. 30 September 2012. Retrieved 2013-12-11.

[238] Jørgensen, Jes K.; Favre, Cécile; Bisschop, Suzanne E.; et al. (20 September 2012). "Detection of the simplest sugar, glycolaldehyde, in a solar-type protostar with ALMA" (PDF). *The Astrophysical Journal Letters* (Bristol, England: IOP Publishing for the American Astronomical Society) **757** (1): L4. arXiv:1208.5498. Bibcode:2012ApJ...757L...4J. doi:10.1088/2041-8205/757/1/L4. ISSN 2041-8213. L4. Retrieved 2015-06-23.

[239] "'Life chemical' detected in comet". *BBC News* (London: BBC). 18 August 2009. Retrieved 2015-06-23.

[240] Thompson, William Reid; Murray, B. G.; Khare, Bishun Narain; Sagan, Carl (30 December 1987). "Coloration and darkening of methane clathrate and other ices by charged particle irradiation: Applications to the outer solar system". *Journal of Geophysical Research* (Washington, D.C.: American Geophysical Union) **92** (A13): 14933–14947. Bibcode:1987JGR....9214933T. doi:10.1029/JA092iA13p14933. ISSN 0148-0227. PMID 11542127.

[241] Stark, Anne M. (5 June 2013). "Life on Earth shockingly comes from out of this world". Livermore, CA: Lawrence Livermore National Laboratory. Retrieved 2015-06-23.

[242] Goldman, Nir; Tamblyn, Isaac (20 June 2013). "Prebiotic Chemistry within a Simple Impacting Icy Mixture". *Journal of Physical Chemistry A* (Washington, D.C.: American Chemical Society) **117** (24): 5124–5131. doi:10.1021/jp402976n. ISSN 1089-5639. PMID 23639050.

[243] Carey, Bjorn (18 October 2005). "Life's Building Blocks 'Abundant in Space'". *Space.com* (Watsonville, CA: Imaginova). Retrieved 2015-06-23.

[244] Hudgins, Douglas M.; Bauschlicher, Charles W., Jr.; Allamandola, Louis J. (10 October 2005). "Variations in the Peak Position of the 6.2 μm Interstellar Emission Feature: A Tracer of N in the Interstellar Polycyclic Aromatic Hydrocarbon Population" (PDF). *The Astrophysical Journal* (Bristol, England: IOP Publishing for the American Astronomical Society) **632** (1): 316–332. Bibcode:2005ApJ...632..316H. doi:10.1086/432495. ISSN 0004-637X.

[245] Des Marais, David J.; Allamandola, Louis J.; Sandford, Scott; et al. (2009). "Cosmic Distribution of Chemical Complexity". *Ames Research Center*. Mountain View, CA: NASA. Retrieved 2015-06-24. See the Ames Research Center 2009 annual team report to the NASA Astrobiology Institute here .

[246] García-Hernández, Domingo. A.; Manchado, Arturo; García-Lario, Pedro; et al. (20 November 2010). "Formation of Fullerenes in H-Containing Planetary Nebulae". *The Astrophysical Journal Letters* (Bristol, England: IOP Publishing for the American Astronomical Society) **724** (1): L39–L43. arXiv:1009.4357. Bibcode:2010ApJ...724L..39G. doi:10.1088/2041-8205/724/1/L39. ISSN 2041-8213.

[247] Atkinson, Nancy (27 October 2010). "Buckyballs Could Be Plentiful in the Universe". *Universe Today* (Courtenay, British Columbia: Fraser Cain). Retrieved 2015-06-24.

[248] Marlaire, Ruth, ed. (3 March 2015). "NASA Ames Reproduces the Building Blocks of Life in Laboratory". *Ames Research Center*. Moffett Field, CA: NASA. Retrieved 2015-03-05.

[249] Lancet, Doron (30 December 2014). "Systems Prebiology-Studies of the origin of Life". *The Lancet Lab*. Rehovot, Israel: Department of Molecular Genetics; Weizmann Institute of Science. Retrieved 2015-06-26.

[250] Segré, Daniel; Ben-Eli, Dafna; Deamer, David W.; Lancet, Doron (February 2001). "The Lipid World" (PDF). *Origins of Life and Evolution of the Biosphere* (Kluwer Academic Publishers) 31 (1–2): 119–145. doi:10.1023/A:1006746807104. ISSN 0169-6149. PMID 11296516. Retrieved 2008-09-11.

[251] Eigen, Manfred; Schuster, Peter (November 1977). "The Hypercycle. A Principle of Natural Self-Organization. Part A: Emergence of the Hypercycle" (PDF). *Naturwissenschaften* (Berlin: Springer-Verlag) 64 (11): 541–565. Bibcode:1977NW.....64..541E. doi:10.1007/bf00450633. ISSN 0028-1042. PMID 593400. Retrieved 2015-06-13.

- Eigen, Manfred; Schuster, Peter (1978). "The Hypercycle. A Principle of Natural Self-Organization. Part B: The Abstract Hypercycle" (PDF). *Naturwissenschaften* (Berlin: Springer-Verlag) 65: 7–41. Bibcode:1978NW.....65....7E. doi:10.1007/bf00420631. ISSN 0028-1042. Retrieved 2015-06-13.

- Eigen, Manfred; Schuster, Peter (July 1978). "The Hypercycle. A Principle of Natural Self-Organization. Part C: The Realistic Hypercycle" (PDF). *Naturwissenschaften* (Berlin: Springer-Verlag) 65 (7): 341–369. Bibcode:1978NW.....65..341E. doi:10.1007/bf00439699. ISSN 0028-1042. Retrieved 2015-06-13.

[252] Markovitch, Omer; Lancet, Doron (Summer 2012). "Excess Mutual Catalysis Is Required for Effective Evolvability" (PDF). *Artificial Life* (Cambridge, MA: MIT Press) 18 (3): 243–266. doi:10.1162/artl_a_00064. ISSN 1064-5462. PMID 22662913. Retrieved 2015-06-26.

[253] Tessera, Marc (2011). "Origin of Evolution *versus* Origin of Life: A Shift of Paradigm". *International Journal of Molecular Sciences* (Basel, Switzerland: MDPI) 12 (6): 3445–3458. doi:10.3390/ijms12063445. ISSN 1422-0067. PMC 3131571. PMID 21747687. Special Issue: "Origin of Life 2011"

[254] Brown, Michael R. W.; Kornberg, Arthur (16 November 2004). "Inorganic polyphosphate in the origin and survival of species". *Proc. Natl. Acad. Sci. U.S.A.* (Washington, D.C.: National Academy of Sciences) 101 (46): 16085–16087. Bibcode:2004PNAS..10116085B. doi:10.1073/pnas.0406909101. ISSN 0027-8424. PMC 528972. PMID 15520374.

[255] Clark, David P. (3 August 1999). "The Origin of Life". *Microbiology 425: Biochemistry and Physiology of Microorganism* (Lecture). Carbondale, IL: College of Science; Southern Illinois University Carbondale. Archived from the original on 2000-10-02. Retrieved 2015-06-26.

[256] Pasek, Matthew A. (22 January 2008). "Rethinking early Earth phosphorus geochemistry". *Proc. Natl. Acad. Sci. U.S.A.* (Washington, D.C.: National Academy of Sciences) 105 (3): 853–858. Bibcode:2008PNAS..105..853P. doi:10.1073/pnas.0708205105. ISSN 0027-8424. PMC 2242691. PMID 18195373.

[257] Witt, Adolf N.; Vijh, Uma P.; Gordon, Karl D. (2003). "Discovery of Blue Fluorescence by Polycyclic Aromatic Hydrocarbon Molecules in the Red Rectangle". *Bulletin of the American Astronomical Society* (Washington, D.C.: American Astronomical Society) 35: 1381. Bibcode:2003AAS...20311017W. Archived from the original on 2003-12-19. Retrieved 2015-06-26. American Astronomical Society Meeting 203, #110.17, January 2004.

[258] "NASA Cooks Up Icy Organics to Mimic Life's Origins". *Space.com*. Ogden, UT: Purch. 20 September 2012. Retrieved 2015-06-26.

[259] Gudipati, Murthy S.; Rui Yang (1 September 2012). "In-situ Probing of Radiation-induced Processing of Organics in Astrophysical Ice Analogs—Novel Laser Desorption Laser Ionization Time-of-flight Mass Spectroscopic Studies". *The Astrophysical Journal Letters* (Bristol, England: IOP Publishing for the American Astronomical Society) 756 (1): L24. Bibcode:2012ApJ...756L..24G. doi:10.1088/2041-8205/756/1/L24. ISSN 2041-8213. L24.

[260] "NASA Ames PAH IR Spectroscopic Database". NASA. Retrieved 2015-06-17.

[261] Dartnell, Lewis (12 January 2008). "Did life begin on a radioactive beach?". *New Scientist* (London: Reed Business Information) (2638): 8. ISSN 0262-4079. Retrieved 2015-06-26.

[262] Adam, Zachary (2007). "Actinides and Life's Origins". *Astrobiology* (New Rochelle, NY: Mary Ann Liebert, Inc.) 7 (6): 852–872. Bibcode:2007AsBio...7..852A. doi:10.1089/ast.2006.0066. ISSN 1531-1074. PMID 18163867.

[263] Parnell, John (December 2004). "Mineral Radioactivity in Sands as a Mechanism for Fixation of Organic Carbon on the Early Earth". *Origins of Life and Evolution of Biospheres* (Kluwer Academic Publishers) 34 (6): 533–547. Bibcode:2004OLEB...34..533P. doi:10.1023/B:ORIG.0000043132.23966.a1. ISSN 0169-6149. PMID 15570707.

[264] Michaelian, Karo (30 June 2009). "Thermodynamic Function of Life". arXiv:0907.0040 [physics.gen-ph].

[265] Michaelian, Karo (25 January 2011). "Biological catalysis of the hydrological cycle: life's thermodynamic function". *Hydrology and Earth System Sciences Discussions* (Göttingen, Germany: Copernicus Publications on behalf of the European Geosciences Union) 8: 1093–1123. Bibcode:2011HESSD...8.1093M. doi:10.5194/hessd-8-1093-2011. ISSN 1812-2116.

[266] Michaelian, Karo (11 March 2011). "Thermodynamic Dissipation Theory for the Origin of Life" (PDF). *Earth System Dynamics* (Göttingen, Germany: Copernicus Publications on behalf of the European Geosciences Union) 2: 37–51. arXiv:0907.0042. Bibcode:2011ESD.....2...37M. doi:10.5194/esd-2-37-2011. ISSN 2190-4987. Retrieved 2015-06-28.

[267] Cnossen, Ingrid; Sanz-Forcada, Jorge; Favata, Fabio; et al. (February 2007). "Habitat of early life: Solar X-ray and UV radiation at Earth's surface 4–3.5 billion years ago". *Journal of Geophysical Research* (Washington, D.C.: American Geophysical Union) 112 (E2): E02008. arXiv:astro-ph/0702529. Bibcode:2007JGRE..112.2008C. doi:10.1029/2006JE002784. ISSN 0148-0227.

[268] Sagan, Carl (April 1973). "Ultraviolet Selection Pressure on the Earliest Organisms". *Journal of Theoretical Biology* (Amsterdam, the Netherlands: Elsevier) 39 (1): 195–200. doi:10.1016/0022-5193(73)90216-6. ISSN 0022-5193. PMID 4741712.

[269] Michaelian, Karo; Simeonov, Aleksander (19 August 2015). "Fundamental molecules of life are pigments which arose and co-evolved as a response to the thermodynamic imperative of dissipating the prevailing solar spectrum". *Biogeosciences* 12: 4913–4937. doi:10.5194/bg-12-4913-2015.

[270] Michaelian, Karo; Simeonov, Aleksandar (16 May 2014). "Fundamental Molecules of Life are Pigments which Arose and Evolved to Dissipate the Solar Spectrum". arXiv:1405.4059 [physics.bio-ph].

[271] Michaelian, Karo (2013). "A non-linear irreversible thermodynamic perspective on organic pigment proliferation and biological evolution" (PDF). *Journal of Physics: Conference Series* (Bristol, England: IOP Publishing) 475 (conference 1): 012010. arXiv:1307.5924. Bibcode:2013JPhCS.475a2010M. doi:10.1088/1742-6596/475/1/012010. ISSN 1742-6596. "4th National Meeting in Chaos, Complex System and Time Series 29 November to 2 December 2011, Xalapa, Veracruz, Mexico"

[272] Knauth 1992, pp. 123–152

[273] Knauth, L. Paul; Lowe, Donald R. (May 2003). "High Archean climatic temperature inferred from oxygen isotope geochemistry of cherts in the 3.5 Ga Swaziland group, South Africa". *Geological Society of America Bulletin* (Boulder, CO: Geological Society of America) 115: 566–580. Bibcode:2003GSAB..115..566K. doi:10.1130/0016-7606(2003)115<0566:hactif>2.0.co;2. ISSN 0016-7606.

[274] Lowe, Donald R.; Tice, Michael M. (June 2004). "Geologic evidence for Archean atmospheric and climatic evolution: Fluctuating levels of CO_2, CH_4, and O_2 with an overriding tectonic control". *Geology* (Boulder, CO: Geological Society of America) 32 (6): 493–496. Bibcode:2004Geo....32..493L. doi:10.1130/G20342.1. ISSN 0091-7613.

[275] Michaelian, Karo; Santillán Padilla, Norberto (24 November 2014). "DNA Denaturing through UV-C Photon Dissipation: A Possible Route to Archean Non-enzymatic Replication" (PDF). *bioRxiv* (Cold Spring Harbor, NY: Cold Spring Harbor Laboratory). doi:10.1101/009126. Retrieved 2015-06-29.

[276] "Massachusetts physicist claims he solved mystery of how life emerged from matter". *RT*. Jan 23, 2014. Retrieved Dec 11, 2014.

[277] Wolchover, Natalie (Jan 28, 2014). "A New Physics Theory of Life". *Scientific American*. Retrieved Dec 11, 2014.

[278] Perunov, Nikolai; Marsland, Robert; England, Jeremy (2014-12-04). "Statistical Physics of Adaptation". arXiv:1412.1875 [physics.bio-ph].

[279] Davies, Paul (December 2007). "Are Aliens Among Us?" (PDF). *Scientific American* (Stuttgart: Georg von Holtzbrinck Publishing Group) 297 (6): 62–69. doi:10.1038/scientificamerican1207-62. ISSN 0036-8733. Retrieved 2015-07-16. ...if life does emerge readily under terrestrial conditions, then perhaps it formed many times on our home planet. To pursue this possibility, deserts, lakes and other extreme or isolated environments have been searched for evidence of "alien" life-forms—organisms that would differ fundamentally from known organisms because they arose independently.

[280] Hartman, Hyman (October 1998). "Photosynthesis and the Origin of Life". *Origins of Life and Evolution of Biospheres* (Kluwer Academic Publishers) 28 (4–6): 515–521. Bibcode:1998OLEB...28..515H. doi:10.1023/A:1006548904157. ISSN 0169-6149. PMID 11536891.

[281] Damer, Bruce; Deamer, David (13 March 2015). "Coupled Phases and Combinatorial Selection in Fluctuating Hydrothermal Pools: A Scenario to Guide Experimental Approaches to the Origin of Cellular Life". *Life* (Basel, Switzerland: MDPI) 5 (1): 872–887. doi:10.3390/life5010872. ISSN 2075-1729. PMC 4390883. PMID 25780958.

39.11 Bibliography

- Altermann, Wladyslaw (2009). "From Fossils to Astrobiology – A Roadmap to Fata Morgana?" (PDF). In Seckbach, Joseph; Walsh, Maud. *From Fossils to Astrobiology: Records of Life on Earth and the*

- *Search for Extraterrestrial Biosignatures*. Cellular Origin, Life in Extreme Habitats and Astrobiology **12**. Dordrecht, the Netherlands; London: Springer Science+Business Media. ISBN 978-1-4020-8836-0. LCCN 2008933212. Retrieved 2015-06-05.

- Bada, Jeffrey L.; Lazcano, Antonio (2009). "The Origin of Life". In Ruse, Michael; Travis, Joseph. *Evolution: The First Four Billion Years*. Foreword by Edward O. Wilson. Cambridge, MA: Belknap Press of Harvard University Press. ISBN 978-0-674-03175-3. LCCN 2008030270. OCLC 225874308.

- Barton, Nicholas H.; Briggs, Derek E. G.; Eisen, Jonathan A.; et al. (2007). *Evolution*. Cold Spring Harbor, NY: Cold Spring Harbor Laboratory Press. ISBN 978-0-87969-684-9. LCCN 2007010767. OCLC 86090399.

- Bastian, H. Charlton (1871). *The Modes of Origin of Lowest Organisms*. London; New York: Macmillan and Company. LCCN 11004276. OCLC 42959303. Retrieved 2015-06-06.

- Bernal, J. D. (1951). *The Physical Basis of Life*. London: Routledge & Kegan Paul. LCCN 51005794.

- Bernal, J. D. (1960). "The Problem of Stages in Biopoesis". In Florkin, M. *Aspects of the Origin of Life*. International Series of Monographs on Pure and Applied Biology. Oxford, UK; New York: Pergamon Press. ISBN 978-1-4831-3587-8. LCCN 60013823.

- Bernal, J. D. (1967) [Reprinted work by A. I. Oparin originally published 1924; Moscow: The Moscow Worker]. *The Origin of Life*. The Weidenfeld and Nicolson Natural History. Translation of Oparin by Ann Synge. London: Weidenfeld & Nicolson. LCCN 67098482.

- Bock, Gregory R.; Goode, Jamie A., eds. (1996). *Evolution of Hydrothermal Ecosystems on Earth (and Mars?)*. Ciba Foundation Symposium **202**. Chichester, UK; New York: John Wiley & Sons. ISBN 0-471-96509-X. LCCN 96031351.

- Bondeson, Jan (1999). *The Feejee Mermaid and Other Essays in Natural and Unnatural History*. Ithaca, NY: Cornell University Press. ISBN 0-8014-3609-5. LCCN 98038295.

- Bryson, Bill (2004). *A Short History of Nearly Everything*. London: Black Swan. ISBN 978-0-552-99704-1. OCLC 55589795.

- Calvin, Melvin (1969). *Chemical Evolution: Molecular Evolution Towards the Origin of Living Systems on the Earth and Elsewhere*. Oxford, UK: Clarendon Press. ISBN 0-19-855342-0. LCCN 70415289. OCLC 25220.

- Chaichian, Masud; Rojas, Hugo Perez; Tureanu, Anca (2014). "Physics and Life". *Basic Concepts in Physics: From the Cosmos to Quarks*. Undergraduate Lecture Notes in Physics. Berlin; Heidelberg: Springer Berlin Heidelberg. doi:10.1007/978-3-642-19598-3_12. ISBN 978-3-642-19597-6. ISSN 2192-4791. LCCN 2013950482. OCLC 900189038.

- Chang, Thomas Ming Swi (2007). *Artificial Cells: Biotechnology, Nanomedicine, Regenerative Medicine, Blood Substitutes, Bioencapsulation, and Cell/Stem Cell Therapy*. Regenerative Medicine, Artificial Cells and Nanomedicine **1**. Hackensack, NJ: World Scientific. ISBN 978-981-270-576-1. LCCN 2007013738. OCLC 173522612.

- Clancy, Paul; Brack, André; Horneck, Gerda (2005). *Looking for Life, Searching the Solar System*. Cambridge, UK: Cambridge University Press. ISBN 978-0-521-82450-7. LCCN 2006271630. OCLC 57574490.

- Dalrymple, G. Brent (2001). "The age of the Earth in the twentieth century: a problem (mostly) solved". In Lewis, C. L. E.; Knell, S. J. *The Age of the Earth: from 4004 BC to AD 2002*. Geological Society Special Publication **190**. London: Geological Society of London. Bibcode:2001GSLSP.190..205D. doi:10.1144/gsl.sp.2001.190.01.14. ISBN 1-86239-093-2. ISSN 0305-8719. LCCN 2003464816. OCLC 48570033.

- Darwin, Charles (1887). Darwin, Francis, ed. *The Life and Letters of Charles Darwin, Including an Autobiographical Chapter* **3** (3rd ed.). London: John Murray. OCLC 834491774.

- Davies, Geoffrey F. (2007). "Chapter 2.3 Dynamics of the Hadean and Archaean Mantle". In van Kranendonk, Martin J.; Smithies, R. Hugh; Bennett, Vickie C. *Earth's Oldest Rocks*. Developments in Precambrian Geology **15**. Amsterdam, the Netherlands; Boston: Elsevier. doi:10.1016/S0166-2635(07)15023-4. ISBN 978-0-444-52810-0. LCCN 2009525003.

- Davies, Paul (1999). *The Fifth Miracle: The Search for the Origin of Life*. London: Penguin Books. ISBN 0-14-028226-2.

- Dawkins, Richard (1996). *The Blind Watchmaker* (Reissue with a new introduction ed.). New York: W. W. Norton & Company. ISBN 0-393-31570-3. LCCN 96229669. OCLC 35648431.

- Dawkins, Richard (2004). *The Ancestor's Tale: A Pilgrimage to the Dawn of Evolution*. Boston, MA: Houghton Mifflin. ISBN 0-618-00583-8. LCCN 2004059864. OCLC 56617123.

- Dobell, Clifford (1960) [Originally published 1932; New York: Harcourt, Brace & Company]. *Antony van Leeuwenhoek and His 'Little Animals'*. New York: Dover Publications. LCCN 60002548.

- Dyson, Freeman (1999). *Origins of Life* (Revised ed.). Cambridge, UK; New York: Cambridge University Press. ISBN 0-521-62668-4. LCCN 99021079.

- Eigen, M.; Schuster, P. (1979). *The Hypercycle: A Principle of Natural Self-Organization*. Berlin; New York: Springer-Verlag. ISBN 0-387-09293-5. LCCN 79001315. OCLC 4665354.

- Fesenkov, V. G. (1959). "Some Considerations about the Primaeval State of the Earth". In Oparin, A. I.; et al. *The Origin of Life on the Earth*. I.U.B. Symposium Series 1. Edited for the International Union of Biochemistry by Frank Clark and R. L. M. Synge (English-French-German ed.). London; New York: Pergamon Press. ISBN 978-1-4832-2240-0. LCCN 59012060. Retrieved 2015-06-03. International Symposium on the Origin of Life on the Earth (held at Moscow, 19–24 August 1957)

- Hazen, Robert M. (2005). *Genesis: The Scientific Quest for Life's Origin*. Washington, D.C.: Joseph Henry Press. ISBN 0-309-09432-1. LCCN 2005012839. OCLC 60321860.

- Huxley, Thomas Henry (1968) [Originally published 1897]. "VIII Biogenesis and Abiogenesis [1870]". *Discourses, Biological and Geological*. Collected Essays VIII (Reprint ed.). New York: Greenwood Press. LCCN 70029958. Retrieved 2014-05-19.

- Kauffman, Stuart (1993). *The Origins of Order: Self-Organization and Selection in Evolution*. New York: Oxford University Press. ISBN 978-0-19-507951-7. LCCN 91011148. OCLC 23253930.

- Kauffman, Stuart (1995). *At Home in the Universe: The Search for Laws of Self-Organization and Complexity*. New York: Oxford University Press. ISBN 0-19-509599-5. LCCN 94025268.

- Klyce, Brig (22 January 2001). Kingsley, Stuart A.; Bhathal, Ragbir, eds. *Panspermia Asks New Questions*. The Search for Extraterrestrial Intelligence (SETI) in the Optical Spectrum III. Bellingham, WA: SPIE. doi:10.1117/12.435366. ISBN 0-8194-3951-7. LCCN 2001279159. Retrieved 2015-06-09. Proceedings of the SPIE held at San Jose, CA, 22–24 January 2001

- Knauth, L. Paul (1992). "Origin and diagenesis of cherts: An isotopic perspective". In Clauer, Norbert; Chaudhuri, Sambhu. *Isotopic Signatures and Sedimentary Records*. Lecture Notes in Earth Sciences 43. Berlin; New York: Springer-Verlag. doi:10.1007/BFb0009863. ISBN 3-540-55828-4. ISSN 0930-0317. LCCN 92025372. OCLC 26262469.

- Lane, Nick (2009). *Life Ascending: The 10 Great Inventions of Evolution* (1st American ed.). New York: W. W. Norton & Company. ISBN 978-0-393-06596-1. LCCN 2009005046. OCLC 286488326.

- Lankenau, Dirk-Henner (2011). "Two RNA Worlds: Toward the Origin of Replication, Genes, Recombination and Repair". In Egel, Richard; Lankenau, Dirk-Henner; Mulkdjanian,, Armen Y. *Origins of Life: The Primal Self-Organization*. Heidelberg: Springer. doi:10.1007/978-3-642-21625-1. ISBN 978-3-642-21624-4. LCCN 2011935879. OCLC 733245537.

- Lennox, James G. (2001). *Aristotle's Philosophy of Biology: Studies in the Origins of Life Science*. Cambridge Studies in Philosophy and Biology. Cambridge, UK; New York: Cambridge University Press. ISBN 0-521-65976-0. LCCN 00026070.

- McKinney, Michael L. (1997). "How do rare species avoid extinction? A paleontological view". In Kunin, William E.; Gaston, Kevin J. *The Biology of Rarity: Causes and consequences of rare—common differences* (1st ed.). London; New York: Chapman & Hall. ISBN 0-412-63380-9. LCCN 96071014. OCLC 36442106.

- Michod, Richard E. (1999). "Darwinian Dynamics: Evolutionary Transitions in Fitness and Individuality". Princeton, NJ: Princeton University Press. ISBN 0-691-02699-8. LCCN 98004166. OCLC 38948118.

- Miller, G. Tyler; Spoolman, Scott E. (2012). *Environmental Science* (14th ed.). Belmont, CA: Brooks/Cole. ISBN 978-1-111-98893-7. LCCN 2011934330. OCLC 741539226.

- Oparin, A. I. (1953) [Originally published 1938; New York: The Macmillan Company]. *The Origin of Life*. Translation and new introduction by Sergius Morgulis (2nd ed.). Mineola, NY: Dover Publications. ISBN 0-486-49522-1. LCCN 53010161.

- Orgel, Leslie E. (1987). "Evolution of the Genetic Apparatus: A Review". *Evolution of Catalytic Function*. Cold Spring Harbor Symposia on Quantitative Biology **52**. Cold Spring Harbor, NY: Cold Spring Harbor Laboratory Press. doi:10.1101/SQB.1987.052.01.004. ISBN 0-87969-054-2. OCLC 19850881. "Proceedings of a symposium held at Cold Spring Harbor Laboratory in 1987"

- Raven, Peter H.; Johnson, George B. (2002). *Biology* (6th ed.). Boston, MA: McGraw-Hill. ISBN 0-07-112261-3. LCCN 2001030052. OCLC 45806501.

- Ross, Alexander (1652). *Arcana Microcosmi*. Book II. London. Retrieved 2015-07-07.

- Shapiro, Robert (1987). *Origins: A Skeptic's Guide to the Creation of Life on Earth*. Toronto; New York: Bantam Books. ISBN 0-553-34355-6.

- Sheldon, Robert B. (22 September 2005). Hoover, Richard B.; Levin, Gilbert V.; Rozanov, Alexei Y.; Gladstone, G. Randall, eds. *Historical Development of the Distinction between Bio- and Abiogenesis* (PDF). Astrobiology and Planetary Missions. Bellingham, WA: SPIE. doi:10.1117/12.663480. ISBN 978-0-8194-5911-4. LCCN 2005284378. Retrieved 2015-04-13. Proceedings of the SPIE held at San Diego, CA, 31 July–2 August 2005

- Stearns, Beverly Peterson; Stearns, Stephen C. (1999). *Watching, from the Edge of Extinction*. New Haven, CT: Yale University Press. ISBN 0-300-07606-1. LCCN 98034087. OCLC 47011675.

- Tyndall, John (1905) [Originally published 1871; London; New York: Longmans, Green & Co.; D. Appleton and Company]. *Fragments of Science* **2** (6th ed.). New York: P.F. Collier & Sons. OCLC 726998155. Retrieved 2015-06-06.

- Vartanian, Aram (1973). "Spontaneous Generation". In Wiener, Philip P. *Dictionary of the History of Ideas* IV. New York: Charles Scribner's Sons. ISBN 0-684-13293-1. LCCN 72007943. Retrieved 2015-06-05.

- Voet, Donald; Voet, Judith G. (2004). *Biochemistry* **1** (3rd ed.). New York: John Wiley & Sons. ISBN 0-471-19350-X. LCCN 2003269978.

- Woodward, Robert J., ed. (1969). *Our Amazing World of Nature: Its Marvels & Mysteries*. Pleasantville, NY: Reader's Digest Association. ISBN 0-340-13000-8. LCCN 69010418.

- Yarus, Michael (2010). *Life from an RNA World: The Ancestor Within*. Cambridge, MA: Harvard University Press. ISBN 978-0-674-05075-4. LCCN 2009044011.

39.12 Further reading

- Arrhenius, Gustaf O.; Sales, Brian C.; Mojzsis, Stephen J.; et al. (21 August 1997). "Entropy and Charge in Molecular Evolution—the Case of Phosphate" (PDF). *Journal of Theoretical Biology* (Amsterdam, the Netherlands: Elsevier) **187** (4): 503–522. doi:10.1006/jtbi.1996.0385. ISSN 0022-5193. PMID 9299295.

- Cavalier-Smith, Thomas (June 2006). "Cell evolution and Earth history: stasis and revolution". *Philosophical Transactions of the Royal Society B* (London: Royal Society) **361** (1470): 969–1006. doi:10.1098/rstb.2006.1842. ISSN 0962-8436. PMC 1578732. PMID 16754610.

- de Duve, Christian (1995). *Vital Dust: Life As A Cosmic Imperative* (1st ed.). New York: Basic Books. ISBN 0-465-09044-3. LCCN 94012964. OCLC 30624716.

- Fernando, Chrisantha T.; Rowe, Jonathan (7 July 2007). "Natural selection in chemical evolution". *Journal of Theoretical Biology* (Amsterdam, the Netherlands) **247** (1): 152–167. doi:10.1016/j.jtbi.2007.01.028. ISSN 0022-5193. PMID 17399743.

- Gribbin, John (1998). *The Case of the Missing Neutrinos: And other Curious Phenomena of the Universe* (1st Fromm International ed.). New York: Fromm International. ISBN 0-88064-199-1. LCCN 98027948. OCLC 39368356.

- Harris, Henry (2002). *Things Come to Life: Spontaneous Generation Revisited*. Oxford, UK; New York: Oxford University Press. ISBN 0-19-851538-3. LCCN 2001054856. OCLC 48100507.

- Horgan, John (February 1991). "In the Beginning...". *Scientific American* (Stuttgart: Georg von Holtzbrinck Publishing Group) **264** (2): 116–125. doi:10.1038/scientificamerican0291-116. ISSN 0036-8733.

- Ignatov, Ignat; Mosin, Oleg V. (2013). "Modeling of Possible Processes for Origin of Life and Living Matter in Hot Mineral and Seawater with Deuterium". *Journal of Environment and Earth Science* (New York: International Institute for Science, Technology and Education) **3** (14): 103–118. ISSN 2224-3216. Retrieved 2015-06-29.

- Jortner, Joshua (October 2006). "Conditions for the emergence of life on the early Earth: summary and reflections". *Philosophical Transactions of the Royal*

- *Society B* (London: Royal Society) **361** (1474): 1877–1891. doi:10.1098/rstb.2006.1909. ISSN 0962-8436. PMC 1664691. PMID 17008225.

- Klotz, Irene (24 February 2012). "Did Life Start in a Pond, Not Oceans?". *Discovery News* (Silver Spring, MD: Discovery Communications). Retrieved 2015-06-29.

- Knoll, Andrew H. (2003). *Life on a Young Planet: The First Three Billion Years of Evolution on Earth*. Princeton, NJ: Princeton University Press. ISBN 0-691-00978-3. LCCN 2002035484. OCLC 50604948.

- Luisi, Pier Luigi (2006). *The Emergence of Life: From Chemical Origins to Synthetic Biology*. Cambridge, UK: Cambridge University Press. ISBN 978-0-521-82117-9. LCCN 2006285720. OCLC 173609999.

- Maynard Smith, John; Szathmáry, Eörs (1999). *The Origins of Life: From the Birth of Life to the Origin of Language*. Oxford, UK; New York: Oxford University Press. ISBN 0-19-850493-4. LCCN 99230990. OCLC 40980149.

- Morowitz, Harold J. (1992). *Beginnings of Cellular Life: Metabolism Recapitulates Biogenesis*. New Haven, CT: Yale University Press. ISBN 0-300-05483-1. LCCN 92006849. OCLC 25316379.

- NASA Astrobiology Institute: Harrison, T. Mark; McKeegan, Kevin D.; Mojzsis, Stephen J. "Earth's Early Environment and Life: When did Earth become suitable for habitation?". Archived from the original on 2012-02-17. Retrieved 2015-06-30.

- NASA Specialized Center of Research and Training in Exobiology: Arrhenius, Gustaf O. (11 September 2002). "Arrhenius". Archived from the original on 2007-12-21. Retrieved 2015-06-30.

- "The physico-chemical basis of life". *What is Life*. Spring Valley, CA: Lukas K. Buehler. Retrieved 27 October 2005.

- Pitsch, Stefan; Krishnamurthy, Ramanarayanan; Arrhenius, Gustaf O. (6 September 2000). "Concentration of Simple Aldehydes by Sulfite-Containing Double-Layer Hydroxide Minerals: Implications for Biopoesis". *Helvetica Chimica Acta* (Hoboken, NJ: John Wiley & Sons) **83** (9): 2398–2411. doi:10.1002/1522-2675(20000906)83:9<2398::AID-HLCA2398>3.0.CO;2-5. ISSN 0018-019X. PMID 11543578.

- Pons, Marie-Laure; Quitté, Ghylaine; Fujii, Toshiyuki; et al. (25 October 2011). "Early Archean Serpentine Mud Volcanoes at Isua, Greenland, as a Niche for Early Life". *Proc. Natl. Acad. Sci. U.S.A.* (Washington, D.C.: National Academy of Sciences) **108** (43): 17639–17643. Bibcode:2011PNAS..10817639P. doi:10.1073/pnas.1108061108. ISSN 0027-8424. PMC 3203773. PMID 22006301.

- Pross, Addy (2012). *What is Life?: How Chemistry Becomes Biology* (1st ed.). Oxford, UK: Oxford University Press. ISBN 978-0-19-964101-7. LCCN 2012538842. OCLC 812020290.

- Roy, Debjani; Schleyer, Paul von Ragué (2010). "Chemical Origin of Life: How do Five HCN Molecules Combine to form Adenine under Prebiotic and Interstellar Conditions". In Matta, Chérif F. *Quantum Biochemistry*. Weinheim, Germany: Wiley-VCH. doi:10.1002/9783527629213.ch6. ISBN 978-3-527-62921-3. LCCN 2011499476. OCLC 905973537.

- Russell, Michael J.; Hall, A. J.; Cairns-Smith, Alexander Graham; et al. (10 November 1988). "Submarine hot springs and the origin of life". *Nature* (London: Nature Publishing Group) **336** (6195): 117. Bibcode:1988Natur.336..117R. doi:10.1038/336117a0. ISSN 0028-0836. PMID 11536607.

- Shock, Everett L. (25 October 1997). "High-temperature life without photosynthesis as a model for Mars" (PDF). *Journal of Geophysical Research* (Washington, D.C.: American Geophysical Union) **102** (E10): 23687–23694. Bibcode:1997JGR...10223687S. doi:10.1029/97je01087. ISSN 0148-0227.

39.13 External links

- "Exploring Life's Origins: A Virtual Exhibit". *Exploring Life's Origins: A Virtual Exhibit*. Arlington County, VA: National Science Foundation. Retrieved 2015-07-02.

- Fields, Helen (October 2010). "The Origins of Life". *Smithsonian* (Washington, D.C.: Smithsonian Institution). ISSN 0037-7333. Retrieved 2015-07-02.

- Fox, Douglas (28 March 2007). "Primordial Soup's On: Scientists Repeat Evolution's Most Famous Experiment". *Scientific American* (Stuttgart: Georg von Holtzbrinck Publishing Group). ISSN 0036-8733. Retrieved 2015-07-02.

- "The Geochemical Origins of Life by Michael J. Russell & Allan J. Hall". Glasgow, Scotland: University of Glasgow. 13 December 2008. Retrieved 2015-07-02.
- Kauffman, Stuart (8 August 1996). "Even peptides do it". *Nature* (London: Nature Publishing Group) **382** (6591): 496–497. Bibcode:1996Natur.382..496K. doi:10.1038/382496a0. ISSN 0028-0836. PMID 8700218. Archived from the original on 2006-10-15. Retrieved 2015-07-02.
- Malory, Marcia. "How life began on Earth". *Earth Facts*. Retrieved 2015-07-02.
- Nowak, Martin A.; Ohtsuki, Hisashi (30 September 2008). "Prevolutionary dynamics and the origin of evolution" (PDF). *Proc. Natl. Acad. Sci. U.S.A.* (Washington, D.C.: National Academy of Sciences) **105** (39): 14924–14927. Bibcode:2008PNAS..10514924N. doi:10.1073/pnas.0806714105. ISSN 0027-8424. PMC 2567469. PMID 18791073.
- "Possible Connections Between Interstellar Chemistry and the Origin of Life on the Earth". *Space Science and Astrobiology at Ames*. NASA. Archived from the original on 2009-07-31. Retrieved 2015-07-02.
- "Research Spotlight: Jack Szostak: Making Life from Scratch". *Origins of Life Initiative*. Cambridge, MA: Harvard University. Retrieved 2015-07-02.
- Schirber, Michael (9 June 2006). "How Life Began: New Research Suggests Simple Approach". *LiveScience* (Ogden, UT: Purch). Retrieved 2015-07-02.
- "Scientists Find Clues That Life Began in Deep Space". *NASA Astrobiology Institute*. Mountain View, CA: NASA. 30 January 2001. Archived from the original on 2013-04-29. Retrieved 2015-07-02.
- "Simple Artificial Cell Created From Scratch To Study Cell Complexity". *Science Daily* (Rockville, MD: ScienceDaily, LLC). 16 May 2008. Retrieved 2015-07-02. Post is reprinted from materials provided by Pennsylvania State University.
- Singer, Emily (19 July 2015). "Chemists Invent New Letters for Nature's Genetic Alphabet". *Wired*. New York: Condé Nast. Retrieved 2015-07-20.
- Swaminathan, Nikhil (10 June 2008). "Scientists Close to Reconstructing First Living Cell". *Scientific American* (News) (Stuttgart: Georg von Holtzbrinck Publishing Group). ISSN 0036-8733. Retrieved 2015-07-02.
- Vasas, Vera; Fernando, Chrisantha; Santos, Mauro; et al. (5 January 2012). "Evolution before genes" (PDF). *Biology Direct* (London: BioMed Central) **7**: 1. doi:10.1186/1745-6150-7-1. ISSN 1745-6150.
- Zlobin, Andrei E. (2013). "Tunguska similar impacts and origin of life". *Modern Scientific Researches and Innovations* (Moscow: International Centre of Science and Innovations Ltd.) (12). Retrieved 2015-07-02.
- Zlobin, Andrei E. (2014). "Symmetry infringement in mathematical metrics of hydrogen atom as illustration of ideas by V.I.Vernadsky concerning origin of life and biosphere" (PDF). *Acta Naturae* (Moscow: Park Media Ltd.) (Special Issue 1): 48. ISSN 2075-8251. Retrieved 2015-07-02.

39.13.1 Video resources

- Hazen, Robert M. (29 April 2014). *The Origins of Life* (Webcast). Baltimore, MD: Space Telescope Science Institute. Retrieved 2015-07-03. — A 2014 Spring Symposium webcast (video; 38 m)
- "The Origin of Life" on YouTube — A Royal Institution Discourse lecture given by John Maynard Smith in 1995 (video; 58 m)
- "Space Experts Discuss the Search for Life in the Universe at NASA" on YouTube — Panel discussion at NASA headquarters on 14 July 2014 (video; 87 m)

Chapter 40

Hypercycle (chemistry)

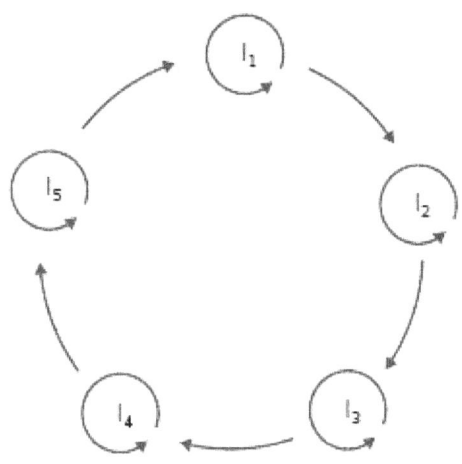

A hypercycle

In chemistry, a **hypercycle** is an abstract model of organization of self-replicating molecules connected in a cyclic, autocatalytic manner. It was introduced in an ordinary differential equation (ODE) form by the Nobel Prize winner Manfred Eigen in 1971[1] and subsequently further extended in collaboration with Peter Schuster.[2][3] It was proposed as a solution to the error threshold problem encountered during modelling of replicative molecules that hypothetically existed on the primordial Earth (see: abiogenesis). As such, it explained how life on Earth could have begun using only relatively short genetic sequences, which in theory were too short to store all essential information.[3] The hypercycle is a special case of the replicator equation.[4] The most important properties of hypercycles are autocatalytic growth competition between cycles, once-for-ever selective behaviour, utilization of small selective advantage, rapid evolvability, increased information capacity, and selection against parasitic branches.[1]

The hypercycle is a cycle of connected, self-replicating macromolecules. In the hypercycle, all molecules are linked such that each of them catalyses the creation of its successor, with the last molecule catalysing the first one. In such a manner, the cycle reinforces itself. Furthermore, each molecule is additionally a subject for self-replication. The resultant system is a new level of self-organization that incorporates both cooperation and selfishness. The coexistence of many genetically non-identical molecules makes it possible to maintain a high genetic diversity of the population. This can be a solution to the error threshold problem, which states that, in a system without ideal replication, an excess of mutation events would destroy the ability to carry information and prevent the creation of larger and fitter macromolecules. Moreover, it has been shown that hypercycles could originate naturally and that incorporating new molecules can extend them. Hypercycles are also subject to evolution and, as such, can undergo a selection process. As a result, not only does the system gain information, but its information content can be improved. From an evolutionary point of view, the hypercycle is an intermediate state of self-organization, but not the final solution.[1]

Over the years, the hypercycle theory has experienced many reformulations and methodological approaches. Among them, the most notable are applications of partial differential equations,[5] cellular automata,[6][7][8][9] and stochastic formulations of Eigen's problem.[10][11] Despite many advantages that the concept of hypercycles presents, there were also some problems regarding the traditional model formulation using ODEs: a vulnerability to parasites and a limited size of stable hypercycles.[6][7][8][9][10][12][13] In 2012, the first experimental proof for the emergence of a cooperative network among fragments of self-assembling ribozymes was published, demonstrating their advantages over self-replicating cycles.[14] However, even though this experiment proves the existence of cooperation among the recombinase ribozyme subnetworks, this cooperative network does not form a hypercycle per se, so we still lack the experimental demonstration of hypercycles.[15]

40.1 Model formulation

40.1.1 Model evolution

1971: Eigen introduces the hypercycle concept[1]

1977: Eigen and Schuster extend the hypercycle concept, propose a hypercycle theory and introduce the concept of quasispecies[2]

1982: Discovery of ribozyme catalytic properties[16][17]

2001: Partial RNA polymerase ribozyme is designed via directed evolution[18]

2012: Experimental demonstration that ribozymes can form collectively autocatalytic sets[15]

40.1.2 Error threshold problem

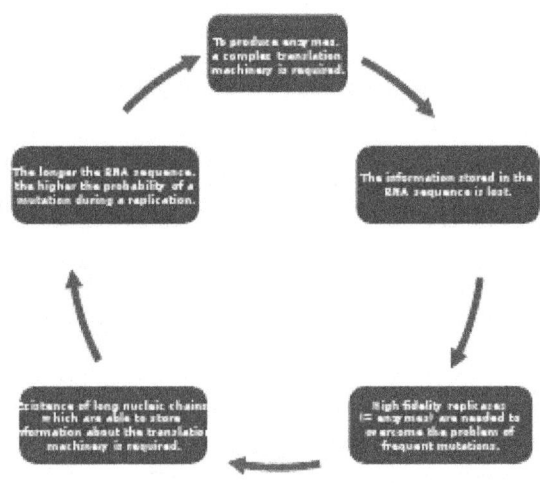

A presentation of the error threshold problem

When a model of replicating molecules was created,[1][2] it was found that, for effective storage of information, macromolecules on prebiotic Earth could not exceed a certain threshold length. This problem is known as the error threshold problem. It arises because replication is an imperfect process, and during each replication event, there is a risk of incorporating errors into a new sequence, leading to the creation of a quasispecies. In a system that is deprived of high-fidelity replicases and error-correction mechanisms, mutations occur with a high probability. As a consequence, the information stored in a sequence can be lost due to the rapid accumulation of errors, a so-called error catastrophe. Moreover, it was shown that the genome size of any organism is roughly equal to the inverse of mutation rate per site per replication.[19][20][21] Therefore, a high mutation rate imposes a serious limitation on the length of the genome.

To overcome this problem, a more specialized replication machinery that is able to copy genetic information with higher fidelity is needed. Manfred Eigen suggested that proteins are necessary to accomplish this task.[1] However, to encode a system as complex as a protein, longer nucleotide sequences are needed, which increases the probability of a mutation even more and requires even more complex replication machinery. This vicious circle is known as Eigen's Paradox.[3]

According to current estimations, the maximum length of a replicated chain that can be correctly reproduced and maintained in enzyme-free systems is about 100 bases, which is assumed to be insufficient to encode replication machinery. This observation was the motivation for the formulation of the hypercycle theory.[22]

40.1.3 Hypercycle models

It was suggested that the problem with building and maintaining larger, more complex, and more accurately replicated molecules can be circumvented if several information carriers, each of them storing a small piece of information, are connected such that they only control their own concentration.[1][2] Studies of the mathematical model describing replicating molecules revealed that to observe a cooperative behaviour among self-replicating molecules, they have to be connected by a positive feedback loop of catalytic actions.[23][24][25] This kind of closed network consisting of self-replicating entities connected by a catalytic positive-feedback loop was named an elementary hypercycle. Such a concept, apart from an increased information capacity, has another advantage. Linking self-replication with mutual catalysis can produce nonlinear growth of the system. This, first, makes the system resistant to so-called parasitic branches. Parasitic branches are species coupled to a cycle that do not provide any advantage to the reproduction of a cycle, which, in turn, makes them useless and decreases the selective value of the system. Secondly, it reinforces the self-organization of molecules into the hypercycle, allowing the system to evolve without losing information, which solves the error threshold problem.[2]

Analysis of potential molecules that could form the first hypercycles in nature prompted the idea of coupling an information carrier function with enzymatic properties. At the time of the hypercycle theory formulation, enzymatic properties were attributed only to proteins, while nucleic acids were recognized only as carriers of information. This led to the formulation of a more complex model of a hypercycle with translation. The proposed model consists of a number of nucleotide sequences I (I stands for intermediate) and the same number of polypeptide chains E (E stands for enzyme). Sequences I have a limited chain length and carry

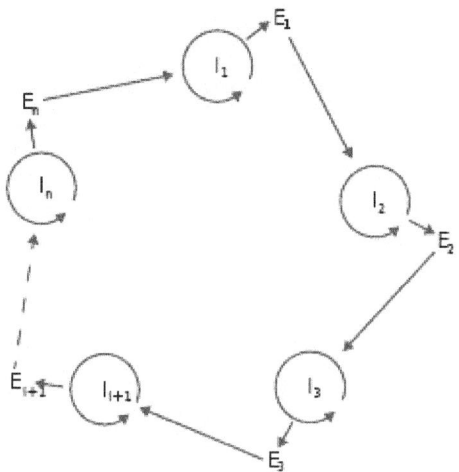

A hypercycle with translation.

the information necessary to build catalytic chains E. The sequence Ii provides the matrix to reproduce itself and a matrix to build the protein Ei. The protein Ei gives the catalytic support to build the next sequence in the cycle, I_{i+1}. The self-replicating sequences I form a cycle consisting of positive and negative strands that periodically reproduce themselves. Therefore, many cycles of the +/− nucleotide collectives are linked together by the second-order cycle of enzymatic properties of E, forming a catalytic hypercycle. Without the secondary loop provided by catalysis, I chains would compete and select against each other instead of cooperating. The reproduction is possible thanks to translation and polymerization functions encoded in I chains. In his principal work, Manfred Eigen stated that the E coded by the I chain can be a specific polymerase or an enhancer (or a silencer) of a more general polymerase acting in favour of formation of the successor of nucleotide chain I. Later, he indicated that a general polymerase leads to the death of the system.[2] Moreover, the whole cycle must be closed, so that En must catalyse I_1 formation for some integer $n > 1$.[1][2]

40.1.4 Alternative concepts

During their research, Eigen and Schuster also considered types of protein and nucleotide coupling other than hypercycles. One such alternative was a model with one replicase that performed polymerase functionality and that was a translational product of one of the RNA matrices existing among the quasispecies. This RNA-dependent RNA polymerase catalysed the replication of sequences that had specific motifs recognized by this replicase. The other RNA matrices, or just one of their strands, provided translational products which had specific anticodons and were responsible for unique assignment and transportation of amino acids.[2]

Another concept devised by Eigen and Schuster was a model in which each RNA template's replication was catalysed by its own translational product; at the same time, this RNA template performed a transport function for one amino acid type. Existence of more than one such RNA template could make translation possible.[2]

Nevertheless, in both alternative concepts, the system will not survive due to the internal competition among its constituents. Even if none of the constituents of such a system is selectively favoured, which potentially allows coexistence of all of the coupled molecules, they are not able to coevolve and optimize their properties. In consequence, the system loses its internal stability and cannot live on. The reason for inability to survive is the lack of mutual control of constituent abundances.[2]

40.2 Mathematical model

40.2.1 Elementary hypercycle

The dynamics of the elementary hypercycle can be modelled using the following differential equation:[3]

$$\dot{x}_i = x_i \left(k_i + \sum_j k_{i,j} x_j - \frac{1}{x}\phi \right)$$

where

$$x = \sum_i x_i,$$
$$k_i = f_i - d_i.$$

In the equation above, xi is the concentration of template Ii; x is the total concentration of all templates; ki is the excess production rate of template Ii, which is a difference between formation fi by self-replication of the template and its degradation di, usually by hydrolysis; ki,j is the production rate of template Ii catalysed by Ij; and φ is a dilution flux; which guarantees that the total concentration is constant. Production and degradation rates are expressed in numbers of molecules per time unit at unit concentration ($xi = 1$). Assuming that at high concentration x the term ki can be neglected, and, moreover, in the hypercycle, a template can be replicated only by itself and the previous member of the cycle, the equation can be simplified to:[3]

$$\dot{x}_i = x_i \left(k_{i,i-1} x_{i-1} - \frac{1}{x} \phi \right)$$

where according to the cyclic properties, it can be assumed that

$$k_{i,0} = k_{i,n},$$
$$x_0 = x_n.$$

40.2.2 Hypercycle with translation

A hypercycle with translation consists of polynucleotides I_i (with concentration x_i) and polypeptides E_i (with concentration y_i). It is assumed that the kinetics of nucleotide synthesis follows a Michaelis–Menten-type reaction scheme in which the concentration of complexes cannot be neglected. During replication, molecules form complexes $I_i E_{i-1}$ (occurring with concentration z_i). Thus, the total concentration of molecules (x_i^0 and y_i^0) will be the sum of free molecules and molecules involved in a complex:

$$x_i^0 = x_i + z_i,$$
$$y_i^0 = y_i + z_{i+1}.$$

The dynamics of the hypercycle with translation can be described using a system of differential equations modelling the total number of molecules:

$$\dot{x}_i^0 = f_i z_i - \frac{x_i^0}{c_I} \phi_x,$$
$$\dot{y}_i^0 = k_i x_i - \frac{y_i^0}{c_E} \phi_y$$

where

$$c_I = \sum_i x_i^0,$$
$$c_E = \sum_i y_i^0.$$

In the above equations, cE and cI are total concentrations of all polypeptides and all polynucleotides, φx and φy are dilution fluxes, ki is the production rate of polypeptide E_i translated from the polynucleotide I_i, and fi is the production rate of polynucleotide I_i synthesised by the complex $I_i E_{i-1}$ (through replication and polymerization).[3]

Coupling nucleic acids with proteins in such a model of hypercycle with translation demanded the proper model for the origin of translation code as a necessary condition for the origin of hypercycle organization. At the time of hypercycle theory formulation, two models for the origin of translation code were proposed by Crick and his collaborators. These were models stating that the first codons were constructed according to either an RRY or an RNY scheme, in which R stands for the purine base, Y for pyrimidine, and N for any base, with the latter assumed to be more reliable. Nowadays, it is assumed that the hypercycle model could be realized by utilization of ribozymes without the need for a hypercycle with translation, and there are many more theories about the origin of the genetic code.[26]

40.3 Evolution of hypercycles

40.3.1 Formation of the first hypercycles

Eigen made several assumptions about conditions that led to the formation of the first hypercycles.[2] Some of them were the consequence of the lack of knowledge about ribozymes, which were discovered a few years after the introduction of the hypercycle concept[16][17] and negated Eigen's assumptions in the strict sense. The primary of them was that the formation of hypercycles had required the availability of both types of chains: nucleic acids forming a quasispecies population and proteins with enzymatic functions. Nowadays, taking into account the knowledge about ribozymes, it may be assumed that a hypercycle's members were selected from the quasispecies population and the enzymatic function was performed by RNA. According to the hypercycle theory, the first primitive polymerase emerged precisely from this population. As a consequence, the catalysed replication could exceed the uncatalysed reactions, and the system could grow faster. However, this rapid growth was a threat to the emerging system, as the whole system could lose control over the relative amount of the RNAs with enzymatic function. The system required more reliable control of its constituents—for example, by incorporating the coupling of essential RNAs into a positive feedback loop. Without this feedback loop, the replicating system would be lost. These positive feedback loops formed the first hypercycles.[2]

In the process described above, the fact that the first hypercycles originated from the quasispecies population (a population of similar sequences) created a significant advantage. One possibility of linking different chains I—which is relatively easy to achieve taking into account the quasispecies properties—is that the one chain I improves the synthesis of the similar chain I'. In this way, the existence of similar sequences I originating from the same quasispecies population promotes the creation of the linkage between molecules I and I'.[2]

40.3.2 Evolutionary dynamics

After formation, a hypercycle reaches either an internal equilibrium or a state with oscillating concentrations of each type of chain I, but with the total concentration of all chains remaining constant. In this way, the system consisting of all chains can be expressed as a single, integrated entity. During the formation of hypercycles, several of them could be present in comparable concentrations, but very soon, a selection of the hypercycle with the highest fitness value will take place.[1] Here, the fitness value expresses the adaptation of the hypercycle to the environment, and the selection based on it is very sharp. After one hypercycle wins the competition, it is very unlikely that another one could take its place, even if the new hypercycle would be more efficient than the winner. Usually, even large fluctuations in the numbers of internal species cannot weaken the hypercycle enough to destroy it. In the case of a hypercycle, we can speak of one-for-ever selection, which is responsible for the existence of a unique translation code and a particular chirality.[2]

The above-described idea of a hypercycle's robustness results from an exponential growth of its constituents caused by the catalytic support. However, Eörs Szathmáry and Irina Gladkih showed that an unconditional coexistence can be obtained even in the case of a non-enzymatic template replication that leads to a subexponential or a parabolic growth. This could be observed during the stages preceding a catalytic replication that are necessary for the formation of hypercycles. The coexistence of various non-enzymatically replicating sequences could help to maintain a sufficient diversity of RNA modules used later to build molecules with catalytic functions.[27]

From the mathematical point of view, it is possible to find conditions required for cooperation of several hypercycles. However, in reality, the cooperation of hypercycles would be extremely difficult, because it requires the existence of a complicated multi-step biochemical mechanism or an incorporation of more than two types of molecules. Both conditions seem very improbable; therefore, the existence of coupled hypercycles is assumed impossible in practice.[2]

Evolution of a hypercycle ensues from the creation of new components by the mutation of its internal species. Mutations can be incorporated into the hypercycle, enlarging it if, and only if, two requirements are satisfied. First, a new information carrier I_{new} created by the mutation must be better recognized by one of the hypercycle's members I_i than the chain I_{i+1} that was previously recognized by it. Secondly, the new member I_{new} of the cycle has to better catalyse the formation of the polynucleotide I_{i+1} that was previously catalysed by the product of its predecessor I_i. In theory, it is possible to incorporate into the hypercycle mutations that do not satisfy the second condition. They would form parasitic branches that use the system for their own replication but do not contribute to the system as a whole. However, it was noticed that such mutants do not pose a threat to the hypercycle, because other constituents of the hypercycle grow nonlinearly, which prevents the parasitic branches from growing.[2]

40.3.3 Evolutionary dynamics: a mathematical model

According to the definition of a hypercycle, it is a nonlinear, dynamic system, and, in the simplest case, it can be assumed that it grows at a rate determined by a system of quadratic differential equations. Then, the competition between evolving hypercycles can be modelled using the differential equation:[3]

$$\dot{C}_l = q_l C_l^2 - C_l \frac{\phi}{C}$$

where

$$C = \sum_l C_l.$$

Here, Cl is the total concentration of all polynucleotide chains belonging to a hypercycle Hl, C is the total concentration of polynucleotide chains belonging to all hypercycles, ql is the rate of growth, and φ is a dilution flux that guarantees that the total concentration is constant. According to the above model, in the initial phase, when several hypercycles exist, the selection of the hypercycle with the largest ql value takes place. When one hypercycle wins the selection and dominates the population, it is very difficult to replace it, even with a hypercycle with a much higher growth rate q.[3]

40.4 Compartmentalization and genome integration

Hypercycle theory proposed that hypercycles are not the final state of organization, and further development of more complicated systems is possible by enveloping the hypercycle in some kind of membrane.[2] After evolution of compartments, a genome integration of the hypercycle can proceed by linking its members into a single chain, which forms a precursor of a genome. After that, the whole individualized and compartmentalized hypercycle can behave like a simple self-replicating entity. Compartmentalization provides some advantages for a system that has already established a linkage between units. Without compartments, genome integration would boost competition by limiting space and resources. Moreover, adaptive evolution requires the package of transmissible information for advantageous mutations in order not to aid less-efficient copies of the gene. The first advantage is that it maintains a high local concentration of molecules, which helps to locally increase the rate of synthesis. Secondly, it keeps the effect of mutations local, while at the same time affecting the whole compartment. This favours preservation of beneficial mutations, because it prevents them from spreading away. At the same time, harmful mutations cannot pollute the entire system if they are enclosed by the membrane. Instead, only the contaminated compartment is destroyed, without affecting other compartments. In that way, compartmentalization allows for selection for genotypic mutations. Thirdly, membranes protect against environmental factors because they constitute a barrier for high-weight molecules or UV irradiation. Finally, the membrane surface can work as a catalyst.[3]

Despite the above-mentioned advantages, there are also potential problems connected to compartmentalized hypercycles. These problems include difficulty in the transport of ingredients in and out, synchronizing the synthesis of new copies of the hypercycle constituents, and division of the growing compartment linked to a packing problem.[2]

In the initial works, the compartmentalization was stated as an evolutionary consequence of the hypercyclic organization. Carsten Bresch and coworkers raised an objection that hypercyclic organization is not necessary if compartments are taken into account.[28] They proposed the so-called package model in which one type of a polymerase is sufficient and copies all polynucleotide chains that contain a special recognition motif. However, as pointed out by the authors, such packages are—contrary to hypercycles—vulnerable to deleterious mutations as well as a fluctuation abyss, resulting in packages that lack one of the essential RNA molecules. Eigen and colleagues argued that simple package of genes cannot solve the information integration problem and hypercycles cannot be simply replaced by compartments, but compartments may assist hypercycles.[29] This problem, however, raised more objections, and Eörs Szathmáry and László Demeter reconsidered whether packing hypercycles into compartments is a necessary intermediate stage of the evolution. They invented a stochastic corrector model[10] that assumed that replicative templates compete within compartments, and selective values of these compartments depend on the internal composition of templates. Numerical simulations showed that when stochastic effects are taken into account, compartmentalization is sufficient to integrate information dispersed in competitive replicators without the need for hypercycle organization. Moreover, it was shown that compartmentalized hypercycles are more sensitive to the input of deleterious mutations than a simple package of competing genes. Nevertheless, package models do not solve the error threshold problem that originally motivated the hypercycle.[30]

40.5 Hypercycles and ribozymes

At the time of the hypercycle theory formulation, ribozymes were not known. After the breakthrough of discovering RNA's catalytic properties in 1982,[16][17] it was realized that RNA had the ability to integrate protein and nucleotide-chain properties into one entity. Ribozymes potentially serving as templates and catalysers of replication can be considered components of quasispecies that can self-organize into a hypercycle without the need to invent a translation process. In 2001, a partial RNA polymerase ribozyme was designed via directed evolution.[18] Nevertheless, it was able to catalyse only a polymerization of a chain having the size of about 14 nucleotides, even though it was 200 nucleotides long. The most up-to-date version of this polymerase was shown in 2013.[31] While it has an ability to catalyse polymerization of longer sequences, even of its own length, it cannot replicate itself due to a lack of sequence generality and its inability to transverse secondary structures of long RNA templates. However, it was recently shown that those limitations could in principle be overcome by the assembly of active polymerase ribozymes from several short RNA strands.[32] In 2014, a cross-chiral RNA polymerase ribozyme was demonstrated.[33] It was hypothesized that it offers a new mode of recognition between an enzyme and substrates, which is based on the shape of the substrate, and allows avoiding the Watson-Crick pairing and, therefore, may provide greater sequence generality. Various other experiments have shown that, besides bearing polymerase properties, ribozymes could have developed other kinds of evolutionarily useful catalytic activity such as synthase, ligase, or aminoacylase activities.[18] Ribozymal aminoacylators and ribozymes with the ability to form peptide bonds might have been crucial to inventing translation. An RNA ligase, in turn, could link various components of quasispecies into one chain, beginning the process of a genome integration. An RNA with a synthase or a synthetase activity could be critical for building compartments and providing building blocks for growing RNA and protein chains as well as other types of molecules. Many examples of this kind of ribozyme are currently known, including a peptidyl transferase ribozyme,[34] a ligase,[35][36] and a nucleotide synthetase.[37] A transaminoacylator described in 2013 has five nucleotides,[38] which is sufficient for a trans-amino acylation reaction and makes it the smallest ribozyme that has been discovered. It supports a peptidyl-RNA synthesis that could be a precursor for the contemporary process of linking amino acids to tRNA molecules. An RNA ligase's catalytic domain, consisting of 93 nucleotides, proved to be sufficient to catalyse a linking reaction between two RNA chains.[39] Similarly, an acyltransferase ribozyme 82 nucleotides long was sufficient to perform an acyltransfer reaction.[40] Altogether, the results concerning the RNA ligase's catalytic domain and the acyltransferase ribozyme are in agreement with the estimated upper limit of 100 nucleotides set by the error threshold problem. However, it was hypothesized that even if the putative first RNA-dependent RNA-polymerases are estimated to be longer—the smallest reported up-to-date RNA-dependent polymerase ribozyme is 165 nucleotides long[18]—they did not have to arise in one step. It is more plausible that ligation of smaller RNA chains performed by the first RNA ligases resulted in a longer chain with the desired catalytically active polymerase domain.[41]

Forty years after the publication of Manfred Eigen's primary work dedicated to hypercycles,[1] Nilesh Vaidya and colleagues showed experimentally that ribozymes can form catalytic cycles and networks capable of expanding their sizes by incorporating new members.[14] However, this is not a demonstration of a hypercycle in accordance with its definition, but an example of a collectively autocatalytic set.[15] Earlier computer simulations showed that molecular networks can arise, evolve and be resistant to parasitic RNA branches.[42] In their experiments, Vaidya et al. used an Azoarcus group I intron ribozyme that, when fragmented, has an ability to self-assemble by catalysing recombination reactions in an autocatalytic manner. They mutated the three-nucleotide-long sequences responsible for recognition of target sequences on the opposite end of the ribozyme (namely, Internal Guide Sequences or IGSs) as well as these target sequences. Some genotypes could introduce cooperation by recognizing target sequences of the other ribozymes, promoting their covalent binding, while other selfish genotypes were only able to self-assemble. In separation, the selfish subsystem grew faster than the cooperative one. After mixing selfish ribozymes with cooperative ones, the emergence of cooperative behaviour in a merged population was observed, outperforming the self-assembling subsystems. Moreover, the selfish ribozymes were integrated into the network of reactions, supporting its growth. These results were also explained analytically by the ODE model and its analysis. They differ substantially from results obtained in evolutionary dynamics.[43] According to evolutionary dynamics theory, selfish molecules should dominate the system even if the growth rate of the selfish subsystem in isolation is lower than the growth rate of the cooperative system. Moreover, Vaidya et al. proved that, when fragmented into more pieces, ribozymes that are capable of self-assembly can not only still form catalytic cycles but, indeed, favour them. Results obtained from experiments by Vaidya et al. gave a glimpse on how inefficient prebiotic polymerases, capable of synthesizing only short oligomers, could be sufficient at the pre-life stage to spark off life. This could happen because coupling the synthesis of short RNA fragments by the first ribozymal polymerases to a system capable of self-assembly not only enables building longer sequences but also allows exploiting the fitness space more efficiently with the use of the recombination process. Another experiment performed by Hannes Mutschler et al.[32] showed that the RNA polymerase ribozyme, which they described, can be synthesized in situ from the ligation of four smaller fragments, akin to a recombination of Azoarcus ribozyme from four inactive oligonucleotide fragments described earlier. Apart from a substantial contribution of the above experiments to the research on the origin of life, they have not proven the existence of hypercycles experimentally.[44]

40.6 Related problems and reformulations

The hypercycle concept has been continuously studied since its origin. Shortly after Eigen and Schuster published their main work regarding hypercycles,[2] John Maynard Smith raised an objection that the catalytic support for the replication given to other molecules is altruistic.[45] Therefore, it cannot be selected and maintained in a system. He also underlined hypercycle vulnerability to parasites, as they are favoured by selection. Later on, Josef Hofbauer and Karl Sigmund[46] indicated that in reality, a hypercycle can maintain only fewer than five members. In agreement with Eigen and Schuster's principal analysis, they argued that systems with five or more species exhibit limited and unstable cyclic behaviour, because some species can die out due to stochastic events and break the positive feedback loop that sustains the hypercycle. The extinction of the hypercycle then follows. It was also emphasized that a hypercycle size of up to four is too small to maintain the amount of information sufficient to cross the information threshold.[2]

Several researchers proposed a solution to these problems by introducing space into the initial model either explicitly[6][13][47][48] or in the form of a spatial segregation within compartments.[10][28] Bresch et al.[28] proposed a package model as a solution for the parasite problem. Later on, Szathmáry and Demeter[10] proposed a stochastic corrector machine model. Both compartmentalized systems proved to be robust against parasites. However, package models do not solve the error threshold problem that originally motivated the idea of the hypercycle. A few years later, Maarten Boerlijst and Paulien Hogeweg, and later Nobuto Takeuchi, studied the replicator equations with the use of partial differential equations[5] and cellular automata models,[6][7][9] methods that already proved to be successful in other applications.[49][50] They demonstrated that spatial self-structuring of the system completely solves the problem of global extinction for large systems and, partially, the problem of parasites.[13] The latter was also analysed by Robert May,[12] who noticed that an emergent rotating spiral wave pattern, which was observed during computational simulations performed on cellular automata, proved to be stable and able to survive the invasion of parasites if they appear at some distance from the wave core. Unfortunately, in this case, rotation decelerates as the number of hypercycle members increases, meaning that selection tends toward decreasing the amount of information stored in the hypercycle. Moreover, there is also a problem with adding new information into the system. In order to be preserved, the new information has to appear near to the core of the spiral wave. However, this would make the system vulnerable to parasites, and, as a consequence, the hypercycle would not be stable. Therefore, stable spiral waves are characterized by once-for-ever selection, which creates the restrictions that, on the one hand, once the information is added to the system, it cannot be easily abandoned; and on the other hand, new information cannot be added.[12]

Another model based on cellular automata, taking into account a simpler replicating network of continuously mutating parasites and their interactions with one replicase species, was proposed by Takeuchi and Hogeweg[7] and exhibited an emergent travelling wave pattern. Surprisingly, travelling waves not only proved to be stable against moderately strong parasites, if the parasites' mutation rate is not too high, but the emergent pattern itself was generated as a result of interactions between parasites and replicase species. The same technique was used to model systems that include formation of complexes.[51] Finally, hypercycle simulation extending to three dimensions showed the emergence of the three-dimensional analogue of a spiral wave, namely, the scroll wave.[52]

40.7 See also

- Chemoton

40.8 References

This article incorporates content from PLoS Computational Biology that is available under the terms of the Creative Commons Attribution 4.0 License.

[1] Eigen, Manfred (October 1971). "Selforganization of matter and the evolution of biological macromolecules". *Die Naturwissenschaften* 58 (10): 465–523. doi:10.1007/BF00623322.

[2] Schuster, M. Eigen ; P. (1979). *The Hypercycle : a principle of natural self-organization* (Reprint. ed.). Berlin [West] [u.a.]: Springer. ISBN 9783540092933.

[3] Eigen, M.; Schuster, P. (1982). "Stages of emerging life — Five principles of early organization". *Journal of Molecular Evolution* 19 (1): 47–61. doi:10.1007/BF02100223. PMID 7161810.

[4] Schuster, Peter; Sigmund, Karl (February 1983). "Replicator dynamics". *Journal of Theoretical Biology* 100 (3): 533–538. doi:10.1016/0022-5193(83)90445-9.

[5] Boerlijst, Maarten C.; Hogeweg, Pauline (November 1995). "Spatial gradients enhance persistence of hypercycles". *Physica D: Nonlinear Phenomena* 88 (1): 29–39. doi:10.1016/0167-2789(95)00178-7.

[6] Boerlijst, M.C.; Hogeweg, P. (February 1991). "Spiral wave structure in pre-biotic evolution: Hypercycles stable against parasites". *Physica D: Nonlinear Phenomena* 48 (1): 17–28. doi:10.1016/0167-2789(91)90049-F.

[7] Hogeweg, Paulien; Takeuchi, Nobuto (2003). "Multilevel Selection in Models of Prebiotic Evolution: Compartments and Spatial Self-organization". *Origins of Life and Evolution of the Biosphere* 33 (4/5): 375–403. doi:10.1023/A:1025754907141.

[8] Takeuchi, Nobuto; Hogeweg, Paulien; Stormo, Gary D. (16 October 2009). "Multilevel Selection in Models of Prebiotic Evolution II: A Direct Comparison of Compartmentalization and Spatial Self-Organization". *PLoS Computational Biology* 5 (10): e1000542. doi:10.1371/journal.pcbi.1000542.

[9] Takeuchi, Nobuto; Hogeweg, Paulien (September 2012). "Evolutionary dynamics of RNA-like replicator systems: A bioinformatic approach to the origin of life". *Physics of Life Reviews* 9 (3): 219–263. doi:10.1016/j.plrev.2012.06.001.

[10] Szathmáry, Eörs; Demeter, László (1987). "Group selection of early replicators and the origin of life". *Journal of Theoretical Biology* 128 (4): 463–486. doi:10.1016/S0022-5193(87)80191-1. PMID 2451771.

[11] Musso, Fabio (16 March 2010). "A Stochastic Version of the Eigen Model". *Bulletin of Mathematical Biology* 73 (1): 151–180. doi:10.1007/s11538-010-9525-4.

[12] May, Robert M. (17 October 1991). "Hypercycles spring to life". *Nature* 353 (6345): 607–608. doi:10.1038/353607a0.

[13] MC Boerlijst, P Hogeweg (1991) "Self-structuring and selection: Spiral Waves as a Substrate for Prebiotic Evolution". Conference Paper *Artificial Life II*

[14] Vaidya, Nilesh; Manapat, Michael L.; Chen, Irene A.; Xulvi-Brunet, Ramon; Hayden, Eric J.; Lehman, Niles (17 October 2012). "Spontaneous network formation among cooperative RNA replicators". *Nature* **491** (7422): 72–77. doi:10.1038/nature11549.

[15] Szathmáry, Eörs (2013). "On the propagation of a conceptual error concerning hypercycles and cooperation". *Journal of Systems Chemistry* **4** (1): 1. doi:10.1186/1759-2208-4-1.

[16] Kruger, Kelly; Grabowski, Paula J.; Zaug, Arthur J.; Sands, Julie; Gottschling, Daniel E.; Cech, Thomas R. (1982). "Self-splicing RNA: Autoexcision and autocyclization of the ribosomal RNA intervening sequence of tetrahymena". *Cell* **31** (1): 147–157. doi:10.1016/0092-8674(82)90414-7. ISSN 0092-8674. PMID 6297745.

[17] Guerrier-Takada, Cecilia; Gardiner, Katheleen; Marsh, Terry; Pace, Norman; Altman, Sidney (1983). "The RNA moiety of ribonuclease P is the catalytic subunit of the enzyme". *Cell* **35** (3): 849–857. doi:10.1016/0092-8674(83)90117-4. ISSN 0092-8674. PMID 6197186.

[18] Johnston, W. K. (18 May 2001). "RNA-Catalyzed RNA Polymerization: Accurate and General RNA-Templated Primer Extension". *Science* **292** (5520): 1319–1325. doi:10.1126/science.1060786.

[19] Drake, J. W. (1 May 1993). "Rates of spontaneous mutation among RNA viruses.". *Proceedings of the National Academy of Sciences* **90** (9): 4171–4175. doi:10.1073/pnas.90.9.4171.

[20] Drake, JW; Charlesworth, B; Charlesworth, D; Crow, JF (April 1998). "Rates of spontaneous mutation.". *Genetics* **148** (4): 1667–86. PMID 9560386.

[21] STADLER, BÄRBEL M. R.; STADLER, PETER F. (March 2003). "MOLECULAR REPLICATOR DYNAMICS". *Advances in Complex Systems* **06** (01): 47–77. doi:10.1142/S0219525903000724.

[22] Eigen, Manfred; Schuster, Peter (1977). "The hypercycle. A principle of natural self-organization. Part A: Emergence of the hypercycle.". *Naturwissenschaften* **64** (11): 541–565. doi:10.1007/BF00450633. PMID 593400.

[23] Hofbauer, J.; Schuster, P.; Sigmund, K.; Wolff, R. (April 1980). "Dynamical Systems Under Constant Organization II: Homogeneous Growth Functions of Degree $p = 2$". *SIAM Journal on Applied Mathematics* **38** (2): 282–304. doi:10.1137/0138025.

[24] Hofbauer, J.; Schuster, P.; Sigmund, K. (February 1981). "Competition and cooperation in catalytic selfreplication". *Journal of Mathematical Biology* **11** (2): 155 168. doi:10.1007/BF00275439.

[25] Schuster, P; Sigmund, K; Wolff, R (June 1979). "Dynamical systems under constant organization. III. Cooperative and competitive behavior of hypercycles". *Journal of Differential Equations* **32** (3): 357–368. doi:10.1016/0022-0396(79)90039-1.

[26] Crick, F. H. C.; Brenner, S.; Klug, A.; Pieczenik, G. (December 1976). "A speculation on the origin of protein synthesis". *Origins of Life* **7** (4): 389–397. doi:10.1007/BF00927934.

[27] Szathmáry, Eörs; Gladkih, Irina (May 1989). "Sub-exponential growth and coexistence of non-enzymatically replicating templates". *Journal of Theoretical Biology* **138** (1): 55–58. doi:10.1016/S0022-5193(89)80177-8.

[28] Bresch, C.; Niesert, U.; Harnasch, D. (1980). "Hypercycles, parasites and packages". *Journal of Theoretical Biology* **85** (3): 399–405. doi:10.1016/0022-5193(80)90314-8. ISSN 0022-5193. PMID 6893729.

[29] Eigen, M.; Gardiner, W.C.; Schuster, P. (August 1980). "Hypercycles and compartments". *Journal of Theoretical Biology* **85** (3): 407–411. doi:10.1016/0022-5193(80)90315-X.

[30] ZINTZARAS, ELIAS; SANTOS, MAURO; SZATHMÁRY, EÖRS (July 2002). '"Living" Under the Challenge of Information Decay: The Stochastic Corrector Model vs. Hypercycles'. *Journal of Theoretical Biology* **217** (2): 167–181. doi:10.1006/jtbi.2002.3026.

[31] Attwater, James; Wochner, Aniela; Holliger, Philipp (20 October 2013). "In-ice evolution of RNA polymerase ribozyme activity". *Nature Chemistry* **5**: 1011–1018. doi:10.1038/nchem.1781.

[32] Mutschler, Hannes; Wochner, Aniela; Holliger, Philipp (4 May 2015). "Freeze–thaw cycles as drivers of complex ribozyme assembly". *Nature Chemistry* **7** (6): 502–508. doi:10.1038/nchem.2251.

[33] Sczepanski, Jonathan T.; Joyce, Gerald F. (29 October 2014). "A cross-chiral RNA polymerase ribozyme". *Nature* **515** (7527): 440–442. doi:10.1038/nature13900.

[34] Cech, Thomas R.; Zhang, Biliang (6 November 1997). "Peptide bond formation by in vitro selected ribozymes". *Nature* **390** (6655): 96–100. doi:10.1038/36375.

[35] Robertson, MP; Hesselberth, JR; Ellington, AD (April 2001). "Optimization and optimality of a short ribozyme ligase that joins non-Watson-Crick base pairings.". *RNA (New York, N.Y.)* **7** (4): 513–23. PMID 11345430.

[36] Paul, N.; Joyce, G. F. (18 September 2002). "A self-replicating ligase ribozyme". *Proceedings of the National Academy of Sciences* **99** (20): 12733–12740. doi:10.1073/pnas.202471099.

[37] Bartel, David P.; Unrau, Peter J. (17 September 1998). "RNA-catalysed nucleotide synthesis". *Nature* **395** (6699): 260–263. doi:10.1038/26193.

[38] Bianconi, Ginestra; Zhao, Kun; Chen, Irene A.; Nowak, Martin A.; Doebeli, Michael (9 May 2013). "Selection for Replicases in Protocells". *PLoS Computational Biology* **9** (5): e1003051. doi:10.1371/journal.pcbi.1003051.

[39] Ekland, E.; Szostak, J.; Bartel, D. (21 July 1995). "Structurally complex and highly active RNA ligases derived from random RNA sequences". *Science* **269** (5222): 364–370. doi:10.1126/science.7618102.

[40] Suga, Hiroaki; Lee, Nick; Bessho, Yoshitaka; Wei, Kenneth; Szostak, Jack W. (1 January 2000). "Ribozyme-catalyzed tRNA aminoacylation". *Nature Structural Biology* **7** (1): 28–33. doi:10.1038/71225.

[41] Costanzo, G.; Pino, S.; Ciciriello, F.; Di Mauro, E. (2 October 2009). "Generation of Long RNA Chains in Water". *Journal of Biological Chemistry* **284** (48): 33206–33216. doi:10.1074/jbc.M109.041905.

[42] Szathmary, E. (29 October 2006). "The origin of replicators and reproducers". *Philosophical Transactions of the Royal Society B: Biological Sciences* **361** (1474): 1761–1776. doi:10.1098/rstb.2006.1912.

[43] Nowak, Martin A. (2006). *Evolutionary dynamics : exploring the equations of life*. Cambridge, Mass. [u.a.]: Belknap Press of Harvard Univ. Press. ISBN 9780674023383.

[44] Hayden, Eric J.; Lehman, Niles (August 2006). "Self-Assembly of a Group I Intron from Inactive Oligonucleotide Fragments". *Chemistry & Biology* **13** (8): 909–918. doi:10.1016/j.chembiol.2006.06.014.

[45] Smith, John Maynard (9 August 1979). "Hypercycles and the origin of life". *Nature* **280** (5722): 445–446. doi:10.1038/280445a0.

[46] Sigmund, Josef Hofbauer; Karl (1992). *The theory of evolution and dynamical systems : mathematical aspects of selection* (Repr. ed.). Cambridge [u.a.]: Cambridge Univ. Press. ISBN 0-521-35838-8.

[47] McCaskill, J.S.; Füchslin, R.M.; Altmeyer, S. (30 January 2001). "The Stochastic Evolution of Catalysts in Spatially Resolved Molecular Systems". *Biological Chemistry* **382** (9). doi:10.1515/BC.2001.167.

[48] Szabó, Péter; Scheuring, István; Czárán, Tamás; Szathmáry, Eörs (21 November 2002). "In silico simulations reveal that replicators with limited dispersal evolve towards higher efficiency and fidelity". *Nature* **420** (6913): 340–343. doi:10.1038/nature01187.

[49] Wasik, Szymon; Jackowiak, Paulina; Krawczyk, Jacek B.; Kedziora, Paweł; Formanowicz, Piotr; Figlerowicz, Marek; Błażewicz, Jacek (2010). "Towards Prediction of HCV Therapy Efficiency". *Computational and Mathematical Methods in Medicine* **11** (2): 185–199. doi:10.1080/17486700903170712.

[50] Wasik, Szymon; Jackowiak, Paulina; Figlerowicz, Marek; Blazewicz, Jacek (February 2014). "Multi-agent model of hepatitis C virus infection". *Artificial Intelligence in Medicine* **60** (2): 123–131. doi:10.1016/j.artmed.2013.11.001.

[51] Takeuchi, Nobuto; Hogeweg, Paulien (23 October 2007). "The Role of Complex Formation and Deleterious Mutations for the Stability of RNA-Like Replicator Systems". *Journal of Molecular Evolution* **65** (6): 668–686. doi:10.1007/s00239-007-9044-6.

[52] S Altmeyer, C Wilke, T Martinetz (2008) "How fast do structures emerge in hypercycle-systems?" in *Third German Workshop on Artificial Life*

40.9 External links

- J. Padgett's Hypercycle model implemented in repast

Chapter 41

Autocatalytic set

An **autocatalytic set** is a collection of entities, each of which can be created catalytically by other entities within the set, such that as a whole, the set is able to catalyze its own production. In this way the set *as a whole* is said to be autocatalytic. Autocatalytic sets were originally and most concretely defined in terms of molecular entities, but have more recently been metaphorically extended to the study of systems in sociology and economics.

Autocatalytic sets also have the ability to replicate themselves if they are split apart into two physically separated spaces. Computer models illustrate that split autocatalytic sets will reproduce all of the reactions of the original set in each half, much like cellular mitosis. In effect, using the principles of autocatalysis, a small metabolism can replicate itself with very little high level organization. This property is why autocatalysis is a contender as the foundational mechanism for complex evolution.

Prior to Watson and Crick, biologists considered autocatalytic sets the way metabolism functions in principle, i.e. one protein helps to synthesize another protein and so on. After the discovery of the double helix, the central dogma of molecular biology was formulated, which is that DNA is transcribed to RNA which is translated to protein. The molecular structure of DNA and RNA, as well as the metabolism that maintains their reproduction, are believed to be too complex to have arisen spontaneously in one step from a soup of chemistry.

Several models of the origin of life are based on the notion that life may have arisen through the development of an initial molecular autocatalytic set which evolved over time. Most of these models which have emerged from the studies of complex systems predict that life arose not from a molecule with any particular trait (such as self-replicating RNA) but from an autocatalytic set. The first empirical support came from Lincoln and Joyce, who obtained autocatalytic sets in which "two [RNA] enzymes catalyze each other's synthesis from a total of four component substrates."[1] Furthermore, an evolutionary process that began with a population of these self-replicators yielded a population dominated by recombinant replicators.

Modern life has the traits of an autocatalytic set, since no particular molecule, nor any class of molecules, is able to replicate itself. There are several models based on autocatalytic sets, including those of Stuart Kauffman[2] and others.

41.1 Formal definition

41.1.1 Definition

Given a set M of molecules, chemical reactions can be roughly defined as pairs r = (A, B) of subsets from M:[3]

$a_1 + a_2 + ... + a_k \to b_1 + b_2 + ... + b_k$

Let R be the set of allowable reactions. A pair (M, R) is a *reaction system* (RS).

Let C be the set of molecule-reaction pairs specifying which molecules can catalyze which reactions:

$C = \{(m, r) \mid m \in M, r \in R\}$

Let $F \subseteq M$ be a set of *food* (small numbers of molecules freely available from the environment) and $R' \subseteq R$ be some subset of reactions. We define a closure of the food set relative to this subset of reactions ClR'(F) as the set of molecules that contains the food set plus all molecules that can be produced starting from the food set and using only reactions from this subset of reactions. Formally ClR'(F) is a minimal subset of M such that $F \subseteq$ ClR'(F) and for each reaction r'(A, B) $\subseteq$ R':

$A \subseteq$ ClR'(F) $\Rightarrow B \subseteq$ ClR'(F)

A reaction system (ClR'(F), R') is *autocatalytic*, if and only if for each reaction r'(A, B) $\subseteq$ R':

1. there exists a molecule $c \subseteq$ ClR'(F) such that (c, r') $\subseteq$ C,

2. $A \subseteq$ ClR'(F).

41.1.2 Example

Let M = {a, b, c, d, f, g} and F = {a, b}. Let the set R contains the following reactions:

a + b → c + d, catalyzed by g a + f → c + b, catalyzed by d
c + b → g + a, catalyzed by d or f

From the F = {a, b} we can produce {c, d} and then from {c, b} we can produce {g, a} so the closure is equal to:

ClR'(F) = {a, b, c, d, g}

According to the definition the maximal autocatalytic subset R' will consists of two reactions:

a + b → c + d, catalyzed by g c + b → g + a, catalyzed by d

The reaction for (a + f) does not belong to R' because f does not belong to closure. Similarly the reaction for (c + b) in the autocatalytic set can only be catalyzed by d and not by f.

41.2 Probability that a random set is autocatalytic

Studies of the above model show that random RS can be autocatalytic with high probability under some assumptions. This comes from the fact that with a growing number of molecules, the number of possible reactions and catalysations grows even larger if the molecules grow in complexity, producing stochastically enough reactions and catalysations to make a part of the RS self-supported.[4] An autocatalytic set then extends very quickly with growing number of molecules for the same reason. These theoretical results make autocatalytic sets attractive for scientific explanation of the very early origin of life.

41.3 Formal limitations

Formally, it is difficult to treat molecules as anything but unstructured entities, since the set of possible reactions (and molecules) would become infinite. Therefore, a derivation of arbitrarily long polymers as needed to model DNA, RNA or proteins is not possible, yet. Studies of the RNA World suffer from the same problem.

41.4 Linguistic aspects

Contrary to the above definition, which applies to the field of Artificial chemistry, no agreed-upon notion of autocatalytic sets exists today.

While above, the notion of catalyst is secondary insofar that only the set as a whole has to catalyse its own production, it is primary in other definitions, giving the term "Autocatalytic Set" a different emphasis. There, *every* reaction (or function, transformation) has to be mediated by a catalyst. As a consequence, while mediating its respective reaction, every catalyst *denotes* its reaction, too, resulting in a self denoting system, which is interesting for two reasons. First, real metabolism is structured in this manner. Second, self denoting systems can be considered as an intermediate step towards self describing systems.

From both a structural and a natural historical point of view, one can identify the ACS as seized in the formal definition the more original concept, while in the second, the reflection of the system in itself is already brought to an explicit presentation, since catalysts represent the reaction induced by them. In ACS literature, both concept are present, but differently emphasised.

To complete the classification from the other side, generalised self reproducing systems move beyond self-denotation. There, no unstructured entities carry the transformations anymore, but structured, described ones. Formally, a generalised self reproducing system consists of two function, u and c, together with their descriptions Desc(u) and Desc(c) along following definition:

u : Desc(X) -> X   c : Desc(X) -> Desc(X)

where the function 'u' is the "universal" constructor, that constructs everything in its domain from appropriate descriptions, while 'c' is a copy function for any description. Practically, 'u' and 'c' can fall apart into many subfunctions or catalysts.

Note that the (trivial) copy function 'c' is necessary because though the universal constructor 'u' would be able to construct any description, too, the description it would base on, would in general be longer than the result, rendering full self replication impossible.

This last concept can be attributed to von Neumann's work on self reproducing automata, where he holds a self description necessary for any nontrivial (generalised) self reproducing system to avoid interferences. Von Neumann planned to design such a system for a model chemistry, too.

41.5 Non-autonomous autocatalytic sets

Virtually all articles on autocatalytic sets leave open whether the sets are to be considered autonomous or not. Often, autonomy of the sets is silently assumed.

Likely, the above context has a strong emphasis on autonomous self replication and early origin of life. But the concept of autocatalytic sets is really more general and in practical use in various technical areas, e.g. where self-sustaining tool chains are handled. Clearly, such sets are not autonomous and are objects of human agency.

Examples of practical importance of non-autonomous autocatalytic sets can be found e.g. in the field of compiler construction and in operating systems, where the self-referential nature of the respective constructions is explicitly discussed, very often in terms of the chicken and egg problem.

41.6 References

[1] Lincoln TA, Joyce GF (February 2009). "Self-sustained replication of an RNA enzyme". *Science* 323 (5918): 1229–32. doi:10.1126/science.1167856. PMC 2652413. PMID 19131595.

[2] Kauffman, Stuart A. (2008) *Reinventing the Sacred: A New View of Science, Reason, and Religion.* [Basic Books] - ISBN 0-465-00300-1, chapter 5, especially pp.59-71

[3] Hordijk W (2013). "Autocatalytic Sets: From the Origin of Life to the Economy". *BioScience* 63 (11): 887–881. doi:10.1525/bio.2013.63.11.6.

[4] Mossel E, Steel M. (2005). "Random biochemical networks and the probability of self-sustaining autocatalysis". *Journal of Theoretical Biology* 233 (3): 327–336. doi:10.1016/j.jtbi.2004.10.011.

41.7 See also

- Autocatalytic reactions and order creation
- Autopoiesis

Chapter 42

Multi-agent system

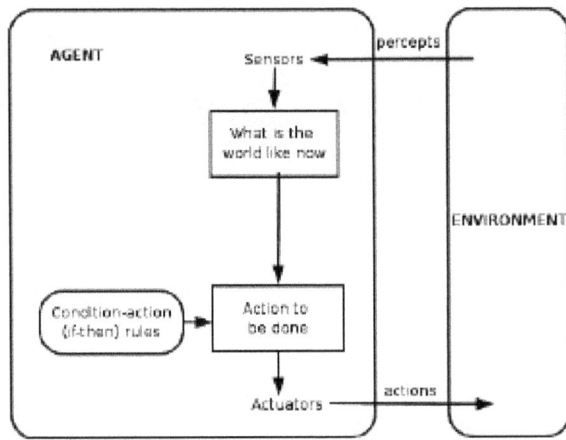

Simple reflex agent

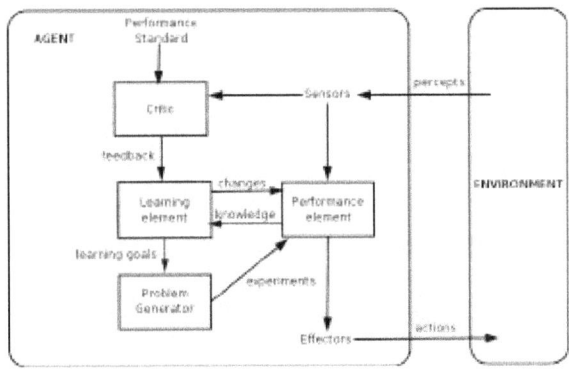

Learning agent

A **multi-agent system** (M.A.S.) is a computerized system composed of multiple interacting intelligent agents within an environment. Multi-agent systems can be used to solve problems that are difficult or impossible for an individual agent or a monolithic system to solve. Intelligence may include some methodic, functional, procedural approach, algorithmic search or reinforcement learning. Although there is considerable overlap, a multi-agent system is not always the same as an agent-based model (ABM). The goal of an ABM is to search for explanatory insight into the collective behavior of agents (which don't necessarily need to be "intelligent") obeying simple rules, typically in natural systems, rather than in solving specific practical or engineering problems. The terminology of ABM tends to be used more often in the sciences, and MAS in engineering and technology.[1] Topics where multi-agent systems research may deliver an appropriate approach include on-line trading,[2] disaster response,[3][4] and modelling social structures.[5]

42.1 Concept

Multi-agent systems consist of agents and their environment. Typically multi-agent systems research refers to software agents. However, the agents in a multi-agent system could equally well be robots,[6] humans or human teams. A multi-agent system may contain combined human-agent teams.

Agents can be divided into different types ranging from simple to complex. Some categories suggested to define these types include:

- **Passive agents**[7] or agent without goals (like obstacle, apple or key in any simple simulation)
- **Active agents**[7] with simple goals (like birds in flocking, or wolf–sheep in prey-predator model)
- Cognitive agents (which contain complex calculations)

Agent environments can be divided into:

- Virtual Environment
- Discrete Environment
- Continuous Environment

Agent environments can also be organized according to various properties like: accessibility (depending on if it is possible to gather complete information about the environment), determinism (if an action performed in the environment causes a definite effect), dynamics (how many entities influence the environment in the moment), discreteness (whether the number of possible actions in the environment is finite), episodicity (whether agent actions in certain time periods influence other periods),[8] and dimensionality (whether spatial characteristics are important factors of the environment and the agent considers space in its decision making).[9] Agent actions in the environment are typically mediated via an appropriate middleware. This middleware offers a first-class design abstraction for multi-agent systems, providing means to govern resource access and agent coordination.[10]

42.1.1 Characteristics

The agents in a multi-agent system have several important characteristics:[11]

- **Autonomy**: the agents are at least partially independent, self-aware, autonomous
- **Local views**: no agent has a full global view of the system, or the system is too complex for an agent to make practical use of such knowledge
- **Decentralization**: there is no designated controlling agent (or the system is effectively reduced to a monolithic system)[12]

42.1.2 Self-organization and self-steering

Multi-agent systems can manifest self-organization as well as self-steering and other control paradigms and related complex behaviors even when the individual strategies of all their agents are simple. When agents can share knowledge using any agreed language, within the constraints of the system's communication protocol, the approach may lead to a common improvement. Example languages are Knowledge Query Manipulation Language (KQML) or FIPA's Agent Communication Language (ACL).

42.1.3 Systems paradigms

Many M.A. systems are implemented in computer simulations, stepping the system through discrete "time steps". The MAS components communicate typically using a weighted request matrix, e.g.

Speed-VERY_IMPORTANT: min=45 mph, Path length-MEDIUM_IMPORTANCE: max=60 expectedMax=40, Max-Weight-UNIMPORTANT Contract Priority-REGULAR

and a weighted response matrix, e.g.

Speed-min:50 but only if weather sunny, Path length:25 for sunny / 46 for rainy Contract Priority-REGULAR note - ambulance will override this priority and you'll have to wait

A challenge-response-contract scheme is common in MAS systems, where

First a "**Who can?**" question is distributed. Only the relevant components respond: "**I can, at this price**". Finally, a contract is set up, usually in several more short communication steps between sides,

also considering other components, evolving "contracts", and the restriction sets of the component algorithms.

Another paradigm commonly used with MAS systems is the pheromone, where components "leave" information for other components "next in line" or "in the vicinity". These "pheromones" may "evaporate" with time, that is their values may decrease (or increase) with time.

42.1.4 Properties

M.A. systems, also referred to as "self-organized systems", tend to find the best solution for their problems "without intervention". There is high similarity here to physical phenomena, such as energy minimizing, where physical objects tend to reach the lowest energy possible within the physically constrained world. For example: many of the cars entering a metropolis in the morning will be available for leaving that same metropolis in the evening.

The main feature which is achieved when developing multi-agent systems, is flexibility, since a multi-agent system can be added to, modified and reconstructed, without the need for detailed rewriting of the application.[13] The systems also tend to prevent propagation of faults, self-recover and be fault tolerant, mainly due to the redundancy of components.

42.2 Study of multi-agent systems

The study of multi-agent systems is "concerned with the development and analysis of sophisticated AI problem-solving and control architectures for both single-agent and multiple-agent systems."[14] Topics of research in MAS include:

- agent-oriented software engineering
- beliefs, desires, and intentions (BDI)
- cooperation and coordination
- distributed constraint optimization (DCOPs)

- organization
- communication
- negotiation
- distributed problem solving
- multi-agent learning
- agent mining
- scientific communities (e.g., on biological flocking, language evolution, and economics)[15][16]
- dependability and fault-tolerance
- robotics,[17] multi-robot systems (MRS), robotic clusters

42.3 Frameworks

While ad hoc multi-agent systems are often created from scratch by researchers and developers, some frameworks have arisen that implement common standards (such as the FIPA agent system platforms and communication languages). These frameworks save developers time and also aid in the standardization of MAS development. One such developmental framework for robotics is given in .[18] See also Comparison of agent-based modeling software.

42.4 Applications in the real world

Multi-agent systems are applied in the real world to graphical applications such as computer games. Agent systems have been used in films.[19] They are also used for coordinated defence systems. Other applications include transportation,[20] logistics,[21] graphics, GIS as well as in many other fields. It is widely being advocated for use in networking and mobile technologies, to achieve automatic and dynamic load balancing, high scalability, and self-healing networks.

42.5 See also

- Comparison of agent-based modeling software
- Agent-based computational economics (ACE)
- Artificial brain
- Artificial intelligence
- Artificial life
- Artificial life framework
- Complex systems
- Discrete event simulation
- Distributed artificial intelligence
- Emergence
- Evolutionary computation
- Human-based genetic algorithm
- Knowledge Query and Manipulation Language (KQML)
- Microbial intelligence
- Multi-agent planning
- Pattern-oriented modeling
- PlatBox Project
- Reinforcement learning
- Scientific community metaphor
- Self-reconfiguring modular robot
- Simulated reality
- Social simulation
- Software agent
- Swarm intelligence

42.6 References

[1] Niazi, Muaz; Hussain, Amir (2011). "Agent-based Computing from Multi-agent Systems to Agent-Based Models: A Visual Survey" (PDF). *Scientometrics* (Springer) **89** (2): 479–499. doi:10.1007/s11192-011-0468-9.

[2] Rogers, Alex; David, E.; Schiff, J.; Jennings, N.R. (2007). "The Effects of Proxy Bidding and Minimum Bid Increments within eBay Auctions". *ACM Transactions on the Web*.

[3] Schurr, Nathan; Marecki, Janusz; Tambe, Milind; Scerri, Paul; Kasinadhuni, Nikhil; Lewis, J.P. (2005). "The Future of Disaster Response: Humans Working with Multiagent Teams using DEFACTO" (PDF).

[4] Genc, Zulkuf; et al. (2013). "Agent-based information infrastructure for disaster management]" (PDF). *Intelligent Systems for Crisis Management*: 349–355.

[5] Sun, Ron; Naveh, Isaac. "Simulating Organizational Decision-Making Using a Cognitively Realistic Agent Model". *Journal of Artificial Societies and Social Simulation*.

[6] Kaminka, G. A. (December 2004). "Robots are Agents, Too!]". *AgentLink News*: 16–17.

[7] Kubera, Yoann; Mathieu, Philippe; Picault, Sébastien (2010), "Everything can be Agent!" (PDF), *Proceedings of the ninth International Joint Conference on Autonomous Agents and Multi-Agent Systems (AAMAS'2010)* (Toronto, Canada): 1547–1548

[8] Russell, Stuart J.; Norvig, Peter (2003), *Artificial Intelligence: A Modern Approach* (2nd ed.), Upper Saddle River, New Jersey: Prentice Hall, ISBN 0-13-790395-2

[9] Salamon, Tomas (2011). *Design of Agent-Based Models*. Repin: Bruckner Publishing. p. 22. ISBN 978-80-904661-1-1.

[10] Weyns, Danny; Omicini, Amdrea; Odell, James (2007). "Environment as a first-class abstraction in multiagent systems" (PDF). *Autonomous Agents and Multi-Agent Systems* **14** (1): 5–30. doi:10.1007/s10458-006-0012-0. Retrieved 2013-05-31.

[11] Wooldridge, Michael (2002). *An Introduction to MultiAgent Systems*. John Wiley & Sons. p. 366. ISBN 0-471-49691-X.

[12] Panait, Liviu; Luke, Sean (2005). "Cooperative Multi-Agent Learning: The State of the Art" (PDF). *Autonomous Agents and Multi-Agent Systems* **11** (3): 387–434. doi:10.1007/s10458-005-2631-2.

[13] Rzevski & Skobelev, Managing Complexity (2014) Wit Press

[14] "The Multi-Agent Systems Lab". University of Massachusetts Amherst. Retrieved Oct 16, 2009.

[15] Cucker, Felipe; Steve Smale (2007). "The Mathematics of Emergence" (PDF). *Japanese Journal of Mathematics* **2**: 197. doi:10.1007/s11537-007-0647-x. Retrieved 2008-06-09.

[16] Shen, Jackie (Jianhong) (2008). "Cucker–Smale Flocking under Hierarchical Leadership". *SIAM J. Applied Math.* **68** (3): 694. doi:10.1137/060673254. Retrieved 2008-06-09.

[17] Ahmed, S.; Karsiti, M.N. (2007). "A testbed for control schemes using multi agent nonholonomic robots". *2007 IEEE International Conference on Electro/Information Technology*. doi:10.1109/EIT.2007.4374547.

[18] Ahmed, Salman; Karsiti, Mohd N.; Agustiawan, Herman (2007). "A development framework for collaborative robots using feedback control".

[19] "Film showcase". MASSIVE. Retrieved 28 April 2012.

[20] Xiao-Feng Xie, S. Smith, G. Barlow. Schedule-driven coordination for real-time traffic network control. International Conference on Automated Planning and Scheduling (ICAPS), Sao Paulo, Brazil, 2012: 323-331.

[21] Máhr, T. S.; Srour, J.; De Weerdt, M.; Zuidwijk, R. (2010). "Can agents measure up? A comparative study of an agent-based and on-line optimization approach for a drayage problem with uncertainty". *Transportation Research Part C: Emerging Technologies* **18**: 99. doi:10.1016/j.trc.2009.04.018.

42.7 Further reading

- Wooldridge, Michael (2002). *An Introduction to MultiAgent Systems*. John Wiley & Sons. p. 366. ISBN 0-471-49691-X.

- Shoham, Yoav; Leyton-Brown, Kevin (2008). *Multiagent Systems: Algorithmic, Game-Theoretic, and Logical Foundations*. Cambridge University Press. p. 496. ISBN 978-0-521-89943-7.

- Mamadou, Tadiou Koné; Shimazu, A.; Nakajima, T. (August 2000). "The State of the Art in Agent Communication Languages (ACL)". *Knowledge and Information Systems Journal (KAIS)* (London: Springer-Verlag) **2** (2): 1–26.

- Hewitt, Carl; Inman, Jeff (Nov–Dec 1991). "DAI Betwixt and Between: From "Intelligent Agents" to Open Systems Science". *IEEE Transactions on Systems, Man, and Cybernetics*.

- *The Journal of Autonomous Agents and Multi-Agent Systems (JAAMAS)*

- Weiss, Gerhard, ed. (1999). *Multiagent Systems, A Modern Approach to Distributed Artificial Intelligence*. MIT Press. ISBN 0-262-23203-0.

- Ferber, Jacques (1999). *Multi-Agent Systems: An Introduction to Artificial Intelligence*. Addison-Wesley. ISBN 0-201-36048-9.

- Sun, Ron (2006). *Cognition and Multi-Agent Interaction*. Cambridge University Press. ISBN 0-521-83964-5.

- Keil, David; Goldin, Dina (2006). Weyns, Danny; Parunak, Van; Michel, Fabien, eds. "Indirect Interaction in Environments for Multiagent Systems" (PDF). *Environments for Multiagent Systems II*. LNCS 3830 (Springer).

- *Whitestein Series in Software Agent Technologies and Autonomic Computing*, published by Springer Science+Business Media Group

- Salamon, Tomas (2011). *Design of Agent-Based Models: Developing Computer Simulations for a Better Understanding of Social Processes*. Bruckner Publishing. ISBN 978-80-904661-1-1.

- Russell, Stuart J.; Norvig, Peter (2003), *Artificial Intelligence: A Modern Approach* (2nd ed.), Upper Saddle River, New Jersey: Prentice Hall, ISBN 0-13-790395-2

- Fasli, Maria (2007). *Agent-technology for E-commerce*. John Wiley & Sons. p. 480. ISBN 978-0-470-03030-1.

- Cao, Longbing, Gorodetsky, Vladimir, Mitkas, Pericles A. (2009). Agent Mining: The Synergy of Agents and Data Mining, IEEE Intelligent Systems, vol. 24, no. 3, 64-72.

42.8 External links

- Random Agent-Based Simulations by Borys Biletskyy – Random agent-base simulations for multi-robot system and Belousov-Zhabotinsky reaction. Java applets available.

- CORMAS (COmmon Resources Multi-Agent System) An open-source framework for Multi-Agent Systems based on SmallTalk. Spatialized, it focuses on issues related to natural resource management and negotiation between stakeholders.

- JaCaMo MAS Platform - An open-source platform for Multi-Agent Systems based on Jason, CArtAgO, and Moise.

- Janus multiagent Platform – Holonic multiagent execution platform (Apache License), written in Java, and directly supporting the SARL agent-oriented programming language.

- HarTech Technologies - HarTech Technologies developed a dedicated Distributed Multi Agent System Framework used in both simulation and large scale command and control system. This unique framework called the Generic Blackboard (GBB) provides a development framework for such systems which is domain independent. Distributed Multi Agent Framework.

- MaDKit is a lightweight open source Java library for designing and simulating Multi-Agent Systems. MaDKit is built upon the AGR (Agent/Group/Role) organizational model: agents are situated in groups and play roles, MAS are conceived as artificial societies.

- Agent Factory Agent Factory is an open source collection of tools, platforms, and languages that support the development and deployment of Multi-Agent Systems.

Chapter 43

Self-organizing network

A **Self-Organizing Network** (SON) is an automation technology designed to make the planning, configuration, management, optimization and healing of mobile radio access networks simpler and faster. SON functionality and behavior has been defined and specified in generally accepted mobile industry recommendations produced by organizations such as 3GPP (3rd Generation Partnership Project) and the NGMN (Next Generation Mobile Networks).

SON has been codified within 3GPP Release 8 and subsequent specifications in a series of standards including 36.902,[1] as well as public white papers outlining use cases from the NGMN.[2] The first technology making use of SON features will be Long Term Evolution (LTE), but the technology has also been retro-fitted to older radio access technologies such as Universal Mobile Telecommunications System (UMTS). The LTE specification inherently supports SON features like Automatic Neighbor Relation (ANR) detection, which is the 3GPP LTE Rel. 8 flagship feature.[3]

Newly added base stations should be self-configured in line with a "plug-and-play" paradigm while all operational base stations will regularly self-optimize parameters and algorithmic behavior in response to observed network performance and radio conditions. Furthermore, self-healing mechanisms can be triggered to temporarily compensate for a detected equipment outage, while awaiting a more permanent solution.

43.1 SON architectural types

Self-organizing networks are commonly divided into three major architectural types.

43.1.1 Distributed SON

In this type of SON (D-SON), functions are distributed among the network elements at the edge of the network, typically the ENodeB elements. This implies a certain degree of localization of functionality and is normally supplied by the network equipment vendor manufacturing the radio cell.

43.1.2 Centralized SON

In centralized SON (C-SON), function is more typically concentrated closer to higher-order network nodes or the network OSS, to allow a broader overview of more edge elements and coordination of e.g. load across a wide geographic area. Due to the need to inter-work with cells supplied by different equipment vendors, C-SON systems are more typically supplied by 3rd parties.

43.1.3 Hybrid SON

Hybrid SON is a mix of centralized and distributed SON, combining elements of each in a hybrid solution.

43.2 SON sub-functions

Self-organizing network functionalities are commonly divided into three major sub-functional groups, each containing a wide range of decomposed use cases.

43.2.1 Self-configuration functions

Self-configuration strives towards the "plug-and-play" paradigm in the way that new base stations shall automatically be configured and integrated into the network. This means both connectivity establishment, and download of configuration parameters are software. Self-configuration is typically supplied as part of the software delivery with each radio cell by equipment vendors. When a new base station is introduced into the network and powered on, it gets immediately recognized and registered by the network. The neighboring base stations then automatically adjust their technical parameters (such as emission power, antenna tilt, etc.) in order to provide the required coverage and capacity, and, in the same time, avoid the interference.

43.2.2 Self-optimization functions

Main article: Self-optimization

Every base station contains hundreds of configuration parameters that control various aspects of the cell site. Each of these can be altered to change network behavior, based on observations of both the base station itself and measurements at the mobile station or handset. One of the first SON features establishes neighbor relations automatically (ANR) while others optimize random access parameters or mobility robustness in terms of handover oscillations. A very illustrative use case is the automatic switch-off of a percent of base stations during the night hours. The neighboring base station would then re-configure their parameters in order to keep the entire area covered by the signal. In case of a sudden growth in connectivity demand for any reason, the "sleeping" base stations "wake up" almost instantaneously. This mechanism leads to significant energy savings for operators.

43.2.3 Self-healing functions

When some nodes in the network become inoperative, self-healing mechanisms aim at reducing the impacts from the failure, for example by adjusting parameters and algorithms in adjacent cells so that other nodes can support the users that were supported by the failing node. In legacy networks, the failing base stations are at times hard to identify and a significant amount of time and resources is required to fix it. This function of SON permits to spot such a failing base stations immediately in order to take further measures, and ensure no or insignificant degradation of service for the users.

See also: Self-healing ring and Resilience (network)

43.3 Introduction of SON

Self-organizing Networks features are being introduced gradually with the arrival of new 4G systems in radio access networks, allowing for the impact of potential 'teething troubles' to be limited and gradually increasing confidence. Self-optimization mechanisms in mobile radio access networks can be seen to have some similarities to automated trading algorithms in financial markets. SON has also been retrofitted to existing 3G networks to help reduce cost and improve service reliability.

The Mobile World Congress trade conference in 2009 saw the first major announcements of SON functionality for LTE mobile networks. First deployments occurred in Japan and USA during 2009/10.[4]

The first commercial live test of one of the official 3GPP SON functions was held 12 November by TeliaSonera and Ericsson.[5] It verified the Automatic Neighbor Relations (ANR) feature in parts of a commercially deployed network in Sweden, using a mobile for a commercial chipset vendor. ANR automatically sets up neighbor relations based on actual radio conditions.

43.4 References

[1] 3GPP

[2] NGMN Downloads

[3] Nomor

[4] Dallas News

[5] Self-Organizing Network solution tested in live LTE Network

43.5 Literature

C. Brunner, D. Flore: *Generation of Pathloss and Interference Maps as SON Enabler in Deployed UMTS Networks.* In: *Proceedings of IEEE Vehicular Technology Conf. (VTC Spring '09).* Barcelona, Spain, April 2009

43.6 External links

- 3GPP
- NGMN
- Nomor Research: White Paper "Self-Organizing Networks (SON) in 3GPP Long Term Evolution"
- Self-Organizing Networks, SON, tutorial covering self configuration, optimization and healing

Chapter 44

Dual-phase evolution

Dual phase evolution (DPE) is a process that drives self-organization within complex adaptive systems.[1] It arises in response to phase changes within the network of connections formed by a system's components. DPE occurs in a wide range of physical, biological and social systems. Its applications to technology include methods for manufacturing novel materials and algorithms to solve complex problems in computation.

44.1 Introduction

Dual phase evolution (DPE) is a process that promotes the emergence of large-scale order in complex systems. It occurs when a system repeatedly switches between various kinds of phases, and in each phase different processes act on the components or connections in the system. DPE arises because of a property of graphs and networks: the connectivity avalanche that occurs in graphs as the number of edges increases.[2]

Social networks provide a familiar example. In a social network the nodes of the network are people and the network connections (edges) are relationships or interactions between people. For any individual, social activity alternates between a *local phase*, in which they interact only with people they already know, and a *global phase* in which they can interact with a wide pool of people not previously known to them. Historically, these phases have been forced on people by constraints of time and space. People spend most of their time in a local phase and interact only with those immediately around them (family, neighbors, colleagues). However, intermittent activities such as parties, holidays, and conferences involve a shift into a global phase where they can interact with different people they do not know. Different processes dominate each phase. Essentially, people make new social links when in the global phase, and refine or break them (by ceasing contact) while in the local phase.

44.2 The DPE mechanism

The following features are necessary for DPE to occur.[1]

44.2.1 Underlying network

DPE occurs where a system has an underlying network. That is, the system's components form a set of nodes and there are connections (edges) that join them. For example, a family tree is a network in which the nodes are people (with names) and the edges are relationships such as "mother of" or "married to". The nodes in the network can take physical form, such as atoms held together by atomic forces, or they may be dynamic states or conditions, such as positions on a chess board with moves by the players defining the edges.

In mathematical terms (graph theory), a graph $G = \langle N, E \rangle$ is a set of nodes N and a set of edges $E \subset \{(x,y) \mid x, y \in N\}$. Each edge (x,y) provides a link between a pair of nodes x and y. A network is a graph in which values are assigned to the nodes and/or edges.

44.2.2 Phase shifts

Graphs and networks have two phases: disconnected (fragmented) and connected. In the connected phase every node is connected by an edge to at least one other node and for any pair of nodes, there is at least one path (sequence of edges) joining them.

The Erdős–Rényi model shows that random graphs undergo a connectivity avalanche as the density of edges in a graph increases.[2] This avalanche amounts to a sudden phase change in the size of the largest connected subgraph. In effect, a graph has two phases: connected (most nodes are linked by pathways of interaction) and fragmented (nodes are either isolated or form small subgraphs). These are often referred to as **global** and **local** phases, respectively.

An essential feature of DPE is that the system undergoes

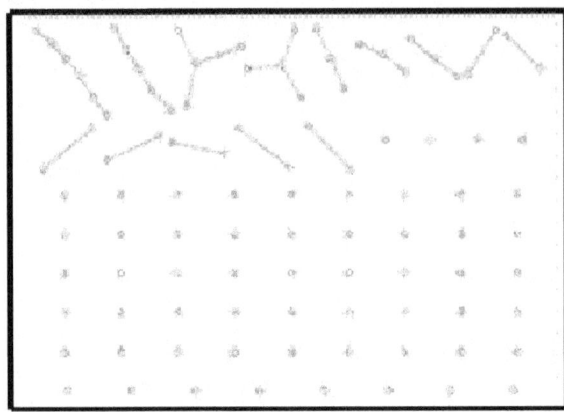

Fragmented graph.

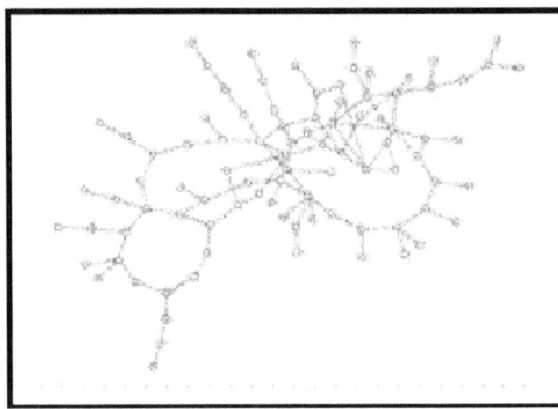

Connected graph.

repeated shifts between the two phases. In many cases, one phase is the system's normal state and it remains in that phase until shocked into the alternate phase by a disturbance, which may be external in origin.

44.2.3 Selection and variation

In each of the two phases, the network is dominated by different processes.[1] In a local phase, the nodes behave as individuals; in the global phase, nodes are affected by interactions with other nodes. Most commonly the two processes at work can be interpreted as *variation* and *selection*. Variation refers to new features, which typically appear in one of the two phases. These features may be new nodes, new edges, or new properties of the nodes or edges. Selection here refers to ways in which the features are modified, refined, selected or removed. A simple example would be new edges being added at random in the global phase and edges being selectively removed in the local phase.

44.2.4 System memory

The effects of changes in one phase carry over into the other phase. This means that the processes acting in each phase can modify or refine patterns formed in the other phase. For instance, in a social network, if a person makes new acquaintances during a global phase, then some of these new social connections might survive into the local phase to become long-term friends. In this way, DPE can create effects that may be impossible if both processes act at the same time.

44.3 Examples

DPE has been found to occur in many natural and artificial systems.[3]

44.3.1 Social networks

DPE is capable of producing social networks with known topologies, notably small-world networks and scale-free networks.[3] Small world networks, which are common in traditional societies, are a natural consequence of alternating *local* and *global* phases: new, long-distance links are formed during the global phase and existing links are reinforced (or removed) during the local phase. The advent of social media has decreased the constraining influence that space used to impose on social communication, so time has become the chief constraint for many people.

The alternation between local and global phases in social networks occurs in many different guises. Some transitions between phases occur regularly, such as the daily cycle of people moving between home and work. This alternation can influence shifts in public opinion.[4] In the absence of social interaction, the uptake of an opinion promoted by media is a Markov process. The effect of social interaction under DPE is to retard the initial takeup until the number converted reaches a critical point, after which takeup accelerates rapidly.

44.3.2 Socio-economics

DPE models of socio-economics interpret the economy as networks of economic agents.[5] Several studies have examined the way socioeconomics evolve when DPE acts on different parts of the network. One model[6] interpreted society as a network of occupations with inhabitants matched to those occupations. In this model social dynamics become a process of DPE within the network, with regular transitions between a development phase, during which the network settles into an equilibrium state, and a mutating phase, during which the network is transformed in random ways by the creation of new occupations.

Another model[7] interpreted growth and decline in socioeconomic activity as a conflict between cooperators and defectors. The cooperators form networks that lead to prosperity. However, the network is unstable and invasions by defectors intermittently fragment the network, reducing prosperity, until invasions of new cooperators rebuild networks again. Thus prosperity is seen as a dual phase process of alternating highly prosperous, connected phases and unprosperous, fragmented phases.

44.3.3 Forest ecology

In a forest, the landscape can be regarded as a network of sites where trees might grow.[8] Some sites are occupied by living trees; others sites are empty. In the local phase, sites free of trees are few and they are surrounded by forest, so the network of free sites is fragmented. In competition for these free sites, local seed sources have a massive advantage, and seeds from distant trees are virtually excluded.[1] Major fires (or other disturbances) clear away large tracts of land, so the network of free sites becomes connected and the landscape enters a global phase. In the global phase, competition for free sites is reduced, so the main competitive advantage is adaptation to the environment.

Most of the time a forest is in the local phase, as described above. The nett effect is that established tree populations largely exclude invading species.[9] Even if a few isolated trees do find free ground, their population is prevented from expanding by established populations, even if the invaders are better adapted to the local environment. A fire in such conditions leads to an explosion of the invading population, and possibly to a sudden change in the character of the entire forest.

This dual phase process in the landscape explains the consist appearance of pollen zones in the postglacial forest history of North America, Europe, as well as the suppression of widespread taxa, such as Beech and Hemlock, followed by huge population explosions. Similar patterns, pollen zones truncated by fire-induced boundaries, have been recorded in most parts of the world

44.3.4 Search algorithms

Dual phase evolution is a family of search algorithms that exploit phase changes in the search space to mediate between local and global search. In this way they control the way algorithms explore a search space, so they can be regarded as a family of meta heuristic methods.

Problems such as optimization can typically be interpreted as finding the tallest peak (optimum) within a search space of possibilities. The task can be approached in two ways: *local search* (e.g. hill climbing) involves tracing a path from point to point, and always moving "uphill". *Global search* involves sampling at wide-ranging points in the search space to find high points.

Many search algorithms involve a transition between phases of global search and local search.[3] A simple example is the Great Deluge algorithm in which the searcher can move at random across the landscape, but cannot enter low-lying areas that are flooded. At first the searcher can wander freely, but rising water levels eventually confine the search to a local area. Many other nature-inspired algorithms adopt similar approaches. Simulated annealing achieves a transition between phases via its cooling schedule. The cellular genetic algorithm places solutions in a pseudo landscape in which they breed only with local neighbours. Intermittent disasters clear patches, flipping the system into a global phase until gaps are filled again.

Some variations on the memetic algorithm involve alternating between selection at different levels. These are related to the Baldwin effect, which arises when processes acting on *phenotypes* (e.g. learning) influence selection at the level of *genotypes*. In this sense, the Baldwin effect alternates between global search (genotypes) an local search (phenotypes).

44.4 Related processes

Dual phase evolution is related to the well-known phenomenon of *self-organized criticality* (SOC). Both concern processes in which critical phase changes promote adaptation and organization within a system. However, SOC differs from DPE in several fundamental ways.[1] Under SOC, a system's natural condition is to be in a critical state; in DPE a system's natural condition is a non-critical state. In SOC the size of disturbances follows a power law; in DPE disturbances are not necessarily distributed the same way. In SOC a system is not necessarily subject to other processes; in DPE different processes (e.g. selection and variation) operate in the two phases.

44.5 See also

- Cellular evolutionary algorithm
- Complex systems
- Erdős–Rényi model
- Graph theory

- Memetic algorithm
- Meta heuristic
- Network theory
- Self-organization
- Self-organized criticality
- Scale-free network
- Simulated annealing
- Small-world network

44.6 References

[1] Green, D.G., Liu, J. and Abbass, H. (2014). *Dual Phase Evolution: from Theory to Practice*. Berlin: Springer. ISBN 978-1441984227.

[2] Erdős, P. and Rényi, A. (1960). "On the evolution of random graphs" (PDF). *Publications of the Mathematical Institute of the Hungarian Academy of Sciences* 5: 17–61.

[3] Paperin, G. and Green, D.G. and Sadedin, S. (2011). "Dual Phase Evolution in Complex Adaptive Systems" (PDF). *Journal of the Royal Society Interface* 8 (58): 609–629. doi:10.1098/rsif.2010.0719.

[4] Stocker, R. and Cornforth, D. and Green, D.G. (2003). "A simulation of the impact of media on social cohesion". *Advances in Complex Systems* 6 (3): 349–359. doi:10.1142/S0219525903000931.

[5] Goodman, J. (2014). "Evidence for ecological learning and domain specificity in rational asset pricing and market efficiency" (PDF). *The Journal of Socio-Economics* 48: 27–39. doi:10.1016/j.socec.2013.10.002.

[6] Xu, G. and Yang, J. and Li, G. (2013). "Simulating society transitions: standstill, collapse and growth in an evolving network model". *PLOS ONE* 8 (9): e75433. Bibcode:2013PLoSO...875433X. doi:10.1371/journal.pone.0075433. PMC 3783390. PMID 24086530.

[7] Cavaliere, M. AND Sedwards, C. AND Tarnita, C.E. AND Nowak, M.A. AND Csikász-Nagy, A. (2012). "Prosperity is associated with instability in dynamical networks" (PDF). *Journal of Theoretical Biology* 299: 126–138. doi:10.1016/j.jtbi.2011.09.005.

[8] Green, David G. (1994). "Connectivity and complexity in ecological systems". *Pacific Conservation Biology* 1 (3): 194–200.

[9] Green, David G (1982). "Fire and stability in the postglacial forests of southwest Nova Scotia". *Journal of Biogeography* 9 (1): 29–40. doi:10.2307/2844728.

Chapter 45

Molecular assembler

A **molecular assembler**, as defined by K. Eric Drexler, is a "proposed device able to guide chemical reactions by positioning reactive molecules with atomic precision". A molecular assembler is a kind of molecular machine. Some biological molecules such as ribosomes fit this definition. This is because they receive instructions from messenger RNA and then assemble specific sequences of amino acids to construct protein molecules. However, the term "molecular assembler" usually refers to theoretical human-made devices.

Beginning in 2007, the British Engineering and Physical Sciences Research Council has funded development of ribosome-like molecular assemblers. Clearly, molecular assemblers are possible in this limited sense. A technology roadmap project, led by the Battelle Memorial Institute and hosted by several U.S. National Laboratories has explored a range of atomically precise fabrication technologies, including both early-generation and longer-term prospects for programmable molecular assembly; the report was released in December, 2007.[1] In 2008 the Engineering and Physical Sciences Research Council provided funding of 1.5 million pounds over six years for research working towards mechanized mechanosynthesis, in partnership with the Institute for Molecular Manufacturing, amongst others.[2]

Likewise, the term "molecular assembler" has been used in science fiction and popular culture to refer to a wide range of fantastic atom-manipulating nanomachines, many of which may be physically impossible in reality. Much of the controversy regarding "molecular assemblers" results from the confusion in the use of the name for both technical concepts and popular fantasies. In 1992, Drexler introduced the related but better-understood term "molecular manufacturing," which he defined as the programmed "chemical synthesis of complex structures by mechanically positioning reactive molecules, not by manipulating individual atoms."[3]

This article mostly discusses "molecular assemblers" in the popular sense. These include hypothetical machines that manipulate individual atoms and machines with organism-like self-replicating abilities, mobility, ability to consume food, and so forth. These are quite different from devices that merely (as defined above) "guide chemical reactions by positioning reactive molecules with atomic precision".

Because synthetic molecular assemblers have never been constructed and because of the confusion regarding the meaning of the term, there has been much controversy as to whether "molecular assemblers" are possible or simply science fiction. Confusion and controversy also stem from their classification as nanotechnology, which is an active area of laboratory research which has already been applied to the production of real products; however, there had been, until recently, no research efforts into the actual construction of "molecular assemblers".

Nonetheless, a 2013 paper published in the journal *Science* details a new method of synthesizing a peptide in a sequence-specific manner by using an artificial molecular machine that is guided by a molecular strand. This functions in the same way as a ribosome building proteins by assembling amino acids according to a messenger RNA blueprint. The structure of the machine is based on a rotaxane, which is a molecular ring sliding along a molecular axle. The ring carries a thiolate group which removes amino acids in sequence from the axle, transferring them to a peptide assembly site.[4]

In another paper published in March 2015, also in *Science*, chemists at the University of Illinois report a platform that automates the synthesis of 14 classes of small molecules, with thousands of compatible building blocks.[5]

45.1 Nanofactories

A **nanofactory** is a proposed system in which nanomachines (resembling molecular assemblers, or industrial robot arms) would combine reactive molecules via mechanosynthesis to build larger atomically precise parts. These, in turn, would be assembled by positioning mechanisms of assorted sizes to build macroscopic (visible) but still atomically-precise products.

A typical nanofactory would fit in a desktop box, in the vision of K. Eric Drexler published in *Nanosystems: Molecular Machinery, Manufacturing and Computation* (1992), a notable work of "exploratory engineering". During the last decade, others have extended the nanofactory concept, including an analysis of nanofactory convergent assembly by Ralph Merkle, a systems design of a replicating nanofactory architecture by J. Storrs Hall, Forrest Bishop's "Universal Assembler", the patented exponential assembly process by Zyvex, and a top-level systems design for a 'primitive nanofactory' by Chris Phoenix (Director of Research at the Center for Responsible Nanotechnology). All of these nanofactory designs (and more) are summarized in Chapter 4 of *Kinematic Self-Replicating Machines* (2004) by Robert Freitas and Ralph Merkle. The Nanofactory Collaboration,[6] founded by Freitas and Merkle in 2000, is a focused ongoing effort involving 23 researchers from 10 organizations and 4 countries that is developing a practical research agenda[7] specifically aimed at positionally-controlled diamond mechanosynthesis and diamondoid nanofactory development.

In 2005, a computer-animated short film of the nanofactory concept was produced by John Burch, in collaboration with Drexler. Such visions have been the subject of much debate, on several intellectual levels. No one has discovered an insurmountable problem with the underlying theories and no one has proved that the theories can be translated into practice. However, the debate continues, with some of it being summarized in the molecular nanotechnology article.

If nanofactories could be built, severe disruption to the world economy would be one of many possible negative impacts, though it could be argued that this disruption would have little negative effect if everyone had such nanofactories. Great benefits also would be anticipated. Various works of science fiction have explored these and similar concepts. The potential for such devices was part of the mandate of a major UK study led by mechanical engineering professor Dame Ann Dowling.

45.2 Self-replication

"Molecular assemblers" have been confused with self-replicating machines. To produce a practical quantity of a desired product, the nanoscale size of a typical science fiction universal molecular assembler requires an extremely large number of such devices. However, a single such theoretical molecular assembler might be programmed to self-replicate, constructing many copies of itself. This would allow an exponential rate of production. Then after sufficient quantities of the molecular assemblers were available, they would then be re-programmed for production of the desired product. However, if self-replication of molecular assemblers were not restrained then it might lead to competition with naturally occurring organisms. This has been called ecophagy or the grey goo problem.[8]

One method to building molecular assemblers is to mimic evolutionary processes employed by biological systems. Biological evolution proceeds by random variation combined with culling of the less-successful variants and reproduction of the more-successful variants. Production of complex molecular assemblers might be evolved from simpler systems since "A complex system that works is invariably found to have evolved from a simple system that worked. . . . A complex system designed from scratch never works and can not be patched up to make it work. You have to start over, beginning with a system that works."[9] However, most published safety guidelines include "recommendations against developing ... replicator designs which permit surviving mutation or undergoing evolution".[10]

Most assembler designs keep the "source code" external to the physical assembler. At each step of a manufacturing process, that step is read from an ordinary computer file and "broadcast" to all the assemblers. If any assembler gets out of range of that computer, or when the link between that computer and the assemblers is broken, or when that computer is unplugged, the assemblers stop replicating. Such a "broadcast architecture" is one of the safety features recommended by the "Foresight Guidelines on Molecular Nanotechnology", and a map of the 137-dimensional replicator design space[11] recently published by Freitas and Merkle provides numerous practical methods by which replicators can be safely controlled by good design.

45.3 Drexler and Smalley debate

Main article: Drexler–Smalley debate on molecular nanotechnology

One of the most outspoken critics of some concepts of "molecular assemblers" was Professor Richard Smalley (1943–2005) who won the Nobel prize for his contributions to the field of nanotechnology. Smalley believed that such assemblers were not physically possible and introduced scientific objections to them. His two principal technical objections were termed the "fat fingers problem" and the "sticky fingers problem". He believed these would exclude the possibility of "molecular assemblers" that worked by precision picking and placing of individual atoms. Drexler and coworkers responded to these two issues[12] in a 2001 publication.

Smalley also believed that Drexler's speculations about apocalyptic dangers of self-replicating machines that have been equated with "molecular assemblers" would threaten the public support for development of nanotechnology. To address the debate between Drexler and Smalley regarding molecular assemblers *Chemical & Engineering News* published a point-counterpoint consisting of an exchange of letters that addressed the issues.[3]

45.4 Regulation

Speculation on the power of systems that have been called "molecular assemblers" has sparked a wider political discussion on the implication of nanotechnology. This is in part due to the fact that nanotechnology is a very broad term and could include "molecular assemblers." Discussion of the possible implications of fantastic molecular assemblers has prompted calls for regulation of current and future nanotechnology. There are very real concerns with the potential health and ecological impact of nanotechnology that is being integrated in manufactured products. Greenpeace for instance commissioned a report concerning nanotechnology in which they express concern into the toxicity of nanomaterials that have been introduced in the environment.[13] However, it makes only passing references to "assembler" technology. The UK Royal Society and Royal Academy of Engineering also commissioned a report entitled "Nanoscience and nanotechnologies: opportunities and uncertainties"[14] regarding the larger social and ecological implications on nanotechnology. This report does not discuss the threat posed by potential so-called "molecular assemblers."

45.5 Formal scientific review

In 2006, U.S. National Academy of Sciences released the report of a study of molecular manufacturing as part of a longer report, *A Matter of Size: Triennial Review of the National Nanotechnology Initiative*[15] The study committee reviewed the technical content of *Nanosystems*, and in its conclusion states that no current theoretical analysis can be considered definitive regarding several questions of potential system performance, and that optimal paths for implementing high-performance systems cannot be predicted with confidence. It recommends experimental research to advance knowledge in this area:

> "Although theoretical calculations can be made today, the eventually attainable range of chemical reaction cycles, error rates, speed of operation, and thermodynamic efficiencies of such bottom-up manufacturing systems cannot be reliably predicted at this time. Thus, the eventually attainable perfection and complexity of manufactured products, while they can be calculated in theory, cannot be predicted with confidence. Finally, the optimum research paths that might lead to systems which greatly exceed the thermodynamic efficiencies and other capabilities of biological systems cannot be reliably predicted at this time. Research funding that is based on the ability of investigators to produce experimental demonstrations that link to abstract models and guide long-term vision is most appropriate to achieve this goal."

45.6 Grey goo

Main article: Grey goo

One potential scenario that has been envisioned is out-of-control self-replicating molecular assemblers in the form of grey goo which consumes carbon to continue its replication. If unchecked such mechanical replication could potentially consume whole ecoregions or the whole Earth (ecophagy), or it could simply outcompete natural lifeforms for necessary resources such as carbon, ATP, or UV light (which some nanomotor examples run on). However, the ecophagy and 'grey goo' scenarios, like synthetic molecular assemblers, are based upon still-hypothetical technologies that have not yet been demonstrated experimentally.

45.7 In fiction

Main article: Nanotechnology in fiction

Molecular assemblers are a popular topic in science fiction, for example, the matter compiler in The Diamond Age and the cornucopia machine in Singularity Sky. The replicator in *Star Trek* might also be considered a molecular assembler. A molecular assembler is also a key element of the plot of the computer game *Deus Ex* (called a "universal constructor" in the game).

In the political sci-fi comic series Transmetropolitan, written by Warren Ellis, machines called "Makers" are used to replicate and reform matter. Each morning, Makers sweep the streets for garbage, gathering the matter to recycle it into more useful objects. The main character also uses a Maker in his apartment to instantly produce a pair of glasses which take photos, as well as other objects such as clothing.

In Dead Money, a DLC of the video game Fallout: New Vegas, the player can obtain useful items from vending machines that use an unknown form of molecular assembly technology to transform casino chips that the player can find into any of several items.

45.8 See also

- Nanotechnology
- Molecular machine
- Bioethics
- Biosafety
- Biosecurity
- Biotechnology
- Ecocide
- Ecophagy
- Santa Claus machine
- 3D Printing

45.9 References

[1] Productive Nanosystems: a technology roadmap

[2] "Grants on the Web". Archived from the original on November 4, 2011.

[3] "C&En: Cover Story - Nanotechnology".

[4] "Sequence-Specific Peptide Synthesis by an Artificial Small-Molecule Machine".

[5] "Synthesis of many different types of organic small molecules using one automated process".

[6] "Nanofactory Collaboration".

[7] "Nanofactory Technical Challenges".

[8] "Nanotechnology: Grey Goo is a Small Issue".

[9] Gall, John, (1986) Systemantics: How Systems Really Work and How They Fail, 2nd ed. Ann Arbor, MI : The General Systemantics Press.

[10] "Foresight Guidelines on Molecular Nanotechnology".

[11] "Kinematic Self-Replicating Machines".

[12] "Institute for Molecular ManufacturingDebate About Assemblers — Smalley Rebuttal".

[13] Future Technologies, Today's Choices Nanotechnology, Artificial Intelligence and Robotics; A technical, political and institutional map of emerging technologies. A report for the Greenpeace Environmental Trust

[14] "Nanoscience and nanotechnologies:opportunities and uncertainties".

[15] "A Matter of Size: Triennial Review of the National Nanotechnology Initiative - The National Academies Press".

45.10 External links

- [Nanoengineer-1] free Open-source multi-scale modeling and simulation program for nano-composites with special support for structural DNA nanotechnology
- Nano-Hive: Nanospace Simulator free software for modeling nanotech entities
- Foresight Institute proposes guidelines for responsible development of molecular manufacturing technologies
- Center for Responsible Nanotechnology
- Molecular Assembler website
- Rage Against the (Green) Machine article originally in Wired
- Government launches nano study UK EducationGuardian, 11 June 2003
- "Unraveling the Big Debate over Small Machines"
- Paper on assembly
- http://www.zyvex.com/nanotech/nano4/merklePaper.html
- Kinematic Self-Replicating Machines <http://www.MolecularAssembler.com/KSRM.htm> online technical book: first comprehensive survey of molecular assemblers (2004) by Robert Freitas and Ralph Merkle
- Design of a Primitive Nanofactory
- Video - Nanofactory in Action
- Nanofactory technology
- Review of Molecular Manufacturing

Chapter 46

Critical mass (sociodynamics)

In social dynamics, **critical mass** is a sufficient number of adopters of an innovation in a social system so that the rate of adoption becomes self-sustaining and creates further growth. It is an aspect of the theory of diffusion of innovations, written extensively on by Everett Rogers in his book *Diffusion of Innovations*.[1] The term is borrowed from nuclear physics and in that field it refers to the amount of a substance needed to start a chain reaction.

Social factors influencing critical mass may involve the size, interrelatedness and level of communication in a society or one of its subcultures. Another is social stigma, or the possibility of public advocacy due to such a factor.

Critical mass may be closer to majority consensus in political circles, where the most effective position is more often that held by the majority of people in society. In this sense, small changes in public consensus can bring about swift changes in political consensus, due to the majority-dependent effectiveness of certain ideas as tools of political debate.

Critical mass is a concept used in a variety of contexts, including physics, group dynamics, politics, public opinion, and technology.

46.1 History

The concept of critical mass was originally created by game theorist Thomas Schelling and sociologist Mark Granovetter to explain the actions and behaviors of a wide range of people and phenomenon. The concept was first established (although not explicitly named) in Schelling's essay about racial segregation in neighborhoods, published in 1971 in the Journal of Mathematical Sociology,[2] and later refined in his book, *Micromotives and Macrobehavior*, published in 1978.[3] He did use the term "critical density" with regard to pollution in his "On the Ecology of Micromotives".[4] Mark Granovetter, in his essay "Threshold models of collective behavior", published in the American Journal of Sociology in 1978[5] worked to solidify the theory.[6] Everett Rogers later cites them both in his important work *Diffusion of Innovations*, in which critical mass plays an important role.

46.1.1 Predecessors

The concept of critical mass had existed before it entered a sociology context. It was an established concept in medicine, specifically epidemiology, since the 1920s, as it helped to explain the spread of illnesses.

It had also been a present, if not solidified, idea in the study of consumer habits and economics, especially in General Equilibrium Theory. In his papers, Schelling quotes the well-known "The Market for Lemons: Quality Uncertainty and the Market Mechanism" paper written in 1970 by George Akerlof.[7] Similarly, Granovetter cited the Nash Equilibrium game in his papers.

Finally, Herbert A. Simon's essay, "Bandwagon and underdog effects and the possibility of election predictions", published in 1954 in Public Opinion Quarterly,[8] has been cited as a predecessor to the concept we now know as critical mass.

46.2 Logic of collective action and common good

Critical mass and the theories behind it help us to understand aspects of humans as they act and interact in a larger social setting. Certain theories, such as Mancur Olson's Logic of Collective Action[9] or Garrett Hardin's Tragedy of the Commons,[10] work to help us understand why humans do or adopt certain things which are beneficial to them, or, more importantly, why they do not. Much of this reasoning has to do with individual interests trumping that which is best for the collective whole, which may not be obvious at the time.

Oliver, Marwell, and Teixeira tackle this subject in relation to critical theory in an 1985 article published in the

American Journal of Sociology.[11] In their essay, they define that action in service of a public good as "collective action". "Collective Action" is beneficial to all, regardless of individual contribution. By their definition, then, "critical mass" is the small segment of a societal system that does the work or action required to achieve the common good. The "Production Function" is the correlation between resources, or what individuals give in an effort to achieve public good, and the achievement of that good. Such function can be decelerating, where there is less utility per unit of resource, and in such a case, resource can taper off. On the other hand, the function can be accelerating, where the more resources that are used the bigger the payback. "Heterogeneity" is also important to the achievement of a common good. Variations (heterogeneity) in the value individuals put on a common good or the effort and resources people give is beneficial, because if certain people stand to gain more, they are willing to give or pay more.

46.3 Gender politics

See also: Women in government (quotas)

Critical mass theory in gender politics and collective political action is defined as the critical number of personnel needed to affect policy and make a change not as the token but as an influential body.[12] This number has been placed at 30%, before women are able to make a substantial difference in politics.[13][14] However, other research suggests lower numbers of women working together in legislature can also affect political change.[15][16] Kathleen Bratton goes so far as to say that women, in legislatures where they make up less than 15% of the membership, may actually be encouraged to develop legislative agendas that are distinct from those of their male colleagues.[17] Others argue that we should look more closely at parliamentary and electoral systems instead of critical mass.[18][19]

46.4 Interactive media

While critical mass can be applied to many different aspects of sociodynamics, it becomes increasingly applicable to innovations in interactive media such as the telephone, fax, or email. With other non-interactive innovations, the dependence on other users was generally sequential, meaning that the early adopters influenced the later adopters to use the innovation. However, with interactive media, the interdependence was reciprocal, meaning both users influenced each other. This is due to the fact that interactive media have high network effect,[1] where in the value and utility of a good or service increases the more users it has.

Thus, the increase of adopters and quickness to reach critical mass can therefore be faster and more intense with interactive media, as can the rate at which previous users discontinue their use. The more people that use it, the more beneficial it will be, thus creating a type of snowball effect, and conversely, if users begin to stop using the innovation, the innovation loses utility, thus pushing more users to discontinue their use.[20]

46.4.1 Markus essay

In M. Lynne Markus' essay in *Communication Research* entitled "Toward a 'Critical Mass' Theory of Interactive Media",[20] several propositions are made that try to predict under what circumstances interactive media is most likely to achieve critical mass and reach universal access, a "common good" using Oliver, *et al.*'s terminology. One proposition states that such media's existence is all or nothing, where in if universal access is not achieved, then, eventually, use will discontinue. Another proposition suggests that a media's ease of use and inexpensiveness, as well as its utilization of an "active notification capability" will help it achieve universal access. The third proposition states that the heterogeneity, as discussed by Oliver, et al. is beneficial, especially if users are dispersed over a larger area, thus necessitating interactivity via media. Fourth, it is very helpful to have highly sought-after individuals to act as early adopters, as their use acts as incentive for later users. Finally, Markus posits that interventions, both monetarily and otherwise, by governments, businesses, or groups of individuals will help a media reach its critical mass and achieve universal access.

46.4.2 Fax machine example

A fax machine

An example put forth by Rogers in *Diffusion of Innovations* was that of the fax machine, which had been around for almost 150 years before it became popular and widely used. It had existed in various forms and for various uses, but with more advancements in the technology of faxes, including the use of existing phone lines to transmit information, coupled with falling prices in both machines and cost per fax, the fax machine reached a critical mass in 1987, when "Americans began to assume that 'everybody else' had a fax machine".[21]

46.5 See also

- Bandwagon effect
- Economics of networks
- Network effect
- One-third hypothesis
- Positive feedback
- Tipping point (sociology)
- Viral phenomenon

46.6 References

[1] Rogers, Everett M. Diffusion of Innovations. New York: Simon & Schuster, 2003. Print.

[2] Thomas Schelling, "Dynamic models of segregation",Journal of Mathematical Sociology, 1971

[3] Schelling, Thomas C. Micromotives and Macrobehavior. New York: Norton, 1978. Print.

[4] Schelling, Thomas C. "On the Ecology of Micromotives," The Public Interest, No. 25, Fall 1971.

[5] Granovetter, Mark. "Threshold Models of Collective Behavior." American Journal of Sociology 83.6 (1978): 1420. Print.

[6] Krauth, Brian. "Notes for a History of the Critical Mass Model." SFU.ca. Web. 29 November 2011. <http://www.sfu.ca/~{}bkrauth/papers/critmass.htm>.

[7] Akerlof, George A. The Market for "lemons": Quality Uncertainty and the Market Mechanism. 2003. Print.

[8] Kuran, Timur. "Chameleon Voters and Public Choice." Public Choice 53.1 (1987): 53-78. Print.

[9] Olson, Mancur. The Logic of Collective Action: Public Goods and the Theory of Groups. Cambridge, MA: Harvard UP, 1971. Print.

[10] "The Tragedy of the Commons." Science 162.3859 (1968): 1243-248. Print.

[11] Oliver, P., G. Marwell, and R. Teixeira. "A Theory of Critical Mass: I. Interdependence, Group Heterogeneity, and the Production of Collective Action." American Journal of Sociology 9.3 (1985): 552-56. Print.

[12] Kanter, Rosabeth Moss (March 1977). "Some effects of proportions on group life: skewed sex ratios and responses to token women". American Journal of Sociology (JSTOR for the University of Chicago Press) 82 (5): 965–990. JSTOR 2777808. Pdf from Norges Handelshøyskole (NHH), the Norwegian School of Economics.

[13] Dahlerup, Drude (December 1988). "From a small to a large minority: women in Scandinavian politics". Scandinavian Political Studies (Wiley) 11 (4): 275–297. doi:10.1111/j.1467-9477.1988.tb00372.x.

[14] Dahlerup, Drude (December 2006). "The story of the theory of critical mass". Politics & Gender (Cambridge Journals) 2 (4): 511–522. doi:10.1017/S1743923X0624114X.

[15] Childs, Sarah; Krook, Mona Lena (December 2006). "Should feminists give up on critical mass? A contingent yes". Politics & Gender (Cambridge Journals) 2 (4): 522–530. doi:10.1017/S1743923X06251146.

[16] Childs, Sarah; Krook, Mona Lena (October 2008). "Critical mass theory and women's political representation". Political Studies (Wiley) 56 (3): 725–736. doi:10.1111/j.1467-9248.2007.00712.x. Pdf.

[17] Bratton, Kathleen A. (March 2005). "Critical mass theory revisited: the behavior and success of token women in state legislatures". Politics & Gender (Cambridge Journals) 1 (1): 97–125. doi:10.1017/S1743923X0505004X.

[18] Tremblay, Manon (December 2006). "The substantive representation of women and PR: some reflections on the role of surrogate representation and critical mass". Politics & Gender (Cambridge Journals) 2 (4): 502–511. doi:10.1017/S1743923X06231143.

[19] Grey, Sandra (December 2006). "Numbers and beyond: the relevance of critical mass in gender research". Politics & Gender (Cambridge Journals) 2 (4): 492–502. doi:10.1017/S1743923X06221147.

[20] M. Lynne Markus (1987). Toward a 'Critical Mass' Theory of Interactive Media: Universal Access, Interdependence and Diffusion, 14:491. Communication Research.

[21] Holmlov, Kramer and Karl-Eric Warneryd (1990). Adoption and Use of Fax in Sweden. Elmservier Science.

46.7 Further reading

- Philip Ball: Critical Mass: How One Thing Leads to Another, Farrar, Straus and Giroux, ISBN 0-374-53041-6

- Mancur Olson: *The Logic of Collective Action*, Harvard University Press, 1971

Chapter 47

Herd behavior

Herd behavior describes how individuals in a group can act collectively without centralized direction. The term can refer to the behavior of animals in herds, packs, bird flocks, fish schools and so on, as well as the behavior of humans in demonstrations, riots and general strikes,[1] sporting events, religious gatherings, episodes of mob violence and everyday decision-making, judgement and opinion-forming.

Raafat, Chater and Frith proposed an integrated approach to herding, describing two key issues, the mechanisms of transmission of thoughts or behavior between individuals and the patterns of connections between them.[2] They suggested that bringing together diverse theoretical approaches of herding behavior illuminates the applicability of the concept to many domains, ranging from cognitive neuroscience to economics.[3]

47.1 In animals

Shimmering behaviour of Apis dorsata *(giant honeybees)*

A group of animals fleeing from a predator shows the nature of herd behavior. In 1971, in the oft cited article "Geometry For The Selfish Herd," evolutionary biologist W. D. Hamilton asserted that each individual group member reduces the danger to itself by moving as close as possible to the center of the fleeing group. Thus the herd appears as a unit in moving together, but its function emerges from the uncoordinated behavior of self-serving individuals.[4]

47.2 Symmetry-breaking

Asymmetric aggregation of animals under panic conditions has been observed in many species, including humans, mice, and ants.[5] Theoretical models have demonstrated symmetry-breaking similar to observations in empirical studies. For example, when panicked individuals are confined to a room with two equal and equidistant exits, a majority will favor one exit while the minority will favor the other.

Possible mechanisms for this behavior include Hamilton's selfish herd theory, neighbor copying, or the byproduct of communication by social animals or runaway positive feedback.

Characteristics of escape panic include:

- Individuals attempt to move faster than normal.
- Interactions between individuals become physical.
- Exits become arched and clogged.
- Escape is slowed by fallen individuals serving as obstacles.
- Individuals display a tendency towards mass or copied behavior.
- Alternative or less used exits are overlooked.[4][6]

47.3 In human societies

The philosophers Søren Kierkegaard and Friedrich Nietzsche were among the first to criticize what they referred to as "the crowd" (Kierkegaard) and "herd morality" and the "herd instinct" (Nietzsche) in human society. Modern psychological and economic research has identified herd behavior in humans to explain the phenomena of large numbers of people acting in the same way at the same time. The British surgeon Wilfred Trotter popularized the "herd behavior" phrase in his book, *Instincts of the Herd in Peace and War* (1914). In *The Theory of the Leisure Class*, Thorstein Veblen explained economic behavior in terms of social influences such as "emulation," where some members of a group mimic other members of higher status. In "The Metropolis and Mental Life" (1903), early sociologist George Simmel referred to the "impulse to sociability in man", and sought to describe "the forms of association by which a mere sum of separate individuals are made into a 'society' ". Other social scientists explored behaviors related to herding, such as Freud (crowd psychology), Carl Jung (collective unconscious), and Gustave Le Bon (the popular mind). Swarm theory observed in non-human societies is a related concept and is being explored as it occurs in human society.

47.3.1 Stock market bubbles

Large stock market trends often begin and end with periods of frenzied buying (bubbles) or selling (crashes). Many observers cite these episodes as clear examples of herding behavior that is irrational and driven by emotion—greed in the bubbles, fear in the crashes. Individual investors join the crowd of others in a rush to get in or out of the market.[7]

Some followers of the technical analysis school of investing see the herding behavior of investors as an example of extreme market sentiment.[8] The academic study of behavioral finance has identified herding in the collective irrationality of investors, particularly the work of Nobel laureates Vernon L. Smith, Amos Tversky, Daniel Kahneman, and Robert Shiller.[9][a]

Hey and Morone (2004) analyzed a model of herd behavior in a market context. Their work is related to at least two important strands of literature. The first of these strands is that on herd behavior in a non-market context. The seminal references are Banerjee (1992) and Bikhchandani, Hirshleifer and Welch (1992), both of which showed that herd behavior may result from private information not publicly shared. More specifically, both of these papers showed that individuals, acting sequentially on the basis of private information and public knowledge about the behavior of others, may end up choosing the socially undesirable option. The second of the strands of literature motivating this paper is that of information aggregation in market contexts. A very early reference is the classic paper by Grossman and Stiglitz (1976) that showed that uninformed traders in a market context can become informed through the price in such a way that private information is aggregated correctly and efficiently. In this strand of the literature, the most commonly used empirical methodologies to test for herding toward the average, are the works of Christie and Huang (1995) and Chang, Cheng and Khorana (2000). Overall, it was shown that it is possible to observe herd-type behavior in a market context. The results refer to a market with a well-defined fundamental value. Even if herd behavior might only be observed rarely, this has important consequences for a whole range of real markets – most particularly foreign exchange markets.

One such herdish incident was the price volatility that surrounded the 2007 Uranium bubble, which started with flooding of the Cigar Lake Mine in Saskatchewan, during the year 2006.[10][11][12]

47.3.2 In crowds

Main article: Crowd psychology

Crowds that gather on behalf of a grievance can involve herding behavior that turns violent, particularly when confronted by an opposing ethnic or racial group. The Los Angeles riots of 1992, New York Draft Riots and Tulsa Race Riot are notorious in U.S. history. The idea of a "group mind" or "mob behavior" was put forward by the French social psychologists Gabriel Tarde and Gustave Le Bon.

47.3.3 Everyday decision-making

"Benign" herding behaviors may occur frequently in everyday decisions based on learning from the information of others, as when a person on the street decides which of two restaurants to dine in. Suppose that both look appealing, but both are empty because it is early evening; so at random, this person chooses restaurant A. Soon a couple walks down the same street in search of a place to eat. They see that restaurant A has customers while B is empty, and choose A on the assumption that having customers makes it the better choice. Because other passersby do the same thing into the evening, restaurant A does more business that night than B. This phenomenon is also referred as an information cascade.[13][14][15][16]

47.4 In Marketing

Herd behavior is often a useful tool in marketing and, if used properly, can lead to increases in sales and changes to the structure of society. Whilst it has been shown that financial incentives cause action in large numbers of people, herd mentality often wins out in a case of "Keeping up with the Jones's."

47.4.1 Herd Behavior in Brand and Product success

Communications technologies have contributed to the proliferation to consumer choice and "the power of crowds," [17] Consumers increasingly have more access to opinions and information from both opinion leaders and formers on platforms that have largely user-generated content, and thus have more tools with which to complete any decision-making process. Popularity is seen as an indication of better quality, and consumers will use the opinions of others posted on these platforms as a powerful compass to guide them towards products and brands that align with their preconceptions and the decisions of others in their peer groups.[18] Taking into account differences in needs and their position in the socialization process, Lessig & Park examined groups of students and housewives and the influence that these reference groups have on one another. By way of herd mentality, students tended to encourage each other towards beer, hamburger and cigarettes, whilst housewives tended to encourage each other towards furniture and detergent. Whilst this particular study was done in 1977, one cannot discount its findings in today's society. A study done by Burke, Leykin, Li and Zhang in 2014 on the social influence on shopper behavior shows that shoppers are influenced by direct interactions with companions, and as a group size grows, herd behaviour becomes more apparent. Discussions that create excitement and interest have greater impact on touch frequency and purchase likelihood grows with greater involvement caused by a large group.[19] Shoppers in this Midwestern American shopping outlet were monitored and their purchases noted, and it was found up to a point, potential customers preferred to be in stores which had moderate levels of traffic. The other people in the store not only served as company, but also provided an inference point on which potential customers could model their behavior and make purchase decisions, as with any reference group or community.

Social media can also be a powerful tool in perpetuating herd behaviour. Its immeasurable amount of user-generated content serves as a platform for opinion leaders to take the stage and influence purchase decisions, and recommendations from peers and evidence of positive online experience all serve to help consumers make purchasing decisions.[20] Gunawan and Huarng's 2015 study concluded that social influence is essential in framing attitudes towards brands, which in turn leads to purchase intention. [21] Influencers form norms which their peers are found to follow, and targeting extroverted personalities increases chances of purchase even further.[20] This is because the stronger personalities tend to be more engaged on consumer platforms and thus spread word of mouth information more efficiently.[22] Many brands have begun to realise the importance of brand ambassadors and influencers, and it is being shown more clearly that herd behaviour can be used to drive sales and profits exponentially in favour of any brand through examination of these instances.

47.4.2 Herd Behavior in Social Marketing

Marketing can easily transcend beyond commercial roots, in that it can be used to encourage action to do with health, environmentalism and general society. Herd mentality often takes a front seat when it comes to social marketing, paving the way for campaigns such as Earth Day, and the variety of anti-smoking and anti-obesity campaigns seen in every country. Within cultures and communities, marketers must aim to influence opinion leaders who in turn influence each other,[23] as it is the herd mentality of any group of people that ensures a social campaign's success. A campaign run by Som la Pera in Spain to combat teenage obesity found that campaigns run in schools are more effective due to influence of teachers and peers, and students' high visibility, and their interaction with one another. Opinion leaders in schools created the logo and branding for the campaign, built content for social media and led in-school presentations to engage audience interaction. It was thus concluded that the success of the campaign was rooted in the fact that its means of communication was the audience itself, giving the target audience a sense of ownership and empowerment.[24] As mentioned previously, students exert a high level of influence over one anothers, and by encouraging stronger personalities to lead opinions, the organizers of the campaign were able to secure the attention of other students who identified with the reference group.

Herd behaviour not only applies to students in schools where they are highly visible, but also amongst communities where perceived action plays a strong role. Between 2003 and 2004, California State University carried out a study to measure household conservation of energy, and motivations for doing so. It was found that factors like saving the environment, saving money or social responsibility did not have as great an impact on each household as the perceived behaviour of their neighbours did.[25] Although the financial incentives of saving money, closely followed by moral incentives of protecting the environment, are often thought of as being a community's greatest guiding compass, more households responded to the encouragement to save energy when they were told that 77% of their neighbours were using fans instead of air conditioning, proving that communities are more likely to engage in a behaviour if they think that everyone else is already taking part.

Herd behaviours shown in the two examples exemplify that it can be a powerful tool in social marketing, and if harnessed correctly, has the potential to achieve great change. It is clear that opinion leaders and their influence achieve huge reach amongst their reference groups and thusly can be used as the loudest voices to encourage others in any collective direction.

47.5 See also

- Anxiety
- Bandwagon effect
- Collective behavior
- Collective consciousness
- Collective effervescence
- Collective intelligence
- Crowd psychology
- Conformity
- Group behavior
- Groupthink
- Herd mentality
- Hive mind
- Informational cascade
- Mass hysteria
- Mean world syndrome
- Meme
- Mob rule
- Moral panic
- Propaganda
- Self-organization
- Sheeple
- Social proof
- Socionomics
- Spontaneous order
- Swarm intelligence
- Team player
- The 2009 Birmingham, Millennium Point stampede
- Riot
- Stampede
- Symmetry breaking of escaping ants

47.6 Notes

a. ^ See for example the Wikipedia article on his book *Irrational Exuberance*.[9]

47.7 References

[1] Braha, D (2012) *Global Civil Unrest: Contagion, Self-Organization, and Prediction*. PLoS ONE 7(10): e48596, article doi:10.1371/journal.pone.0048596

[2] Raafat, R. M.; Chater, N.; Frith, C. (2009). "Herding in humans". *Trends in Cognitive Sciences* 13 (10): 420–428. doi:10.1016/j.tics.2009.08.002.

[3] Burke, C. J.; Tobler, P. N.; Schultz, W.; Baddeley, M. (2010). "Striatal BOLD response reflects the impact of herd information on financial decisions". *Frontiers in Human Neuroscience* 4: 48. doi:10.3389/fnhum.2010.00048. PMID 20589242.

[4] Hamilton, W. D. (1971). "Geometry for the Selfish Herd". *Journal of Theoretical Biology* 31 (2): 295–311. doi:10.1016/0022-5193(71)90189-5. PMID 5104951.

[5] Altshuler, E.; Ramos, O.; Nuñez, Y.; Fernández, J. "Panic-induced symmetry breaking in escaping ants" (PDF). University of Havana, Havana, Cuba. Retrieved 2011-05-18.

[6] Altshuler, E.; Ramos, O.; Núñez, Y.; Fernández, J.; Batista-Leyva, A. J.; Noda, C. (2005). "Symmetry Breaking in Escaping Ants". *The American Naturalist* 166 (6): 643–649. doi:10.1086/498139.

[7] Markus K. Brunnermeier, Asset *Pricing under Asymmetric Information: Bubbles, Crashes, Technical Analysis, and Herding*, Oxford University Press (2001).

[8] Robert Prechter, *The Wave Principle of Human Social Behavior*, New Classics Library (1999), pp. 152–153.

[9] Shiller, Robert J. (2000). *Irrational Exuberance*. Princeton University Press. pp. 149–153. Retrieved 4 March 2013.

[10] In Focus article (8 June 2012), "WNFM: A Focus on Fundamentals One Year After Fukushima", *Reproduced article from Nuclear Market Review*, TradeTech, retrieved 4 March 2013 There are several reproduced In Focus articles on this page. The relevant one is near the bottom, under the title in this reference

[11] UraniumSeek.com, Gold Seek LLC (2008-08-22). "Uranium Has Bottomed: Two Uranium Bulls to Jump on Now". UraniumSeek.com. Retrieved 2011-09-19.

[12] "Uranium Bubble & Spec Market Outlook". News.goldseek.com. Retrieved 2011-09-19.

[13] Banerjee, Abhijit V. (1992). "A Simple Model of Herd Behavior". *Quarterly Journal of Economics* **107** (3): 797–817. doi:10.2307/2118364.

[14] Bikhchandani, Sushil; Hirshleifer, David; Welch, Ivo (1992). "A Theory of Fads, Fashion, Custom, and Cultural Change as Informational Cascades". *Journal of Political Economy* **100** (5): 992–1026. doi:10.1086/261849.

[15] Froot, K; Schaferstein, DS; Jeremy Stein, J (1992). "Herd on the street: Informational inefficiencies in a market with short-term speculation" (PDF). *Journal of Finance* **47**: 1461–1484. doi:10.1111/j.1540-6261.1992.tb04665.x.

[16] Hirshleifer, D; Teoh, SH (2003). "Herd behaviour and cascading in capital markets: A review and synthesis" (PDF). *European Financial Management* **9** (1): 25–66. doi:10.1111/1468-036X.00207.

[17] Chen, Yi-Fen (2008-09-01). "Herd behavior in purchasing books online". *Computers in Human Behavior*. Including the Special Issue: Internet Empowerment **24** (5): 1977–1992. doi:10.1016/j.chb.2007.08.004.

[18] Lessig, V (1977). "Students and Housewives: Differences in Susceptibility to Reference Group Influence". *Journal of Consumer Research*.

[19] Zhang, Xiaoling; Li, Shibo; Burke, Raymond R.; Leykin, Alex (2014-05-13). "An Examination of Social Influence on Shopper Behavior Using Video Tracking Data". *Journal of Marketing* **78** (5): 24–41. doi:10.1509/jm.12.0106. ISSN 0022-2429.

[20] Dhar, Joydip; Jha, Abhishek Kumar (2014-10-03). "Analyzing Social Media Engagement and its Effect on Online Product Purchase Decision Behavior". *Journal of Human Behavior in the Social Environment* **24** (7): 791–798. doi:10.1080/10911359.2013.876376. ISSN 1091-1359.

[21] Gunawan, Dedy Darsono; Huarng, Kun-Huang (2015-11-01). "Viral effects of social network and media on consumers' purchase intention". *Journal of Business Research* **68** (11): 2237–2241. doi:10.1016/j.jbusres.2015.06.004.

[22] Cheung, Christy M. K.; Xiao, Bo Sophia; Liu, Ivy L. B. (2014-09-01). "Do actions speak louder than voices? The signaling role of social information cues in influencing consumer purchase decisions". *Decision Support Systems*. Crowdsourcing and Social Networks Analysis **65**: 50–58. doi:10.1016/j.dss.2014.05.002.

[23] James M. Cronin; Mary B. McCarthy (2011-07-12). "Preventing game over: A study of the situated food choice influences within the videogames subculture". *Journal of Social Marketing* **1** (2): 133–153. doi:10.1108/20426761111141887. ISSN 2042-6763.

[24] Lozano, Natàlia; Prades, Jordi; Montagut, Marta (2015-10-01). "Som la Pera: How to develop a social marketing and public relations campaign to prevent obesity among teenagers in Catalonia". *Catalan Journal of Communication & Cultural Studies* **7** (2): 251–259. doi:10.1386/cjcs.7.2.251_1.

[25] Nolan, Jessica M.; Schultz, P. Wesley; Cialdini, Robert B.; Goldstein, Noah J.; Griskevicius, Vladas (2008-07-01). "Normative Social Influence is Underdetected". *Personality and Social Psychology Bulletin* **34** (7): 913–923. doi:10.1177/0146167208316691. ISSN 0146-1672. PMID 18550863.

47.8 Further reading

- Bikhchandani, Sushil; Hirshleifer, David; Welch, Ivo (1992). "A Theory of Fads, Fashion, Custom, and Cultural Change as Informational Cascades". *Journal of Political Economy* **100** (5): 992–1026. doi:10.1086/261849. JSTOR 2138632.

- Trotter, Wilfred (1914). *The Instincts of the Herd in Peace and War*.

- Brunnermeier, Markus Konrad (2001). *Asset Pricing under Asymmetric Information: Bubbles, Crashes, Technical Analysis, and Herding*. Oxford, UK ; New York: Oxford University Press.

- Rook, Laurens (2006). "An Economic Psychological Approach to Herd Behavior". *Journal of Economic Issues* **40** (1): 75–95.

- Hamilton, W. D. (1970). *Geometry for the Selfish Herd*. Diss. Imperial College.

- Stanford, Craig B. (2001). "Avoiding Predators: Expectations and Evidence in Primate Antipredator Behaviour". *International Journal of Primatology* **23**: 741–757. doi:10.1023/A:1015572814388. Ebsco. Fall. Keyword: Herd Behaviour.

- Ottaviani, Marco; Sorenson, Peter (2000). "Herd Behavior and Investment: Comment". *American Economic Review* **90** (3): 695–704. doi:10.1257/aer.90.3.695. JSTOR 117352.

- Altshuler, E.; et al. (2005). "Symmetry Breaking in Escaping Ants". *The American Naturalist* **166**: 643–649. doi:10.1086/498139.

- Hey, John D.; Morone, Andrea (2004). "Do Markets Drive out Lemmings—or Vice Versa?". *Economica* **71** (284): 637–659. doi:10.1111/j.0013-0427.2004.00392.x. JSTOR 3548984.

Chapter 48

Groupthink

Groupthink is a psychological phenomenon that occurs within a group of people in which the desire for harmony or conformity in the group results in an irrational or dysfunctional decision-making outcome. Group members try to minimize conflict and reach a consensus decision without critical evaluation of alternative viewpoints by actively suppressing dissenting viewpoints, and by isolating themselves from outside influences.

Groupthink requires individuals to avoid raising controversial issues or alternative solutions, and there is loss of individual creativity, uniqueness and independent thinking. The dysfunctional group dynamics of the "ingroup" produces an "illusion of invulnerability" (an inflated certainty that the right decision has been made). Thus the "ingroup" significantly overrates its own abilities in decision-making and significantly underrates the abilities of its opponents (the "outgroup"). Furthermore, groupthink can produce dehumanizing actions against the "outgroup".

Antecedent factors such as group cohesiveness, faulty group structure, and situational context (e.g., community panic) play into the likelihood of whether or not groupthink will impact the decision-making process.

Groupthink is a construct of social psychology but has an extensive reach and influences literature in the fields of communication studies, political science, management, and organizational theory,[1] as well as important aspects of deviant religious cult behaviour.[2][3]

Groupthink is sometimes stated to occur (more broadly) within natural groups within the community, for example to explain the lifelong different mindsets of conservatives versus liberals,[4] or the solitary nature of introverts.[5] However, this conformity of viewpoints within a group does not mainly involve deliberate group decision-making, and might be better explained by the collective confirmation bias of the individual members of the group.

Most of the initial research on groupthink was conducted by Irving Janis, a research psychologist from Yale University.[6] Janis published an influential book in 1972, which was revised in 1982.[7][8] Janis used the Bay of Pigs disaster (the failed invasion of Castro's Cuba in 1961) and the Japanese attack on Pearl Harbor in 1941 as his two prime case studies. Later studies have evaluated and reformulated his groupthink model.[9][10]

48.1 History

From "Groupthink" by William H. Whyte, Jr. in Fortune *magazine, March 1952*

William H. Whyte, Jr. coined the term in 1952 in *Fortune* magazine:

> Groupthink being a coinage - and, admittedly, a loaded one - a working definition is in order. We are not talking about mere instinctive conformity - it is, after all, a perennial failing of mankind. What we are talking about is a *rationalized* conformity - an open, articulate philosophy which holds that group values are not only expedient but right and good as well.[11][12]

Irving Janis pioneered the initial research on the groupthink theory. He does not cite Whyte, but coined the term by analogy with "doublethink" and similar terms that were part of the newspeak vocabulary in the novel *Nineteen Eighty-Four* by George Orwell. He initially defined groupthink as follows:

> I use the term groupthink as a quick and easy way to refer to the mode of thinking that persons engage in when *concurrence-seeking* becomes so dominant in a cohesive ingroup that it tends to override realistic appraisal of alternative courses of action. Groupthink is a term of the same order as the words in the newspeak vocabulary George Orwell used in his dismaying world of *1984*. In that context, groupthink takes on an invidious connotation. Exactly such a connotation is intended, since the term refers to a deterioration in mental efficiency, reality testing and moral judgments as a result of group pressures.[6]:43

He went on to write:

> The main principle of groupthink, which I offer in the spirit of Parkinson's Law, is this: *The more amiability and esprit de corps there is among the members of a policy-making ingroup, the greater the danger that independent critical thinking will be replaced by groupthink, which is likely to result in irrational and dehumanizing actions directed against outgroups.*[6]:44

Janis set the foundation for the study of groupthink starting with his research in the American Soldier Project where he studied the effect of extreme stress on group cohesiveness. After this study he remained interested in the ways in which people make decisions under external threats. This interest led Janis to study a number of "disasters" in American foreign policy, such as failure to anticipate the Japanese attack on Pearl Harbor (1941); the Bay of Pigs Invasion fiasco (1961); and the prosecution of the Vietnam War (1964–67) by President Lyndon Johnson. He concluded that in each of these cases, the decisions occurred largely because of groupthink, which prevented contradictory views from being expressed and subsequently evaluated.

After the publication of Janis' book *Victims of Groupthink* in 1972,[7] and a revised edition with the title *Groupthink: Psychological Studies of Policy Decisions and Fiascoes* in 1982,[8] the concept of groupthink was used to explain many other faulty decisions in history. These events included Nazi Germany's decision to invade the Soviet Union in 1941, the Watergate Scandal and others. Despite the popularity of the concept of groupthink, fewer than two dozen studies addressed the phenomenon itself following the publication of *Victims of Groupthink*, between the years 1972 and 1998.[1]:107 This is surprising considering how many fields of interests it spans, which include political science, communications, organizational studies, social psychology, management, strategy, counseling, and marketing. One can most likely explain this lack of follow-up in that group research is difficult to conduct, groupthink has many independent and dependent variables, and it is unclear "how to translate [groupthink's] theoretical concepts into observable and quantitative constructs."[1]:107–108

Nevertheless, outside research psychology and sociology, wider culture has come to detect groupthink (somewhat fuzzily defined) in observable situations, for example:

- " [...] critics of Twitter point to the predominance of the hive mind in such social media, the kind of groupthink that submerges independent thinking in favor of conformity to the group, the collective"[13]

- "[...] leaders often have beliefs which are very far from matching reality and which can become more extreme as they are encouraged by their followers. The predilection of many cult leaders for abstract, ambiguous, and therefore unchallengeable ideas can further reduce the likelihood of reality testing, while the intense milieu control exerted by cults over their members means that most of the reality available for testing is supplied by the group environment. This is seen in the phenomenon of 'groupthink', alleged to have occurred, notoriously, during the Bay of Pigs fiasco."[14]

- "Groupthink by Compulsion [...] [G]roupthink at least implies voluntarism. When this fails, the organization is not above outright intimidation. [...] In [a nationwide telecommunications company], refusal by the new hires to cheer on command incurred consequences not unlike the indoctrination and brainwashing techniques associated with a Soviet-era gulag."[15]

48.2 Symptoms

To make groupthink testable, Irving Janis devised eight symptoms indicative of groupthink.

Type I: Overestimations of the group — its power and morality

1. *Illusions of invulnerability* creating excessive optimism and encouraging risk taking.
2. *Unquestioned belief* in the morality of the group, causing members to ignore the consequences of their actions.

Type II: Closed-mindedness

1. *Rationalizing warnings* that might challenge the group's assumptions.
2. *Stereotyping* those who are opposed to the group as weak, evil, biased, spiteful, impotent, or stupid.

Type III: Pressures toward uniformity

1. *Self-censorship* of ideas that deviate from the apparent group consensus.
2. *Illusions of unanimity* among group members, silence is viewed as agreement.
3. *Direct pressure* to conform placed on any member who questions the group, couched in terms of "disloyalty"
4. *Mindguards*— self-appointed members who shield the group from dissenting information.

48.3 Causes

Janis prescribed three antecedent conditions to groupthink.[7]:9

1. High group cohesiveness
   - deindividuation: group cohesiveness becomes more important than individual freedom of expression
2. Structural faults:
   - insulation of the group
   - lack of impartial leadership
   - lack of norms requiring methodological procedures
   - homogeneity of members' social backgrounds and ideology
3. Situational context:
   - highly stressful external threats
   - recent failures
   - excessive difficulties on the decision-making task
   - moral dilemmas

Although it is possible for a situation to contain all three of these factors, all three are not always present even when groupthink is occurring. Janis considered a high degree of cohesiveness to be the most important antecedent to producing groupthink and always present when groupthink was occurring; however, he believed high cohesiveness would not always produce groupthink. A very cohesive group abides to all group norms; whether or not groupthink arises is dependent on what the group norms are. If the group encourages individual dissent and alternative strategies to problem solving, it is likely that groupthink will be avoided even in a highly cohesive group. This means that high cohesion will lead to groupthink only if one or both of the other antecedents is present, situational context being slightly more likely than structural faults to produce groupthink.[16]

48.4 Prevention

As observed by Aldag & Fuller (1993), the groupthink phenomenon seems to rest on a set of unstated and generally restrictive assumptions:[17]

1. The purpose of group problem solving is mainly to improve decision quality
2. Group problem solving is considered a rational process.
3. Benefits of group problem solving:
   - variety of perspectives
   - more information about possible alternatives
   - better decision reliability
   - dampening of biases
   - social presence effects
4. Groupthink prevents these benefits due to structural faults and provocative situational context
5. Groupthink prevention methods will produce better decisions
6. An illusion of well-being is presumed to be inherently dysfunctional.
7. Group pressures towards consensus lead to concurrence-seeking tendencies.

It has been thought that groups with the strong ability to work together will be able to solve dilemmas in a quicker and more efficient fashion than an individual. Groups have a greater amount of resources which lead them to be able to store and retrieve information more readily and come up with more alternative solutions to a problem. There was a recognized downside to group problem solving in that it takes groups more time to come to a decision and requires that people make compromises with each other. However, it was not until the research of Janis appeared that anyone really considered that a highly cohesive group could impair the group's ability to generate quality decisions. Tight-knit groups may appear to make decisions better because they can come to a consensus quickly and at a low energy cost; however, over time this process of decision-making may decrease the members' ability to think critically. It is, therefore, considered by many to be important to combat the effects of groupthink.[16]

According to Janis, decision-making groups are not necessarily destined to groupthink. He devised ways of preventing groupthink:[7]:209–215

1. Leaders should assign each member the role of "critical evaluator". This allows each member to freely air objections and doubts.

2. Leaders should not express an opinion when assigning a task to a group.

3. Leaders should absent themselves from many of the group meetings to avoid excessively influencing the outcome.

4. The organization should set up several independent groups, working on the same problem.

5. All effective alternatives should be examined.

6. Each member should discuss the group's ideas with trusted people outside of the group.

7. The group should invite outside experts into meetings. Group members should be allowed to discuss with and question the outside experts.

8. At least one group member should be assigned the role of Devil's advocate. This should be a different person for each meeting.

By following these guidelines, groupthink can be avoided. After the Bay of Pigs invasion fiasco, President John F. Kennedy sought to avoid groupthink during the Cuban Missile Crisis using "vigilant appraisal."[8]:148–153 During meetings, he invited outside experts to share their viewpoints, and allowed group members to question them carefully. He also encouraged group members to discuss possible solutions with trusted members within their separate departments, and he even divided the group up into various subgroups, to partially break the group cohesion. Kennedy was deliberately absent from the meetings, so as to avoid pressing his own opinion.

48.5 Empirical findings and meta-analysis

Testing groupthink in a laboratory is difficult because synthetic settings remove groups from real social situations, which ultimately changes the variables conducive or inhibitive to groupthink.[18] Because of its subjective nature, researchers have struggled to measure groupthink as a complete phenomenon, instead frequently opting to measure its particular factors. These factors range from causal to effectual and focus on group and situational aspects.[19][20]

Park (1990) found that "only 16 empirical studies have been published on groupthink," and concluded that they "resulted in only partial support of his [Janis's] hypotheses."[21]:230 Park concludes, "despite Janis' claim that group cohesiveness is the major necessary antecedent factor, no research has showed a significant main effect of cohesiveness on groupthink."[21]:230 Park also concludes that research on the interaction between group cohesiveness and leadership style does not support Janis' claim that cohesion and leadership style interact to produce groupthink symptoms.[21] Park presents a summary of the results of the studies analyzed. According to Park, a study by Huseman and Drive (1979) indicates groupthink occurs in both small and large decision-making groups within businesses.[21] This results partly from group isolation within the business. Manz and Sims (1982) conducted a study showing that autonomous work groups are susceptible to groupthink symptoms in the same manner as decisions making groups within businesses.[21][22] Fodor and Smith (1982) produced a study revealing that group leaders with high power motivation create atmospheres more susceptible to groupthink.[21][23] Leaders with high power motivation possess characteristics similar to leaders with a "closed" leadership style—an unwillingness to respect dissenting opinion. The same study indicates that level of group cohesiveness is insignificant in predicting groupthink occurrence. Park summarizes a study performed by Callaway, Marriott, and Esser (1985) in which groups with highly dominant members "made higher quality decisions, exhibited lowered state of anxiety, took more time to reach a decision, and made more statements of disagreement/agreement."[21]:232[24] Overall, groups with highly dominant members expressed characteristics inhibitory to groupthink. If highly dominant members are considered equivalent to leaders with high power motivation, the results of Callaway, Marriott, and Esser contradict the results of Fodor and Smith. A study by Leana (1985) indicates the interaction between level of group cohesion and leadership style is completely insignificant in predicting groupthink.[21][25] This finding refutes Janis' claim that the factors of cohesion and leadership style interact to produce groupthink. Park summarizes a study by McCauley (1989) in which structural conditions of the group were found to predict groupthink while situational conditions did not.[10][21] The structural conditions included group insulation, group homogeneity, and promotional leadership. The situational conditions included group cohesion. These findings refute Janis' claim about group cohesiveness predicting groupthink.

Overall, studies on groupthink have largely focused on the factors (antecedents) that predict groupthink. Groupthink occurrence is often measured by number of ideas/solutions generated within a group, but there is no uniform, concrete standard by which researchers can objectively conclude groupthink occurs.[18] The studies of groupthink and groupthink antecedents reveal a mixed body of results. Some studies indicate group cohesion and leadership style to be powerfully predictive of groupthink, while other studies indicate the insignificance of these factors. Group homogeneity and group insulation are generally supported as factors predictive of groupthink.

48.6 Case studies

48.6.1 Politics and military

Groupthink can have a strong hold on political decisions and military operations, which may result in enormous wastage of human and material resources. Highly qualified and experienced politicians and military commanders sometimes make very poor decisions when in a suboptimal group setting. Scholars such as Janis and Raven attribute political and military fiascoes, such as the Bay of Pigs Invasion, Vietnam War, and the Watergate scandal, to the effect of groupthink.[8][26] More recently, Dina Badie argued that groupthink was largely responsible for the shift in the U.S. administration's view on Saddam Hussein that eventually led to the 2003 invasion of Iraq by the United States.[27] After 9/11, "stress, promotional leadership, and intergroup conflict" were all factors that gave rise to the occurrence of groupthink.[27]:283 Political case studies of groupthink serve to illustrate the impact that the occurrence of groupthink can have in today's political scene.

Bay of Pigs invasion and the Cuban Missile Crisis

The United States Bay of Pigs Invasion of April 1961 was the primary case study that Janis used to formulate his theory of groupthink.[6] The invasion plan was initiated by the Eisenhower administration, but when the Kennedy White House took over, it "uncritically accepted" the CIA's plan.[6]:44 When some people, such as Arthur M. Schlesinger, Jr. and Senator J. William Fulbright, attempted to present their objections to the plan, the Kennedy team as a whole ignored these objections and kept believing in the morality of their plan.[6]:46 Eventually Schlesinger minimized his own doubts, performing self-censorship.[6]:74 The Kennedy team stereotyped Castro and the Cubans by failing to question the CIA about its many false assumptions, including the ineffectiveness of Castro's air force, the weakness of Castro's army, and the inability of Castro to quell internal uprisings.[6]:46

Janis claimed the fiasco that ensued could have been prevented if the Kennedy administration had followed the methods to preventing groupthink adopted during the Cuban Missile Crisis, which took place just one year later in October 1962. In the latter crisis, essentially the same political leaders were involved in decision-making, but this time they learned from their previous mistake of seriously under-rating their opponents.[6]:76

Pearl Harbor

The attack on Pearl Harbor on December 7, 1941 is a prime example of groupthink. A number of factors such as shared illusions and rationalizations contributed to the lack of precaution taken by Naval officers based in Hawaii. The United States had intercepted Japanese messages and they discovered that Japan was arming itself for an offensive attack *somewhere* in the Pacific. Washington took action by warning officers stationed at Pearl Harbor, but their warning was not taken seriously. They assumed that Japan was taking measures in the event that their embassies and consulates in enemy territories were usurped.

The Navy and Army in Pearl Harbor also shared rationalizations about why an attack was unlikely. Some of them included:[8]:83,85

- "The Japanese would never dare attempt a full-scale surprise assault against Hawaii because they would realize that it would precipitate an all-out war, which the United States would surely win."

- "The Pacific Fleet concentrated at Pearl Harbor was a major deterrent against air or naval attack."

- "Even if the Japanese were foolhardy to send their carriers to attack us [the United States], we could certainly detect and destroy them in plenty of time."

- "No warships anchored in the shallow water of Pearl Harbor could ever be sunk by torpedo bombs launched from enemy aircraft."

48.6.2 Corporate world

In the corporate world, ineffective and suboptimal group decision-making can negatively affect the health of a company and cause a considerable amount of monetary loss.

Swissair

Aaron Hermann and Hussain Rammal illustrate the detrimental role of groupthink in the collapse of Swissair, a Swiss airline company that was thought to be so financially stable that it earned the title the "Flying Bank."[28] The authors argue that, among other factors, Swissair carried two symptoms of groupthink: the belief that the group is invulnerable and the belief in the morality of the group.[28]:1056 In addition, before the fiasco, the size of the company board was reduced, subsequently eliminating industrial expertise. This may have further increased the likelihood of groupthink.[28]:1055 With the board members lacking expertise in the field and having somewhat similar background, norms, and values, the pressure to conform may have become more prominent.[28]:1057 This phenomenon is called group homogeneity, which is an antecedent to groupthink. Together, these conditions may have contributed to the poor decision-making process that eventually led to Swissair's collapse.

Marks & Spencer and British Airways

Another example of groupthink from the corporate world is illustrated in the UK based companies, Marks & Spencer and British Airways. The negative impact of groupthink took place during the 1990s as both companies released globalization expansion strategies. Researcher Jack Eaton's content analysis of media press releases revealed that all eight symptoms of groupthink were present during this period. The most predominant symptom of groupthink was the illusion of invulnerability as both companies underestimated potential failure due to years of profitability and success during challenging markets. Up until the consequence of groupthink erupted they were considered blue chips and darlings of the London Stock Exchange. During 1998 - 1999 the price of Marks & Spencer shares fell from 590 to less than 300 and that of British Airways from 740 to 300. Both companies had already featured prominently in the UK press and media for more positive reasons, to do with national pride in their undoubted sector-wide performance.[29]

48.6.3 Sports

Recent literature of groupthink attempts to study the application of this concept beyond the framework of business and politics. One particularly relevant and popular arena in which groupthink is rarely studied is sports. The lack of literature in this area prompted Charles Koerber and Christopher Neck to begin a case-study investigation that examined the effect of groupthink on the decision of the Major League Umpires Association (MLUA) to stage a mass resignation in 1999. The decision was a failed attempt to gain a stronger negotiating stance against Major League Baseball.[30]:21 Koerber and Neck suggest that three groupthink symptoms can be found in the decision-making process of the MLUA. First, the umpires overestimated the power that they had over the baseball league and the strength of their group's resolve. The union also exhibited some degree of closed-mindedness with the notion that MLB is the enemy. Lastly, there was the presence of self-censorship; some umpires who disagreed with the decision to resign failed to voice their dissent.[30]:25 These factors, along with other decision-making defects, led to a decision that was suboptimal and ineffective.

48.7 Recent developments

48.7.1 Ubiquity model

Researcher Robert Baron (2005) contends that the connection between certain antecedents Janis believed necessary have not been demonstrated by the current collective body of research on groupthink. He believes that Janis' antecedents for groupthink is incorrect and argues that not only are they "not necessary to provoke the symptoms of groupthink, but that they often will not even amplify such symptoms."[31] As an alternative to Janis' model, Baron proposed a ubiquity model of groupthink. This model provides a revised set of antecedents for groupthink, including social identification, salient norms, and low self-efficacy.

General group problem-solving (GGPS) model

Aldag and Fuller (1993) argue that the groupthink concept was based on a "small and relatively restricted sample" that became too broadly generalized.[17] Furthermore, the concept is too rigidly staged and deterministic. Empirical support for it has also not been consistent. The authors compare groupthink model to findings presented by Maslow and Piaget; they argue that, in each case, the model incites great interest and further research that, subsequently, invalidate the original concept. Aldag and Fuller thus suggest a new model called the general group problem-solving (GGPS) model, which integrates new findings from groupthink literature and alters aspects of groupthink itself.[17]:534 The primary difference between the GGPS model and groupthink is that the former is more value neutral and more political.[17]:544

48.7.2 Reexamination

Other scholars attempt to assess the merit of groupthink by reexamining case studies that Janis had originally used to buttress his model. Roderick Kramer (1998) believed that, because scholars today have a more sophisticated set of ideas about the general decision-making process and because new and relevant information about the fiascos have surfaced over the years, a reexamination of the case studies is appropriate and necessary.[32] He argues that new evidence does not support Janis' view that groupthink was largely responsible for President Kennedy's and President Johnson's decisions in the Bay of Pigs Invasion and U.S. escalated military involvement in the Vietnam War, respectively. Both presidents sought the advice of experts outside of their political groups more than Janis suggested.[32]:241 Kramer also argues that the presidents were the final decision-makers of the fiascos; while determining which course of action to take, they relied more heavily on their own construals of the situations than on any group-consenting decision presented to them.[32]:241 Kramer concludes that Janis' explanation of the two military issues is flawed and that groupthink has much less influence on group decision-making than is popularly believed to be.

48.7.3 Reformulation

Whyte (1998) suggests that collective efficacy plays a large role in groupthink because it causes groups to become less vigilant and to favor risks, two particular factors that characterize groups affected by groupthink.[33] McCauley recasts aspects of groupthink's preconditions by arguing that the level of attractiveness of group members is the most prominent factor in causing poor decision-making.[34] The results of Turner's and Pratkanis' (1991) study on social identity maintenance perspective and groupthink conclude that groupthink can be viewed as a "collective effort directed at warding off potentially negative views of the group."[3] Together, the contributions of these scholars have brought about new understandings of groupthink that help reformulate Janis' original model.

48.7.4 Sociocognitive theory

According to a new theory many of the basic characteristics of groupthink - e.g., strong cohesion, indulgent atmosphere, and exclusive ethos - are the result of a special kind of mnemonic encoding (Tsoukalas, 2007). Members of tightly knit groups have a tendency to represent significant aspects of their community as episodic memories and this has a predictable influence on their group behavior and collective ideology.[35]

48.8 See also

- Tuckman's stages of group development
- Spiral of silence
- Asch conformity experiments
- Abilene paradox
- Bandwagon effect
- Conformity (psychology)
- Deindividuation
- Emotional contagion
- Group behaviour
- Group flow
- Group narcissism
- Group polarization
- Group-serving bias
- Groupshift
- Herd behaviour
- Homophily
- In-group favoritism
- Mob rule
- No soap radio
- Organizational dissent
- Peer pressure
- Risky shift
- Scapegoating
- Social comparison theory

- System justification
- Three men make a tiger
- Vendor lock-in
- Woozle effect

Diversity

- Cultural diversity
- Multiculturalism

48.9 References

[1] Turner, M. E.; Pratkanis, A. R. (1998). "Twenty-five years of groupthink theory and research: lessons from the evaluation of a theory" (PDF). *Organizational Behavior and Human Decision Processes* 73: 105–115. doi:10.1006/obhd.1998.2756.

[2] Wexler, Mark N. (1995). "Expanding the groupthink explanation to the study of contemporary cults". *Cultic Studies Journal* 12 (1): 49–71.

[3] Turner, M.; Pratkanis, A. (1998). "A social identity maintenance model of groupthink". *Organizational Behavior and Human Decision Processes* 73: 210–235. doi:10.1006/obhd.1998.2757.

[4] "Does Liberal Truly Mean Open-Minded?". psychologytoday.com.

[5] Cain, Susan (January 13, 2012). "The rise of the new groupthink". *New York Times.*.

[6] Janis, I. L. (November 1971). "Groupthink" (PDF). *Psychology Today* 5 (6): 43–46, 74–76. Archived from the original on April 1, 2010.

[7] Janis, I. L. (1972). *Victims of Groupthink: a Psychological Study of Foreign-Policy Decisions and Fiascoes*. Boston: Houghton Mifflin. ISBN 0-395-14002-1.

[8] Janis, I. L. (1982). *Groupthink: Psychological Studies of Policy Decisions and Fiascoes*. Boston: Houghton Mifflin. ISBN 0-395-31704-5.

[9] 't Hart, P. (1998). "Preventing groupthink revisited: evaluating and reforming groups in government". *Organizational Behavior and Human Decision Processes* 73: 306–326. doi:10.1006/obhd.1998.2764.

[10] McCauley, C. (1989). "The nature of social influence in groupthink: compliance and internalization". *Journal of Personality and Social Psychology* 57: 250–260. doi:10.1037/0022-3514.57.2.250.

[11] Whyte, W. H., Jr. (March 1952). "Groupthink". *Fortune*. pp. 114–117, 142, 146.

[12] Safire, W. (August 8, 2004). "Groupthink". *New York Times*. Retrieved February 2, 2012. If the committee's other conclusions are as outdated as its etymology, we're all in trouble. 'Groupthink' (one word, no hyphen) was the title of an article in Fortune magazine in March 1952 by William H. Whyte Jr. ... Whyte derided the notion he argued was held by a trained elite of Washington's 'social engineers.'

[13] Cross, Mary (2011). *Bloggerati, Twitterati: How Blogs and Twitter are Transforming Popular Culture*. ABC-CLIO. p. 62. ISBN 9780313384844. Retrieved 2013-11-17. [...] critics of twitter point to the predominance of the hive mind in such social media, the kind of groupthink that submerges independent thinking in favor of conformity to the group, the collective.

[14] Taylor, Kathleen (2006). *Brainwashing: The Science of Thought Control*. Oxford University Press. p. 42. ISBN 9780199204786. Retrieved 2013-11-17. [...] leaders often have beliefs which are very far from matching reality and which can become more extreme as they are encouraged by their followers. The predilection of many cult leaders for abstract, ambiguous, and therefore unchallengeable ideas can further reduce the likelihood of reality testing, while the intense milieu control exerted by cults over their members means that most of the reality available for testing is supplied by the group environment. This is seen in the phenomenon of 'groupthink', alleged to have occurred, notoriously, during the Bay of Pigs fiasco.

[15] Jonathan I., Klein (2000). *Corporate Failure by Design: Why Organizations are Built to Fail*. Greenwood Publishing Group. p. 145. ISBN 9781567202977. Retrieved 2013-11-17. Groupthink by Compulsion [...] [G]roupthink at least implies voluntarism. When this fails, the organization is not above outright intimidation. [...] In [a nationwide telecommunications company], refusal by the new hires to cheer on command incurred consequences not unlike the indoctrination and brainwashing techniques associated with a Soviet-era gulag.

[16] Hart, Paul't (1991). "Irving L. Janis' Victims of Groupthink". *Political Psychology* 2: 247–278. doi:10.2307/3791464.

[17] Aldag, R. J.; Fuller, S. R. (1993). "Beyond fiasco: A reappraisal of the groupthink phenomenon and a new model of group decision processes" (PDF). *Psychological Bulletin* 113 (3): 533–552. doi:10.1037/0033-2909.113.3.533.

[18] Flowers, M.L. (1977). "A laboratory test of some implications of Janis's groupthink hypothesis". *Journal of Personality and Social Psychology* 35 (12): 888–896. doi:10.1037/0022-3514.35.12.888.

[19] Schafer, M.; Crichlow, S. (1996). "Antecedents of groupthink: a quantitative study". *Journal of Conflict Resolution* 40 (3): 415–435. doi:10.1177/0022002796040003002.

[20] Cline, R. J. W. (1990). "Detecting groupthink: methods for observing the illusion of unanimity". *Communication Quarterly* 38 (2): 112–126. doi:10.1080/01463379009369748.

[21] Park, W.-W. (1990). "A review of research on Groupthink" (PDF). *Journal of Behavioral Decision Making* 3 (4): 229–245. doi:10.1002/bdm.3960030402.

[22] Manz, C. C.; Sims, H. P. (1982). "The Potential for "Groupthink" in Autonomous Work Groups". *Human Relations* 35 (9): 773–784. doi:10.1177/001872678203500906.

[23] Fodor, Eugene M.; Smith, Terry, Jan 1982, The power motive as an influence on group decision making, Journal of Personality and Social Psychology, Vol 42(1), 178-185. doi: 10.1037/0022-3514.42.1.178

[24] Callaway, Michael R.; Marriott, Richard G.; Esser, James K., Oct 1985, Effects of dominance on group decision making: Toward a stress-reduction explanation of groupthink, Journal of Personality and Social Psychology, Vol 49(4), 949-952. doi: 10.1037/0022-3514.49.4.949

[25] Carrie, R. Leana (1985). A partial test of Janis' Groupthink Model: Effects of group cohesiveness and leader behavior on defective decision making, "Journal of Management", vol. 11(1), 5-18. doi: 10.1177/014920638501100102

[26] Raven, B. H. (1998). "Groupthink: Bay of Pigs and Watergate reconsidered". *Organizational Behavior and Human Decision Processes* 73 (2/3): 352–361. doi:10.1006/obhd.1998.2766.

[27] Badie, D. (2010). "Groupthink, Iraq, and the War on Terror: explaining US policy shift toward Iraq". *Foreign Policy Analysis* 6 (4): 277–296. doi:10.1111/j.1743-8594.2010.00113.x.

[28] Hermann, A.; Rammal, H. G. (2010). "The grounding of the "flying bank"". *Management Decision* 48 (7): 1051. doi:10.1108/00251741011068761.

[29] Eaton, Jack (2001). "Management communication: the threat of groupthink". *Corporate Communications: An International Journal* 6 (4): 183–192. doi:10.1108/13563280110409791.

[30] Koerber, C. P.; Neck, C. P. (2003). "Groupthink and sports: an application of Whyte's model". *International Journal of Contemporary Hospitality Management* 15: 20–28. doi:10.1108/09596110310458954.

[31] Baron, R. (2005). "So right it's wrong: Groupthink and the ubiquitous nature of polarized group decision making". *Advances in Experimental Social Psychology* 37: 35. doi:10.1016/s0065-2601(05)37004-3.

[32] Kramer, R. M. (1998). "Revisiting the Bay of Pigs and Vietnam decisions 25 years later: How well has the groupthink hypothesis stood the test of time?". *Organizational Behavior & Human Decision Processes* 73 (2/3): 238. doi:10.1006/obhd.1998.2762.

[33] Whyte, G. (1998). "Recasting Janis's Groupthink model: The key role of collective efficacy in decision fiascoes". *Organization Behavior and Human Decision Processes* 73 (2/3): 185–209. doi:10.1006/obhd.1998.2761.

[34] McCauley, C. (1998). "Group dynamics in Janis's theory of groupthink: Backward and forward". *Organizational Behavior and Human Decision Processes* 73 (2/3): 142–162. doi:10.1006/obhd.1998.2759.

[35] Tsoukalas, I. (2007). "Exploring the microfoundations of group consciousness". *Culture and Psychology* 13 (1): 39–81. doi:10.1177/1354067x07073650.

48.10 Further reading

- Baron, R. S. (2005). "So right it's wrong: groupthink and the ubiquitous nature of polarized group decision making". *Advances in Experimental Social Psychology* 37: 219–253. doi:10.1016/S0065-2601(05)37004-3.

- Ferraris, C.; Carveth, R. (2003). "NASA and the Columbia disaster: decision-making by groupthink?" (PDF). *Proceedings of the 2003 Association for Business Communication Annual Convention*.

- Esser, J. K. (1998). "Alive and well after 25 years: a review of groupthink research" (PDF). *Organizational Behavior and Human Decision Processes* 73 (2–3): 116–141. doi:10.1006/obhd.1998.2758.

- Hogg, M. A.; Hains, S. C. (1998). "Friendship and group identification: a new look at the role of cohesiveness in groupthink". *European Journal of Social Psychology* 28 (3): 323–341. doi:10.1002/(SICI)1099-0992(199805/06)28:3<323::AID-EJSP854>3.0.CO;2-Y.

- Klein, D. B.; Stern, C. (Spring 2009). "Groupthink in academia: majoritarian departmental politics and the professional pyramid". *The Independent Review: A Journal of Political Economy (Independent Institute)* 13 (4): 585–600.

- Kowert, P. (2002). *Groupthink or Deadlock: When do Leaders Learn from their Advisors?*. Albany: State University of New York Press. ISBN 0-7914-5250-6.

- Mullen, B.; Anthony, T.; Salas, E.; Driskell, J. E. (1994). "Group cohesiveness and quality of decision making: an integration of tests of the groupthink hypothesis". *Small Group Research* 25 (2): 189–204. doi:10.1177/1046496494252003.

- Moorhead, G.; Ference, R.; Neck, C. P. (1991). "Group decision fiascoes continue: Space Shuttle Challenger and a revised groupthink framework" (PDF). *Human Relations* 44 (6): 539–550. doi:10.1177/001872679104400601.

- O'Connor, M. A. (Summer 2003). "The Enron board: the perils of groupthink". *University of Cincinnati Law Review* **71** (4): 1233–1320.

- Packer, D. J. (2009). "Avoiding groupthink: whereas weakly identified members remain silent, strongly identified members dissent about collective problems" (PDF). *Psychological Science* **20** (5): 546–548. doi:10.1111/j.1467-9280.2009.02333.x. PMID 19389133.

- Rose, J. D. (Spring 2011). "Diverse perspectives on the groupthink theory – a literary review" (PDF). *Emerging Leadership Journeys* **4** (1): 37–57.

- Schafer, M.; Crichlow, S. (2010). *Groupthink versus High-Quality Decision Making in International Relations*. New York: Columbia University Press. ISBN 978-0-231-14888-7.

- 't Hart, P. (1990). *Groupthink in Government: a Study of Small Groups and Policy Failure*. Amsterdam; Rockland, MA: Swets & Zeitlinger. ISBN 90-265-1113-2.

- 't Hart, P.; Stern, E. K.; Sundelius, B. (1997). *Beyond Groupthink: Political Group Dynamics and Foreign Policy-Making*. Ann Arbor: University of Michigan Press. ISBN 0-472-09653-2.

- Tetlock, P. E. (1979). "Identifying victims of groupthink from public statements of decision makers" (PDF). *Journal of Personality and Social Psychology* **37** (8): 1314–1324. doi:10.1037/0022-3514.37.8.1314.

- Tetlock, P. E.; Peterson, R. S.; McGuire, C.; Chang, S. J.; Feld, P. (1992). "Assessing political group dynamics: a test of the groupthink model" (PDF). *Journal of Personality and Social Psychology* **63** (3): 403–425. doi:10.1037/0022-3514.63.3.403.

- Turner, M. E.; Pratkanis, A. R.; Probasco, P.; Leve, C. (1992). "Threat, cohesion, and group effectiveness: Testing a social identity maintenance perspective on groupthink" (PDF). *Journal of Personality and Social Psychology* **63** (5): 781–796. doi:10.1037/0022-3514.63.5.781.

- Whyte, G. (1989). "Groupthink reconsidered". *Academy of Management Review* **14** (1): 40–56. doi:10.2307/258190.

Chapter 49

Joint attention

A parent and child engage in joint attention

Joint attention (also: *shared attention*) is the shared focus of two individuals on an object. It is achieved when one individual alerts another to an object by means of eye-gazing, pointing or other verbal or non-verbal indications. An individual gazes at another individual, points to an object and then returns their gaze to the individual. Scaife and Bruner were the first researchers to present a cross-sectional description of children's ability to follow eye gaze in 1975. They found that most eight- to ten-month-old children followed a line of regard, and that all 11- to 14-month-old children did so. This early research showed it was possible for an adult to bring certain objects in the environment to an infant's attention using eye gaze.[1]

Subsequent research demonstrates that two important skills in joint attention are following eye gaze and identifying intention. The ability to share gaze with another individual is an important skill in establishing reference. The ability to identify intention is important in a child's ability to learn language and direct the attention of others. Joint attention is important for many aspects of language development including comprehension, production and word learning. Episodes of joint attention provide children with information about their environment, allowing individuals to establish reference from spoken language and learn words. Socio-emotional development and the ability to take part in normal relationships are also influenced by joint attention abilities. The ability to establish joint attention may be negatively affected by deafness, blindness, and developmental disorders such as autism.

Other animals such as great apes, orangutans, chimpanzees, dogs, and horses also show some elements of joint attention.

49.1 Humans

49.1.1 Levels of joint attention

Defining levels of joint attention is important in determining if children are engaging in age-appropriate joint attention. There are three levels of joint attention: triadic, dyadic, and shared gaze.

Triadic joint attention is the highest level of joint attention and involves two individuals looking at an object.[2] Each individual must understand that the other individual is looking at the same object and realize that there is an element of shared attention.[3] For an instance of social engagement to count as triadic joint attention it requires at least two individuals attending to an object or focusing their attention on each other.[4] Additionally, the individual must display awareness that focus is shared between himself or herself and another individual.[4] Triadic attention is marked by the individual looking back to the other individual after looking at the object.

Dyadic joint attention is a conversation-like behavior that individuals engage in. This is especially true for human adults and infants, who engage in this behavior starting at two months of age.[2] Adults and infants take turns exchanging facial expressions, noises, and in the case of the adult, speech.

Shared gaze occurs when two individuals are simply looking at an object.[5] Shared gaze is the lowest level of joint attention.

Individuals who engage in triadic joint attention must understand both gaze and intention to establish common reference. Gaze refers to a child's understanding of the link between mental activity and the physical act of seeing. Intention refers to the child's ability to understand the goal of another person's mental processes.

49.1.2 Gaze

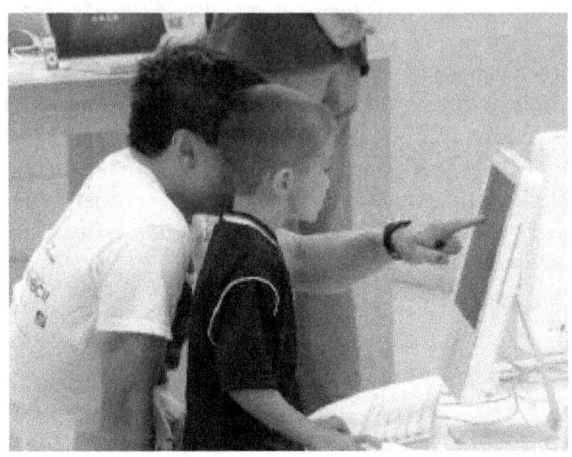

A parent and child engage in joint attention through pointing

For an individual to engage in joint attention they must establish reference.[6] Following the gaze or directive actions (such as pointing) of others is a common way of establishing reference.[6] For an individual to understand that following gaze establishes reference the individual must display:

- Recognition that looking is intentional behavior directed to external objects and events. Following gaze serves the purpose of establishing reference.[6]

- An understanding that looking results in the mental experience of seeing an object or event.[6]

- Recognition that eyes are responsible for seeing.[6]

- Recognition that others share in the capacity to see things.[6]

- An understanding that voice direction helps determine whether the speaker is talking to them and what he or she is referring to or focused on.[7]

Gaze becomes more complex with age and practice.[8][9] As gaze increases in complexity, individuals are better able to discriminate what others are referring to.[10] Joint attention is also important for social learning. Gaze following reflects an expectation-based type of orienting in which an individual's attention is cued by another's head turn or eye turn.[11] Individuals are motivated to follow another's gaze and engage in joint attention because gaze is a cue for which rewarding events occur.[11]

49.1.3 Intention

The ability to identify intention is critical to joint attention. When individuals understand that others have goals, intentions, and attentional states, they are able to enter into and direct another's attention.[6] Joint attention promotes and maintains dyadic exchanges and learning about the nature of social partners.[6] The ability to engage in joint attention is crucial for language development.[12][13]

Individuals who are intentional in their actions display regularity in their behavior.[14] Individuals locate objects with their eyes, move towards the object, and then use hands to make contact with and manipulate the object.[14] Change in gaze direction is one of several behavioral cues that individuals use in combination with changes in facial and vocal displays and body posture to mark the intention to act on an object.[14] Individuals who seek or follow a joint focus of attention display knowledge that what is in their awareness is also in another's awareness.[3] They believe that they are experiencing the same world as others.[3]

Joint attention plays an important role in the development of theory of mind. Theory of mind and joint attention are important precursors to a fully developed grasp of another individual's mental activity.[10]

49.1.4 Language comprehension

Two teenagers engage in joint attention by reading a book.

The ability of children to extract information from their environment rests on understandings of attentional behaviors such as pointing.[8] Episodes of joint attention provide children with a great deal of information about objects by establishing reference and intention.[8] Joint attention occurs within particular environments. The items and events in that environment provide a context that enables the child to associate meaning with a particular utterance.[15] Joint attention makes relevant aspects of the context salient, helping children comprehend what is taking place.

49.1.5 Language production

An infant's social environment relates to his or her later language development.[16] Children's first words are closely linked to their early language experience.[2] For children with typically developing language skills, there is a close match between maternal speech and their environment: up to 78% of maternal speech is matched to the object the child is focusing on.[2] In children with delayed language development, only 50% of maternal speech is matched to the object the infant is focusing on.[2] Infants are more likely to engage in joint attention when the parent talks about an object that the child is attending to as opposed to an object outside of the infant's attention.[16] This increased level of joint attention aids in encouraging normal language development, including word comprehension and production.[16] When joint attention is present, it plays an important role in word learning, a crucial aspect of language development.[17]

49.1.6 Relationship to socio-emotional development

Joint attention and the ability to attend to an aspect of one's environment are fundamental to normal relationships that rely on the sharing of experience and knowledge.[11] Infants are highly motivated to share experience. An infant's motivation to engage in joint attention is strong enough that infants voluntarily turn away from interesting sights to engage in joint attention with others.[9]

As described in attachment theory, infants need to develop a relationship with a primary caregiver to achieve normal social and emotional development. A key part of the ability to develop this relationship may be joint attention. In addition to language development, joint attention serves the function of preparing infants for more complex social structures involved in adult conversation. Children's skills in initiating and responding to joint attention predict their social competence at 30 months of age.[18] Anticipatory smiling (a low level form of joint attention involving smiling at an object then turning the smile to one's communicative partner) at 9 months positively predicts parent-rated social competence scores at 30 months in infants.[19] Early joint attention abilities account for differences in social and emotional abilities in later life.[19]

49.1.7 Developmental markers in infancy

At the age of 2 months, children engage in dyadic joint attention and conversation-like exchanges with adults during which each is the focus of the other's attention and they take turns exchanging looks, noises and mouth movements.[20] At age 3 months, children display joint attention skills by calling to a caregiver when they are not perceivable.[3] When caregiver does not respond in a similar manner, child exhibits a series of responses that were first studied in early 1970s by Edward Tronick[21] in collaboration with pediatrician T. Berry Brazelton at the time when the latter was creating the Neonatal Behavioral Assessment Scale. At age 6 months, infants display joint attentional skills by:

- Orienting themselves in the same general direction (in their visual field) as another person.[20] Infants also cease to focus on the first interesting (salient) object they encounter.[20]

- Following outward directed gaze of adults.[20]

- Extending more sophisticated behaviors, such as gaze checking, when initial gaze following is not successful.[10]

- Paying more attention to eyes, responding to shifts in eye gaze direction, and directing their own attention based on another's gaze.[8]

At age 8 months, infants demonstrate joint attention through proto-declarative pointing, particularly in girls.[20] At 9 months of age, infants begin to display triadic joint attention.[2] Infants also will display joint attention activities, such as communicative gestures, social referencing, and using the behavior of others to guide response to novel things.[20]

At one year of age, joint attention is displayed through a child's understanding of pointing as an intentional act.[20] One-year-olds also establish joint attention for objects within their visual field before objects beyond their current visual field. At this age, infants are not yet able to represent their entire environment, only what they can see.[20] At age 15 months, children recognize the minds of others.[20] At this age, children also recognize the importance of eyes for seeing and that physical objects can block sight.[8] At age 18 months, infants are capable of following an individual's gaze to outside their visual field and establishing (representative) joint attention.[20] 18-month-olds also grasp the intentional, referential nature of looking, the mentalistic experience of seeing and the role of eyes[8] and are skilled at following both gaze and pointing with precision.[8] At two years of age, children display joint attention by extending attention beyond the present and understanding that the targets of other's attention extends to the past as well.[3] Two-year-olds are also capable of representational thought or increased memory.[3]

49.1.8 Individuals with disabilities

Several studies have shown that problems with joint attention are associated with developmental processes.[22] Difficulties in establishing joint attention may partially account for differences in social abilities of children with developmental disorders (i.e. Autism spectrum disorders).[22] A core deficit noted in autism is eye gaze.[23] Autistic children have difficulty alternating their attention towards a partner and third object.[23] This difficulty is attributed to their deficiencies in following gaze, resulting in difficulty initiating and maintaining joint attention.[23] Deaf infants are able to engage in joint attention similar to hearing infants; however, the time spent engaged in joint attention is often reduced in deaf infants born to hearing parents.[24] Hearing parents of deaf infants often are less likely to respond and expand on their deaf infants' initiative and communicative acts.[24] Deaf infants of deaf parents do not show reduced time spent in joint attention.[24] Auditory input is not critical to joint attention but similar modes of communication and understanding are vital.[24] Furthermore, mothers who are unable to successfully establish regular joint attention with their child rate that infant lower on scales of social competence.[24] Judgement of low social competence can be made as early as 18 months of age.[24] In blind infants, joint attention is established by means of auditory input or feeling another person's hand on an object and may be delayed compared to sighted infants.[25]

49.2 Other animals

49.2.1 Definitions in non-human animals

Triadic joint attention is the highest level of joint attention and involves two individuals looking at an object.[2] Each individual must understand that the other individual is looking at the same object and realize that there is an element of shared attention.[3][4] Triadic attention is marked by the individual looking back to the other individual after looking at the object.[5] Dyadic joint attention involves mutual gaze between the parent and infant.[5] Mutual gaze is marked by both the parent and infant looking at each other's face.[26] If two individuals are simply looking at an object, it is referred to as shared gaze.[5]

49.2.2 Dyadic joint attention

Infant and parent chimpanzees show dyadic joint attention in an affectionate manner by looking at each other's eyes[27] Non-human animals such as Japanese monkeys, baboons, and other Old World monkeys seldom engage in dyadic joint attention.[27] For these animals, the eye contact involved in dyadic joint attention is deemed threatening.[27]

49.2.3 Shared gaze

Gaze following, or shared gaze, can be found in a number of primates.[5]:155–71 Domesticated animals such as dogs and horses also demonstrate shared gaze.[28] This type of joint attention is important for animals because gaze shifts serve as indicators alerting the animal to the location of predators, mates, or food.[5]

Chimpanzees are capable of actively locating objects that are the focus of another individual's attention by tracking the gaze of others.[29] They are not limited to following eye gaze to the first interesting object in their view.[29] They use a number of different cues to engage in shared focus, including head movement and eye gaze.[5] Infant chimpanzees start to follow tap, point, and head turn cues of an experimenter by nine months of age.[5] By 13 months of age, they show following responses to glance cues without a head turn.[5] There is no evidence to support that infant chimpanzees are able to use eye gaze alone as a cue for following responses.[5] By 20 months of age, infant chimpanzees are able to follow an experimenter's cues to a target behind the chimpanzee but infant chimpanzees do not look back to the experimenter after looking at the target.[5] Moving targets are more salient than stationary targets for infant chimpanzees.[5] Chimpanzee infants are sensitive to faces which are gazing at them, but chimpanzees less than three to four years old only look within their visual field when using the experimenter's head turn as their cue.[5]

49.3 See also

- Asperger syndrome
- Cooperative eye hypothesis
- Grounding in communication
- Vocabulary development

49.4 References

[1] Moore, C.; Dunham, P (1995). *Joint Attention: Its Origins and Role in Development*. Lawrence Erlbaum Associates.

ISBN 0-8058-1437-X.

[2] Oates J, & Grayson A. (2004). *Cognitive and Language Development in Children*. Oxford, UK: Blackwell Publishing. ISBN 1-4051-1045-7

[3] Reddy, V. (2005). Before the third element: Understanding attention to self. In N. Eilan, C. Hoerl, T. McCormack & J. Roessler (Eds.), *Joint attention: Communication and other minds* (pp. 85–109). Oxford, UK: Oxford University Press.ISBN 0-19-924563-0

[4] Hobson, R.P. (2005). "What puts the jointness in joint attention?". In Roessler, J. *Joint attention: Communication and other minds*. Oxford University Press. pp. 185–204. ISBN 0-19-924563-0.

[5] Okamoto-Barth, S.; Tomonaga, M. (2006). "Development of Joint Attention in Infant Chimpanzees". In Tanaka, M. *Cognitive Development in Chimpanzees*. Spinger-Verlag. pp. 155–171. ISBN 4-431-30246-8.

[6] D'Entremont, B., Yazbeck, A., Morgan, A. & MacAulay, S. (2007). Early gaze-following and the understanding of others. In R. Flomm, K. Lee & D. Muir (Eds.), *Gaze-Following: Its Development and Significance* (pp. 77–94). Mahwah, NJ: Lawrence Erlbaum Associates. ISBN 0-19-924563-0

[7] Rossano, Federico; Malinda Carpenter; Michael Tomasello (2012). "One-Year-Old Infants Follow Others' Voice Direction". *Psychological Science* 23: 1298–1302. doi:10.1177/0956797612450032.

[8] Woodward, A. (2005). Infants' understanding of the actions involved in joint attention. In N. Eilan, C. Hoerl, T. McCormack & J. Roessler (Eds.), *Joint attention: Communication and other minds* (pp. 110–128). Oxford, UK: Oxford University Press.ISBN 0-19-924563-0

[9] Carpenter, M. (2011). Social Cognition and Social Motivations in Infancy. In U. Goswami (Eds.), *The Wiley-Blackwell handbook of childhood cognitive development* (pp. 106–128). West Sussex, UK: Blackwell Publishing Ltd.ISBN 1-4051-9116-3

[10] Heal, J. (2005). "Joint attention and understanding the mind". In Roessler, J. *Joint attention: Communication and other minds*. Oxford University Press. pp. 34–44.

[11] MacPherson, A. C. & Moore, C. (2007). Attentional control by gaze cues in infancy. In R. Flomm, K. Lee & D. Muir (Eds.), *Gaze-Following: Its Development and Significance* (pp. 53–76). Mahwah, NJ: Lawrence Erlbaum Associates.

[12] Tomasello, M.; Farrar, J. (1986). "Joint attention and early language". *Child Development* 57: 1454–1463. doi:10.1111/j.1467-8624.1986.tb00470.x.

[13] Baldwin, D.A. (1995). Understanding the link between joint attention and language. In C. Moore & P.J. Dunham (Eds.) *Joint attention: Its origins and role in development* (pp.131-158). Hillsdale, NJ: Erlbaum.

[14] Mumme, D., Bushnell, E., DiCorcia, J. & Lariviere, L. (2007). Infants' use of gaze cues to interpret others' actions and emotional reactions. In R. Flomm, K. Lee & D. Muir (Eds.), *Gaze-Following: Its Development and Significance* (pp. 143–170). Mahwah, NJ: Lawrence Erlbaum Associates. ISBN 0-8058-4750-2

[15] Bruner, J. (1983). *Child's talk: Learning to use language*. (pp. 67–88). New York, NY: W.W. Norton & Co.

[16] Rollins, P. R. (2003). "Caregivers' contingent comments to 9-month-old infants: Relationships with later language". *Applied Psycholinguistics* 24: 221–234. doi:10.1017/S0142716403000110.

[17] Hirotani, M.; Stets, M.; Striano, T.; Friederic, A.D. (2009). "Joint attention helps infants learn new words: event-related potential evidence". *Developmental Neuroscience* 20: 600–605. doi:10.1097/WNR.0b013e32832a0a7c.

[18] Van Hecke, Amy Vaughan; Mundy, Peter C.; Acra, C.Franciose; Block, Jessica J.; Delgado, Christine E.F.; Parlade, Meaghan V.; Myers, Jessica A.; Neal, A. Rebecca; Popmares, Yuly B. (2007). "Infant Joint Attention, Temperament, and Social Competence in Preschool Children". *Child Development* 78 (1): 53–69. doi:10.1111/j.1467-8624.2007.00985.x.

[19] Paralade, Meaghan Venezia; Messinger, Daniel S.; Delgado, Christine E.F.; Kaiser, Marygrace Yale; Van Hecke, Amy Vaughan; Mundy, Peter C. (2009). "Anticipatory smiling: Linking early affective communication and social outcome". *Infant Behavior and Development* 32: 33–43. doi:10.1016/j.infbeh.2008.09.007.

[20] Goswami, Usha (2008)*Cognitive Development: The Learning Brain*. New York, NY:Psychology.

[21] Trusting Relationships Are Central to Children's Learning -- Lessons From Edward Tronick, Huffington Post, 31 January 2011

[22] Bhat, AN; Galloway, JC; Landa, RJ. "Social and non-social visual attention patterns and associative learning in infants at risk for autism". *Journal of Child Psychology and Psychiatry* 51 (9): 989–997. doi:10.1111/j.1469-7610.2010.02262.x.

[23] Bruinsma, Y.; Koegel, R.; Koegel, L. (2004). "Joint attention and children with autism: A review of the literature". *Mental Retardation and Development Disabilities* 10: 169–175. doi:10.1002/mrdd.20036. PMID 15611988.

[24] Nowakowski, Matlida E.; Tasker, Susan L.; Schmidt, Louis A. (2009). "Establishment of joint attention in dyads involving hearing mothers of deaf and hearing children and its relation to adaptive social behavior". *American Annals of the Deaf* 154 (1): 15–29. doi:10.1353/aad.0.0071.

[25] Bigelow, A. E. (2003). "The development of joint attention in blind infants". *Development and Psychopathology* 15: 259–275. doi:10.1017/S0954579403000142.

[26] Social Cognition Development in First 2 Years. In T. Matsuzawa M. Tomonaga & M. Tanaka (Eds.), *Cognitive Development in Chimpanzees* (pp.182-197). Tokyo: Springer-Verlag.

[27] Evolutionary Origins of Mother-Infant Relationship. In T. Matsuzawa M. Tomonaga & M. Tanaka (Eds.), *Cognitive Development in Chimpanzees* (pp.127-141). Tokyo: Springer-Verlag.

[28] Itakura, S. (2004). "Gaze Following and Joint Visual Attention in Nonhuman Animals". *Japanese Psychological Research* **46** (3): 216–226. doi:10.1111/j.1468-5584.2004.00253.x.

[29] Tomasello et al. & Emery et al. as cited in (2006). Development of Joint Attention in Infant Chimpanzees. In T. Matsuzawa M. Tomonaga & M. Tanaka (Eds.), *Cognitive Development in Chimpanzees* (pp.155–171). Tokyo: Springer-Verlag.

49.5 Text and image sources, contributors, and licenses

49.5.1 Text

- **Self-organization** *Source:* https://en.wikipedia.org/wiki/Self-organization?oldid=724091941 *Contributors:* The Anome, Miguel~enwiki, Tedernst, Edward, Michael Hardy, Lexor, Kku, MartinHarper, EntmootsOfTrolls, Charles Matthews, Dysprosia, Nickg, Robbot, Fredrik, Rursus, Moink, Michael Snow, Mu6, Dina, Snobot, Ancheta Wis, Alensha, Pcarbonn, Margana, Karol Langner, The Land, Elektron, Pgreenfinch, Robin klein, Andreas Kaufmann, RevRagnarok, Chris Howard, Jwdietrich2, Ronaldo~enwiki, MiddleOfNowhere, Rich Farmbrough, Avriette, Vsmith, Wk muriithi, Smyth, Dave souza, JimR, Dmr2, Bender235, FirstPrinciples, Shrike, Zenohockey, Alex Kosorukoff, RoyBoy, Cretog8, Smalljim, Viriditas, .:Ajvol:., Physicistjedi, Ire and curses, Mdd, HasharBot~enwiki, Jheald, RJII, DV8 2XL, Sylvainremy, Rvanschaik, BryanKaplan, Grammarbot, Rjwilmsi, KYPark, Pleiotrop3, ElKevbo, The wub, Jeffmcneill, Mathbot, Diza, Hamidifar, YurikBot, Wavelength, Mukkakukaku, Duracell~enwiki, Pseudomonas, CLW, Curpsbot-unicodify, KnightRider~enwiki, SmackBot, Stpalli, WebDrake, Vald, Pokipsy76, M stone, Skizzik, Mobius27, Thumperward, Complexica, Colonies Chris, Royboycrashfan, Fotoguzzi, Cícero, Ericbritton, Will Beback, Eliyak, Nick Green, JoseREMY, Camazine, Kerbii, Dave Runger, Mr3641, Zarex, N2e, Pfhenshaw, Cydebot, Krauss, Gmusser, Skittleys, Miguel de Servet, Oszillodrom, Letranova, Kilva, Noclevername, Luna Santin, Rudick.JG, Davedrh, Smartse, Phanerozoic, JAnDbot, Narssarssuaq, Athkalani~enwiki, Gerculanum, Freshacconci, GrahameKing, Vernanimalcula, Economizer, Snowded, KConWiki, Dirac66, David Eppstein, User A1, Rvsole, Masaki K, Jim.henderson, Emathematica, Pilgaard, Keesiewonder, Grosscha, Crakkpot, 1000Faces, Korotkikh, Elizabeth McMillan, Pleasantville, Dggreen, Crscrs, Rollo44, Vipinhari, AllGloryToTheHypnotoad, Ordermaven, Northfox, Gbawden, SieBot, Thehotelambush, GeneCallahan, Adelanwar, Der Golem, Techdoer, Synergier, Gulmammad, Rhododendrites, Sun Creator, EhJJ, Bracton, Schreiber-Bike, Adriansrfr, Life of Riley, Koumz, Xiaoju zheng, Dthomsen8, Cyberoo, Fd42, WikHead, Thomas h ray, Addbot, USchick, Unesn6iduja, MrOllie, LarryJeff, Lightbot, Mcamus, Jarble, بسم, Luckas-bot, Yobot, II MusLiM HyBRiD II, Azcolvin429, AnomieBOT, Jim1138, Phantom Hoover, Materialscientist, Citation bot, LilHelpa, The Banner, Omnipaedista, Sahehco, Chjoaygame, FrescoBot, TheSen, Citation bot 1, Winterst, Gray1, Charbee, Regular Polyhedron, Jandalhandler, Ambarsande, Trappist the monk, Reflexinio, Barryclemson, We system, Blueshifting, Noresponse, Lithistman, Hhhippo, Quickmute, JuanCano, Cymru.lass, Carl Wivagg, Allanwik, Robbiemorrison, Ems2715, NinjaQuick, TuxFighter, Jrichardliston, ClueBot NG, Fgunnars, Panleek, Joel B. Lewis, MerlIwBot, Helpful Pixie Bot, Richardjb25, Revisor2011, RogerBF, BG19bot, GlaedrH, DPL bot, Terrykel, Kfriston, Soler99, Zach Lipsitz, Khazar2, Nathanielfirst, IjonTichyIjonTichy, Dexbot, Makecat-bot, BurritoBazooka, Mre env, Samotny Wędrowiec, Andy Quarry, Duchifat, Otherocketman, FrB.TG, Monkbot, 我, Mit0126, Asuscreative, Isambard Kingdom, KasparBot, Jman9058, Sangqiu5, Robcduk and Anonymous: 136

- **Statistical fluctuations** *Source:* https://en.wikipedia.org/wiki/Statistical_fluctuations?oldid=670246277 *Contributors:* Pol098, Malcolma, SmackBot, Melcombe, GoingBatty, RockMagnetist, Tianbai Cui and Anonymous: 2

- **Positive feedback** *Source:* https://en.wikipedia.org/wiki/Positive_feedback?oldid=724568935 *Contributors:* Derek Ross, DavidLevinson, Anthere, Edward, Patrick, Lexor, Ahoerstemeier, William M. Connolley, Angela, Julesd, Glenn, Dcoetzee, Bhuston, Pstudier, Denelson83, Altenmann, Mattflaschen, Alan Liefting, Giftlite, Wolfkeeper, CyborgTosser, Cantus, Beland, Thorsten1, Bender235, Cafzal, Nigelj, Spalding, Arcadian, Mdd, Mduvekot, Wtmitchell, Wtshymanski, RyanGerbil10, Joriki, Eleassar777, Gimboid13, DaveApter, Marudubshinki, Mandarax, BD2412, Rjwilmsi, NeonMerlin, Nihiltres, Nivix, Ewlyahoocom, Nimur, Bgwhite, NSR, YurikBot, Wavelength, Jenks1987, Member, Wimt, Jpbowen, DeadEyeArrow, Light current, ReCover, C-randles, SmackBot, InverseHypercube, McGeddon, Kslays, Indium, Andy M. Wang, KD5TVI, Oli Filth, Nbarth, Scwlong, RandyBurge, Richard001, Ryan Roos, Sporkot, Byelf2007, Dudecon, Harryboyles, Mike1901, Slakr, Rogerbrent, Dicklyon, Geologyguy, Novangelis, Onionmon, Kvng, Philmcrae, Igoldste, Tawkerbot2, Chetvorno, Xcentaur, Peter1c, CmdrObot, Circuit dreamer, Seven of Nine, Penbat, Herenthere, AndrewHowse, Cydebot, Peripitus, Epbr123, Headbomb, Id447, Luigifan, RichardVeryard, Dawnseeker2000, Jj137, LibLord, Phanerozoic, Athkalani~enwiki, Yill577, Alta-Snowbird, Mluo2010, JaGa, Squidonius, Nikoladie~enwiki, Roger Mexico, Anaxial, Glrx, R'n'B, Erkan Yilmaz, Maurice Carbonaro, Ginsengbomb, Andareed, Winderful1, Brendan19, DorganBot, Hulten, AndreasJSbot, Hammersoft, Mark v1.0, Fredrick day, Mannafredo, Every name is taken12345, Lova Falk, Spinningspark, FlyingLeopard2014, Pmarshal, SieBot, Mikemoral, Caltas, Oxymoron83, Rinconsoleao, Jonathanstray, Sokari, ClueBot, ImperfectlyInformed, WMCEREBELLUM, Mild Bill Hiccup, VandalCruncher, Yuckhil, Riking8, Djr32, Magwo, Vivio Testarossa, Sun Creator, Calrosfing, BOTarate, Johnuniq, Editor2020, Nathan Johnson, Thinboy00P, Addbot, Emiehling, Drrdudley, Gizziiusa, Download, Semiwiki, Swarm, Luckas-bot, Yobot, AnomieBOT, Trevithj, LilHelpa, Gap9551, Ktpeil, GrouchoBot, Omnipaedista, Chjoaygame, GliderMaven, Recognizance, Skyerise, USMCSoftLove, Allthingstoallpeople, Wjomlex, Wotnow, Nascar1996, RjwilmsiBot, Kakuchirana, EmausBot, Denniskakai, Jopienator, Dcirovic, NESFreak92, Donner60, ModManDan, ChuispastonBot, JanetteDoe, Teapeat, ClueBot NG, Robthepiper, O.Koslowski, Widr, Helpful Pixie Bot, Nashhinton, BG19bot, Cisba, Walk&check, Bowser423, Ugncreative Usergname, Robertmacl, John1988cassidy, Rob Hurt, Dchandran1, Pawlowiec, Mbharat23, Sapientsage, Kernsters, Greenteawhitetea, EvergreenFir, Sam Sailor, Monkbot, Loraof, MurderByDeadcopy, Klsyhncck, Chaithanya Prabhu M, Coolbreeze19876, Trev74 and Anonymous: 201

- **Chaos theory** *Source:* https://en.wikipedia.org/wiki/Chaos_theory?oldid=726457269 *Contributors:* AxelBoldt, Tobias Hoevekamp, Sodium, Mav, Zundark, Gareth Owen, Arvindn, Roadrunner, SimonP, David spector, Heron, Gumpu, Edward, Michael Hardy, Tez, Lexor, Isomorphic, Chinju, Karada, Iluvcapra, Ahoerstemeier, William M. Connolley, Snoyes, Darkwind, Kevin Baas, Evercat, Smack, Schneelocke, Charles Matthews, Adam Bishop, Dino, Dysprosia, Jitse Niesen, Doradus, Munford, K1Bond007, Jose Ramos, Fairandbalanced, Bevo, Traroth, Banno, JorgeGG, Phil Boswell, Robbot, Bernhard Bauer, Goethean, Gandalf61, Chopchopwhitey, MathMartin, Sverdrup, Academic Challenger, Ojigiri~enwiki, Zubras, Paul Murray, Dave Bass, Dbroadwell, Wile E. Heresiarch, Tea2min, Enochlau, Decumanus, Giftlite, Smjg, Fennec, Gene Ward Smith, Vir4030, Kim Bruning, Everyking, Curps, Sunny256, Pucicu, Chowbok, Utcursch, LucasVB, Antandrus, Mako098765, Quarl, Vanished user 1234567890, Karol Langner, Rdsmith4, Oneiros, Pmanderson, Zfr, Sam Hocevar, Lumidek, Jmeppley, Joyous!, Barnaby dawson, TheObtuseAngleOfDoom, Shiftchange, Discospinster, Rich Farmbrough, TedPavlic, Avriette, Guanabot, Vsmith, Dave souza, Lulu of the Lotus-Eaters, Fluzwup, Paul August, Bender235, Neurophyre, Loren36, Fenice, Brian0918, El C, Pjrich, Alereon, AJP, Rwh, Semper discens, Billymac00, John Vandenberg, Thomas G Graf, Flammifer, Obradovic Goran, Mdd, Cyrloc, Msh210, Defunkt, Prashmail, Alansohn, Eric Kvaalen, Arthena, Keenan Pepper, CommodoreMan, Lectonar, WhiteC, BryanD, Sligocki, Hu, Bart133, PaePae, Helixblue, HenkvD, Evil Monkey, Cal 1234, RainbowOfLight, DV8 2XL, Embryomystic, Kazvorpal, Dan100, OleMaster, Simetrical, Linas, Ramsremedies, Scriberius, Igny, VanFullOfMidgets, LOL, Scid, Guardian of Light, KickAir8P~, Ruud Koot, MONGO, Kelisi, GregorB, XaosBits, Graham87, Magister Mathematicae, Anarchivist, Jorunn, Rjwilmsi, Joakim Munkhammar, KYPark, XP1, TheRingess, Brighterorange, Scartol, The wub, Bhadani,

49.5. TEXT AND IMAGE SOURCES, CONTRIBUTORS, AND LICENSES

Yamamoto Ichiro, Mathbot, Greg321, Sunayana, Nivix, RexNL, Nabarry, Incompetnce, Smithbrenon, Nicholasink, Chobot, Evilphoenix, Bgwhite, Cactus.man, Gwernol, YurikBot, Wavelength, Deeptrivia, Pmg, Hillman, Nmondal, Splash, JabberWok, Prokaryote1234, Stephenb, Jugander, Chaos, Alex Bakharev, Rsrikanth05, David R. Ingham, Dtrebbien, Grafen, Winonanick, JocK, Dhollm, Raven4x4x, Moe Epsilon, Zwobot, Epipelagic, Romarin, Dlyons493, Suso, Bota47, Dan131m, Cat2020, Zunaid, WAS 4.250, Phgao, Ninly, Imaninjapirate, Arthur Rubin, GraemeL, DGaw, Madrazz, Vicarious, Reject, Kungfuadam, DVD R W, Soir, Benjamindees, Marquez~enwiki, SmackBot, 4dhayman, ManaUser, Maksim-e~enwiki, Sethmasters, Stellea, The hoodie, InverseHypercube, KnowledgeOfSelf, Unyoyega, C.Fred, Rosaak, Thunderboltz, Flux.books, PeterSymonds, Gilliam, Sbonsib, Skizzik, GwydionM, Izehar, Bluebot, Persian Poet Gal, RDBrown, Telempe, Alex brollo, SchfiftyThree, GabrielPere, Complexica, Bazonka, Sudharsansn, CSWarren, DHN-bot~enwiki, Jdthood, Yanksox, Hellfire81, QuimGil, Gorgeorgeus, Can't sleep, clown will eat me, MyNameIsVlad, Jahiegel, Rrburke, Spectrogram, Nakon, Anmnd, Mini-Geek, Thismarty, Profyorke, Wybot, DMacks, SashatoBot, Lambiam, Mukadderat, Luigi-ish, Kuru, Lakinekaki, Lapaz, Buchanan-Hermit, Joshua Andersen, Chodorkovskiy, JorisvS, Dumelow, Jim.belk, IronGargoyle, Mosgiel, Atomic Duck!, Brazucs, Dicklyon, Xiaphias, Invisifan, Candybars, Dr.K., Dfred, Inquisitus, Rlinfinity, Xionbox, Asyndeton, Mdanziger, PSOfan2000, Iridescent, Shoeofdeath, Cumi~enwiki, Rhetth, Daveyork, Experiment123, Tawkerbot2, Chetvorno, Timrem, PurpleRain, CRGreathouse, Crownjewel82, Aherunar, Avanu, TheTito, Neelix, Grein, Mct mht, CX, Yaris678, Gogo Dodo, Lugnuts, Pascal.Tesson, Alpharius~enwiki, Tawkerbot4, DumbBOT, Chrislk02, Romon, Letranova, Thijs!bot, Epbr123, Hervegirod, UXs, Sagaciousuk, Scientio, Oliver202, Headbomb, Zardoze, Perrygogas, West Brom 4ever, James086, Nezzadar, Charukesi, Universal Hero, Widenet, Gfalco, Northumbrian, AntiVandalBot, Devanshi.shah, Ben pcc, Doc Tropics, Jcsellak, Jj137, JAnDbot, Ashishval44, Husond, Gandhi gaurav, MER-C, Sophie means wisdom, Igodard, Hut 8.5, MSBOT, Kirrages, Captain head, Peteymills, Coffee2theorems, Jill.marleigh, Magioladitis, Diderot7, VoABot II, Catslash, JamesBWatson, Mbc362, Carlylecastle, Brewhaha@edmc.net, Brother Francis, Catgut, Ensign beedrill, Mjkelley79, David Eppstein, Kotinopoulos, Vssun, JoergenB, DerHexer, JaGa, Bryt, Falcor84, Waitati, Cocytus, Stephenchou0722, DancingPenguin, MartinBot, Arjun01, Poeloq, InnocuousPseudonym, Tomasao, Ayonbd2000, Erkan Yilmaz, J.delanoy, Oshron, Trusilver, AstroHurricane001, MikeBaharmast, Maurice Carbonaro, Zakholdsworth, Thegreenj, Ian.thomson, JAK2112, Salih, Katalaveno, Enuja, Quasi-Abstract, V.V zzzzz, Coppertwig, Chiswick Chap, NewEnglandYankee, Policron, MKoltnow, Zojj, MetsFan76, TottyBot, Ahshabazz, Lamp90, Prot D, Yodler, JavierMC, Nnnagig, Cmarnold, Ogunjosam, Idioma-bot, JLBernstein, Funandtrvl, Phlounder, Yoeb137, Torcini, Mimigary, Pleasantville, DSRH, Tunnels of Set, Jeff G., JohnBlackburne, AlnoktaBOT, HeckXX, Richardseel, DancingMan, Philip Trueman, TXiKiBoT, Oshwah, Ggggdxn, Red Act, A4bot, Tagalong99, IPSOS, Voorlandt, Magmi, Corvus cornix, Garravogue, Rubseb, PDFbot, Katimawan2005, 3p1416, Kızılsungur, Inductiveload, Kaiketsu, Kilmer-san, Wolfrock, Jacob501, Sheildofthunder, The The Fool on the Hill, Blazen nite, HiDrNick, Symane, SamuraiGabe, Radagast3, Maxlittle2007, SieBot, Tosun, Cwkmail, This, that and the other, Zsniew, Revent, Vanished User 8a9b4725f8376, Africangenesis, Warhammer 8, Somecreepyoldguy, Prestonmag, Trang Oul, Oxymoron83, AngelOfSadness, Hello71, KoshVorlon, Michael Courtney, Fratrep, Convictionist, StaticGull, Szalagloria, Mike2vil, Abmcdonald, Tojuro, Tommi Ronkainen, Wikiskimmer, SUPERSONICOOHHHOHOH, Escape Orbit, Stu, Francvs, Apsimpson02, Axel-Rega, ClueBot, Avenged Eightfold, The Thing That Should Not Be, Sijokjoseph, Plastikspork, Ribbon Salminen, Herakles01, Abrfreek777, Der Golem, Gommert, Ksmadden, Niceguyedc, JJIG, LizardJr8, ChandlerMapBot, Lbertolotti, Paulcmnt, Djr32, Eboyjr, Feline Hymnic, IPrussian, Echion2, Jmlipton, Mikaey, Niyse, La Pianista, Flower Priest, Versus22, SoxBot III, Vanished user uih38riiw4hjlsd, Nori Llane, Un Piton, Wbblaze4, Golddragon24, XLinkBot, Jovianeye, Rror, Colliric, Addbot, Mortense, Rakeshfern, TheDestitutionOfOrganizedReligion, Melab-1, The Equilibrium, Otisjimmy1, DougsTech, Fgnievinski, Ronhjones, Funky Fantom, SomeUsr, Glane23, Nutter13, Ytbau, Debresser, Favonian, XFreakonaLeashX, SpBot, LinkFA-Bot, Lipehauss~enwiki, Freakonaleashnj, Tassedethe, Bwrs, LarryJeff, Lightbot, Gail, Zorrobot, Jarble, Jamesevi, Megaman en m, Vicky sekar, CS2020, Legobot, Luckas-bot, Yobot, Dgurubaran, AnomieBOT, Kristen Eriksen, IRP, Collieuk, Aeortiz, Kingpin13, Flewis, Materialscientist, Jacksonroberts25, To Fight a Vandal, Citation bot, Srinivas, Onesius, Ruby2010, Spidermanizdabest, Xqbot, TitusCarus, CathNek, GrouchoBot, Damienivan, 7h3 3L173, RibotBOT, SassoBot, Energybender, Smallman12q, Elizabeth Linden Rahway, A. di M., Frozenevolution, FrescoBot, Justinodem, Thayts, Sławomir Biały, Argumzio, Kwiki, Citation bot 1, Milly.mortimer, Theory2reality, Pinethicket, SimmonsNorwood, Therealfozzy, MastiBot, FoxBot, Koolguy1029, Anonwhymus, Trappist the monk, Wotnow, Willihans, Redfan45x, Math.geek3.1415926, Inferior Olive, Duoduoduo, Arided, DARTH SIDIOUS 2, RjwilmsiBot, DSP-user, CanadianPenguin, John of Reading, Orphan Wiki, Karsh07007, KurtLC, Jaguar6cy, Perfect Introvert, AppuruPan, Slightsmile, AgRince, Mussermaster, Dcirovic, K6ka, Koryds2008, Earthh, Hhhippo, Ida Shaw, Hugo.cavalcante, Shuipzv3, Askedonty, JPfreak, Arbnos, Wayne Slam, Music Sorter, Donner60, Inka 888, Bill william compton, Subanm, AnthonyMarkes, Mr Schneebly, Support.and.Defend, Mikhail Ryazanov, ClueBot NG, Guswfla1, Marechal Ney, Davidcarfi, Timflutre, Minki6656, Helpful Pixie Bot, Richardjb25, Lottsy, Bibcode Bot, Jeraphine Gryphon, Rhysjeans, BG19bot, LangdonAlger29, Cispyre, Birdtracks, FiveColourMap, Yowanvista, Falkirks, Sf jeff2, Joshua Jonathan, HMman, Westcoastg24, Brad7777, Packman744, Randomguess, BattyBot, Ema--or, DIY Sunrise, Ruidilao, Cjripper, Prayforrain, Hamzaata, Trololol115, Zirconzx, Illia Connell, Dexbot, Theillusionking, CuriousMind01, Pal.bjartan, Sundarsharath, TheKing44, Patrick.knoll96, Anthaceorote, William2001, NerdGirl1988, Ydoc52, Ufoneda, Paulpgh, Myconix, Penitence, Docirish7, VAggarwal, Nigellwh, Francois-Pier, Irte, Anrnusna, TheSawyerBean, Sheddow, JaconaFrere, Masdpofham, Jsmk, Monkbot, HowlingAngel, Yikkayaya, Rebusch, ChaoticPoet, Garfield Garfield, Sajidiqbal14, JC713, Purgy Purgatorio, Loraof, TheOddsMaker, Kidsankyran, SageGreenRider, Wikplan, Shifra987, Fustbariclation, Anomalistic, TheDarkMessiah, Sladegrantham, LolShrek, Fmadd and Anonymous: 1018

- **Extremal principles in non-equilibrium thermodynamics** Source: https://en.wikipedia.org/wiki/Extremal_principles_in_non-equilibrium_thermodynamics?oldid=716466881 Contributors: Rich Farmbrough, Jheald, SmackBot, Kmarinas86, Myasuda, Headbomb, RebelRobot, Kbrose, Rl1rl1, Qgil-WMF, TutterMouse, Yobot, AnomieBOT, Omnipaedista, Nerdseeksblonde, Chjoaygame, Buddy23Lee, ThePowerofX, Helpful Pixie Bot, BG19bot and Anonymous: 9

- **Self-assembly** Source: https://en.wikipedia.org/wiki/Self-assembly?oldid=725151447 Contributors: William Avery, Lexor, Kku, Omegatron, Robbot, Fuelbottle, DavidCary, Dratman, Beland, Karol Langner, Icairns, Rich Farmbrough, NeuronExMachina, JoeSmack, Cristofv, Mdd, Passw0rd, Triddle, CharlesC, JohnJohn, Rjwilmsi, Wavelength, Aeusoes1, Daniel Mietchen, SmackBot, M stone, JonHarder, Smokefoot, Rspanton, Vina-iwbot~enwiki, Iridescent, PaddyM, Harold f, N2e, Neelix, Cydebot, Skittleys, Headbomb, Athkalani~enwiki, Nanotrix, Antony-22, Ryarod, Nopetro, Lightmouse, Pelesko, Acannon2, Ste nohype, SchreiberBike, XLinkBot, Koumz, Addbot, LaaknorBot, Aviados, Yobot, Materialscientist, Citation bot, Euanmc, Carlog3, Citation bot 1, Tom.Reding, Gray1, Jandalhandler, SchreyP, Chemengine, John of Reading, Dcirovic, AManWithNoPlan, Swarm123, Maxkingesq, Iqcp, Helpful Pixie Bot, BG19bot, UAwiki, Pfd1986, Wegallagher, Dexbot, Mogism, Marcela louis, Fasantos, Monkbot, Fulticethu, Faraday Telsa, Joao.justo001, Jrlaw001, GoldCar, Ta 91 and Anonymous: 43

- **Cellular automaton** Source: https://en.wikipedia.org/wiki/Cellular_automaton?oldid=726448862 Contributors: AxelBoldt, LC~enwiki, Bryan Derksen, XJaM, PierreAbbat, Ezubaric, Hephaestos, Jose Icaza, Hfastedge, Michael Hardy, Jdandr2, Kku, Eric119, Angela, Error, AugPi,

Samw, Schneelocke, Iseeaboar, Dysprosia, Jogloran, Wik, Zoicon5, Selket, Saltine, Bevo, Kizor, Kyber~enwiki, Chopchopwhitey, P0lyglut, TittoAssini, Tea2min, Giftlite, Smjg, Curps, MingMecca, Guanaco, Jasper Chua, LucasVB, Watcher, Joseph Myers, Elektron, Robin klein, Chmod007, Yonkeltron, Balsarxml, Imroy, Peak Freak, On you again, ZeroOne, Ben Standeven, Kb, Edward Z. Yang, Dalf, Erauch, Christian Kreibich, ACW, Photonique, Ferkel, Perceval, Keenan Pepper, Benjah-bmm27, Kotasik, Alexwg, Samohyl Jan, LukeSurl, Oleg Alexandrov, Marasmusine, Zorbid, Woohookitty, Mihai Damian, Kzollman, Lgallindo, -Ril-, GregorB, CharlesC, Graham87, Deltabeignet, Rjwilmsi, MarSch, Sbp, Pygy, FlaBot, Mathbot, RexNL, Orborde, Quuxplusone, Mahlon, Srleffler, Kri, Ahunt, Visor, DVdm, Hmonroe, Wavelength, Karlscherer3, RussBot, Allister MacLeod, Xihr, ENeville, Welsh, R.e.s., Lpdurocher, JocK, Chakazul, Pcorteen, Raven4x4x, Scs, William R. Buckley, Iztok.jeras~enwiki, Silverhill, Ninly, Bhumiya, MaNeMeBasat, Curpsbot-unicodify, Ilmari Karonen, Tropylium, Banus, Nekura, Mosiah~enwiki, SmackBot, RDBury, PEHowland, InverseHypercube, K-UNIT, Axd, Hmains, Chris the speller, Dra, Zom-B, Felicity Knife, Froese, Gragus, Crazilla, Phaedriel, Radagast83, Spectrogram, Nakon, Jon Awbrey, Sadi Carnot, Joeyramoney, Sam Tobar, SashatoBot, Metric, Acidburn24m, AnonEMouse, Mgiganteus1, Alpha Omicron, Ckatz, 524, The Temple Of Chuck Norris, Warrado, EmreDuran, Iridescent, JoeBot, Mudd1, Xerophytes, Argon233, Requestion, Cydebot, DumbBOT, Scolobb, Mattisse, Headbomb, Oubiwann, I do not exist, Ideogram, Dawnseeker2000, Navigatr85, Geneffects, AntiVandalBot, Dhushara, Gioto, Caileagleisg, Hannes Eder, Pixelface, Kaini, JAnDbot, Asmeurer, Avaya1, Hillgentleman, Baccyak4H, EagleFan, Torchiest, David Eppstein, Cypherzero0, JaGa, Hiner, NAHID, J.delanoy, Marcus Wilkinson, Chiswick Chap, DadaNeem, B.huseini, Kneb~enwiki, Mydogtrouble, Dcornforth, Torcini, Pleasantville, TXiKiBoT, Yugsdrawkcabeht, Mbaudier, Calwiki, Chuckwolber, RyanB88, Anonymous Dissident, Akramm1, Bearian, Lamro, SQL, Dmcq, AlleborgoBot, EmxBot, AdRock, SieBot, Setoodehs, BotMultichill, Gerakibot, Pi is 3.14159, Lightmouse, Sph110, JL-Bot, FerrenMacI, Beeblebrox, ClueBot, NickCT, Justin W Smith, HairyFotr, Sun Creator, Gleishma, Oliviersc2, Cstheoryguy, Versus22, DumZiBoT, Gthen, XLinkBot, Pichpich, Jytdog, Mandalaschmandala, Dekart, Bprentice, Shoemaker's Holiday, Addbot, Wli625, MrOllie, LinkFA-Bot, Hope09, ScAvenger, Jarble, Yobot, AnomieBOT, JackieBot, Flewis, JuliusCarver, Citation bot, Watertree, Loveless, Nippashish, Artem M. Pelenitsyn, Sharkyangliu, OgreBot, Citation bot 1, Kiefer.Wolfowitz, MondalorBot, Dave Feldman, Throwaway85, Jonkerz, BorysB, RjwilmsiBot, Bento00, Laesod, EmausBot, Johnhwynne, Svrist, Dcirovic, Sumanafsu, GSM83~enwiki, Agora2010, Grondilu, ZéroBot, PBS-AWB, Ὁ οἶστρος, Beddowve, Tijfo098, ChuispastonBot, Wolfpax50, Mishael1, ClueBot NG, Deer*lake, LunchboxGuy, Helpful Pixie Bot, Jlopez1967, Bibcode Bot, Alvyray, BG19bot, Pasicles, Tklauser, Dexbot, Ashleydan, Luanoz, Frizzil, 420mysteryman69, Mark viking, Captain Ford, PierreYvesLouis, Pdecalculus, Genaro.juarez.martinez, Andy Quarry, Sharkyangliu916, Nigellwh, Ginsuloft, JeremyLThompson, Prestigiouzman, Lev Kalmykov, Todd Rowland, Monktues, Monkbot, Joseph2302, KasparBot, Mhanga parto, Nicole tylor, Fahkr smith, Socaacos, Gulumeemee and Anonymous: 277

- **Emergence** *Source:* https://en.wikipedia.org/wiki/Emergence?oldid=723735203 *Contributors:* CYD, The Anome, WillWare, ChangChienFu, Heron, Bdesham, Michael Hardy, Owl, Lexor, Pnm, Kku, Karada, Ronz, Angela, Andres, Palfrey, Pipis, TonyClarke, Technopilgrim, Ec5618, RodC, Charles Matthews, Nickg, Greenrd, Jeffrey Smith, Jerzy, Banno, Tlogmer, Vespristiano, Chopchopwhitey, Steeev, Rursus, Blainster, Wikibot, Aetheling, Paul Murray, Aknxy, Jleedev, Stirling Newberry, Ancheta Wis, Giftlite, Gwalla, Tom harrison, SantiagoGala~enwiki, Henry Flower, Leonard G., Finn-Zoltan, Edcolins, John Abbe, Andycjp, Loremaster, Karol Langner, BookgirlST, Histrion, Talrias, Jmeppley, IcycleMort, Robin klein, Andreas Kaufmann, Chris Howard, MiddleOfNowhere, Rich Farmbrough, Cagliost, Dbachmann, Pavel Vozenilek, Goochelaar, Bender235, ESkog, Ben Standeven, El C, Vipul, Aaronbrick, Ray Dassen, Mike Schwartz, C S, Teorth, Viriditas, Tmh, JavOs, Mdd, HasharBot~enwiki, Kitoba, Mote, Silver hr, Diego Moya, Minority Report, Hu, Radical Mallard, ClockworkSoul, Zenter~enwiki, Cburnett, Stephan Leeds, Cal 1234, Eternal March, Drat, Acadac, Kazvorpal, Oleg Alexandrov, Woohookitty, PoccilScript, Kzollman, Jeff3000, Abu ari, Ludocrat, Ziji, Christianjb, DaveApter, Marudubshinki, Ashmoo, Rjwilmsi, Mayumashu, Gohn, Nightscream, Koavf, Dudegalea, Krash, JFromm, Sydbarrett74, Pe3~enwiki, Diza, Chobot, Fourdee, Bgwhite, Adonisick, YurikBot, Wavelength, Flameviper, RussBot, John2000, Rintrah, Ksyrie, Arkapravo, DarkFireTaker, BlackAndy, Thiseye, Slarson, Adamrush, Rbarreira, Shadowfax0, Larry laptop, Moe Epsilon, LodeRunner, Epipelagic, MBDowd, WAS 4.250, HereToHelp, Raveled, Curpsbot-unicodify, MagneticFlux, Bwiki, Luk, KnightRider~enwiki, SmackBot, Moxon, Saravask, ElectricRay, Tomdw, Peteresch, ZS, Cazort, Ohnoitsjamie, Betacommand, Chaojoker, Isaac Dupree, Grokmoo, Ben.c.roberts, Fuzzform, Nbarth, Hongooi, Toomuchnoise, Gbuffett, OrphanBot, Xyzzyplugh, Cybercobra, Pwjb, Richard001, NickPenguin, Jon Awbrey, A.W.Shred, Dr. Gabriel Gojon, Vina-iwbot~enwiki, Cast, Bcasterline, Prionesse, Harryboyles, Kreb Dragonrider, John, Rigadoun, Writtenonsand, Tktktk, Physis, Dchudz, Tasc, Wmattis, Olag, Nabeth, Tones, Papertiger, Asatruer, Joseph Solis in Australia, Antonio Prates, GDallimore, ChrisCork, Ripounet, CmdrObot, CBM, USMCM1A1, N2e, AshLin, Pfhenshaw, Emesghali, ONUnicorn, John courtneidge, Arnold.Sikkema, Logicombat, Myasuda, Gregbard, CX, Phatom87, Fyrius, Cydebot, Clappingsimon, Steel, Peterdjones, Anthonyhcole, Mirrormundo, Studerby, Skittleys, Shirulashem, L7HOMAS, Krylonblue83, Trev M, Letranova, Thijs!bot, Wikid77, ConceptExp, D4g0thur, Headbomb, Pjvpjv, Marek69, Mr pand, Dfrg.msc, Muaddeeb, Nick Number, Timf1234, Majorly, Dougher, Athkalani~enwiki, Davemarshall04, Albany NY, Andonic, Nessman, Psychohistorian, Aka042, LookingGlass, JaGa, Rickard Vogelberg, Profitip, Logan1939, Geoinmn, CommonsDelinker, Fixaller, Erkan Yilmaz, AstroHurricane001, Rlsheehan, BillWSmithJr, Alexjryan, Soiducked, Maurice Carbonaro, Lantonov, BobEnyart, Grosscha, Chiswick Chap, Aquaepulse, Tgooding, Halrhp, Jknd, Hammersoft, VolkovBot, Pleasantville, Dggreen, Toddy1, LuckyInWaco, Rollo44, VivekVish, Karmela, Rei-bot, Lordvolton, Sjeng, Littlealien182, Sintaku, Dendodge, JhsBot, Don4of4, BL2593, Myscience, Andrewaskew, Lova Falk, SieBot, Sweetp80, Djayjp, Scorpion451, Lord Phat, Sunrise, Emptymountains, Mr. Granger, Rowmn, Rojorulet, ClueBot, Kai-Hendrik, WurmWoode, Napzilla, Der Golem, Alexbot, Brews ohare, SchreiberBike, Bbbeard, Jmanigold, JKeck, XLinkBot, Saurus68, Ecolabs, Rreagan007, MystBot, Jonathanmoyer, Anticipation of a New Lover's Arrival, The, Svea Kollavainen, Addbot, Xp54321, Claudio Gnoli~enwiki, MrOllie, Dyaa, SimonB1710, Mjhunton, Zorrobot, Jarble, Ben Ben, Luckas-bot, Yobot, Isotelesis, IW.HG, Examtester, AnomieBOT, 1exec1, Trevithj, Galoubet, 90 Auto, MorgothX, Citation bot, ArthurBot, Carbaholic, Tomwsulcer, Srich32977, Omnipaedista, RibotBOT, Friesin76, SchnitzelMannGreek, Constructive editor, FrescoBot, LucienBOT, Dwightfowler, Machine Elf 1735, Journalmuncher, Diavel, DivineAlpha, Citation bot 1, Cbarlow, Pinethicket, Exjhawk, Aizquier, Filthylaugh, Jandalhandler, Sroel, Mjs1991, Pollinosisss, Jonkerz, LilyKitty, Inferior Olive, Reaper Eternal, Catcamus, Bento00, Djjr, EmausBot, Rusfuture, Irvbesen, GoingBatty, Dcirovic, Tuxedo junction, PBS-AWB, Alpha Quadrant, SporkBot, Libertaar, Providus, Ricardsolewiki, RockMagnetist, Just granpa, Spicemix, ClueBot NG, MohamedBishr, BarryKayton, Frietjes, SpaniardGR, Panleek, Tr00rle, Helpful Pixie Bot, Calgg, Bibcode Bot, BG19bot, Rosalegria, Dr. Whooves, Manjusri Wickramasinghe, Michaelweinstock, Joshua Jonathan, MHeder, Run to the hills, cos the end of the world is soon!, Warmtub, Symphonic Spenguin, Dexbot, ZutZut, Polyrahul, Limit theorem, Danny Sprinkle, Georgeandrews, I am One of Many, Alfy32, Igjohnston, Pamphilia, Dsomers74, Aubreybardo, Francois-Pier, SJ Defender, Deegeejay333, Peter Corning, Occurring, Saectar, Chaya5260, TheEpTic, Loraof, Social Theory, You better look out below!, Kew8888 and Anonymous: 296

- **Gauss's principle of least constraint** *Source:* https://en.wikipedia.org/wiki/Gauss's_principle_of_least_constraint?oldid=669054430 *Con-*

49.5. TEXT AND IMAGE SOURCES, CONTRIBUTORS, AND LICENSES

tributors: Michael Hardy, SebastianHelm, Bender235, Noetica, Anima Rytak, WillowW, Haseldon, Hertz1888, PixelBot, Addbot, EmausBot, Mattte and Anonymous: 1

- **Self-organized criticality** *Source:* https://en.wikipedia.org/wiki/Self-organized_criticality?oldid=723876716 *Contributors:* Fnielsen, Michael Hardy, Kku, Charles Matthews, Herbee, Karol Langner, Ivn~enwiki, Pgreenfinch, Vsmith, JimR, Iamunknown, I9Q79oL78KiL0QTFHgyc, Mdd, Embryomystic, Linas, Ruud Koot, Kelisi, Rjwilmsi, Nihiltres, Artgirl88, Salsb, Shadowfax0, SmackBot, WebDrake, Badjeros, Chris the speller, Good Intentions, Lapaz, Robofish, Rhetth, Sonswoo, Headbomb, JustAGal, Magioladitis, BigrTex, Hugh Hudson, Westfall3, Dggreen, Cwkmail, AnonyScientist, Addbot, DOI bot, Lightbot, Jarble, Yobot, Citation bot, Omnipaedista, FrescoBot, Satrapa, Citation bot 1, Dimo400, Nastasyuk v, Rayman60, Benlansdell, Dcirovic, Crbazevedo, Vacation9, Helpful Pixie Bot, Bibcode Bot, BG19bot, Jcphillips8, Compsim, AppliedMathematics, Marknew12, Otherocketman, Self Order, Monkbot, Meforwk, Isambard Kingdom and Anonymous: 50

- **Spontaneous order** *Source:* https://en.wikipedia.org/wiki/Spontaneous_order?oldid=718491179 *Contributors:* Michael Hardy, Kku, Big iron, Nickg, Owen, Goethean, Nagelfar, Nikodemos, Ubernetizen, Everyking, Ravn, Christofurio, Turion, Karol Langner, Rich Farmbrough, Cfailde, Bender235, Cretog8, Viriditas, Generalebriety, Mdd, Gary, Sstoneb, RJII, Siafu, NotSuper, Rjwilmsi, Phileas, Crazynas, Jrtayloriv, Russ-Bot, PEZ, Briaboru, Pigman, Stijn Calle, RL0919, Anclation~enwiki, RG2, NetRolller 3D, SmackBot, Zazaban, Brianski, Bluebot, Persian Poet Gal, D-Rock, Xchbla423, NickDupree, PrometheusX303, LoveMonkey, Afaus, Byelf2007, Rigadoun, SparksWillFly, Vision Thing, RelicLord, N2e, Tzalumen, Cydebot, JamesAM, Thijs!bot, Biruitorul, PHaze, Carolmooredc, Athkalani~enwiki, Skomorokh, Alastair Haines, Freshacconci, Daniel Cordoba-Bahle, Tonyfaull, KConWiki, SlamDiego, Anarcho-capitalism, Teardrop onthefire, Working Poor, Crakkpot, Ontarioboy, Childhoodsend, TXiKiBoT, Rollo44, Malinaccier, Esotericengineer, Jesin, Austriacus, Macdonald-ross, Pointsmyth, Dstlascaux, Bombastus, Operation Spooner, Der Golem, Alexbot, Byates5637, 1OwenJones, Addbot, Elsendero, USchick, 5 albert square, Tassedethe, Luckas-bot, Yobot, AnomieBOT, PublicSquare, Aeortiz, 90 Auto, Citation bot, Kaoruchan21, Xqbot, Srich32977, Eisfbnore, Skyerise, Dmitry St, Jonkerz, Libertatis, Thermoworld, Roastedpepper, Jbradley904, Xerographica, Financestudent, Helpsome, Benjamin9832, Thomask0, Rurik the Varangian, Helpful Pixie Bot, Revisor2011, BG19bot, Geistcj, PhnomPencil, Wodrow, Iansha, Platospigmonster, Fajrbot, Austrartsua, Monkbot, Loraof and Anonymous: 82

- **Metastability** *Source:* https://en.wikipedia.org/wiki/Metastability?oldid=724094269 *Contributors:* Damian Yerrick, AxelBoldt, William Avery, Heron, Michael Hardy, Ixfd64, GCarty, RodC, Jitse Niesen, Colin Marquardt, Robbot, DavidCary, Beland, Vsmith, Brim, AngryParsley, Linas, ScottDavis, Calréfa Wéná, Kushboy, BorgHunter, Wavelength, Jimp, Hede2000, CarlHewitt, Xorx, KnightRider~enwiki, SmackBot, Eskimbot, Pwjb, DMacks, John, JorisvS, Wizard191, George100, Nick Number, SeanTater, Adrian J. Hunter, Damuna, Rustyfence, LordAnubisBOT, VolkovBot, OlavN, !dea4u, SieBot, Kotabatubara, Istaro, Commutator, Jmcclare, DragonBot, Djr32, Coinmanj, 1ForTheMoney, DumZiBoT, Tullywinters, Addbot, Download, SpBot, Luckas-bot, Yobot, Ptbotgourou, AnomieBOT, LouriePieterse, ArthurBot, Xqbot, Mladjowie, Mcoupal, FrescoBot, Waldemahr, RjwilmsiBot, ElPeste, EmausBot, GoingBatty, Slightsmile, ZéroBot, GianniG46, RockMagnetist, Masssly, George8211, Wikifan2744, Dough34, JonahSpars and Anonymous: 55

- **Butterfly effect** *Source:* https://en.wikipedia.org/wiki/Butterfly_effect?oldid=727079365 *Contributors:* The Cunctator, Lee Daniel Crocker, Bryan Derksen, DonDaMon, Chuq, Mrwojo, D, Michael Hardy, Lexor, Dominus, Wwwwolf, Ixfd64, William M. Connolley, Darrell Greenwood, BenKovitz, Andres, Evercat, Raven in Orbit, Ventura, Przepla, Peregrine981, Big Bob the Finder, Furrykef, Jusjih, PuzzletChung, Robbot, Lowellian, Gandalf61, Sverdrup, Litefantastic, Blainster, Hadal, Wereon, Jholman, Dina, Giftlite, JamesMLane, Philwiki, Paul Richter, Wolfkeeper, Lee J Haywood, Snowdog, SanderSpek~enwiki, DemonThing, ChicXulub, Bact, Phe, Karol Langner, DragonflySixtyseven, Esperant, Kate, Gazpacho, Discospinster, Vsmith, Lulu of the Lotus-Eaters, User2004, Eric Shalov, Paul August, Bender235, Duemellon, El C, MPS, Viriditas, Elipongo, Thomas G Graf, Mdd, Storm Rider, Alansohn, Me at work, DanielLC, Arthena, Plumbago, Kotasik, BryanD, Mysdaao, Malo, Wtmitchell, Velella, Helixblue, Tony Sidaway, Peedl, Jabernal, Henry W. Schmitt, Itsmine, MIT Trekkie, Dismas, AustinZ, OleMaster, Richard Arthur Norton (1958-), Linas, Mu301, Mazca, MattGiuca, Male1979, Hughcharlesparker, XaosBits, Stefanomione, Graham87, Galwhaa, Deadcorpse, Rjwilmsi, Lars T., Koavf, Vary, XP1, Moorlock, The wub, Latka, MosheZadka, BjKa, Fieryfaith, Whodunit, King of Hearts, Matsi, YurikBot, Wavelength, Hawaiian717, RobotE, ~Viper~, Bhny, Bergsten, Zlobny, PepsiMax181, Yamara, Gaius Cornelius, Bovineone, NawlinWiki, Dysmorodrepanis~enwiki, Trovatore, Zwobot, Epipelagic, Asarelah, Pegship, Cdiggins, 2over0, Arthur Rubin, Canley, Mobius288, Emc2, Ilmari Karonen, Allens, Ephilei, Katieh5584, NeilN, GrinBot~enwiki, SmackBot, Prodego, InverseHypercube, K-UNIT, Alksub, Srnelson, Ema Zee, Gilliam, Pfulgione, Valley2city, Supersox, Thumperward, Dawd, Kotra, Sneltrekker, Max David, Chlewbot, Sommers, Jjjsixsix, Cybercobra, Dvc214, Mini-Geek, Dacxjo, Kukini, Masterpjz9, Z-d, ArglebargleIV, Lapaz, Markdr, EsterhDG, Majorclanger, A. Parrot, Tarcieri, Comicist, Akerensky99, Hypnosifl, InedibleHulk, TheOtherStephan, Mego'brien, Hu12, Tó campos~enwiki, Lovelight, Svlad Jelly, CRGreathouse, SerenadeOp24, BeenAroundAWhile, BGH122, Joelvanatta, KBuck, Lentower, Denstat, Neelix, Gregbard, CX, Spenser~enwiki, JFreeman, Kiske, Omicronpersei8, Bulmabriefs144, Letranova, RobbieG, Epbr123, Vinoduec, JDK77590, Commonlaw504, Headbomb, Marek69, Electron9, Bward1, Leon7, Dfrg.msc, 2QT2BSTR8, Mentifisto, AntiVandalBot, TTN, Hannes Eder, SkoreKeep, Pixelface, Storkk, Uusitunnus, Narssarssuaq, Quentar~enwiki, Mkch, Darkxavier, MSBOT, Y2kcrazyjoker4, Stardotboy, Seanette, S0uj1r0, Bongwarrior, VoABot II, Nyttend, Johnnybee13, Theroadislong, Hiplibrarianship, Captin Shmit, Cmf576, Boffob, David Eppstein, Spellmaster, DerHexer, Stephenchou0722, Osquar, MartinBot, Barnjo, RP88, Kostisl, CommonsDelinker, AlexiusHoratius, LittleOldMe old, Erkan Yilmaz, J.delanoy, AstroHurricane001, Maurice Carbonaro, DeadGuy, Dimentico, Daniel Berwick, Chiswick Chap, Touch Of Light, Kernel NickM, DorganBot, Idioma-bot, Funandtrvl, X!, JohnBlackburne, VasilievVV, Vlmastra, QuackGuru, Greggreggreg, Anna Lincoln, Fizzackerly, Fgsherrill, Wiae, Rprpsych, Rjgodoy, Leviathan~enwiki, Gillyweed, Insanity Incarnate, The Blizzard King, D. Recorder, BotMultichill, Thelegendarystm, Yintan, JabbaTheBot, Zbvhs, Flyer22 Reborn, Oxymoron83, SoulEspresso, BrightRoundCircle, Ks0stm, Michael Courtney, OKBot, Ralph-Michael von Khuon-Wildegg, Tcrow777, EveryDayJoe45, Pinkadelica, Teh.cmn, ClueBot, Snigbrook, The Thing That Should Not Be, Pan narrans, Ndenison, Meekywiki, DanielDeibler, CounterVandalismBot, Blanchardb, Bob bobato, Toad of Steel, JFlav, Thejoshwolfe, Grandpallama, Lame Name, DragonBot, No such user, Excirial, Luligal, Artisticked, Gtstricky, Sophiemeloy, 7&6=thirteen, JasonAQuest, Aitias, Sucnidea, Chasecarter, DumZiBoT, Vanished user k3rmwkdmn4tjna3d, Nepenthes, Little Mountain 5, WillOakland, ZooFari, Omnisage, Stupubstupub, Djdisconess, DOI bot, Blake de doosten, Betterusername, Egg1234, Capouch, Download, Morning277, Davemontes, Deamon138, Helskaill, OffsBlink, Krano, Zorrobot, Asad828, Legobot, Ladnscience, Luckas-bot, Imemadhu, Yobot, Bkrishnan, 2D, TaBOT-zerem, Karsachin, Amirobot, Dzied Bulbash, Eric-Wester, AnomieBOT, Jim1138, IRP, Abstruce, Shock Brigade Harvester Boris, Mintrick, Piano non troppo, Materialscientist, Citation bot, Oxygenator, BarrelRollZRTwice, GB fan, ArthurBot, Xqbot, Intelati, Melmann, Wasseem2100, A dullard, Marius.graur, A Quest For Knowledge, 4RugbyRd, Drdpw, Shadowjams, Spellage, Cekli829, Hushpuckena, Undsoweiter, GliderMaven, D'ohBot, Nightgleam, Haeinous, Citation bot 1, Pinethicket, I dream of horses, Notedgrant, 10metreh, Rameshngbot, MJ94, Banej, Reconsider the static, Trappist the monk, Buddy23Lee, DriveMySol, Diannaa, DARTH SIDIOUS 2, Mean as custard, EmausBot,

- **Self-organized criticality control** *Source:* https://en.wikipedia.org/wiki/Self-organized_criticality_control?oldid=690906084 *Contributors:* Michael Hardy, Rjwilmsi, RadioFan, Revent, GorillaWarfare, Yobot, Bibcode Bot, BG19bot, ArticlesForCreationBot, Mark viking, Daviscientist, Acetoyce, Jodosma, Otherocketman, Ginsuloft, Lovegoodscience and Anonymous: 2

- **Spontaneous magnetization** *Source:* https://en.wikipedia.org/wiki/Spontaneous_magnetization?oldid=618965629 *Contributors:* Keoki, CiaPan, Aeusoes1, Roberto DR, Chaiken, SmackBot, Bluebot, Yevgeny Kats, Venny85, Addbot, Yobot, RjwilmsiBot, RockMagnetist, ChrisGualtieri, Mogism and Anonymous: 4

- **Crystallization** *Source:* https://en.wikipedia.org/wiki/Crystallization?oldid=726233918 *Contributors:* Michael Hardy, Darkwind, Charles Matthews, Hyacinth, Omegatron, Samsara, Giftlite, Eequor, Andycjp, Icairns, Sam Hocevar, M1ss1ontomars2k4, Discospinster, Vsmith, Violetriga, Fenice, Shanes, Femto, Smalljim, Nk, 99of9, Pearle, Siim, Alansohn, Bhupesh mishra, Pion, Melaen, RHaworth, Jeff3000, BD2412, Rjwilmsi, Quiddity, FlaBot, Physchim62, Mallocks, YurikBot, Wavelength, RussBot, Tony1, BenBildstein, Pb30, SmackBot, UbUb, Weiguxp, Zephyris, TantalumTelluride, Gilliam, ERcheck, Rolinator, DMacks, A5b, Mbeychok, Physis, Wizard191, Nádvorník, Tawkerbot2, CmdrObot, Tanthalas39, Chrumps, Van helsing, Neelix, CJBot, Cydebot, Krauss, Rifleman 82, Anonymi, Calvero JP, Thijs!bot, Ezkerraldean, Sagaciousuk, Headbomb, Marek69, AntiVandalBot, Tangerines, Dougher, JAnDbot, Igoruha, Husond, Bahar, Elmschrat, Lost-theory, Deathstar 80, Karlhahn, VoABot II, Jeff Dahl, Quantockgoblin, Avicennasis, Adrian J. Hunter, 28421u2232nfenfcenc, DerHexer, JaGa, Pri Lalli, Schmloof, Echis, J.delanoy, Numbo3, Wandering Ghost, Gavinayling, SJP, Toon05, Steve Lund, KylieTastic, Bob, STBotD, Pdcook, ABiochemist, Deor, VolkovBot, Jeff G., DoorsAjar, TXiKiBoT, Oshwah, BuickCenturyDriver, Judge Nutmeg, Themcman1, TBChem, Falcon8765, AlleborgoBot, Logan, Rocky717717, Winchelsea, Kopeliovich, KoshVorlon, Steven Crossin, Techman224, Kudret abi, ClueBot, Mild Bill Hiccup, Niceguyedc, Alexbot, Rubin joseph 10, Estirabot, Irishmik22, Razorflame, Thehelpfulone, Banano03, Nitech2008, SoxBot III, Darkicebot, Anjelgurl202, Kittilhagen, Addbot, Orci, Fieldday-sunday, Flufferputter, MrOllie, Numbo3-bot, Tide rolls, Lightbot, Gail, Zorrobot, Peko, PlankBot, Luckasbot, AnomieBOT, Daniele Pugliesi, JackieBot, Franksucks, Bluerasberry, Materialscientist, MauritsBot, Xqbot, Mrba70, J04n, Valerychani, قلی زادگان, Logger9, Editor182, Eframgoldberg, Josemanimala, Fotaun, Prari, FrescoBot, Basher018, Anny1121, Pinethicket, LittleWink, Btilm, MondalorBot, Felix0411, Lightlowemon, Lotje, Diannaa, Marie Poise, DARTH SIDIOUS 2, Difu Wu, EmausBot, John of Reading, RenamedUser01302013, Solarra, K6ka, ZéroBot, White Trillium, Michel Awkal, Rahulthe1, Ocaasi, Carmichael, ChuispastonBot, ClueBot NG, This lousy T-shirt, Satellizer, Emcdonld, Swallow2011, PhnomPencil, MusikAnimal, Gorthian, Hiberni, Twyegimp, Bcary, BattyBot, Jeremy112233, Webclient101, G.Kiruthikan, Samaam, Zlelik2000, Delunaluno, IvanaMadzarevic, Briefexact32, HMSLavender, Nudroid, Hamster27, Elisavir, JohnCrow47, Tropicalkitty and Anonymous: 247

- **Phase transition** *Source:* https://en.wikipedia.org/wiki/Phase_transition?oldid=724121103 *Contributors:* CYD, Mav, Bryan Derksen, Olof, Roadrunner, Peterlin~enwiki, Patrick, Michael Hardy, Tim Starling, Kku, TakuyaMurata, Ahoerstemeier, Glenn, Charles Matthews, Terse, Phys, Baffclan, Chuunen Baka, Lzur, Dave6, Giftlite, Djinn112, BenFrantzDale, Zigger, Curps, Ezhiki, Steuard, Antandrus, Aulis Eskola, Karol Langner, Zfr, Pgreenfinch, Revised~enwiki, Mike Simpson, Deglr6328, CALR, Varada, Vsmith, JimR, Ponder, Dmr2, Bender235, Evice, Nabla, Brian0918, El C, Rgdboer, I9Q79oL78KiL0QTFHgyc, Slinky Puppet, Kjkolb, Ynhockey, RJFJR, Oleg Alexandrov, Brookie, CoolMike, Linas, V8rik, Zeroparallax, Nanite, Rjwilmsi, Nneonneo, FlaBot, John Baez, Mathbot, GünniX, Gurch, Srleffler, YurikBot, Ugha, Wavelength, Hairy Dude, Supasheep, Chuck Carroll, Salsb, Janke, Tony1, Bota47, Kkmurray, Vicarious, Cmglee, Sbyrnes321, Pankkake, SmackBot, WebDrake, Unyoyega, Gilliam, Kdliss, Joseph Stalin~enwiki, Richfife, Kmarinas86, MalafayaBot, Complexica, Colonies Chris, Wiki me, MureninC, SkanderH, Akriasas, DMacks, Dave314159, Brennan Milligan, JorisvS, Bjankuloski06en~enwiki, FrostyBytes, Soulkeeper, JHunterJ, Inquisitus, JarahE, Wizard191, Japhet, JRSpriggs, Ouishoebean, CmdrObot, Tarchon, Chrumps, Van helsing, Myasuda, Hga, Icek~enwiki, Cydebot, Perfect Proposal, Kweeket, Rifleman 82, JFreeman, JohnInDC, Thijs!bot, Epbr123, TDF, Headbomb, E. Ripley, Nick Number, JAnDbot, Penubag, Primephear, Magioladitis, JamesBWatson, Kaiserkarl13, Dhk, User A1, Jtw11, Hans Dunkelberg, Icseaturtles, TomyDuby, DadaNeem, Forevaclevah, Cuzkatzimhut, Hershen, Nikthestunned, VolkovBot, Oshwah, A4bot, Judge Nutmeg, ChooseAnother, Brianga, Neparis, Kbrose, Barkeep, SieBot, I Like Cheeseburgers, Trumpsternator, Sean.hoyland, Dolphin51, WikiBotas, ClueBot, LAX, Rodhullandemu, Brettstrawbridge, Lame Name, Ktr101, CohesionBot, Tbagasaurausrex21, Thehelpfulone, Sdrtirs, RexxS, Matthieumarechal, Avoided, Hess88, Addbot, AkhtaBot, Download, EconoPhysicist, Alelima, TStein, Tide rolls, Yoavd, Snaily, Luckas-bot, Yobot, Pteradactyle, AnomieBOT, Hunnjazal, ImperatorExercitus, Citation bot, Betsim, Madbard, DSisyphBot, Paula Pilcher, Mvpranav, Omnipaedista, SassoBot, Logger9, FrescoBot, Hornord, Kmdouglass, DivineAlpha, Wdcf, Jonesey95, Dr-b-m, Σ, Gryllida, Trappist the monk, Puzl bustr, Wotnow, Lotje, Marie Poise, Oakycoppice, Ripchip Bot, Rafmag, Super48paul, Slightsmile, Olaw3, K6ka, ZéroBot, Joshlepaknpsa, Mattedia, Octopusbr, Donner60, RockMagnetist, ClueBot NG, Gareth Griffith-Jones, Anagogist, Nscozzaro, Pvnuffel, Rezabot, Pluma, MerllwBot, Helpful Pixie Bot, KLBot2, Bibcode Bot, BG19bot, Bmusician, PhnomPencil, Indah blestari, Jedharris, GenBiorics, Webclient101, JGTZ, Sphelps 9312, Prokaryotes, Stemcells90, Eigenbra, Wireless erlang, Kkosman, Monkbot, Yikkayaya, Briefexact32, Rakeshyashroy, Ppdouble, Cyclingralph, See996able, PDcanonical, KasparBot, Лагічна рэвалюцыйны, Baba Arouj, Delamotteb and Anonymous: 222

- **Critical opalescence** *Source:* https://en.wikipedia.org/wiki/Critical_opalescence?oldid=672402043 *Contributors:* Ravn, Natrij, Jag123, Physicistjedi, Hooperbloob, Papeschr, Linas, David Haslam, Nanite, Salsb, SmackBot, Wen D House, Mion, Cydebot, Magioladitis, Athaenara, Lamro, Phe-bot, Lanomenklatura, Addbot, Rudolf.hellmuth, Felix0411, Zephyrus Tavvier and Anonymous: 13

- **Percolation** *Source:* https://en.wikipedia.org/wiki/Percolation?oldid=719655273 *Contributors:* DavidLevinson, RTC, Michael Hardy, Lexor, Delirium, Dgrant, Docu, Snoyes, Doradus, Steinsky, Robbot, Giftlite, Lgarcia, Pgreenfinch, Antaeus Feldspar, Brian0918, Mdd, Guy Harris, Melaen, Rafti Institute, Uffish, Oleg Alexandrov, Linas, Ruud Koot, BD2412, Rjwilmsi, Windchaser, Vonkje, YurikBot, TexasAndroid, RadioFan, Shell Kinney, NawlinWiki, Welsh, BazookaJoe, CedricVonck, 2over0, SmackBot, Stepa, Mhym, Mwtoews, BranStark, Iridescent, JMK, PetaRZ, Tawkerbot2, Nunquam Dormio, Neelix, Bjordan555, Edchi, Escarbot, F.Shelley, VoABot II, The Big Man, Seba5618, J.delanoy, Idunno271828, Mikael Häggström, Bonadea, Signalhead, Lights, SCriBu, Biscuittin, THEMONKEYZIP, Anton Petrov, Flyer22

Reborn, Oda Mari, Bakashi10, Svick, Water and Land, JohnnyMrNinja, Rziff, Tanvir Ahmeed, ClueBot, Wikeepedian, SilvonenBot, Addbot, Jacopo Werther, Jarble, Yobot, E mraedarab, SwisterTwister, AnomieBOT, Kingpin13, Gggmaster, Dr. Perfessor, Omnipaedista, Mmru, Pinethicket, Felix0411, Heracles31, Erianna, Sitic, Berberisb, ClueBot NG, Frietjes, Bibcode Bot, BG19bot, BendelacBOT, BattyBot, Austen-Head, ChrisGualtieri, CarrieVS, Compsim, Nunoaraujo, Derosigur, KasparBot, Shifra987, Boomer Vial, Weiner18 and Anonymous: 97

- **Structure formation** *Source:* https://en.wikipedia.org/wiki/Structure_formation?oldid=715060074 *Contributors:* Zundark, Edward, Rtfisher, Everyking, Grm wnr, FT2, Vsmith, Nabla, El C, Cmdrjameson, I9Q79oL78KiL0QTFHgyc, RHaworth, Mu301, Jeff3000, Joke137, Drbogdan, Rjwilmsi, R Lee E, Wavelength, Gaius Cornelius, Redgolpe, That Guy, From That Show!, SmackBot, Ashill, MalafayaBot, Droll, Colonies Chris, J 1982, Myasuda, BobQQ, Corpx, Thijs!bot, Headbomb, Lfstevens, Magioladitis, R'n'B, Warut, Sheliak, AlnoktaBOT, SwordSmurf, Lamro, SieBot, Shahidur Rahman, Jonmtkisco, Panos84, Dana boomer, Tailedkupo, DumZiBoT, RexxS, DCCougar, Hess88, Addbot, DOI bot, AkhtaBot, Samiswicked, Glane23, Yobot, Amirobot, AnomieBOT, Christopher.Gordon3, Citation bot, Icosmology, FrescoBot, Paine Ellsworth, Kikuyu3, Citation bot 1, Berkeleyjess, Tom.Reding, RockSolidCosmo, Puzl bustr, JLincoln, RjwilmsiBot, TjBot, John of Reading, Italia2006, RaptureBot, Brownie Charles, EdoBot, Machina Lucis, Lincoln Josh, Bibcode Bot, Khazar2, Tzymne, Penitence, PirtleShell, Jwratner1, Kogge, Xibalban Alchemist, Cosmic connection, Monkbot, Tetra quark, KasparBot, TychosElk, Youknowwhatimsayin and Anonymous: 16

- **De Sitter universe** *Source:* https://en.wikipedia.org/wiki/De_Sitter_universe?oldid=726486700 *Contributors:* Stevertigo, Boud, Palfrey, Timwi, Robbot, Barbara Shack, ShaunMacPherson, LeYaYa, Fropuff, Eequor, SWAdair, Icairns, Lumidek, Cacycle, Pjacobi, Pt, Jag123, I9Q79oL78KiL0QTFHgyc, Falcorian, Pol098, Mpatel, Christopher Thomas, Mattmartin, Carrionluggage, Hillman, Arado, Salsb, Larsobrien, SmackBot, Ashill, Rentier, Eskimbot, Bluebot, Silly rabbit, Scwlong, QFT, Jmnbatista, Lpgeffen, Mindnumbed, Harold f, Cydebot, Michael C Price, Thijs!bot, Marek69, Pervect, R'n'B, Ontarioboy, Dmcq, Vanished User 8902317830, Bobathon71, Excirial, Addbot, SpBot, AnomieBOT, Jim1138, Dogbert66, RedBot, Lotje, EmausBot, Quondum, ClueBot NG, Raidr, Frietjes, Helpful Pixie Bot, Monkbot, Thundergodz, Tetra quark and Anonymous: 20

- **Diffusion-limited aggregation** *Source:* https://en.wikipedia.org/wiki/Diffusion-limited_aggregation?oldid=683685672 *Contributors:* Michael Hardy, Hyacinth, Omegatron, HangingCurve, Jason Quinn, Mach535, Bert Hickman, Macowell, Oleg Alexandrov, Jeremy Young, Hynespm, Gaius Cornelius, SmackBot, Sam Tobar, Kevin R Johnson, CBM, Thijs!bot, BananaSlug, Mgmirkin, STBot, WingkeeLEE, Melcombe, Maniac18, Meigel, Marklar2007, Addbot, Jockelinde, Yobot, Dreamer08, Xqbot, FrescoBot, BenzolBot, EmausBot, Toxmeister, Pruess, Accelerometer, Akarpe and Anonymous: 18

- **Reaction–diffusion system** *Source:* https://en.wikipedia.org/wiki/Reaction%E2%80%93diffusion_system?oldid=714436181 *Contributors:* Michael Hardy, Jitse Niesen, Markhurd, Aetheling, Giftlite, Rich Farmbrough, Oleg Alexandrov, Spencerk, Borgx, Welsh, Allens, SmackBot, Complexica, Neelix, Kupirijo, Oink54321, R'n'B, Chiswick Chap, Danwills, Jmath666, Arcfrk, Tiddly Tom, Huboedeker, Vagogan, XLinkBot, Addbot, Mjl2008, MrOllie, Jockelinde, سعى, Yobot, Ptbotgourou, KamikazeBot, Geek1337~enwiki, Rhodydog, FrescoBot, Petrelharp, TharsHammar, D'ohBot, Wrettyfugi, EdoBot, Xanchester, Vaulttech, Kryomaxim, Bhavya1333, Jasherratt, Awliehr, Andy Quarry and Anonymous: 41

- **Molecular self-assembly** *Source:* https://en.wikipedia.org/wiki/Molecular_self-assembly?oldid=723554328 *Contributors:* Mdd, Rjwilmsi, Maurog, SmackBot, TestPilot, M stone, Satish.murthy~enwiki, N2e, EdJohnston, Rudick.JG, Dirac66, Clipjoint, Nanotrix, Antony-22, Netmonger, Oshwah, Northfox, XenonX3, El bot de la dieta, Addbot, Abduallah mohammed, Yobot, LithoGuy, Materialscientist, Citation bot, Citation bot 1, Mikespedia, Trappist the monk, SchreyP, Dcirovic, RaptureBot, Iqcp, MerllwBot, Bibcode Bot, Dexbot, Fasantos, Monkbot, Vikingo999 and Anonymous: 24

- **Autocatalysis** *Source:* https://en.wikipedia.org/wiki/Autocatalysis?oldid=724193320 *Contributors:* Lexor, Shyamal, Raven in Orbit, Dino, Randomm832, Azra99, RedWolf, UtherSRG, Jason Quinn, D3, Xenoglossophobe, TedPavlic, Ebradsha, Ceyockey, Rjwilmsi, Misternuvistor, Butonic~enwiki, Scope creep, Cedar101, Eno-ja, Limhes, Yakudza, SmackBot, Thumperward, Complexica, Mion, Olin, Tarcieri, Knights who say ni, Imsobored, Ebaskerv, John Riemann Soong, Rifleman 82, Christian75, Thijs!bot, Dirac66, User A1, TechnoFaye, CommonsDelinker, Daniele.tampieri, STBotD, RedAndr, Ordermaven, SieBot, Liamstone, Sunrise, Alexbot, Addbot, EconoPhysicist, Lightbot, LilHelpa, GrouchoBot, DerryTaylor, Nathanielvirgo, Citation bot 1, Dinamik-bot, Dcirovic, Bollyjeff, Surya Prakash.S.A., RockMagnetist, Helpful Pixie Bot, Khazar2, Jamesx12345, Kernsters, Yahadzija, Alékos Elefthérios, 好好好, Skylord a52 and Anonymous: 24

- **Liquid crystal** *Source:* https://en.wikipedia.org/wiki/Liquid_crystal?oldid=725136577 *Contributors:* William Avery, Caltrop, Youandme, Michael Hardy, Dominus, SebastianHelm, Александр, Yoshitaka Mieda~enwiki, Glenn, Smack, Tantalate, Stone, Radiojon, Topbanana, Phil Boswell, Rogper~enwiki, ZimZalaBim, Dusik, Giftlite, BenFrantzDale, Bensaccount, Eequor, Bobblewik, Karol Langner, JohnArmagh, Guanabot, MuDavid, Sergei Frolov, Bender235, Mykhal, Jaberwocky6669, Plugwash, Rgdboer, Femto, Tomgally, Chiral, Rumikohorie, Panjasan, Jared81, Andrewpmk, Lithian, Rwendland, RJFJR, Gene Nygaard, Nick Mks, Rvlaw, Mindmatrix, Wacko, David Haslam, Cbhiii, Jeff3000, Jwanders, GregorB, Eras-mus, CharlesC, V8rik, Chenxlee, Tlroche, Saperaud~enwiki, Rjwilmsi, Koavf, Eptalon, Numa, RexNL, Kebes, Wars, BjKa, Thecurran, Srleffler, Chobot, Bgwhite, DerrickOswald, YurikBot, Armistej, Epolk, Hellbus, Gaius Cornelius, Shaddack, Prime Entelechy, Dysmorodrepanis~enwiki, Dhollm, Misza13, SFC9394, 2over0, Equack, LeonardoRob0t, Claudiozzbo, Ilmari Karonen, SmackBot, AnOddName, M stone, Gilliam, Hugo-cs, Bluebot, Bduke, Martin Jambon, DHN-bot~enwiki, Descalzo, JustUser, Chlewbot, Waskyo, VMS Mosaic, Iamthealchemist, Akriasas, Smokefoot, DMacks, Ohconfucius, SashatoBot, DavidCooke, JorisvS, Mxreb0, Arkrishna, Kvng, Xionbox, JoeBot, Tawkerbot2, WolfgangFaber, CmdrObot, Dgw, Rifleman 82, Raul1022, Epbr123, Barticus88, HappyInGeneral, Headbomb, WVhybrid, Sijarvis, AntiVandalBot, Guy Macon, Emmelie, Sbarnard, Markthemac, JAnDbot, Tsuji, Sangak, Magioladitis, Charlesreid1, Fallschirmjäger, David Eppstein, LorenzoB, DerHexer, MartinBot, Byzantime, AgarwalSumeet, MrBell, Dagar1989, Pmench, 272727, Acalamari, Jcwf, IanusVentus, Liveste, Milogardner, Vanished user 39948282, Hbayat, Odysseyontario, Deor, TXiKiBoT, Oshwah, Showjumpersam, GroveGuy, Rei-bot, Minutemen, Laserspec, I'm with gerrit, Pitel, Rikuharts, Billygettothechopper, RJaguar3, The Unknown Hitchhiker, Harry-, Easamy, Kumioko (renamed), BlnLiCr, ClueBot, Capture88, GregVolk, Toxonomy, Gulmammad, California847, Versus22, SoxBot III, Chhe, Hotcrocodile, Wikijpp, Ngebbett, Addbot, DOI bot, TutterMouse, Chamal N, Remini2, Tassedethe, SPat, Rascallgh, Luckas-bot, Yobot, AnomieBOT, Nintend06, Prof.kulkarni, Law, Jeff Muscato, Materialscientist, 90 Auto, Citation bot, JJagomagi, MauritsBot, Brycker, Gaujmalnieks, GrouchoBot, Omnipaedista, Jezhotwells, A. di M., Dave3457, Antunbalaz, Tobby72, Sky Attacker, Pratik.mallya, Louperibot, Citation bot 1, Pinethicket, Edderso, Tom.Reding, Darkwolf800, Igfmnbo, Bhxho, Σ, Pjvbra, Felix0411, Jsbeeckm, TobeBot, ItsZippy, Rnakatsuji, Callanecc, Marie Poise, NerdyScienceDude, Dcirovic, Cogiati, Wikfr, Timeloop, RockMagnetist, MidasDoas, ClueBot NG, Starshipenterprise, Noulin, ScottSteiner, Widr, Helpful Pixie Bot, Bibcode Bot, Mayast, BattyBot, Arr4, Drjcast, Proffuwec, Z11o22, SRaemiA, Garuda0001,

Mark viking, OhVerr Aiteen, Matchbox30, Ugog Nizdast, Anrnusna, ThatDandyMan, Polikpolik890, Monkbot, Liquid-Crystal-Lens, Maodit, KasparBot, Namelessbj, Allenengel, GkTheMass, Soumyadeep333 and Anonymous: 210

- **Grid complex** *Source:* https://en.wikipedia.org/wiki/Grid_complex?oldid=696746007 *Contributors:* Sahehco
- **Colloidal crystal** *Source:* https://en.wikipedia.org/wiki/Colloidal_crystal?oldid=718812489 *Contributors:* Stone, Altenmann, Vsmith, Cedders, Kjkolb, 99of9, Woohookitty, BD2412, Rjwilmsi, Nihiltres, Bullzeye, 2over0, Banus, ChemGardener, SmackBot, Zephyris, Whpq, Jwy, Ohconfucius, Igoldste, Valoem, Mbell, T@nn, Swpb, R'n'B, CommonsDelinker, Acalamari, M-le-mot-dit, Squids and Chips, Davehi1, Lamro, Northfox, Lightmouse, Dolphin51, Tegrenath, Mild Bill Hiccup, Niceguyedc, Addbot, DougsTech, Fgnievinski, Luckas-bot, Materialscientist, Paula Pilcher, J04n, Abce2, Logger9, Locobot, LucienBOT, Stephen Morley, Tom.Reding, Marie Poise, John of Reading, GoingBatty, Danielacruz, Scientific29, Pierce.Schiller, Helpful Pixie Bot, Calabe1992, Bibcode Bot, AdventurousSquirrel, S Larctia, Makecat-bot, Amr94, Shearflyer and Anonymous: 8
- **Self-assembled monolayer** *Source:* https://en.wikipedia.org/wiki/Self-assembled_monolayer?oldid=708940920 *Contributors:* Rich Farmbrough, Mdd, Gene Nygaard, Dandv, V8rik, BD2412, Rjwilmsi, TheRingess, Physchim62, Jaraalbe, YurikBot, Wavelength, Shaddack, Chase me ladies, I'm the Cavalry, CWenger, Fram, SmackBot, Shalom Yechiel, N2e, Kupirijo, Mystic force, Barticus88, Dwmosley, Headbomb, Mah159, R'n'B, Shanata, Northfox, Moletrouser, ImageRemovalBot, SkeletorUK, Mild Bill Hiccup, ColinBelyea, Sun Creator, AHDGraham, Addbot, Anthonydelaware, Yobot, Elizgoiri, Shootbamboo, Materialscientist, Citation bot, Klaflin, Abnak, Carlog3, FrescoBot, Citation bot 1, DrilBot, Trappist the monk, Mmnbv21, Crazedcatatonic5, RjwilmsiBot, Srujan89, Ovoznyy, ChezChemistry, Bibcode Bot, AdventurousSquirrel, Hamish59, Aisteco, TheElectroChemist, Khazar2, Samorion, Vlakvlak, Ph.D. Chemical Engineering, Monkbot, Yongan Tang and Anonymous: 32
- **Micelle** *Source:* https://en.wikipedia.org/wiki/Micelle?oldid=726082321 *Contributors:* The Anome, Deb, Michael Hardy, Lexor, PolymerTim, Robbot, Modeha, Giftlite, Karol Langner, DanielCD, Pilatus, Alansohn, Jared81, Bart133, Dan100, Eras-mus, LadyofHats, Rjwilmsi, SeanMack, DannyWilde, Solidpeg, Iridos~enwiki, Roboto de Ajvol, YurikBot, Padraig, Gaius Cornelius, Dysmorodrepanis~enwiki, Biopresto, Wojdyr, Zzuuzz, SmackBot, Hugo-cs, Tamfang, Gimeral, Rsm99833, Tinctorius, Clicketyclack, Ocee, SpyMagician, Extremophile, Knights who say ni, Plcoffey, Ubiq, Mikebrand, Christian75, Thijs!bot, Opabinia regalis, Macilent.Flage, AntiVandalBot, Jj137, Farosdaughter, Sophosmoros, JamesBWatson, BatteryIncluded, Dirac66, Oren0, Hodja Nasreddin, Ibrmrn3000, Rod57, Mj sklar, Gregp99, Antony-22, Wmpremed, Showjumpersam, Lamro, Nosferatütr, Alhead, Redfiona99, ClueBot, Mild Bill Hiccup, Alandmanson, Kperkin, Chhe, Addbot, Quercus solaris, Zorrobot, Luckas-bot, Yobot, TaBOT-zerem, Choij, Materialscientist, Gsmgm, RibotBOT, FrescoBot, MBChandl, Pokyrek, I dream of horses, Joonkimphd, Klbrain, Dcirovic, Wikfr, L Kensington, ClueBot NG, PaulRupar, Chameleonwang, Neøn, Sobarwiki, Jacknunn, Robcicle, BattyBot, ChrisGualtieri, HH neuron, Prof.Phillips, Zaparah, JamesMoose, MSWAL, Amr94, EtymAesthete, Elephantsofearth, Ljenningsics and Anonymous: 99
- **Copolymer** *Source:* https://en.wikipedia.org/wiki/Copolymer?oldid=719309695 *Contributors:* Bryan Derksen, Wtanaka, DocWatson42, Leonard G., Rubicon, Walkerma, HenkvD, Ceyockey, Woohookitty, Cruccone, Mandarax, V8rik, Rjwilmsi, HappyCamper, Rolva, YurikBot, Shaddack, WAS 4.250, Doktorbuk, Zerodamage, ChemGardener, SmackBot, Radak, Nathaniel, Gilliam, Folajimi, Kuttanns05, Radagast83, Smokefoot, Disavian, JorisvS, Peterlewis, Ahering@cogeco.ca, DabMachine, Tanthalas39, Odie5533, Sdurf, Headbomb, Chimaeridae, .anacondabot, David Eppstein, Schmloof, Vinograd19, Trusilver, Bidabidabee, Bdodo1992, Irene Ringworm, Kiessling~enwiki, AllGloryToTheHypnotoad, BotKung, Lamro, Locke9k, Zenek.k, Northfox, ClueBot, Vengolis, SoxBot III, NellieBly, Good Olfactory, Addbot, Favonian, Servolo, Zielu20, Yobot, Rogerb67, Daniele Pugliesi, Materialscientist, Citation bot, Maniadis, LilHelpa, Elvim, Ludx, FrescoBot, Citation bot 1, Jujutacular, Oktanyum, Hirsutism, Michel Awkal, Zvelindovsky, Vanischenu, Prasanna Kumara Welahetti, ~riley, Marcela louis, Faizan, YiFeiBot, Amr94, Dough34, Apogalacticon, Nanoeclipse, Gonzalo.villarreal, 115.241.241.2d, CV9933 and Anonymous: 46
- **Langmuir-Blodgett film** *Source:* https://en.wikipedia.org/wiki/Langmuir%E2%80%93Blodgett_film?oldid=704566678 *Contributors:* Giftlite, DragonflySixtyseven, V8rik, BD2412, Srleffler, YurikBot, Shaddack, Anastasius zwerg, SmackBot, Hmains, Sandycx, TenPoundHammer, JorisvS, Ckatz, Dicklyon, Wikiauthor, Christian75, AntiVandalBot, CosineKitty, Jingxin, Shatnerrocks, CattleGirl, Uvainio, Sockettome, ClueBot, XLinkBot, Addbot, Jncraton, Yobot, Jesi, Materialscientist, Citation bot, Major measure, Sfingo, FrescoBot, Nmedard, PJ Geest, DrilBot, Tom.Reding, SchreyP, GoingBatty, ChrisGualtieri, Yadsalohcin, Epigogue and Anonymous: 32
- **Biological organisation** *Source:* https://en.wikipedia.org/wiki/Biological_organisation?oldid=720793520 *Contributors:* Edward, Greenrd, Graeme Bartlett, Rich Farmbrough, Vsmith, SteinbDJ, RichardWeiss, Grafen, Malcolma, Daniel Mietchen, Deville, SmackBot, Kazkaskazkasako, T-borg, Mrdthree, RekishiEJ, Krauss, Skittleys, Krylonblue83, Barticus88, Phopon, Edokter, Gökhan, Noobeditor, PhilKnight, Falcor84, Tgeairn, Kesal, AntiSpamBot, Whiteandnerdy52, Funandtrvl, VolkovBot, JohnBlackburne, TXiKiBoT, Lambyte, Falcon8765, Accounting4Taste, Oldag07, Mimihitam, JL-Bot, Addbot, Luckas-bot, Yobot, Amirobot, AnomieBOT, Materialscientist, WikiKherad, Crzer07, Aaron Kauppi, Citation bot 1, Ambarsande, Trappist the monk, Animalparty, EmausBot, Orphan Wiki, WikitanvirBot, Jager008, ZéroBot, Cameron11598, ClueBot NG, Jack Greenmaven, Helpful Pixie Bot, Faus, Itzuvit, BG19bot, Mark Arsten, AdventurousSquirrel, Meclee, Zaramela, BattyBot, Urmelii, Makecat-bot, Alphama, Refeline, MarqFJA87, I am One of Many, Forever elementary school student, Hendrick 99, Mikala14, Kahtar, Fixuture, BiologicalMe, Loraof, Liance, The Quixotic Potato, 19chasitymarie and Anonymous: 49
- **Protein folding** *Source:* https://en.wikipedia.org/wiki/Protein_folding?oldid=725844078 *Contributors:* Magnus Manske, Bryan Derksen, The Anome, Taw, Malcolm Farmer, Awaterl, Tijmz, Netesq, DennisDaniels, Dwmyers, Lir, Smelialichu, Michael Hardy, Lexor, Ixfd64, MockAE, 168..., Julesd, Peter Kaminski, Lfh, Fuzheado, Zoicon5, Donarreiskoffer, Robbot, Blainster, Intangir, Giftlite, Bfinn, Eequor, ChicXulub, Toytoy, LiDaobing, Piotrus, Klemen Kocjancic, Kevyn, Fawzin, Discospinster, Rich Farmbrough, Cacycle, Bender235, ESkog, Konstantin~enwiki, RoyBoy, WorldDownInFire, Harley peters, Arcadian, Ferkel, Minghong, Kierano, לערי, Rebroad, Cburnett, Jheald, WadeSimMiser, Cyberman, V8rik, Rjwilmsi, Trenchcoatjedi, Computor, DannyWilde, Sharkman~enwiki, Nihiltres, Terrace4, Bkhouser, Roboto de Ajvol, YurikBot, Wavelength, Cathalgarvey, Richyfp, Splette, SpuriousQ, Burrito, Kkmurray, Leptictidium, FF2010, JeramieHicks, Banus, SmackBot, TestPilot, Stepa, Kjaergaard, Tommstein, M stone, Gilliam, Bluebot, RDBrown, Miguel Andrade, Kostmo, Snowmanradio, Sunny17152, Gowantervo, Lostart, Drphilharmonic, Fuzzypeg, Clicketyclack, Ohconfucius, Madeleine Price Ball, Gobonobo, Ben Moore, Bendzh, Adriferr, Erencexor, Dhatz, Macintosh10000, Herd of Swine, WillowW, Myscrnnm, Christian75, SirHaddock, Calvero JP, Barticus88, Opabinia regalis, Headbomb, Nick Number, Kevphenry, Waerlueg, Pro crast in a tor, TimVickers, Tomixdf, Arch dude, Sangak, Magioladitis, WolfmanSF, Grimlock, Hooplehead, David Eppstein, Davepntr, Otvaltak, Jack007, Player 03, Lucaaah~enwiki, Trusilver, Jacobsman, Boghog, Hodja Nasreddin, Andraaide, Antony-22, Jammedshut, Gcrossan, Xnuala, Erwinser, Blooooo, Brianga, AlleborgoBot, D. Recorder, Macdonald-ross, SieBot, Czhangrice, Louismaddox, Movalley, Yves-henri, Ptr123, Dabomb87, Agilemolecule, LarRan, RunninRiot, ClueBot, Ideal gas equation, Bioinfo177, Mild Bill Hiccup,

49.5. TEXT AND IMAGE SOURCES, CONTRIBUTORS, AND LICENSES

Jlalbee, Akane700, Jytdog, Addbot, 5beta5, DOI bot, Samrat Mukhopadhyay, Totakeke423, JJ TKOB, Movado73, Legobot, Luckas-bot, Yobot, Ptbotgourou, TaBOT-zerem, 哇哇, Royote, Materialscientist, Citation bot, Bci2, BlurTento, Biophysik, Xqbot, DSisyphBot, P99am, Jü, GrouchoBot, Shmedia, RibotBOT, Dave3457, Mark Renier, Citation bot 1, Pinethicket, Katherine Folsom, The Great Cook, My very best wishes, Trappist the monk, Jesse V., Mittinatten, RjwilmsiBot, Artoannila, Conjugado, Clarker1, Dcirovic, H3llBot, Khaydock, ClueBot NG, Aejohnst, Memories of lost time, BarryRobson, Lifeonahilltop, Dopeytaylor, X-men2011, Behinddemlune, Lalo1121, Bibcode Bot, Natriumchloratum, Nescio vos, BG19bot, Kalekundert, Annavaughan, Tbonefrog, Vxn002, Scfbio, Cyberbot II, Sk177, Dexbot, Holger87, Proteinfolder, Makecatbot, Leprof 7272, David P Minde, Randykitty, TR13, Doctor Procto, TMHD, Sunny0208, Fisicorina blackhawk, Mmyjak, Jamal Ahmed GM, Monkbot, 17caespa, CV9933, KasparBot, Spinrade, JosiahWilard, WandaLan, Valerieap, GreenC bot and Anonymous: 178

- **Lipid bilayer** *Source:* https://en.wikipedia.org/wiki/Lipid_bilayer?oldid=725437132 *Contributors:* Magnus Manske, Peterlin~enwiki, Heron, Hfastedge, Lir, Lexor, 168..., Raven in Orbit, MichaK, RodC, Stone, Maximus Rex, Ark30inf, Robbot, Meelar, Seth Ilys, GreatWhiteNortherner, Bensaccount, JuanitaJP, Jfdwolff, Chowbok, Pgan002, Danko Georgiev, Antandrus, Karol Langner, Rich Farmbrough, Vsmith, MBisanz, Marx Gomes, Arcadian, Nachocab, Alansohn, Jared81, Monado, Geraldshields11, Bsdlogical, Ceyockey, Eleassar777, LadyofHats, Plw, Magister Mathematicae, Teuteul, Rjwilmsi, Arisa, Ian Pitchford, Xenobog, Kinglz, NawlinWiki, Bota47, Kkmurray, Tetracube, 2over0, Airconswitch, Melchoir, Od Mishehu, EncycloPetey, Jrockley, Durova, RDBrown, DMS, Tree Biting Conspiracy, Chelsea99, Claire van der Meer, Artie p, Clicketyclack, Visium, Winwin, Bendzh, Shinryuu, Arne Saknussemm~enwiki, Figbush1, PaddyM, Switchercat, Scohoust, Was a bee, Skittleys, JamesAM, Headbomb, AntiVandalBot, TimVickers, Smartse, Chill doubt, JAnDbot, Mattert, Sangak, Sophosmoros, Meredyth, Bulbeck, Olavrg, Adrian J. Hunter, Emw, DerHexer, JaGa, Lkc159, CommonsDelinker, Nono64, Nbauman, Catmoongirl, Hodja Nasreddin, Sunset's Light, VolkovBot, Oshwah, Showjumpersam, Guillaume2303, Sodapopinski, 1yesfan, Duncan.Hull, Jdude89, Shanata, Michaeldsuarez, Lamro, Synthebot, Monty845, Northfox, Logan, Steven.Noyce, Emfetz, Yerpo, Abdoulwahid, Verytas, Fratrep, Dlrohrer2003, Animeronin, ClueBot, Yurko~enwiki, Mild Bill Hiccup, Diderot's dreams, Drahkrub, Dderkits, Jytdog, Anticipation of a New Lover's Arrival, The, Addbot, Bouncingball2, Lightbot, Jarble, Luckas-bot, Yobot, Wikipedian Penguin, IW.HG, Backslash Forwardslash, AnomieBOT, Templatehater, Materialscientist, Citation bot, Neurolysis, Richard Jay Morris, Biophysik, LilHelpa, Xqbot, Pontificalibus, MDougM, J04n, Amaury, Miyagawa, D'ohBot, Citation bot 1, I dream of horses, My very best wishes, TobeBot, Codygerman, Jesse V., NEW Rome, J36miles, Dewritech, GoingBatty, Sesamehoneytart, Slightsmile, Dcirovic, JSquish, ZéroBot, Wiensgov, Hazard-SJ, Dclipp, ClueBot NG, Surgdx, Frietjes, Genetics1809, Helpful Pixie Bot, Bibcode Bot, MusikAnimal, Cyberbot II, Dexbot, DaanFloid, Webclient101, Mogism, InstantNull, Kjmarjun, Frosty, Mcdelo4256, Epicgenius, Lemnaminor, Iztwoz, DavidLeighEllis, Babitaarora, Jianhui67, Jlmalcos, Elephantsofearth, Monkbot, NewEnglandDr, Rakeshyashroy, Spacedandy838, CLCStudent, Centurionjoel, Day77 and Anonymous: 190

- **Homeostasis** *Source:* https://en.wikipedia.org/wiki/Homeostasis?oldid=727281584 *Contributors:* Kpjas, Marj Tiefert, Vicki Rosenzweig, Mav, Bryan Derksen, RK, Rmhermen, SimonP, Anthere, Stevertigo, Lir, Lexor, Isomorphic, Kku, Wwwwolf, Ixfd64, Looxix~enwiki, Ahoerstemeier, JWSchmidt, BigFatBuddha, Ijon, Александр, LittleDan, Julesd, Andres, Mxn, Smack, Raven in Orbit, Hashar, Charles Matthews, Timwi, Steinsky, Saltine, Fvw, Francs2000, Shantavira, Joi, Robbot, AlainV, RedWolf, Naddy, Academic Challenger, Hadal, Isopropyl, Alan Liefting, Marc Venot, Gobeirne, Giftlite, Mintleaf~enwiki, Kenny sh, COMPATT, Everyking, No Guru, Michael Devore, Bensaccount, Jfdwolff, Radius, Tom-, Decoy, J. 'mach' wust, Antandrus, Ravikiran r, MacGyverMagic, Jmeppley, Neutrality, Trevor MacInnis, Canterbury Tail, Forschung, Jwdietrich2, Freakofnurture, DanielJanzon~enwiki, CALR, Discospinster, Rich Farmbrough, Inkypaws, Vsmith, David Schaich, AlanBarrett, Paul August, Bender235, Jaberwocky6669, Cyclopia, Kbh3rd, Mashford, JoeSmack, Brian0918, El C, RoyBoy, Bobo192, Spalding, Smalljim, Viriditas, Adrian~enwiki, Aquillion, TheProject, Physicistjedi, Ben@liddicott.com, 99of9, Mdd, Googie man, Orangemarlin, Jumbuck, Alansohn, Gary, Arthena, Burn, Marianocecowski, Danntm, Jheald, Jackerhack, Vuo, Netkinetic, Ceyockey, Bytesmiths, Kamezuki, Ron Ritzman, Woohookitty, Sylvainremy, Camw, JBellis, MONGO, Mpatel, Sir Lewk, Huhsunqu, Wayward, 哇哇哇哇哇, DaveApter, Dysepsion, V8rik, BD2412, FreplySpang, Vanderdecken, Elmers, Mlewan, Sjö, Rjwilmsi, Koavf, Erebus555, Jake Wartenberg, PinchasC, Tangotango, Daniel Collins, Sferrier, Bhadani, Nigosh, Yamamoto Ichiro, RainR, Musical Linguist, Apollo the Archer, RexNL, Ewlyahoocom, Wongm, OrbitOne, Jorgesalgueiro~enwiki, Alphachimp, Chobot, Nastajus, Mhking, Bgwhite, YurikBot, Wavelength, Pip2andahalf, Arado, Nesbit, Russoc4, Shell Kinney, Rsrikanth05, NawlinWiki, Wiki alf, Nutiketaiel, Brandon, Daniel Mietchen, Veracious Rey, Raven4x4x, Moe Epsilon, Misza13, BOT-Superzerocool, DeadEyeArrow, Jeh, Stefeyboy, Kkmurray, Caerwine, Chanueting, Lt-wiki-bot, Arthur Rubin, Pb30, Josh3580, JoanneB, Whouk, Allens, NeilN, Zvika, Lomacar, Xtraeme, Crystallina, SmackBot, NickyMcLean, Eperotao, Dweller, Aim Here, Jclerman, Hydrogen Iodide, Lantianer, Pgk, Blue520, Davewild, Eskimbot, Frymaster, Yamaguchi哇哇, Gilliam, Eug, Cabe6403, Tyciol, Chris the speller, Rkitko, NCurse, Jprg1966, Miquonranger03, SchfiftyThree, Deli nk, Miguel Andrade, DHN-bot~enwiki, QuimGil, KaiserbBot, MBlume, Snowmanradio, Nakon, VegaDark, John D. Croft, Hgilbert, Drphilharmonic, BrotherFlounder, DDima, Dankonikolic, L337p4wn, SashatoBot, Nick Green, John, Edwy, Mgiganteus1, Optakeover, Waggers, Dr.K., Cerealkiller13, Xionbox, D12000, Iridescent, Younesmaia, RokasT~enwiki, Tony Fox, Courcelles, Dartelaar, Tawkerbot2, Lahiru k, JForget, Vaughan Pratt, CmdrObot, Tobes00, Limno, Chrumps, Ninetyone, JohnCD, Devis, Im.a.lumberjack, Dgw, Pajast, Tim1988, Flying Saucer, Nilfanion, Slazenger, Michfan2123, Cydebot, Gogo Dodo, Anthonyhcole, JFreeman, ST47, Adolphus79, Tawkerbot4, DumbBOT, Sweikart, Mattisse, Thijs!bot, Epbr123, DulcetTone, Mathmoclaire, Daniel, Mungomba, Luigifan, Sobreira, Marek69, Woody, Tellyaddict, Ketan Kapoor, Kborer, CharlotteWebb, David D., KrakatoaKatie, AntiVandalBot, Luna Santin, Bas Kooijman, TimVickers, Ashleyy osaurus, Res2216firestar, JAnDbot, Husond, Athkalani~enwiki, Kaobear, Skomorokh, Nthep, Matthew Fennell, Myaca, Ph.eyes, Leolaursen, Rothorpe, Jarkeld, Ariaconditzione, Bongwarrior, VoABot II, AuburnPilot, Wikidudeman, Oliver.nissen, Paroswiki, Arrowcatcher, Nutscode, Animum, Cgingold, BatteryIncluded, Matsumuraseito, Heliac, DerHexer, Hbent, Squidonius, Hannah dh, 1salam1, Yobol, MartinBot, Axlq, Nono64, AgarwalSumeet, Xargque, Tgeairn, Erkan Yilmaz, St475353825, J.delanoy, Richlv, Nbauman, Jrsnbarn, Cmghim925, Michael Daly, Hydroflexology, DarkFalls, McSly, Mikael Häggström, LittleHow, DJbuddy16, Krasniy, Richard D. LeCour, NewEnglandYankee, In Transit, Zerokitsune, Sunderland06, Jackacon, Cometstyles, DorganBot, Treisijs, V. berus, Specter01010, DebateKid, Lwalt, Montchav, Pleclech, VolkovBot, Hibbity Dibbity, Macedonian, Flyingidiot, Soliloquial, Dom Kaos, Barneca, Philip Trueman, DoorsAjar, TXiKiBoT, Oshwah, Technopat, A4bot, Ndaniels, Ewart7034, Crucius~enwiki, Melsaran, Brunton, Leafyplant, PaulTanenbaum, AllGloryToTheHypnotoad, HuskyHuskie, DieBuche, Mishlai, Wiae, Kevens7, Davwillev, Maxim, WinTakeAll, Madhero88, Andy Dingley, Manic medic101, Dirkbb, Lova Falk, Friedsotong, Enviroboy, MCTales, Mjminc, Brianga, Monty845, Petergans, SieBot, Alessgrimal, Timb66, Tiddly Tom, VVVBot, Jacotto, Blakeeatscake, Svenjense, Flyer22 Reborn, Ssands, Radon210, Bookermorgan, Dangerousnerd, Cori Valet, Oxymoron83, Omurphy, Fratrep, Macy, Thomas5436, Anchor Link Bot, Vanished User 8902317830, Hordaland, Denisarona, TheCatalyst31, ClueBot, Philip Sutton, Hippo99, Healthwise, The Thing That Should Not Be, Seth3481, Nnemo, Wysprgr2005, Franamax, Delta40, TheOldJacobite, 3grayb, 3russella, CounterVandalismBot, TheSmuel, Phenylalanine, Mkativerata, Excirial, Nipper94, Nick721, Muhandes, SpikeToronto, Pjeh, Brews ohare, Tyler, Neucleon, Tillyiscool, Calrosfing, Dekisugi, Aleksd, La Pianista, Awesome93, Aitias,

Versus22, Dusen189, SoxBot III, Chrissie2100, Htmlcoderexe, JKeck, Spitfire, Jytdog, WikHead, Nicolae Coman, Qgil-WMF, NellieBly, Enylius, Eleven even, Gazimoff, Izahia, Jojodaho, Mojska, Addbot, Willking1979, Some jerk on the Internet, DOI bot, Non-dropframe, Atethnekos, Zefryl, Montgomery '39, AkhtaBot, Ronhjones, Fieldday-sunday, SoSaysChappy, Glane23, Baffle gab1978, West.andrew.g, Quercus solaris, Pince Nez, Tide rolls, Teles, Ayacop, Zorrobot, 04Patrickg, Pinus jeffreyi, Luckas-bot, Yobot, Ayrton Prost, IW.HG, Samtar, محبوب عالم, AnomieBOT, Tryptofish, Rubinbot, Jim1138, Pyrrhus16, Mbunaman, Merube 89, Law, 05borehama, Godofalltheworld, Materialscientist, Swithrow2546, Citation bot, Naj-GMU, Skhanal-GMU, Rosc702, Frankenpuppy, Obersachsebot, Xqbot, Jonathanmcguinness, Sionus, Fshaikh GMU, Mononomic, NFD9001, RadiX, Omnipaedista, 78.26, Uloggonitor, SD5, Fdardel, Ryryrules100, Tobby72, Pepper, Hirpex, Citation bot 1, Pinethicket, I dream of horses, A412, SpaceFlight89, Tommya182123, Qzqzqzqz, Fumitol, December21st2012Freak, FoxBot, NIHKZ, Mercy11, Juankfe, Etincelles, Ekul81, Vrenator, Daggad11, Zink Dawg, Eddturtle, Jeffrd10, Diannaa, Suffusion of Yellow, Lilleskvat, Sirkablaam, Tbhotch, Sideways713, DARTH SIDIOUS 2, Bernd.Brincken, Beyond My Ken, Mirilikesbpi, Pitlane02, EmausBot, ImprovingWiki, WikitanvirBot, Immunize, Ajraddatz, Hilly8, RA0808, Yowife, Slightsmile, Fromriri, Janedoe52, Wikipelli, Dcirovic, Josve05a, Imperial Monarch, Arbnos, A930913, Rails, Jay-Sebastos, Jesanj, L Kensington, Donner60, Jbergste, Orange Suede Sofa, TheObsidianFriar, Wakebrdkid, DASHBotAV, Ustates2119, Petrb, Xanchester, Minnsurfur2, ClueBot NG, Gareth Griffith-Jones, MelbourneStar, Gilderien, Bodyworxs, Chazede, KnowledgeisPOWA, -sche, Snotbot, Mesoderm, Widr, AlexB531, Mbsciencegeek, Vibhijain, Helpful Pixie Bot, Homeostasis111, Ramaksoud2000, BG19bot, 7204daniel, Hallows AG, MusikAnimal, Mark Arsten, Altafr, Anonymousplus, Wannabemodel, Agent 78787, Samein50, BattyBot, Tutelary, Walrustree, The Illusive Man, GoShow, Silk666, Rinkle gorge, BrightStarSky, Louisfuture, Hbasketball7, Mogism, Indiferente1, Wikilove719, Princess1234567890, Hillbillyholiday, Justicekiller, Eyesnore, Tentinator, Cuzimcoolandall, Ugog Nizdast, CloudStrifeNBHM, Ginsuloft, D Eaketts, Manul, PierreFG5, EtymAesthete, Anila Bakhtawar, Skr15081997, Hayley kadic, Csutric, Kiran cb, Wrae Ann, Awesome2435, BethNaught, PooBlaster3000, Toadface123, Thereppy, SongofSol, Matiia, Freshabautista, 384400km, FourViolas, Eathansmart, Alexander villatoro, Devildog27, Wishva de Silva, Cruithne9, XXMcFabioXx, Charlotte135, Rhansen64, Allan Thomas Needle, Fmadd and Anonymous: 975

- **Pattern formation** Source: https://en.wikipedia.org/wiki/Pattern_formation?oldid=723738682 Contributors: Kku, Nina, Pharos, Wtmitchell, Oleg Alexandrov, BD2412, Josh Parris, Rjwilmsi, DonSiano, JarrahTree, Daniel Mietchen, Reyk, JesseStone, Berland, LeoNomis, Wizard191, Miketwardos, Narayanese, Utopiah, Ste4k, Lopkiol, JaGa, R'n'B, Dr d12, Chiswick Chap, The enemies of god, Danwills, LucDecker, Paradoctor, Forluvoft, SchreiberBike, DumZiBoT, Addbot, Rascallgh, Yobot, FrescoBot, Reirobros, Algorithmgeek, Blueshifting, GoingBatty, WarEqualsPeace, Dcirovic, Danielse, Gongoozler123, Vaulttech, BattyBot, ChrisGualtieri, Cerabot~enwiki, Mk85 2, Monkbot and Anonymous: 19

- **Morphogenesis** Source: https://en.wikipedia.org/wiki/Morphogenesis?oldid=726540629 Contributors: AxelBoldt, Magnus Manske, Malcolm Farmer, Andre Engels, Enchanter, AdamRetchless, Zocky, Alan Peakall, Lexor, Shyamal, Lquilter, 168..., JWSchmidt, Frazzydee, Rhys~enwiki, Sjorford, Jmabel, Romanm, Matty j, Bfinn, Electric goat, Dullhunk, Arminius, Lulu of the Lotus-Eaters, Dmr2, Bender235, Cayte, Scentoni, Kocio, Julien Tuerlinckx, BD2412, Rjwilmsi, Mohawkjohn, Daycd, Moriane, Bgwhite, Hede2000, Arrt-932~enwiki, Chaos, Jknabe, Mccready, Gadget850, Attilios, Stepa, Habibkoite, RDBrown, Cybercobra, Hgilbert, Drphilharmonic, Celefin, Robofish, Mgiganteus1, Lauriec, DabMachine, Phantom mafia, IronChris, Eubanks718, Gogo Dodo, Khatru2, Thijs!bot, Epbr123, Darklilac, Lfstevens, Gcm, Igodard, Dmmd123, Srice13, MartinBot, Ledfloyd, J.delanoy, MikeBaharmast, Dr d12, Chiswick Chap, DadaNeem, VolkovBot, AlnoktaBOT, Fanatix, BartekChom, ClueBot, Quinxorin, Fjellstad, BOTarate, DumZiBoT, XLinkBot, Falco528, SilvonenBot, Anwormy, Eddovar, Addbot, Morphogenesis2008, Lightbot, Arbitrarily0, HerculeBot, Legobot, Luckas-bot, Yobot, EdwardLane, Ptbotgourou, Xqbot, WellsPedia, SassoBot, Izvora, Trkiehl, Pinethicket, Wotnow, RjwilmsiBot, JamesHilt62, Dcirovic, ZéroBot, Bibcode Bot, Barney the barney barney, Drmaitre, Me, Myself, and I are Here, Hkouros, Luvlethalwhites, IAmNitpicking, Anrnusna, Monkbot, Ianwyosnick, JMWSlack, KasparBot and Anonymous: 47

- **Abiogenesis** Source: https://en.wikipedia.org/wiki/Abiogenesis?oldid=726999594 Contributors: Damian Yerrick, AxelBoldt, Joao, Bryan Derksen, The Anome, Sjc, -- April, Ed Poor, SimonP, Maury Markowitz, AdamRetchless, Zadcat, Mjb, Heron, Someone else, Lexor, Gabbe, Martin BENOIT~enwiki, Bobby D. Bryant, Ixfd64, Cyde, Sannse, Mcarling, Ihcoyc, Ellywa, Mdebets, Cyp, JWSchmidt, Julesd, Raven in Orbit, Norwikian, Ec5618, Charles Matthews, Timwi, Steinsky, Foodman, Maximus Rex, David Shay, Populus, Omegatron, Samsara, Jackson~enwiki, Raul654, Johnleemk, Finlay McWalter, Skaffman, Twang, Jason Potter, Robbot, Fredrik, Goethean, Altenmann, Nurg, Rursus, Rebrane, Sheridan, Hadal, Wereon, Raeky, Xanzzibar, Xyzzyva, Giftlite, Mshonle~enwiki, Polsmeth, Pretzelpaws, Everyking, Curps, Solipsist, Bobblewik, Pgan002, Andycjp, Keith Edkins, J.'mach' wust, Sonjaaa, Quadell, Beland, Onco p53, Nograpes, Savant1984, JohnArmagh, Deglr6328, Flex, Lacrimosus, Mike Rosoft, Ta bu shi da yu, Rfl, Discospinster, Rich Farmbrough, Vsmith, ArnoldReinhold, Dave souza, Paul August, Bender235, ESkog, Srbauer, RJHall, Mr. Billion, Crunchy Frog, José Gnudista, Lycurgus, Kwamikagami, Liberatus, Sietse Snel, Art LaPella, RoyBoy, Fufthmin, Guettarda, Causa sui, Bobo192, John Vandenberg, Enric Naval, Viriditas, .:Ajvol:., ZayZayEM, I9Q79oL78KiL0QTFHgyc, VBGFscJUn3, Sulai~enwiki, Hob Gadling, A Karley, Orangemarlin, Marwood, DanielVallstrom, Darrelljon, Psychofox, SlimVirgin, Ferrierd, Kocio, InShaneee, Wtmitchell, Velella, Darco, XB-70, Knowledge Seeker, Pauli133, Tainter, BerndH, Bdrasin, Linas, Mindmatrix, Anilocra, LOL, Rocastelo, Schultz.Ryan, Tabletop, Grace Note, Sadettin, GregorB, CharlesC, Wdanwatts, Essjay, Palica, Gerbrant, GSlicer, RichardWeiss, Alienus, V8rik, BD2412, Rkevins, Sjö, Drbogdan, Rjwilmsi, Mayumashu, Nightscream, Koavf, Zbxgscqf, OneWeirdDude, Bob A, XP1, Crazynas, Mikedelsol, Bfigura, SLi, Duaglooth, Margosbot~enwiki, Nihiltres, Ahlutch, Geologist~enwiki, Vanished user psdfiwnef3niurunfiuh234ruhfwdb7, WhyBeNormal, Knoma Tsujmai, Truthteller, Chobot, DVdm, Bgwhite, Poorsod, YurikBot, Spacepotato, RadioFan2 (usurped), GPS Pilot, The Hokkaido Crow, NawlinWiki, Rick Norwood, DragonHawk, Dysmorodrepanis~enwiki, Uberisaac, Dtrebbien, Seirscius, Zarel, SAE1962, RecSpecz, Apokryltaros, Nick, Kdbuffalo, E rulez, Crasshopper, Kortoso, Stefan Udrea, WAS 4.250, 2over0, Encephalon, Bhumiya, Smoggyrob, Davril2020, Petri Krohn, Red Jay, Fram, DisambigBot, JDspeeder1, NeilN, CIreland, Victor falk, NetRoller 3D, Quadpus, KnightRider~enwiki, SmackBot, Eperotao, PiCo, John Croft, Rtc, TestPilot, David Shear, Lankenau, Bmearns, BiT, Edgar181, Yamaguchi先生, Macintosh User, Gilliam, Portillo, Betacommand, Skizzik, Eloy, Chris the speller, Kaylus, RDBrown, Davep.org, Jprg1966, Thumperward, Silly rabbit, Hibernian, Complexica, Jeff5102, Scwlong, John Hyams, JoelWhy, Jefffire, Viperphantom, Vanished User 0001, Avb, Cfassett, Ines it, Khukri, John D. Croft, Richard001, Archgoon, Smokefoot, Greg.collver, The PIPE, DMacks, Sammy1339, Daniel.Cardenas, Denise from the Cosby Show, Alan G. Archer, Ohconfucius, SashatoBot, Danielrcote, Technocratic, Gloriamarie, Attys, Atkinson 291, Khazar, John, Writtenonsand, J 1982, Butko, JoshuaZ, JorisvS, Robert Stevens, Mgiganteus1, Olin, Scetoasus, Fig wright, Extremophile, 041744, Robbins, A. Parrot, Tarcieri, Smith609, Makyen, Stevebritgimp, Tac2z, Mr Stephen, Xiaphias, Larrymcp, NJA, Novangelis, LenW, Dau Glück, Nehrams2020, Clarityfiend, ShyK, Twas Now, Lent, The Letter J, George100, Chris55, VinnieCool, DangerousPanda, CRGreathouse, Ale jrb, Memetics, BeenAroundAWhile, Runningonbrains, RoliSoft, ButFli, WeggeBot, Moreschi, Richard Keatinge, Nnp, Myasuda, Ciyean, Abeg92, Peterdjones, Cyhawk, Hughgr, Michael C Price, Doug Weller, DumbBOT, Narayanese, DnimrevO, Ebyabe, Crum375, PKT, Thijs!bot, Barticus88, Ryansca, Pstanton, Mojo Hand, Mungomba, Headbomb, James086, Astrobiologist, Davidhorman, Chandler, Gossamers, AntiVandalBot, Luna

49.5. TEXT AND IMAGE SOURCES, CONTRIBUTORS, AND LICENSES

Santin, Guy Macon, Dbrodbeck, Gnixon, TimVickers, Cstreet, Smartse, Fluffy654, Danny lost, Princeofexcess, JAnDbot, XyBot, GromXXVII, MER-C, The Transhumanist, Matthew Fennell, Mildly Mad, Andonic, Xeno, Panarjedde, TAnthony, Tstrobaugh, Rothorpe, Kornbelt888, Magioladitis, WolfmanSF, Carlwev, Sushant gupta, JNW, CattleGirl, Harelx, Trishm, Hubbardaie, Mark PEA, Recurring dreams, Zephyr2k~enwiki, Theroadislong, Cgingold, BatteryIncluded, Allstarecho, Lyoncc, DerHexer, Edward321, Urco, JohanViklund, Mdsats, Robin S, Drm310, Tsinoyboi, Keith D, R'n'B, CommonsDelinker, Verdatum, Leyo, Mzaki, Player 03, PhageRules1, Ulisse0, Ifomichev~enwiki, AstroHurricane001, Avkulkarni, Rlsheehan, Hans Dunkelberg, Sidhekin, AmagicalFishy, Dispenser, It Is Me Here, Enuja, McSly, Tarotcards, Janet1983, Davy p, RobinGrant, Lbeaumont, Jorfer, Cmichael, KylieTastic, AzureCitizen, IceDragon64, Funandtrvl, Novernae, Jamiejoseph, Speaker to wolves, Philip Trueman, Sub-life, Vipinhari, GcSwRhIc, Charlesdrakew, Matthewrossing, Littlealien182, Steven J. Anderson, Awl, AllGloryToTheHypnotoad, Noformation, MacFodder, Mannafredo, Mishlai, Gibson Flying V, Wiae, Maxim, Shanata, WinTakeAll, Distinguisher, SheffieldSteel, Wolfrock, Lamro, Synthebot, Zarcoen, Omermar, Northfox, Rep07, Planet-man828, Hrafn, Nachohosking, EGMAG, Tczuel, Macdonaldross, Gnocchi, Carny, KatieandHandy, Nihil novi, ToePeu.bot, Meldor, Dawn Bard, ConfuciusOrnis, Odd nature, Yintan, 0xFFFF, Abhishikt, Chhandama, Oda Mari, Jc-S0CO, Oxymoron83, Lightmouse, Heliкophis, RW Marloe, Manifolds~enwiki, Jruderman, RyanParis, Sunrise, Diego Grez-Cañete, Skeptical scientist, StaticGull, Mos bratrud, Tesi1700, Hamiltondaniel, Driftwood87, Kalidasa 777, Marmenta, Lucius Sempronius Turpio, Twinsday, Sfan00 IMG, ClueBot, Tmol42, Fyyer, The Thing That Should Not Be, AstroMark, Sexiestjen4u, Desoto10, Pi zero, Unbuttered Parsnip, Jumacdon, Canopus1, Polyamorph, Timberframe, Tfpsly, Niceguyedc, Baegis, Alexis Brooke M, Rotational, Jandew, Paulcmnt, Excirial, Gustavocarra, Winston365, Vital Forces, Shinkolobwe, Abeo iniuria, Sun Creator, Eznight, Coinmanj, NuclearWarfare, SchreiberBike, Audaciter, BOTarate, Truth is relative, understanding is limited, Thusled, Thingg, Aitias, AC+79 3888, Johnuniq, Egmontaz, Editor2020, Goodvac, Bentheadvocate, Darkicebot, CaptainVideo890, XLinkBot, Roxy the dog, Jytdog, Jovianeye, Rror, Bradv, Elfgeek, Ost316, Jungfruchallan, Aloboof123, Opaq87, Aunt Entropy, Virajelix, Thatguyflint, Janisterzaj, Addbot, Roentgenium111, DOI bot, Landon1980, Swissmeister, Ronhjones, CanadianLinuxUser, Dsmith77, Lindert, Download, Redheylin, Bernstein0275, Camedit, Blade13125, Wildreceleste, Polyp2, LinkFA-Bot, Quietmarc, Partofwhole, U3190, Tide rolls, TL782, Romaioi, Nase, Yobot, StarTroll, Scepticus2, Yngvadottir, The Earwig, Punu, 489thCorsica, Cseppala, CinchBug, Dr.Buttons, AnomieBOT, Brroga, Mike Hayes, Kerfuffler, JWSurf, Trabucogold, Csigabi, Mann jess, Materialscientist, Citation bot, Quebec99, Romandoggie, LilHelpa, FreeRangeFrog, Xqbot, Sventington the Second, Blorblowthno, JimVC3, Wapondaponda, Mnnlaxer, Δζ, Nasnema, Mononomic, Turk oğlan, J JMesserly, Crzer07, 7h3 3L173, DerryTaylor, ProtectionTaggingBot, Gui le Roi, Conquistador, Sophus Bie, Ramssiss, Shadowjams, Methcub, Eugene-elgato, Joaquin008, Biem, FrescoBot, Finstergeist, Yanima, Hoffmannrungethailand, Krj373, Riventree, BKMBC3, Machine Elf 1735, Trkiehl, Citation bot 1, Redrose64, ANDROBETA, DrilBot, Winterst, Gravityguy, WaveRunner85, Gamocamo, Jonesey95, Helzrule19, Tom.Reding, Deleteduser2015, Hoo man, SpaceFlight89, FormerIP, Tanzania, Jerrywickey, Mikespedia, Jandalhandler, Kibi78704, MichaelExe, Fartherred, SkyMachine, IVAN3MAN, Trappist the monk, Silenceisgod, MEPK, Fama Clamosa, Comet Tuttle, Mcfl116, Vrenator, Victorfrogg, Jimmetry, Clarkcj12, Bcoolsdad, Diannaa, 564dude, Jynto, Gregrutz, Myrmidon1, DARTH SIDIOUS 2, Tor1714, Onel5969, RjwilmsiBot, Apotheosa, Hppa, Plommespiser, WildBot, Tesseract2, I belong to Jesus Christ, EmausBot, Jeffhughes22, Immunize, Dominus Vobisdu, Niluop, Dewritech, Ibbn, RespoonsibileSQ, Tamtrible, Jmv2009, Pboehnke, Slightsmile, Tommy2010, Kiran Gopi, Mmeijeri, Dcirovic, Solomonfromfinland, Ofekalef, Gershake, Kiwi128, H3llBot, Wayne Slam, David J Johnson, Ksarasofi, Korztin, Jesanj, Brandmeister, L Kensington, Jess, Scientific29, Ego White Tray, Tanoan, Renji911, SemanticMantis, Dr. Hipopotamo, JanetteDoe, Sven Manguard, JonRichfield, Ldvhl, Zuky79, Gary Dee, AUN4, ClueBot NG, Don Para, E3cubestore, Afterrock81, Colin Fredericks, Rainbowwrasse, Jorge 2701, Joefromrandb, Sketchup123, DonaldRichardSands, DS Belgium, Sjmantyl, Asukite, Widr, Telpardec, Keenedged, Wikiwiki180, MerIlwBot, Lotterox, Michaeltdeans, Helpful Pixie Bot, Elefnose, Cinnaplum, Anentiresleeve, Curb Chain, Bibcode Bot, Mwregehr, BG19bot, Lebs27, Expewikiwriter, Vevanpelt, Karmstrong909, Knowledge Examiner, Halstedcw, Mark Arsten, IraChesterfield, Drewrainey, Dkspartan1, Cauhtcoatl, Գարիկ Ավագյան Արցախ, MLearry, Cadiomals, Gorthian, Mthoodhood, Ghostsarememories, Harizotoh9, Blackstar167, HMman, Dontreader, Zedshort, Zetazeros, Dontshootimgay, Benyboy2, BattyBot, Decruft, Sfarney, Hghyux, Marc Tessera, Jimw338, SkepticalRaptor, David B Stephens, Soulbust, TheJJJunk, Garamond Lethe, Tanookiinashu, Khazar2, Ekren, Nathanielfirst, Elfinanciero222, Cmw255, Соляристъ, Pterodactyloid, RGA1980, Dexbot, Webclient101, Jinx69, Cerabot~enwiki, TippyGoomba, Mbreht, TheTahoeNatrLuvnYaho, CuriousMind01, Saehry, Leptus Froggi, 93, TruthOrTruthy, Corinne, Frivolous Consultant, HerbertHuey, FlaviusFerry, Tjmiler, Reatlas, Anastronomer, Bret palmer, Faizan, ICameHereToEdit, Surfer43, KnowledgeIncreases07, Analiticus, StewartGriffiths, Nirendeka, DavidLeighEllis, Ronaldo Laranja, Nigellwh, AbioScientistGenesis, SzostakJack, Mj12hoaxwriter, MDPub13, PubMed2015, Andreas.Geisler, EunuchRU, PrivateMasterHD, SpazAbiogenesis, Leptinresistinadiponectin, NottNott, SuperFreakCell, Anmusna, Stamptrader, Sstur, Dodi 8238, Suelru, Chaya5260, †thing goes, Inphynite, Kkosman, Baltazorgue, Johngraybosch, Monkbot, BethNaught, Acagastya, Garfield Garfield, Shandck, Signedzzz, Brianbleakley, Pombrand, Ruwdaman, Fried Vegetables, BicelPhD, Yazan atheos, BlueFenixReborn, Strongjam, Sarr Cat, Imradinmyownway, Washington Charter, Chemistryorigin, Michaelo1019, One sanguin, KasparBot, Fernando orrego, MusikBot, Kanashimi, Atchoum, Ktns, Paula NK, Joholub123, Shadowblade001, Bik0ser, GoldCar, Dylangenetic, Allthefoxes, Gongwool, Oucherowl and Anonymous: 747

- **Hypercycle (chemistry)** *Source:* https://en.wikipedia.org/wiki/Hypercycle_(chemistry)?oldid=720868710 *Contributors:* Smalljim, Szymcio2001, The Rambling Man, Yoninah, Malcolma, Daniel Mietchen, Nikkimaria, Racklever, Curly Turkey, Tktktk, Utopiah, Perohanych, Sunrise, Elmerfadd, Addbot, Yobot, AnomieBOT, Rubinbot, Project Osprey, FourViolas and Anonymous: 5

- **Autocatalytic set** *Source:* https://en.wikipedia.org/wiki/Autocatalytic_set?oldid=656681667 *Contributors:* AdamRetchless, Heron, Lexor, Raven in Orbit, Emperorbma, Fuelbottle, Dittrich, Woohookitty, Rjwilmsi, Szymcio2001, Gaius Cornelius, Luk, SmackBot, Chris the speller, Complexica, Knights who say ni, CmdrObot, Lewallen, Vancouverbagel, Falcor84, STBot, Selfsame, VolkovBot, ThomHImself, Capitalismojo, Niceguyedc, Jemmy Button, Addbot, Hakan Kayı, Yobot, Xqbot, AvicAWB, Khazar2, Mathmomike, Monkbot and Anonymous: 15

- **Multi-agent system** *Source:* https://en.wikipedia.org/wiki/Multi-agent_system?oldid=717197978 *Contributors:* The Anome, Michael Hardy, Lexor, Haakon, Jll, BAxelrod, Wlievens, Zigger, Gdm, Zootalures, Beyer, Eyrian, Rich Farmbrough, Sladen, Michal Jurosz, Fenice, Alex Kosorukoff, Mdd, CyberSkull, Alai, Emergence~enwiki, Mindmatrix, Mihai Damian, Male1979, Waldir, Merlin.The, Marudubshinki, Rjwilmsi, Quiddity, Intgr, Pouac, KenBailey, Luigi.bozzo, PeWu~enwiki, Gaius Cornelius, CarlHewitt, Gelderlander1, Rbarreira, Rjlabs, Thijswijs, SmackBot, Mneser, KnowledgeOfSelf, Mctoyama, Tomdw, Pmkpmk, Colonies Chris, Jasonb05, Frap, Zvar, Radagast83, Robofish, Christian Guttmann, Cmh, Beetstra, Dan1679, Dpag, Ishnid, Neuknecht, Peterdjones, M.nikraz, Thijs!bot, Headbomb, Peoppenheimer, I am neuron~enwiki, Clan-destine, MSBOT, Americanhero, Gwern, R'n'B, Jiuguang Wang, Gwkronenberger, Chris.selwyn, Aprodan, Lordvolton, Tesfatsion, 450w, AlleborgoBot, Frederick.stanichev, CharlesGillingham, Benno Overeinder, Rinconsoleao, XDanielx, Alexbot, PixelBot, SchreiberBike, Hasanadnantaha, Chaosdruid, MairAW, WikHead, Eric Catoire, Itrebax, Addbot, Jncraton, Pgautier-neuze, Rosemartin, Sdeloach, Wireless friend, Yobot, AnomieBOT, ChristianG2, Xqbot, Charvest, Lbcao, FrescoBot, Cedric71, Ssnestinger, Bassanesi, BorysB, Amandus74,

- **Self-organizing network** *Source:* https://en.wikipedia.org/wiki/Self-organizing_network?oldid=711542608 *Contributors:* Pnm, Kku, Closeapple, RJFJR, Avalon, SmackBot, JanCeuleers, Stim371, Misarxist, Qu3a, Moonriddengirl, Niceguyedc, Muhandes, MatthewVanitas, Addbot, Umbeebmu, MrOllie, Yobot, Eikoseidel, Xqbot, Isheden, Tumbledown, Pinethicket, LittleWink, Micraboy, John of Reading, AsceticRose, Bubble456, Bomazi, Fgunnars, BG19bot, EricEnfermero, Samwalton9, Mogism, Tator2, Claudiahernandez123, Yaacovcohen, Miki Weiser-Padova, Hakanakgunonline, Aosevim and Anonymous: 31

- **Dual-phase evolution** *Source:* https://en.wikipedia.org/wiki/Dual-phase_evolution?oldid=722257320 *Contributors:* Michael Hardy, Bearcat, Ruud Koot, BD2412, Rjwilmsi, Gaius Cornelius, Headbomb, Magioladitis, Dggreen, Wiae, Yobot, Citation bot, BG19bot, Northamerica1000, BattyBot, Me, Myself, and I are Here, Pratyush Sagiraju, Joseph2302 and Anonymous: 1

- **Molecular assembler** *Source:* https://en.wikipedia.org/wiki/Molecular_assembler?oldid=720942231 *Contributors:* Bryan Derksen, Ahoerstemeier, Ehn, Smith03, Omegatron, King Art, Sappe, Vespristiano, Fuelbottle, David Gerard, Curps, Guanaco, Hannes Karnoefel, JRR Tolkien, Apotheon, CesarFelipe, Sam Hocevar, DMG413, Brianhe, Rich Farmbrough, Dmeranda, Bender235, Kjoonlee, Kanzure, Etxrge, Zxcvbnm, Alai, Zbxgscqf, Rsmith, Benlisquare, WriterHound, Hairy Dude, Arado, Mike Treder, Kkmurray, Dieseldrinker, NHSavage, Petri Krohn, SmackBot, M stone, Betacommand, Ppntori, Toughpigs, DéRahier, KaiserbBot, GVnayR, Polonium, Vampus, Paul venter, Harold f, Basawala, Cydebot, KrakatoaKatie, Knotwork, Kawaputra, Oicumaybenght, Drjem3, Pekaje, Antony-22, Sbierwagen, Netmonger, VolkovBot, TXiKiBoT, Karjam, Plastikspork, Kwizy, Coccyx Bloccyx, DumZiBoT, XLinkBot, Addbot, Anthonydelaware, Yobot, Ptbotgourou, Amirobot, AnomieBOT, Materialscientist, Carl086, Xqbot, FrescoBot, Tom.Reding, Skyerise, Alph Bot, ZéroBot, Kni2, สไลห้, Virtualerian, Star A Star, Cyberbot II, Leafonesky, Spacelion88, Comp.arch, Dersman, Fixuture, SoerenMind, Maplestrip, Spectra239 and Anonymous: 62

- **Critical mass (sociodynamics)** *Source:* https://en.wikipedia.org/wiki/Critical_mass_(sociodynamics)?oldid=726621631 *Contributors:* Kku, Bearcat, Chealer, Tomchiukc, Eep², Rich Farmbrough, Mdd, Sin-man, BD2412, Rjwilmsi, Benlisquare, Wavelength, Aeusoes1, ColdFusion650, Jogers, Jtneill, Cybercobra, Sasata, JohnCD, Cydebot, Meno25, Soetermans, Mkdw, Dextrase, Fair Alienor, Useight, Ale2006, RingtailedFox, Ori, Binksternet, Samer.hc, WikHead, Daniel Lièvre, Addbot, Luckas-bot, Yobot, KamikazeBot, Martin scharrer, AnomieBOT, Fritzesumc, LilHelpa, Omnipaedista, Je ne détiens pas la vérité universelle, Alvin Seville, Black Yoshi, Cstorm44, Helpful Pixie Bot, BG19bot, MusikAnimal, Brad7777, Witny23, Khazar2, The Vintage Feminist, Polipassion, Phoenix 123 abc and Anonymous: 14

- **Herd behavior** *Source:* https://en.wikipedia.org/wiki/Herd_behavior?oldid=722876258 *Contributors:* R Lowry, DennisDaniels, Edward, Mic, Cyde, Dwo, Wetman, Secretlondon, Hajor, PBS, Altenmann, Seglea, AnchoriteBuran, Graeme Bartlett, Andries, Everyking, Snowdog, Michael Devore, Stevietheman, Piotrus, Pgreenfinch, WOT, Abdull, Miborovsky, Rich Farmbrough, Bender235, Dr. Colossus, Illuvatar,, Johnkarp, Reinyday, Ilse@, Animakitty, BlankVerse, Teknic, Jshadias, Jaxhere, Rjwilmsi, Jweiss11, Nneonneo, SchuminWeb, Nihiltres, Amyloo, Psantora, Wzk, VolatileChemical, Phantomsteve, RussBot, Bdoxey, NawlinWiki, Korny O'Near, Rjlabs, Epipelagic, Gadget850, WAS 4.250, Johnny waz, SmackBot, SmartGuy Old, Gilliam, Portillo, Betacommand, Squiddy, Chris the speller, Bluebot, MartinPoulter, Scwlong, WinstonSmith, Kcordina, Radagast83, Lambiam, Adagio Cantabile, ML5, JoshuaZ, JorisvS, Nagle, Santa Sangre, RichardF, JForget, N2e, Penbat, Odie5533, Mattisse, Ikhan85, SnoopJeDi, FirefoxRocks, Alphachimpbot, Davemarshall04, Mcorazao, Jackmass, Rgfolsom, Cpl Syx, Neoprote, Maurice Carbonaro, Memestream, Lawlzors, Xnuala, Jwezel, JohnBlackburne, Fences and windows, Ekac, DSGruss, AllGloryToTheHypnotoad, Wassermann~enwiki, Earthdirt, Tracerbullet11, Goldneje, Spzcb10, Rinconsoleao, Clank.r, Ministry of random walks, ClueBot, SummerWithMorons, Bettinakatz, Auntof6, Aitias, Torrentweb, DumZiBoT, Ost316, Addbot, Some jerk on the Internet, Protonk, Redheylin, Mctp1111, Gail, Konstock, Yobot, Andreasmperu, AnomieBOT, ProtectionTaggingBot, Some standardized rigour, FrescoBot, Markeilz, Trappist the monk, Wotnow, Alph Bot, AvicBot, Scythia, ChuispastonBot, Nirakka, ClueBot NG, MelbourneStar, Widr, Helpful Pixie Bot, Schrödinger's Neurotoxin, M0rphzone, Northamerica1000, Rongrong.shu, Mattgnd4, Praxiphenes, EricEnfermero, BattyBot, Timothy j. thomas, Marknew12, Ginsuloft, Monkbot, Gladamas, WyattAlex, Lynnefox, JRFurmanski and Anonymous: 127

- **Groupthink** *Source:* https://en.wikipedia.org/wiki/Groupthink?oldid=722346942 *Contributors:* Hajhouse, Fubar Obfusco, M~enwiki, Roadrunner, Heron, R Lowry, Mrwojo, Michael Hardy, Mahjongg, MartinHarper, Graue, Lquilter, 6birc, Kosebamse, HarmonicSphere, TUF-KAT, LittleDan, Pizza Puzzle, Mydogategodshat, Dcoetzee, Dino, Hyacinth, Cleduc, Omegatron, Bevo, Vardion, Sander123, GreatWhiteNortherner, Matthew Stannard, Andries, Laudaka, Nat Krause, Fudoreaper, Leflyman, Andris, Taak, Khalid hassani, Andycjp, Popefauvexxiii, The Trolls of Navarone, Quickwik, Augur, MFNickster, Kuralyov, Pgreenfinch, WOT, Robin klein, DMG413, Random account 47, Zondor, Lucidish, Sysy, DanielCD, Macrowiz, Quiensabe, D-Notice, Antaeus Feldspar, Pavel Vozenilek, Bender235, Kelly Ramsey, Shanes, Art LaPella, Bookofjude, Femto, Kremit, Dub4u, Marblespire, Johnkarp, Reinyday, BrokenSegue, Mdd, Mareino, 927, Deboerjo, Arnold S. Truman, ClockworkSoul, Tony Sidaway, Mattbrundage, ZOP, Richard Arthur Norton (1958-), StradivariusTV, Rukkyg, James Kemp, Before My Ken, Benking, Male1979, Stefanomione, Rufous, John Hubbard, BD2412, Dpr, ConradKilroy, CheshireKatz, Rjwilmsi, Ghepeu, Keimzelle, GregAsche, FlaBot, Loplin, Twipley, Nihiltres, JdforresterBot, Pharzo, Eldred, Alexjohnc3, Alphachimp, Bgwhite, Wavelength, Jadon, Djcartwright, Matt Fitzpatrick, Stephenb, Gaius Cornelius, CambridgeBayWeather, NawlinWiki, Deodar~enwiki, D. F. Schmidt, Zwobot, Xompanthy, Roid, Maelgwn, Paul Magnussen, Lindentree, Adamkolson, SMcCandlish, Meegs, Victor falk, That Guy, From That Show!, Sardanaphalus, KnightRider~enwiki, A bit iffy, SmackBot, BluePlatypus, Mercifull, Midway, Brick Thrower, Eskimbot, Ck4829, Wittylama, The Rhymesmith, Tnkr111, Portillo, Frédérick Lacasse, Isnoop, Jbroomfield, Chris the speller, MartinPoulter, Colonies Chris, Antonrojo, Gracenotes, TidyCat, Muboshgu, Jefffire, Anthon.Eff, R!ch, Bigturtle, Film.addict, "alyosha", Paroxysm, Hygelac-enwiki, Kukini, Will Beback, The undertow, Mchavez, Z-d, Acebrock, Stavr0~enwiki, Loodog, Jaganath, Gregorydavid, 16@r, Dezro, Santa Sangre, Peterbr~enwiki, MrArt, Mathsci, Dr.K., Iridescent, Bohica1971, Gungasdindin, Sabina F, Mellery, Penbat, Liu Bei, Gregbard, Treybien, Mike Christie, MC10, Was a bee, Agentilini, Prof. Harris, ThatPeskyCommoner, The Real Jean-Luc, NaLalina, Thijs!bot, Beefpelican, Al Lemos, Deathbunny, Wmconlon, Zachary, CharlotteWebb, Laualoha, Escarbot, The Person Who Is Strange, Yonatan, Luna Santin, Widefox, Turlo Lomon, Miker@sundialservices.com, Jj137, Alphachimpbot, Barek, MLilburne, Albany NY, Igodard, SteveSims, Io Katai, Uniлrab, SHCarter, Father Goose, MatthewJS, Rowsdower45, Cgingold, Ensign beedrill, Acornwithwings, A3nm, ArmadilloFromHell, DerHexer, Dontdoit, Mattjs, TylerQC, KTo288, Lksd~enwiki, Adavidb, USN1977, Cyanolinguophile, MistyMorn, Maurice Carbonaro, NaomiRG, JayJasper, BoredTerry, Resonanteye, Fijahh, Kenneth M Burke, Boombaard, M. Frederick, Gersh uwec, VolkovBot, Gsmcghee, Don Quixote de la Mancha, Daustins, Wassermann~enwiki, B. Jennings Perry, Robertekraut, BigDunc, Andrewaskew, Enigmaman, Groundswell, Bporopat, Kchampcal, Rypcord,

49.5. TEXT AND IMAGE SOURCES, CONTRIBUTORS, AND LICENSES

Newbyguesses, Jean-Louis Swiners, Dogah, Ivan Štambuk, YonaBot, BotMultichill, Leejasonc, Alberjohns, Butters7, Flyer22 Reborn, Tiptoety, Waves00, Smilo Don, Escape Orbit, Gopalkrishnan83, ClueBot, SummerWithMorons, Kai-Hendrik, USAjp22, The Thing That Should Not Be, Padler, Jusdafax, Thrdchip, Wndl42, Rhododendrites, SunnyDisp, Bracton, Eustress, Aprock, GFHandel, Paralipsis, NJGW, Appicharlask, DumZiBoT, XLinkBot, Jytdog, Snowmonster, Borock, PeterWD, M4390116, Addbot, DOI bot, AlbinoFerret, Zellfaze, Martindo, LaaknorBot, SamatBot, Iluvbk, Jarble, Brainstewn, Luckas-bot, Yobot, Gongshow, TestEditBot, Eric-Wester, AnomieBOT, Shootbamboo, Captain Quirk, Panther991, Citation bot, Didaktron, ArthurBot, Bandizzle, Gilo1969, Petropoxy (Lithoderm Proxy), Internoob, ProtectionTaggingBot, 4RugbyRd, Mouseodoom, PigFlu Oink, Winterst, Mimzy1990, Denkealsobin, Serols, Crayon101, Qwertywiki77, Lotje, Swimer91, Awesome763, Clayacan, Klindseth, Kookaburra11, RjwilmsiBot, Loziosam, DASHBot, EmausBot, Orphan Wiki, Rowland606, Mr. Paramecium, Solomonfromfinland, ZéroBot, Pherm, SporkBot, TyA, Morgan Hauser, ChuispastonBot, U3964057, YieldNowOrLose, Miradre, ClueBot NG, JohnsonL623, Snotbot, Manpoi, O.Koslowski, Mr Sheep Measham, Edakavlakoglu, Erinrdelaney, Helpful Pixie Bot, Popcornduff, Wiki Groupthink, Lesterhm, Iconoclastik, Kurt.lockhart, BG19bot, Northamerica1000, Pacerier, PhnomPencil, Planetary Chaos Redux, Cadiomals, Tbirch, IndividualHerd, Ellencavanaugh, Dosakata, Euromatty2, Media-hound- thethird, Cyberbot II, Khazar2, IjonTichyIjonTichy, Raymond1922A, Gunko13, SoledadKabocha, Mogism, Berudagon, RayTayMiht, Paul1andrews, Timothy j. thomas, Poornimacvd, Angiez628, GabeIglesia, Isaacsaccount, Waliway, Toksoz, Myconix, Matthewcplowman, NottNott, AlainCo, Monkbot, Oldbizguy, Lightingbolt50, HaleyB3, LauraC1360, Ihaveacatonmydesk, Cusku'i, Stefani518, Scatsloyal078, Jdw702, PraiseTheShroom and Anonymous: 308

- **Joint attention** *Source:* https://en.wikipedia.org/wiki/Joint_attention?oldid=719042513 *Contributors:* Kku, Bearcat, Bumm13, Chris Howard, Miranche, Fritzpoll, Woohookitty, Rjwilmsi, Abyssal, MrRadioGuy, Mr Stephen, Kencf0618, Falott, Roypea, Neelix, Mirrormundo, Headbomb, JustAGal, Mgierdal, Fabrictramp, Maproom, RiverStyx23, Lova Falk, Doc James, DancingPhilosopher, Mr. Stradivarius, Drmies, Blue bear sd, Epigenetic1, Maky, Addbot, Ironholds, AnomieBOT, Citation bot, LilHelpa, Jeffrey Mall, Tbarrett027, Rotideypoc41352, HRoestBot, Koi.lover, Trappist the monk, Komipfeiffer, Aircorn, Lemieu, Chris857, Pyrsmis, CaroleHenson, Helpful Pixie Bot, Marentette, Canoe1967, Amae2, MathewTownsend, NadRose, LianneAnna, Crella09, Star767, Xiauhhuh, Fafnir1, CV9933 and Anonymous: 7

49.5.2 Images

- File:080205_Brusselator_picture.jpg *Source:* https://upload.wikimedia.org/wikipedia/commons/a/a4/080205_Brusselator_picture.jpg *License:* Public domain *Contributors:* Own work (Original text: *self-made*) *Original artist:* Complexica at English Wikipedia
- File:1-cooling-crystallizer-schladen.JPG *Source:* https://upload.wikimedia.org/wikipedia/commons/e/ee/1-cooling-crystallizer-schladen.JPG *License:* CC BY-SA 3.0 *Contributors:* Own work *Original artist:* Elmschrat Coaching38
- File:10_small_subunit.gif *Source:* https://upload.wikimedia.org/wikipedia/commons/3/3d/10_small_subunit.gif *License:* Public domain *Contributors:* <a data-x-rel='nofollow' class='external text' href='http://www.pdb.org/pdb/static.do?p=education_discussion/molecule_of_the_month/pdb10_1.html'>*Molecule of the Month*</a> at the RCSB Protein Data Bank *Original artist:* Animation by David S. Goodsell, RCSB Protein Data Bank
- File:1r3j.png *Source:* https://upload.wikimedia.org/wikipedia/commons/1/18/1r3j.png *License:* CC-BY-SA-3.0 *Contributors:* Own work *Original artist:* Andrei Lomize
- File:7TM4_(GPCR).png *Source:* https://upload.wikimedia.org/wikipedia/commons/6/60/7TM4_%28GPCR%29.png *License:* Public domain *Contributors:* Transferred from en.wikipedia to Commons by valeryns. *Original artist:* Bensaccount at English Wikipedia
- File:ACBP_MSM_from_Folding@home.tiff *Source:* https://upload.wikimedia.org/wikipedia/commons/b/b9/ACBP_MSM_from_Folding%40home.tiff *License:* CC BY-SA 3.0 *Contributors:* Sent to the uploader personally *Original artist:* Vincent Voelz
- File:Adam_Savage_HOPE.jpg *Source:* https://upload.wikimedia.org/wikipedia/commons/4/46/Adam_Savage_HOPE.jpg *License:* CC BY-SA 3.0 *Contributors:* Transferred from en.wikipedia *Original artist:* Original uploader was Porkrind at en.wikipedia
- File:Aegopodium_podagraria1_ies.jpg *Source:* https://upload.wikimedia.org/wikipedia/commons/b/bf/Aegopodium_podagraria1_ies.jpg *License:* CC-BY-SA-3.0 *Contributors:* Own work *Original artist:* Frank Vincentz
- File:Airplane_vortex_edit.jpg *Source:* https://upload.wikimedia.org/wikipedia/commons/f/fe/Airplane_vortex_edit.jpg *License:* Public domain *Contributors:* This image or video was catalogued by Langley Research Center of the United States National Aeronautics and Space Administration (NASA) under **Photo ID**: EL-1996-00130 AND **Alternate ID**: L90-5919.
Original artist: NASA Langley Research Center (NASA-LaRC), Edited by Fir0002
- File:Aleksandr_Oparin_and_Andrei_Kursanov_in_enzymology_laboratory_1938.jpg *Source:* https://upload.wikimedia.org/wikipedia/commons/f/f4/Aleksandr_Oparin_and_Andrei_Kursanov_in_enzymology_laboratory_1938.jpg *License:* Public domain *Contributors:* ? *Original artist:* ?
- File:Ambox_important.svg *Source:* https://upload.wikimedia.org/wikipedia/commons/b/b4/Ambox_important.svg *License:* Public domain *Contributors:* Own work, based off of Image:Ambox scales.svg *Original artist:* Dsmurat (talk · contribs)
- File:Annular_Gap_Junction_Vesicle.jpg *Source:* https://upload.wikimedia.org/wikipedia/commons/7/7b/Annular_Gap_Junction_Vesicle.jpg *License:* Public domain *Contributors:* Own work *Original artist:* Sandraamurray
- File:ArealVelocity.svg *Source:* https://upload.wikimedia.org/wikipedia/commons/9/9b/ArealVelocity.svg *License:* CC-BY-SA-3.0 *Contributors:* Transferred from en.wikipedia to Commons. *Original artist:* The original uploader was Xyzzy n at English Wikipedia
- File:Argon_ice_1.jpg *Source:* https://upload.wikimedia.org/wikipedia/commons/0/0d/Argon_ice_1.jpg *License:* CC-BY-SA-3.0 *Contributors:* No machine-readable source provided. Own work assumed (based on copyright claims). *Original artist:* No machine-readable author provided. Deglr6328~commonswiki assumed (based on copyright claims).
- File:Auklet_flock_Shumagins_1986.jpg *Source:* https://upload.wikimedia.org/wikipedia/commons/5/5e/Auklet_flock_Shumagins_1986.jpg *License:* Public domain *Contributors:* images.fws.gov ([1]) *Original artist:* D. Dibenski
- File:BML_N=200_P=32.png *Source:* https://upload.wikimedia.org/wikipedia/commons/f/f6/BML_N%3D200_P%3D32.png *License:* CC BY-SA 3.0 *Contributors:* Own work *Original artist:* Purpy Pupple

- File:Bacillus_subtilis.jpg *Source:* https://upload.wikimedia.org/wikipedia/commons/9/91/Bacillus_subtilis.jpg *License:* Public domain *Contributors:* Transferred from en.wikipedia to Commons by alnokta. *Original artist:* Allonweiner at English Wikipedia
- File:Bangkok_skytrain_sunset.jpg *Source:* https://upload.wikimedia.org/wikipedia/commons/f/f6/Bangkok_skytrain_sunset.jpg *License:* CC-BY-SA-3.0 *Contributors:* Own work *Original artist:* User:Diliff
- File:Barnsley_fern_plotted_with_VisSim.PNG *Source:* https://upload.wikimedia.org/wikipedia/commons/7/76/Barnsley_fern_plotted_with_VisSim.PNG *License:* CC BY-SA 3.0 *Contributors:* Own work, using model written by Mike Borrello *Original artist:* DSP-user
- File:Bellcurve.svg *Source:* https://upload.wikimedia.org/wikipedia/commons/d/df/Bellcurve.svg *License:* Copyrighted free use *Contributors:* ? *Original artist:* ?
- File:Bilayer_AFM_schematic.png *Source:* https://upload.wikimedia.org/wikipedia/commons/4/44/Bilayer_AFM_schematic.png *License:* Public domain *Contributors:* Own work *Original artist:* MDougM
- File:Bilayer_hydration_profile.svg *Source:* https://upload.wikimedia.org/wikipedia/commons/e/ed/Bilayer_hydration_profile.svg *License:* Public domain *Contributors:* Own work *Original artist:* MDougM
- File:Birmingham_Northern_Rock_bank_run_2007.jpg *Source:* https://upload.wikimedia.org/wikipedia/commons/e/e5/Birmingham_Northern_Rock_bank_run_2007.jpg *License:* CC BY-SA 2.0 *Contributors:* Flickr *Original artist:* Lee Jordan
- File:Blacksmoker_in_Atlantic_Ocean.jpg *Source:* https://upload.wikimedia.org/wikipedia/commons/6/6f/Blacksmoker_in_Atlantic_Ocean.jpg *License:* Public domain *Contributors:* NOAA Photo Library *Original artist:* P. Rona
- File:Br4Py_self-assembly_on_Au.jpg *Source:* https://upload.wikimedia.org/wikipedia/commons/0/0a/Br4Py_self-assembly_on_Au.jpg *License:* CC BY 4.0 *Contributors:* [1] (article supplement) *Original artist:* Tuan Anh Pham et al.
- File:Br4Py_self-assembly_on_Au_2.jpg *Source:* https://upload.wikimedia.org/wikipedia/commons/1/13/Br4Py_self-assembly_on_Au_2.jpg *License:* CC BY 4.0 *Contributors:* [1] (article supplement) *Original artist:* Tuan Anh Pham et al.
- File:Brownian_tree_vertical_large.png *Source:* https://upload.wikimedia.org/wikipedia/commons/2/2d/Brownian_tree_vertical_large.png *License:* CC-BY-SA-3.0 *Contributors:* ? *Original artist:* ?
- File:Buckminsterfullerene-perspective-3D-balls.png *Source:* https://upload.wikimedia.org/wikipedia/commons/0/0f/Buckminsterfullerene-perspective-3D-balls.png *License:* Public domain *Contributors:* Own work *Original artist:* Benjah-bmm27
- File:CA-Moore.png *Source:* https://upload.wikimedia.org/wikipedia/en/d/d2/CA-Moore.png *License:* CC0 *Contributors:*

    Own work

    Original artist:

    Torchiest (talk) (Uploads)
- File:CA-von-Neumann.png *Source:* https://upload.wikimedia.org/wikipedia/en/5/56/CA-von-Neumann.png *License:* CC0 *Contributors:*

    Own work

    Original artist:

    Torchiest (talk) (Uploads)
- File:CA_rule110s.png *Source:* https://upload.wikimedia.org/wikipedia/commons/f/fa/CA_rule110s.png *License:* CC0 *Contributors:* Own work by the original uploader *Original artist:* Grondilu (talk) (Uploads)
- File:CA_rule30s.png *Source:* https://upload.wikimedia.org/wikipedia/commons/9/9d/CA_rule30s.png *License:* CC-BY-SA-3.0 *Contributors:* ? *Original artist:* ?
- File:Causeway-code_poet-4.jpg *Source:* https://upload.wikimedia.org/wikipedia/commons/c/c0/Causeway-code_poet-4.jpg *License:* CC BY-SA 2.0 *Contributors:* http://www.flickr.com/photos/alphageek/20005235/ *Original artist:* code poet on flickr.
- File:CentralTendencyLV.jpg *Source:* https://upload.wikimedia.org/wikipedia/en/6/60/CentralTendencyLV.jpg *License:* PD *Contributors:*

    Own work

    Original artist:

    Elb2000 (talk) (Uploads)
- File:Champagne_vent_white_smokers.jpg *Source:* https://upload.wikimedia.org/wikipedia/commons/a/aa/Champagne_vent_white_smokers.jpg *License:* Public domain *Contributors:* http://oceanexplorer.noaa.gov/explorations/04fire/logs/hirez/champagne_vent_hirez.jpg *Original artist:* NOAA
- File:Chaos_Sensitive_Dependence.svg *Source:* https://upload.wikimedia.org/wikipedia/commons/8/8e/Chaos_Sensitive_Dependence.svg *License:* Public domain *Contributors:* Own work *Original artist:* Radagast3
- File:Chaos_Topological_Mixing.png *Source:* https://upload.wikimedia.org/wikipedia/commons/d/dc/Chaos_Topological_Mixing.png *License:* Public domain *Contributors:* Own work *Original artist:* Radagast3
- File:ChemSepProcDiagram.svg *Source:* https://upload.wikimedia.org/wikipedia/commons/f/f9/ChemSepProcDiagram.svg *License:* Public domain *Contributors:* No machine-readable source provided. Own work assumed (based on copyright claims). *Original artist:* No machine-readable author provided. Slashme assumed (based on copyright claims).
- File:Cholesterinisch.png *Source:* https://upload.wikimedia.org/wikipedia/commons/d/d2/Cholesterinisch.png *License:* CC-BY-SA-3.0 *Contributors:* retirved form german Wikipedia originaly uploaded on 00:55, 19. Jan 2005 *Original artist:* de:Benutzer:Heimoponnath
- File:Cholesteryl_benzoate.png *Source:* https://upload.wikimedia.org/wikipedia/commons/9/9a/Cholesteryl_benzoate.png *License:* Public domain *Contributors:* ? *Original artist:* ?
- File:Circle_map_bifurcation.jpeg *Source:* https://upload.wikimedia.org/wikipedia/commons/a/a5/Circle_map_bifurcation.jpeg *License:* CC-BY-SA-3.0 *Contributors:* ? *Original artist:* ?
- File:CitricAcid_Crystalisation_Timelapse.ogg *Source:* https://upload.wikimedia.org/wikipedia/commons/9/96/CitricAcid_Crystalisation_Timelapse.ogg *License:* CC BY-SA 3.0 *Contributors:* Own work *Original artist:* Zephyris

49.5. TEXT AND IMAGE SOURCES, CONTRIBUTORS, AND LICENSES

- File:Click_Here._No,_Here.jpg *Source:* https://upload.wikimedia.org/wikipedia/commons/1/14/Click_Here._No%2C_Here.jpg *License:* CC BY 2.0 *Contributors:* Click Here. No, Here: 23/09/06 *Original artist:* Ken Banks from Los Altos, USA
- File:ColloidCrystal_40xBrightField_GlassInWater.jpg *Source:* https://upload.wikimedia.org/wikipedia/commons/5/52/ColloidCrystal_40xBrightField_GlassInWater.jpg *License:* CC BY-SA 3.0 *Contributors:* Own work *Original artist:* Zephyris
- File:ColloidCrystal_40xBrightField_GlassInWater_Connectivity.png *Source:* https://upload.wikimedia.org/wikipedia/commons/c/c4/ColloidCrystal_40xBrightField_GlassInWater_Connectivity.png *License:* CC BY-SA 3.0 *Contributors:* Own work *Original artist:* Zephyris
- File:Commons-logo.svg *Source:* https://upload.wikimedia.org/wikipedia/en/4/4a/Commons-logo.svg *License:* CC-BY-SA-3.0 *Contributors:* ? *Original artist:* ?
- File:Comparison_carbon_dioxide_water_phase_diagrams.svg *Source:* https://upload.wikimedia.org/wikipedia/commons/4/40/Comparison_carbon_dioxide_water_phase_diagrams.svg *License:* CC BY-SA 3.0 *Contributors:* Own work *Original artist:* Cmglee
- File:Complex-adaptive-system.jpg *Source:* https://upload.wikimedia.org/wikipedia/commons/0/00/Complex-adaptive-system.jpg *License:* Public domain *Contributors:* Own work by Acadac : Taken from en.wikipedia.org, where Acadac was inspired to create this graphic after reading: *Original artist:* Acadac
- File:ConvectionCells.svg *Source:* https://upload.wikimedia.org/wikipedia/commons/f/f5/ConvectionCells.svg *License:* CC-BY-SA-3.0 *Contributors:* Own work *Original artist:* Eyrian Con-struct
- File:Copolymers.svg *Source:* https://upload.wikimedia.org/wikipedia/commons/d/d6/Copolymers.svg *License:* CC-BY-SA-3.0 *Contributors:* Uppladdarens egna verk (uploader's own work); SVG-version of Image:Copolymers.png; original version at en:Image:Copolymers.png *Original artist:* Mankash (<a href='//commons.wikimedia.org/wiki/User_talk:Mankash' title='User talk:Mankash'>talk</a>); original image by en:User:V8rik
- File:Crab_Nebula.jpg *Source:* https://upload.wikimedia.org/wikipedia/commons/0/00/Crab_Nebula.jpg *License:* Public domain *Contributors:* HubbleSite: gallery, release. *Original artist:* NASA, ESA, J. Hester and A. Loll (Arizona State University)
- File:CriticalPointMeasurementEthane.jpg *Source:* https://upload.wikimedia.org/wikipedia/commons/e/e5/CriticalPointMeasurementEthane.jpg *License:* CC BY-SA 3.0 *Contributors:* Persönlich übergeben von Sven Horstmann. In ähnlicher Form verwendet in: Horstmann S., "Theoretische und experimentelle Untersuchungen zum Hochdruckphasengleichgewichtsverhalten fluider Stoffgemische für die Erweiterung der PSRK-Gruppenbeitragszustandsgleichung", Doktorabeit, C.-v.-O. Universität Oldenburg, 2000 *Original artist:* Dr. Sven Horstmann
- File:Crystal_growth.PNG *Source:* https://upload.wikimedia.org/wikipedia/commons/a/a5/Crystal_growth.PNG *License:* CC-BY-SA-3.0 *Contributors:* ? *Original artist:* ?
- File:Crystallized_honey.jpg *Source:* https://upload.wikimedia.org/wikipedia/commons/3/38/Crystallized_honey.jpg *License:* Public domain *Contributors:* Own work *Original artist:* Stevo-88
- File:Cône_textileII.png *Source:* https://upload.wikimedia.org/wikipedia/commons/a/ae/C%C3%B4ne_textileII.png *License:* CC BY-SA 3.0 *Contributors:* Own work *Original artist:* Didier Descouens
- File:DLA_Cluster.JPG *Source:* https://upload.wikimedia.org/wikipedia/commons/b/b8/DLA_Cluster.JPG *License:* CC BY 2.5 *Contributors:* See Author *Original artist:* Kevin R Johnson
- File:DLA_spiral.png *Source:* https://upload.wikimedia.org/wikipedia/commons/5/5b/DLA_spiral.png *License:* CC BY 2.0 *Contributors:* http://toxiclibs.org/2010/02/new-package-simutils/ *Original artist:* Karsten Schmidt
- File:DNA_nanostructures.png *Source:* https://upload.wikimedia.org/wikipedia/commons/5/55/DNA_nanostructures.png *License:* CC BY 2.5 *Contributors:* Strong M: *Protein Nanomachines.* PLoS Biol 2/3/2004: e73. http://dx.doi.org/10.1371/journal.pbio.0020073 *Original artist:* (Images were kindly provided by Thomas H. LaBean and Hao Yan.)
- File:DTB_2.PNG *Source:* https://upload.wikimedia.org/wikipedia/commons/e/e6/DTB_2.PNG *License:* CC-BY-SA-3.0 *Contributors:* Unknown *Original artist:* Ruben Castelnuovo (myself)
- File:DTB_Xls.png *Source:* https://upload.wikimedia.org/wikipedia/commons/b/b0/DTB_Xls.png *License:* CC-BY-SA-3.0 *Contributors:* ? *Original artist:* ?
- File:Darwin_restored2.jpg *Source:* https://upload.wikimedia.org/wikipedia/commons/b/b6/Darwin_restored2.jpg *License:* Public domain *Contributors:* Library of Congress[1] *Original artist:* Elliott & Fry
- File:Diagram_of_the_Monitor-Analyse-Record-Reflect-Reconstruct-Review-Spiral_algorithm.jpg *Source:* https://upload.wikimedia.org/wikipedia/commons/b/b8/Diagram_of_the_Monitor-Analyse-Record-Reflect-Reconstruct-Review-Spiral_algorithm.jpg *License:* CC BY-SA 3.0 *Contributors:* Created by Laurie F. Thomas and published in the book "Learning Conversations" and published by Routledge, copyright has now returned to the Author(s) *Original artist:* Soler99
- File:Double-compound-pendulum.gif *Source:* https://upload.wikimedia.org/wikipedia/commons/4/45/Double-compound-pendulum.gif *License:* Public domain *Contributors:* Own work *Original artist:* Catslash
- File:Drugroutemap.gif *Source:* https://upload.wikimedia.org/wikipedia/commons/6/64/Drugroutemap.gif *License:* Public domain *Contributors:* CIA Employee *Original artist:* CIA Employee
- File:Edit-clear.svg *Source:* https://upload.wikimedia.org/wikipedia/en/f/f2/Edit-clear.svg *License:* Public domain *Contributors:* The *Tango! Desktop Project.* *Original artist:*
The people from the Tango! project. And according to the meta-data in the file, specifically: "Andreas Nilsson, and Jakub Steiner (although minimally)."
- File:Emblem-money.svg *Source:* https://upload.wikimedia.org/wikipedia/commons/f/f3/Emblem-money.svg *License:* GPL *Contributors:* http://www.gnome-look.org/content/show.php/GNOME-colors?content=82562 *Original artist:* perfectska04
- File:Error-threshold.svg *Source:* https://upload.wikimedia.org/wikipedia/commons/8/87/Error-threshold.svg *License:* CC BY-SA 4.0 *Contributors:* Own work *Original artist:* Nszostak
- File:Ethanol-3D-balls.png *Source:* https://upload.wikimedia.org/wikipedia/commons/b/b0/Ethanol-3D-balls.png *License:* Public domain *Contributors:* ? *Original artist:* ?

- File:Exocytosis-machinery.jpg *Source:* https://upload.wikimedia.org/wikipedia/commons/9/90/Exocytosis-machinery.jpg *License:* CC-BY-SA-3.0 *Contributors:* http://en.wikipedia.org/wiki/Image:Exocytosis-machinery.jpg *Original artist:* Danko Dimchev Georgiev, M.D.

- File:Figure1a.gif *Source:* https://upload.wikimedia.org/wikipedia/en/a/a5/Figure1a.gif *License:* CC-BY-SA-3.0 *Contributors:* ? *Original artist:* ?

- File:Figure1b.gif *Source:* https://upload.wikimedia.org/wikipedia/en/0/01/Figure1b.gif *License:* CC-BY-SA-3.0 *Contributors:*

    Own work

    Original artist:

    Dggreen (talk) (Uploads)

- File:Folder_Hexagonal_Icon.svg *Source:* https://upload.wikimedia.org/wikipedia/en/4/48/Folder_Hexagonal_Icon.svg *License:* Cc-by-sa-3.0 *Contributors:* ? *Original artist:* ?

- File:Formation_of_Glycolaldehyde_in_star_dust.png *Source:* https://upload.wikimedia.org/wikipedia/commons/4/46/Formation_of_Glycolaldehyde_in_star_dust.png *License:* Public domain *Contributors:* NASA *Original artist:* Lara Clemence

- File:Fractal_fern_explained.png *Source:* https://upload.wikimedia.org/wikipedia/commons/4/4b/Fractal_fern_explained.png *License:* Public domain *Contributors:* Own work *Original artist:* António Miguel de Campos

- File:Fugle._oruso_073.jpg *Source:* https://upload.wikimedia.org/wikipedia/commons/d/d6/Fugle%2C_%C3%B8rns%C3%B8_073.jpg *License:* Public domain *Contributors:* Own work *Original artist:* Christoffer A Rasmussen (Rasmussen29892 at da.wikipedia)

- File:Fullerene_Nanogears_-_GPN-2000-001535.jpg *Source:* https://upload.wikimedia.org/wikipedia/commons/b/b6/Fullerene_Nanogears_-_GPN-2000-001535.jpg *License:* Public domain *Contributors:* Great Images in NASA: Home - info - pic *Original artist:* NASA

- File:Garni_Gorge3.jpg *Source:* https://upload.wikimedia.org/wikipedia/commons/3/32/Garni_Gorge3.jpg *License:* CC BY-SA 3.0 *Contributors:* for-wikimedia.wowarmenia.ru *Original artist:* uncredited

- File:Gospers_glider_gun.gif *Source:* https://upload.wikimedia.org/wikipedia/commons/e/e5/Gospers_glider_gun.gif *License:* CC-BY-SA-3.0 *Contributors:* Own work *Original artist:* Kieff

- File:Herdwick_Stampede.jpg *Source:* https://upload.wikimedia.org/wikipedia/commons/e/e8/Herdwick_Stampede.jpg *License:* CC BY 2.0 *Contributors:* Stampede! *Original artist:* Andy Docker from England

- File:Homebrew_reaction_diffusion_example_512iter.jpg *Source:* https://upload.wikimedia.org/wikipedia/en/6/67/Homebrew_reaction_diffusion_example_512iter.jpg *License:* PD *Contributors:* ? *Original artist:* ?

- File:Horizonte_inflacionario.svg *Source:* https://upload.wikimedia.org/wikipedia/commons/b/b4/Horizonte_inflacionario.svg *License:* CC-BY-SA-3.0 *Contributors:* Transferred from en.wikipedia to Commons.; original: *I created this work in Adobe Illustrator*. *Original artist:* Joke137 at English Wikipedia

- File:Hypercycle-eigen.svg *Source:* https://upload.wikimedia.org/wikipedia/commons/c/c3/Hypercycle-eigen.svg *License:* CC BY-SA 4.0 *Contributors:* Own work *Original artist:* Nszostak

- File:Hypercycle2.svg *Source:* https://upload.wikimedia.org/wikipedia/commons/9/96/Hypercycle2.svg *License:* CC BY-SA 4.0 *Contributors:* Own work *Original artist:* Nszostak

- File:Hysteresis_sharp_curve.svg *Source:* https://upload.wikimedia.org/wikipedia/commons/a/aa/Hysteresis_sharp_curve.svg *License:* CC-BY-SA-3.0 *Contributors:* Own work *Original artist:* Alessio Damato

- File:Ideal_feedback_model.svg *Source:* https://upload.wikimedia.org/wikipedia/commons/e/ed/Ideal_feedback_model.svg *License:* Public domain *Contributors:* Own work *Original artist:* Me (Intgr)

- File:Ilc_9yr_moll4096.png *Source:* https://upload.wikimedia.org/wikipedia/commons/3/3c/Ilc_9yr_moll4096.png *License:* Public domain *Contributors:* http://map.gsfc.nasa.gov/media/121238/ilc_9yr_moll4096.png *Original artist:* NASA / WMAP Science Team

- File:IntelligentAgent-Learning.png *Source:* https://upload.wikimedia.org/wikipedia/commons/0/09/IntelligentAgent-Learning.png *License:* Public domain *Contributors:* Author *Original artist:* Utkarshraj Atmaram

- File:IntelligentAgent-SimpleReflex.png *Source:* https://upload.wikimedia.org/wikipedia/commons/3/3f/IntelligentAgent-SimpleReflex.png *License:* Public domain *Contributors:* Author *Original artist:* Utkarshraj Atmaram

- File:Iron_oxide_nanocube.jpg *Source:* https://upload.wikimedia.org/wikipedia/commons/0/02/Iron_oxide_nanocube.jpg *License:* CC BY 3.0 *Contributors:* http://iopscience.iop.org/1468-6996/15/5/055010/article *Original artist:* Erik Wetterskog et al.

- File:Issoria_lathonia.jpg *Source:* https://upload.wikimedia.org/wikipedia/commons/2/2d/Issoria_lathonia.jpg *License:* CC-BY-SA-3.0 *Contributors:* ? *Original artist:* ?

- File:JerkCircuit01.png *Source:* https://upload.wikimedia.org/wikipedia/commons/8/87/JerkCircuit01.png *License:* Public domain *Contributors:* No machine-readable source provided. Own work assumed (based on copyright claims). *Original artist:* No machine-readable author provided. PAR~commonswiki assumed (based on copyright claims).

- File:John_von_Neumann_ID_badge.png *Source:* https://upload.wikimedia.org/wikipedia/commons/d/d9/John_von_Neumann_ID_badge.png *License:* Public domain *Contributors:* ? *Original artist:* ?

- File:Kristalizacija.jpg *Source:* https://upload.wikimedia.org/wikipedia/commons/d/d9/Kristalizacija.jpg *License:* CC BY-SA 4.0 *Contributors:* Own work *Original artist:* Taki Jo

- File:LCDM.jpg *Source:* https://upload.wikimedia.org/wikipedia/commons/7/7d/LCDM.jpg *License:* CC BY-SA 3.0 *Contributors:* Own work *Original artist:* Michael L. Umbricht

- File:LCD_layers.svg *Source:* https://upload.wikimedia.org/wikipedia/commons/d/dc/LCD_layers.svg *License:* CC-BY-SA-3.0 *Contributors:* No machine-readable source provided. Own work assumed (based on copyright claims). *Original artist:* No machine-readable author provided. Ed g2s assumed (based on copyright claims).

49.5. TEXT AND IMAGE SOURCES, CONTRIBUTORS, AND LICENSES

- File:LT-SEM_snow_crystal_magnification_series-3.jpg *Source:* https://upload.wikimedia.org/wikipedia/commons/a/a8/LT-SEM_snow_crystal_magnification_series-3.jpg *License:* Public domain *Contributors:* ? *Original artist:* ?
- File:Levels_of_Organization.svg *Source:* https://upload.wikimedia.org/wikipedia/commons/3/38/Levels_of_Organization.svg *License:* CC BY-SA 3.0 *Contributors:* Own work *Original artist:* Mikala14
- File:Lfg_polymer_milk_(cropped).jpg *Source:* https://upload.wikimedia.org/wikipedia/commons/f/fa/Lfg_polymer_milk_%28cropped%29.jpg *License:* CC BY 3.0 *Contributors:* File:Lfg polymer milk.jpg *Original artist:* Achim Hering
- File:Lipid_bilayer_and_micelle.svg *Source:* https://upload.wikimedia.org/wikipedia/commons/c/c8/Lipid_bilayer_and_micelle.svg *License:* CC-BY-SA-3.0 *Contributors:*
- Lipid_bilayer_and_micelle.png *Original artist:* Lipid_bilayer_and_micelle.png: en:User:Stephen Gilbert
- File:Lipid_bilayer_fusion.svg *Source:* https://upload.wikimedia.org/wikipedia/commons/0/05/Lipid_bilayer_fusion.svg *License:* Public domain *Contributors:* Own work *Original artist:* MDougM
- File:Lipid_bilayer_section.gif *Source:* https://upload.wikimedia.org/wikipedia/commons/f/f0/Lipid_bilayer_section.gif *License:* Public domain *Contributors:* http://en.wikipedia.org/wiki/Image:Lipid_bilayer_section.gif *Original artist:* Bensaccount
- File:Lipid_unsaturation_effect.svg *Source:* https://upload.wikimedia.org/wikipedia/commons/2/2e/Lipid_unsaturation_effect.svg *License:* Public domain *Contributors:* Own work *Original artist:* MDougM
- File:LiquidCrystal-MesogenOrder-ChiralPhases.jpg *Source:* https://upload.wikimedia.org/wikipedia/commons/b/bf/LiquidCrystal-MesogenOrder-ChiralPhases.jpg *License:* CC BY-SA 3.0 *Contributors:* Own work *Original artist:* Kebes
- File:LiquidCrystal-MesogenOrder-Nematic.jpg *Source:* https://upload.wikimedia.org/wikipedia/commons/8/80/LiquidCrystal-MesogenOrder-Nematic.jpg *License:* CC BY-SA 3.0 *Contributors:* Own work *Original artist:* Kebes
- File:LiquidCrystal-MesogenOrder-SmecticPhases.jpg *Source:* https://upload.wikimedia.org/wikipedia/commons/f/f2/LiquidCrystal-MesogenOrder-SmecticPhases.jpg *License:* CC BY-SA 3.0 *Contributors:* Own work *Original artist:* Kebes
- File:LogisticMap_BifurcationDiagram.png *Source:* https://upload.wikimedia.org/wikipedia/commons/7/7d/LogisticMap_BifurcationDiagram.png *License:* Public domain *Contributors:* Own work *Original artist:* PAR
- File:Logo_sociology.svg *Source:* https://upload.wikimedia.org/wikipedia/commons/a/a6/Logo_sociology.svg *License:* Public domain *Contributors:* Own work *Original artist:* Tomeq183
- File:LorenzCoordinatesSmall.jpg *Source:* https://upload.wikimedia.org/wikipedia/commons/1/12/LorenzCoordinatesSmall.jpg *License:* CC BY-SA 2.5 *Contributors:* ? *Original artist:* ?
- File:Lorenz_attractor_yb.svg *Source:* https://upload.wikimedia.org/wikipedia/commons/5/5b/Lorenz_attractor_yb.svg *License:* CC-BY-SA-3.0 *Contributors:* Own work based on images Image:Lorenz system r28 s10 b2-6666.png by User:Wikimol and Image:Lorenz attractor.svg by User:Dschwen *Original artist:* User:Wikimol, User:Dschwen
- File:MBBA.svg *Source:* https://upload.wikimedia.org/wikipedia/commons/f/f0/MBBA.svg *License:* CC0 *Contributors:* Own work *Original artist:* Jkwchui
- File:Manual_coffee_preperation.jpg *Source:* https://upload.wikimedia.org/wikipedia/commons/9/92/Manual_coffee_preperation.jpg *License:* CC BY-SA 2.0 *Contributors:* Flickr *Original artist:* miheco from California, USA
- File:Membrane_fusion_via_stalk_formation.jpg *Source:* https://upload.wikimedia.org/wikipedia/commons/1/1c/Membrane_fusion_via_stalk_formation.jpg *License:* Public domain *Contributors:* Own work *Original artist:* a
- File:Mergefrom.svg *Source:* https://upload.wikimedia.org/wikipedia/commons/0/0f/Mergefrom.svg *License:* Public domain *Contributors:* ? *Original artist:* ?
- File:Meta-stability.svg *Source:* https://upload.wikimedia.org/wikipedia/commons/a/a0/Meta-stability.svg *License:* CC-BY-SA-3.0 *Contributors:* self made drawing *Original artist:* Georg Wiora (Dr. Schorsch)
- File:Methane-2D-stereo.svg *Source:* https://upload.wikimedia.org/wikipedia/commons/9/92/Methane-2D-stereo.svg *License:* Public domain *Contributors:* Own work *Original artist:* SVG version by Patricia.fidi
- File:Micelle_scheme-en.svg *Source:* https://upload.wikimedia.org/wikipedia/commons/4/4d/Micelle_scheme-en.svg *License:* CC BY-SA 3.0 *Contributors:* Own work *Original artist:* SuperManu
- File:MtBaker-Chair8-Top.jpg *Source:* https://upload.wikimedia.org/wikipedia/commons/a/a9/MtBaker-Chair8-Top.jpg *License:* CC BY 3.0 *Contributors:* Own work *Original artist:* Wavepacket
- File:NTCDI_AFM2a.jpg *Source:* https://upload.wikimedia.org/wikipedia/commons/0/0c/NTCDI_AFM2a.jpg *License:* CC BY 3.0 *Contributors:* http://www.nature.com/ncomms/2014/140530/ncomms4931/full/ncomms4931.html *Original artist:* A. M. Sweetman et al.
- File:Nb3O7(OH)_self-organization2.jpg *Source:* https://upload.wikimedia.org/wikipedia/commons/3/3f/Nb3O7%28OH%29_self-organization2.jpg *License:* CC BY 3.0 *Contributors:* http://pubs.rsc.org/en/content/articlehtml/2014/ta/c4ta02202e *Original artist:* Sophia B. Betzler et al.
- File:Nematic-Director.png *Source:* https://upload.wikimedia.org/wikipedia/commons/e/e7/Nematic-Director.png *License:* CC BY-SA 3.0 *Contributors:* Own work *Original artist:* Panjasan
- File:Nematische_Phase_Schlierentextur.jpg *Source:* https://upload.wikimedia.org/wikipedia/commons/e/ec/Nematische_Phase_Schlierentextur.jpg *License:* CC-BY-SA-3.0 *Contributors:* Own work *Original artist:* Minutemen
- File:Nitrous-oxide-3D-balls.png *Source:* https://upload.wikimedia.org/wikipedia/commons/9/93/Nitrous-oxide-3D-balls.png *License:* Public domain *Contributors:* Own work *Original artist:* Ben Mills
- File:Nuvola_apps_edu_mathematics_blue-p.svg *Source:* https://upload.wikimedia.org/wikipedia/commons/3/3e/Nuvola_apps_edu_mathematics_blue-p.svg *License:* GPL *Contributors:* Derivative work from Image:Nuvola apps edu mathematics.png and Image:Nuvola apps edu mathematics-p.svg *Original artist:* David Vignoni (original icon); Flamurai (SVG convertion); bayo (color)

- File:OMV-macrophage99.jpg *Source:* https://upload.wikimedia.org/wikipedia/commons/4/44/OMV-macrophage99.jpg *License:* CC BY-SA 3.0 *Contributors:* Own work *Original artist:* Rakeshyashroy
- File:Of7_p0001_15h.jpg *Source:* https://upload.wikimedia.org/wikipedia/commons/3/3d/Of7_p0001_15h.jpg *License:* Public domain *Contributors:* Own work *Original artist:* WingkLEE
- File:Office-book.svg *Source:* https://upload.wikimedia.org/wikipedia/commons/a/a8/Office-book.svg *License:* Public domain *Contributors:* This and myself. *Original artist:* Chris Down/Tango project
- File:Op-Amp_Schmitt_Trigger.svg *Source:* https://upload.wikimedia.org/wikipedia/commons/6/64/Op-Amp_Schmitt_Trigger.svg *License:* Public domain *Contributors:* Own work *Original artist:* Inductiveload
- File:Open_Access_logo_PLoS_transparent.svg *Source:* https://upload.wikimedia.org/wikipedia/commons/7/77/Open_Access_logo_PLoS_transparent.svg *License:* CC0 *Contributors:* http://www.plos.org/ *Original artist:* art designer at PLoS, modified by Wikipedia users Nina, Beao, and JakobVoss
- File:Oscillator.gif *Source:* https://upload.wikimedia.org/wikipedia/commons/8/86/Oscillator.gif *License:* CC-BY-SA-3.0 *Contributors:* Transferred from en.wikipedia to Commons. Self-made with Java program. *Original artist:* Grontesca at English Wikipedia
- File:Otto_Lehmann.jpg *Source:* https://upload.wikimedia.org/wikipedia/commons/5/51/Otto_Lehmann.jpg *License:* Public domain *Contributors:* Von Einfällen und Zufällen. 2008, ISBN 978-3-9812294-1-7 Seite 17 *Original artist:* Unknown<a href='//www.wikidata.org/wiki/Q4233718' title='wikidata:Q4233718'><img alt='wikidata:Q4233718' src='https://upload.wikimedia.org/wikipedia/commons/thumb/f/ff/Wikidata-logo.svg/20px-Wikidata-logo.svg.png' width='20' height='11' srcset='https://upload.wikimedia.org/wikipedia/commons/thumb/f/ff/Wikidata-logo.svg/30px-Wikidata-logo.svg.png 1.5x, https://upload.wikimedia.org/wikipedia/commons/thumb/f/ff/Wikidata-logo.svg/40px-Wikidata-logo.svg.png 2x' data-file-width='1050' data-file-height='590' /></a>
- File:P-A-Char_surfactant.jpg *Source:* https://upload.wikimedia.org/wikipedia/commons/7/7a/P-A-Char_surfactant.jpg *License:* Public domain *Contributors:* Own work *Original artist:* Major measure
- File:P19_cell_sorting_out.png *Source:* https://upload.wikimedia.org/wikipedia/commons/2/26/P19_cell_sorting_out.png *License:* GFDL *Contributors:* Own work *Original artist:* JWSchmidt
- File:People_icon.svg *Source:* https://upload.wikimedia.org/wikipedia/commons/3/37/People_icon.svg *License:* CC0 *Contributors:* OpenClipart *Original artist:* OpenClipart
- File:Pfeil_SO.svg *Source:* https://upload.wikimedia.org/wikipedia/commons/a/a1/Pfeil_SO.svg *License:* Public domain *Contributors:* made by me (Inkscape or Corel-Draw or Flash) *Original artist:* user:Mjchael
- File:Phanerozoic_Biodiversity.svg *Source:* https://upload.wikimedia.org/wikipedia/commons/4/4d/Phanerozoic_Biodiversity.svg *License:* CC-BY-SA-3.0 *Contributors:* Phanerozoic_Biodiversity.png *Original artist:* SVG version by Albert Mestre
- File:Phase-diag2.svg *Source:* https://upload.wikimedia.org/wikipedia/commons/3/34/Phase-diag2.svg *License:* CC-BY-SA-3.0 *Contributors:* SVG conversion from raster image Image:Phase-diag.png; some additions from Image:Phase diagram.png *Original artist:* me
- File:Phase_change_-_en.svg *Source:* https://upload.wikimedia.org/wikipedia/commons/0/0b/Phase_change_-_en.svg *License:* Public domain *Contributors:* Own work *Original artist:* F l a n k e r, penubag
- File:Phospholipids_aqueous_solution_structures.svg *Source:* https://upload.wikimedia.org/wikipedia/commons/c/c6/Phospholipids_aqueous_solution_structures.svg *License:* Public domain *Contributors:* Own work *Original artist:* Mariana Ruiz Villarreal ,LadyofHats
- File:Phylogenic_Tree-en.svg *Source:* https://upload.wikimedia.org/wikipedia/commons/5/58/Phylogenic_Tree-en.svg *License:* CC BY-SA 3.0 *Contributors:* This file was derived from Phylogenic Tree.jpg: <a href='//commons.wikimedia.org/wiki/File:Phylogenic_Tree.jpg' class='image'><img alt='Phylogenic Tree.jpg' src='https://upload.wikimedia.org/wikipedia/commons/thumb/d/de/Phylogenic_Tree.jpg/50px-Phylogenic_Tree.jpg' width='50' height='33' srcset='https://upload.wikimedia.org/wikipedia/commons/thumb/d/de/Phylogenic_Tree.jpg/75px-Phylogenic_Tree.jpg 1.5x, https://upload.wikimedia.org/wikipedia/commons/thumb/d/de/Phylogenic_Tree.jpg/100px-Phylogenic_Tree.jpg 2x' data-file-width='1069' data-file-height='713' /></a>
Original artist: Phylogenic Tree.jpg: John D. Croft
- File:Pinocytosis.svg *Source:* https://upload.wikimedia.org/wikipedia/commons/b/b7/Pinocytosis.svg *License:* Public domain *Contributors:* modified Image: Endocytosis types.svg, author Mariana Ruiz Villarreal LadyofHats *Original artist:* Jacek FH
- File:Polycyclic_Aromatic_Hydrocarbons.png *Source:* https://upload.wikimedia.org/wikipedia/commons/c/c0/Polycyclic_Aromatic_Hydrocarbons.png *License:* Public domain *Contributors:* Own work by uploader, Accelrys DS Visualizer *Original artist:* Inductiveload
- File:Pore_schematic.svg *Source:* https://upload.wikimedia.org/wikipedia/commons/c/cc/Pore_schematic.svg *License:* Public domain *Contributors:* Own work *Original artist:* MDougM
- File:Portal-puzzle.svg *Source:* https://upload.wikimedia.org/wikipedia/en/f/fd/Portal-puzzle.svg *License:* Public domain *Contributors:* ? *Original artist:* ?
- File:Positive_Feedback-_Childbirth_(1).svg *Source:* https://upload.wikimedia.org/wikipedia/commons/0/0e/Positive_Feedback-_Childbirth_%281%29.svg *License:* CC BY-SA 4.0 *Contributors:* Own work *Original artist:* Hannah.gray05
- File:Positive_feedback_bistable_switch.svg *Source:* https://upload.wikimedia.org/wikipedia/en/f/fd/Positive_feedback_bistable_switch.svg *License:* PD *Contributors:* ? *Original artist:* ?
- File:Protein_folding.png *Source:* https://upload.wikimedia.org/wikipedia/commons/a/a9/Protein_folding.png *License:* Public domain *Contributors:* No machine-readable source provided. Own work assumed (based on copyright claims). *Original artist:* No machine-readable author provided. DrKjaergaard assumed (based on copyright claims).
- File:Protein_folding_schematic.png *Source:* https://upload.wikimedia.org/wikipedia/commons/c/c5/Protein_folding_schematic.png *License:* Public domain *Contributors:* Own work (Original text: self-made) *Original artist:* Tomixdf (talk)
- File:Protein_structure.png *Source:* https://upload.wikimedia.org/wikipedia/commons/0/05/Protein_structure.png *License:* CC BY-SA 3.0 *Contributors:* Own work *Original artist:* Holger87

49.5. TEXT AND IMAGE SOURCES, CONTRIBUTORS, AND LICENSES

- File:Psi2.svg *Source:* https://upload.wikimedia.org/wikipedia/commons/6/6c/Psi2.svg *License:* Public domain *Contributors:* No machine-readable source provided. Own work assumed (based on copyright claims). *Original artist:* No machine-readable author provided. Gdh~commonswiki assumed (based on copyright claims).
- File:Question_book-new.svg *Source:* https://upload.wikimedia.org/wikipedia/en/9/99/Question_book-new.svg *License:* Cc-by-sa-3.0 *Contributors:* Created from scratch in Adobe Illustrator. Based on Image:Question book.png created by User:Equazcion *Original artist:* Tkgd2007
- File:Question_dropshade.png *Source:* https://upload.wikimedia.org/wikipedia/commons/d/dd/Question_dropshade.png *License:* Public domain *Contributors:* Image created by JRM *Original artist:* JRM
- File:R-S_mk2.gif *Source:* https://upload.wikimedia.org/wikipedia/commons/c/c6/R-S_mk2.gif *License:* CC BY 2.0 *Contributors:* Modification of Wikimedia Commons file R-S.gif (shown below) *Original artist:* Napalm Llama
- File:Rec8_3kc2p.jpg *Source:* https://upload.wikimedia.org/wikipedia/commons/c/c4/Rec8_3kc2p.jpg *License:* Public domain *Contributors:* Own work *Original artist:* WingkLEE
- File:Regenerartive_Receiver-S7300056.JPG *Source:* https://upload.wikimedia.org/wikipedia/commons/8/81/Regenerartive_Receiver-S7300056.JPG *License:* Public domain *Contributors:* English WP w:en:File:S7300056.JPG *Original artist:* Ozguy89
- File:SAM_schematic.jpeg *Source:* https://upload.wikimedia.org/wikipedia/commons/2/28/SAM_schematic.jpeg *License:* Public domain *Contributors:* Own work *Original artist:* Abnak
- File:SBSstructure.svg *Source:* https://upload.wikimedia.org/wikipedia/commons/8/86/SBSstructure.svg *License:* Public domain *Contributors:* en:File:SBSstructure.jpg *Original artist:*
- original: User:Peterlewis
- File:Samfax.jpg *Source:* https://upload.wikimedia.org/wikipedia/commons/1/1c/Samfax.jpg *License:* Public domain *Contributors:* Transferred from en.wikipedia to Commons by Quadell using CommonsHelper. *Original artist:* Jonnyt at English Wikipedia (Later version(s) were uploaded by Bility, Csyria, Aezram at en.wikipedia.)
- File:Sand_dune_ripples.jpg *Source:* https://upload.wikimedia.org/wikipedia/commons/c/cd/Sand_dune_ripples.jpg *License:* CC BY-SA 2.0 *Contributors:* http://www.flickr.com/photos/shirazc/3387882509/ *Original artist:* Shiraz Chakera http://www.flickr.com/photos/shirazc/
- File:Sarfus.stearic_acid_one_monolayer.jpg *Source:* https://upload.wikimedia.org/wikipedia/commons/9/9b/Sarfus.stearic_acid_one_monolayer.jpg *License:* GFDL *Contributors:* Nicolas Medard *Original artist:* Nanolane
- File:Sbs_block_copolymer.jpg *Source:* https://upload.wikimedia.org/wikipedia/commons/9/9a/Sbs_block_copolymer.jpg *License:* Public domain *Contributors:* Own work by the original uploader *Original artist:* Peter R Lewis
- File:Science.jpg *Source:* https://upload.wikimedia.org/wikipedia/commons/5/54/Science.jpg *License:* Public domain *Contributors:* ? *Original artist:* ?
- File:Sedimented_red_blood_cells.jpg *Source:* https://upload.wikimedia.org/wikipedia/commons/c/c9/Sedimented_red_blood_cells.jpg *License:* Public domain *Contributors:* Own work *Original artist:* MDougM
- File:Self-assembly_of_iron_oxide_nanocrystals2.jpg *Source:* https://upload.wikimedia.org/wikipedia/commons/d/d5/Self-assembly_of_iron_oxide_nanocrystals2.jpg *License:* CC BY 3.0 *Contributors:* http://iopscience.iop.org/1468-6996/15/5/055010/article *Original artist:* Erik Wetterskog et al.
- File:Self-organizing-Mechanism-for-Development-of-Space-filling-Neuronal-Dendrites-pcbi.0030212.sv003.ogv *Source:* https://upload.wikimedia.org/wikipedia/commons/0/0e/Self-organizing-Mechanism-for-Development-of-Space-filling-Neuronal-Dendrites-pcbi.0030212.sv003.ogv *License:* CC BY 2.5 *Contributors:* Video S3 from Sugimura K, Shimono K, Uemura T, Mochizuki A (2007). "Self-organizing Mechanism for Development of Space-filling Neuronal Dendrites". PLOS Computational Biology. DOI:10.1371/journal.pcbi.0030212. PMID 18020700. PMC: 2077899. *Original artist:* Sugimura K, Shimono K, Uemura T, Mochizuki A
- File:Shimmering_bees_drive_hornet_away.ogg *Source:* https://upload.wikimedia.org/wikipedia/commons/d/de/Shimmering_bees_drive_hornet_away.ogg *License:* CC BY 2.5 *Contributors:* Movie S1 of Kastberger G, Schmelzer E, Kranner I (2008) Social Waves in Giant Honeybees Repel Hornets. PLoS ONE 3(9): e3141. doi:10.1371/journal.pone.0003141 *Original artist:* Kastberger G, Schmelzer E, Kranner I (2008) Social Waves in Giant Honeybees Repel Hornets. PLoS ONE 3(9): e3141. doi:10.1371/journal.pone.0003141
- File:Sigmoid_curve_for_an_autocatalytical_reaction.jpg *Source:* https://upload.wikimedia.org/wikipedia/commons/5/55/Sigmoid_curve_for_an_autocatalytical_reaction.jpg *License:* CC-BY-SA-3.0 *Contributors:* English Wikipedia *Original artist:* Knights who say ni
- File:Smectic_nematic.jpg *Source:* https://upload.wikimedia.org/wikipedia/commons/6/67/Smectic_nematic.jpg *License:* CC-BY-SA-3.0 *Contributors:* ? *Original artist:* ?
- File:Smitt_hysteresis_graph.svg *Source:* https://upload.wikimedia.org/wikipedia/commons/a/a9/Smitt_hysteresis_graph.svg *License:* CC-BY-SA-3.0 *Contributors:* Own work *Original artist:* FDominec
- File:Snow_crystallization_in_Akureyri_2005-02-26_19-03-37.jpeg *Source:* https://upload.wikimedia.org/wikipedia/commons/d/d3/Snow_crystallization_in_Akureyri_2005-02-26_19-03-37.jpeg *License:* Public domain *Contributors:* Own work *Original artist:* Ævar Arnfjörð Bjarmason
- File:SnowflakesWilsonBentley.jpg *Source:* https://upload.wikimedia.org/wikipedia/commons/c/c2/SnowflakesWilsonBentley.jpg *License:* Public domain *Contributors:* Plate XIX of "Studies among the Snow Crystals ... " by Wilson Bentley, "The Snowflake Man." From Annual Summary of the "Monthly Weather Review" for 1902. *Original artist:* Wilson Bentley
- File:Solubilita_Na2SO4.png *Source:* https://upload.wikimedia.org/wikipedia/commons/2/2e/Solubilita_Na2SO4.png *License:* Public domain *Contributors:* No machine-readable source provided. Own work assumed (based on copyright claims). *Original artist:* No machine-readable author provided. Ub assumed (based on copyright claims).
- File:Sound-icon.svg *Source:* https://upload.wikimedia.org/wikipedia/commons/4/47/Sound-icon.svg *License:* LGPL *Contributors:* Derivative work from Silsor's versio *Original artist:* Crystal SVG icon set

- File:Square1.jpg *Source:* https://upload.wikimedia.org/wikipedia/commons/5/55/Lichtenberg_figure_in_block_of_Plexiglas.jpg *License:* Attribution *Contributors:* http://www.capturedlightning.com *Original artist:* Bert Hickman
- File:Stampede_loop.png *Source:* https://upload.wikimedia.org/wikipedia/commons/b/b8/Stampede_loop.png *License:* CC BY-SA 3.0 *Contributors:* Own work *Original artist:* Trevithj
- File:Stereobl.png *Source:* https://upload.wikimedia.org/wikipedia/commons/5/5f/Stereobl.png *License:* CC BY-SA 3.0 *Contributors:* Own work *Original artist:* Materialscientist
- File:Stereoscopic-motion-analysis-in-densely-packed-clusters-3D-analysis-of-the-shimmering-behaviour-in-1742-9994-8-3-S1.ogv *Source:* https://upload.wikimedia.org/wikipedia/commons/b/b5/Stereoscopic-motion-analysis-in-densely-packed-clusters-3D-analysis-of-the-shimmering-behaviour-in-1742-9994-8-3-S1.ogv *License:* CC BY 2.0 *Contributors:* Kastberger G, Maurer M, Weihmann F, Ruether M, Hoetzl T, Kranner I, Bischof H (2011). " Stereoscopic motion analysisin densely packed clusters: 3D analysis of the shimmering behaviour in Giant honey bees". *Frontiers in Zoology*.DOI:10.1186/1742-9994-8-3.PMID21303539.PMC:3050815. *Original artist:* Kastberger G, Maurer M, Weihmann F, Ruether M, Hoetzl T, Kranner I, Bischof H
- File:Stromatolites.jpg *Source:* https://upload.wikimedia.org/wikipedia/commons/c/c0/Stromatolites.jpg *License:* Public domain *Contributors:* National Park Service - http://www.nature.nps.gov/geology/cfprojects/photodb/Photo_Detail.cfm?PhotoID=204 *Original artist:* P. Carrara, NPS
- File:Structure_mode_history.svg *Source:* https://upload.wikimedia.org/wikipedia/commons/e/ef/Structure_mode_history.svg *License:* CC-BY-SA-3.0 *Contributors:* This is my work, calculated using a code I wrote myself, plotted in Gnuplot and edited in Adobe Illustrator. *Original artist:* Joke137 at English Wikipedia
- File:Supported_Lipid_Bilayer_and_Nanoparticles_AFM.png *Source:* https://upload.wikimedia.org/wikipedia/commons/3/34/Supported_Lipid_Bilayer_and_Nanoparticles_AFM.png *License:* CC BY-SA 3.0 *Contributors:* Own work *Original artist:* Yurko
- File:Supramolecular_Assembly_Lehn.jpg *Source:* https://upload.wikimedia.org/wikipedia/commons/7/74/Supramolecular_Assembly_Lehn.jpg *License:* CC-BY-SA-3.0 *Contributors:* Transferred from en.wikipedia to Commons by satish.murthy. *Original artist:* M stone at English Wikipedia
- File:Supramolecular_assembly_of_micelles6.jpg *Source:* https://upload.wikimedia.org/wikipedia/commons/b/b5/Supramolecular_assembly_of_micelles6.jpg *License:* CC BY 4.0 *Contributors:* http://www.nature.com/ncomms/2015/150904/ncomms9127/full/ncomms9127.html *Original artist:* Xiaoyu Li et al
- File:Surfactant.jpg *Source:* https://upload.wikimedia.org/wikipedia/commons/0/03/Surfactant.jpg *License:* Public domain *Contributors:* Own work *Original artist:* Major measure
- File:Symbol_book_class2.svg *Source:* https://upload.wikimedia.org/wikipedia/commons/8/89/Symbol_book_class2.svg *License:* CC BY-SA 2.5 *Contributors:* Mad by Lokal_Profil by combining: *Original artist:* Lokal_Profil
- File:Symbol_list_class.svg *Source:* https://upload.wikimedia.org/wikipedia/en/d/db/Symbol_list_class.svg *License:* Public domain *Contributors:* ? *Original artist:* ?
- File:Technics_SL-1210MK2.jpg *Source:* https://upload.wikimedia.org/wikipedia/commons/0/01/Technics_SL-1210MK2.jpg *License:* CC-BY-SA-3.0 *Contributors:* Self-photographed *Original artist:* Cschirp
- File:Termite_Cathedral_DSC03570.jpg *Source:* https://upload.wikimedia.org/wikipedia/commons/7/73/Termite_Cathedral_DSC03570.jpg *License:* CC-BY-SA-3.0 *Contributors:* [1] *Original artist:* taken by w:User:Yewenyi
- File:Textile_cone.JPG *Source:* https://upload.wikimedia.org/wikipedia/commons/7/7d/Textile_cone.JPG *License:* CC-BY-SA-3.0 *Contributors:* Location: Cod Hole, Great Barrier Reef, Australia *Original artist:* Photographer: Richard Ling (richard@research.canon.com.au)
- File:The_Earth_seen_from_Apollo_17_with_transparent_background.png *Source:* https://upload.wikimedia.org/wikipedia/commons/4/43/The_Earth_seen_from_Apollo_17_with_transparent_background.png *License:* Public domain *Contributors:* http://nssdc.gsfc.nasa.gov/imgcat/html/object_page/a17_h_148_22727.html *Original artist:* NASA
- File:The_Systems7_Diagram.jpg *Source:* https://upload.wikimedia.org/wikipedia/commons/en/5/59/The_Systems7_Diagram.jpg *License:* CC-BY-3.0 *Contributors:* Created by myself (Laurie F. Thomas) during a research project and published in the book 'Learning Conversation' publishd by Routledge 1985 and the copyright has returned to myself

  Original artist:

  Soler99
- File:Tiger_Bush_Niger_Corona_1965-12-31.jpg *Source:* https://upload.wikimedia.org/wikipedia/commons/1/1b/Tiger_Bush_Niger_Corona_1965-12-31.jpg *License:* Public domain *Contributors:* Data available from the U.S. Geological Survey *Original artist:* US Agency
- File:Torus.png *Source:* https://upload.wikimedia.org/wikipedia/commons/1/17/Torus.png *License:* Public domain *Contributors:* This image was created with POV-Ray *Original artist:* LucasVB
- File:Travelling_wave_for_Fisher_equation.svg *Source:* https://upload.wikimedia.org/wikipedia/commons/7/78/Travelling_wave_for_Fisher_equation.svg *License:* Public domain *Contributors:* Own work, created using Matlab *Original artist:* Jitse Niesen
- File:Tree_of_life.svg *Source:* https://upload.wikimedia.org/wikipedia/commons/0/09/Tree_of_life.svg *License:* CC-BY-SA-3.0 *Contributors:* No machine-readable source provided. Own work assumed (based on copyright claims). *Original artist:* No machine-readable author provided. Vanished user fijtji34toksdcknqrjn54yoimascj assumed (based on copyright claims).
- File:Tree_of_life_by_Haeckel.jpg *Source:* https://upload.wikimedia.org/wikipedia/commons/d/de/Tree_of_life_by_Haeckel.jpg *License:* Public domain *Contributors:* First version from en.wikipedia; description page was here. Later versions derived from this scan, from the American Philosophical Society Museum. *Original artist:* Ernst Haeckel
- File:TwoLorenzOrbits.jpg *Source:* https://upload.wikimedia.org/wikipedia/commons/4/44/TwoLorenzOrbits.jpg *License:* CC BY 2.5 *Contributors:* ? *Original artist:* ?
- File:US_Army_52300_\char"0022\relax{}Tell_Me_A_Story\char"0022\relax{}_promotes_academic,_emotional_connections.jpg *Source:* https://upload.wikimedia.org/wikipedia/commons/9/96/US_Army_52300_%22Tell_Me_A_Story%22_promotes_academic%2C_emotional_connections.jpg *License:* Public domain *Contributors:* United States Army *Original artist:* Sgt. Maj. Terry Anderson, 8th Theater Sustainment Command Public Affairs

- File:Visualization_of_wiki_structure_using_prefuse_visualization_package.png *Source:* https://upload.wikimedia.org/wikipedia/commons/9/90/ Visualization_of_wiki_structure_using_prefuse_visualization_package.png *License:* CC BY-SA 3.0 *Contributors:* Own work - I (Mr3641 (talk)) created this work entirely by myself.
  Original artist: Chris Davis at en.wikipedia
- File:Water_Crystals_on_Mercury_20Feb2010_CU1.jpg *Source:* https://upload.wikimedia.org/wikipedia/commons/7/77/Water_Crystals_on_Mercury_ 20Feb2010_CU1.jpg *License:* CC BY-SA 3.0 *Contributors:* I photographed a car window with my Kodak digital camera
  Previously published: Published on Wikipedia, deleted by a vandal, unfortunately *Original artist:* Rusfuture
- File:Wiki_letter_w.svg *Source:* https://upload.wikimedia.org/wikipedia/en/6/6c/Wiki_letter_w.svg *License:* Cc-by-sa-3.0 *Contributors:* ? *Original artist:* ?
- File:Wiki_letter_w_cropped.svg *Source:* https://upload.wikimedia.org/wikipedia/commons/1/1c/Wiki_letter_w_cropped.svg *License:* CC-BY-SA-3.0 *Contributors:* This file was derived from Wiki letter w.svg: <a href='//commons.wikimedia.org/wiki/File:Wiki_letter_w.svg' class='image'><img alt='Wiki letter w.svg' src='https://upload.wikimedia.org/wikipedia/commons/thumb/6/6c/Wiki_letter_w.svg/50px-Wiki_letter_w.svg.png' width='50' height='50' srcset='https://upload. wikimedia.org/wikipedia/commons/thumb/6/6c/Wiki_letter_w.svg/75px-Wiki_letter_w.svg.png 1.5x, https://upload.wikimedia.org/wikipedia/commons/thumb/ 6/6c/Wiki_letter_w.svg/100px-Wiki_letter_w.svg.png 2x' data-file-width='44' data-file-height='44' /></a>
  Original artist: Derivative work by Thumperward
- File:Wiki_tarantula.jpg *Source:* https://upload.wikimedia.org/wikipedia/commons/4/40/Wiki_tarantula.jpg *License:* CC-BY-SA-3.0 *Contributors:* www.nutscode.com *Original artist:* Arno / Coen
- File:Wikibooks-logo-en-noslogan.svg *Source:* https://upload.wikimedia.org/wikipedia/commons/d/df/Wikibooks-logo-en-noslogan.svg *License:* CC BY-SA 3.0 *Contributors:* Own work *Original artist:* User:Bastique, User:Ramac et al.
- File:Wikipedia_Liquid_Crystal_Display_Arduino.jpg *Source:* https://upload.wikimedia.org/wikipedia/commons/1/12/Wikipedia_Liquid_Crystal_ Display_Arduino.jpg *License:* CC BY-SA 4.0 *Contributors:* Own work *Original artist:* <a href='//commons.wikimedia.org/w/index.php?title=User: Ctg4Rahat,&,,action=edit,&,,redlink=1' class='new' title='User:Ctg4Rahat (page does not exist)'>Rahat</a> (Talk * Contributions) 07:56, 26 November 2014 (UTC)
- File:Wikiquote-logo.svg *Source:* https://upload.wikimedia.org/wikipedia/commons/f/fa/Wikiquote-logo.svg *License:* Public domain *Contributors:* Own work *Original artist:* Rei-artur
- File:Wiktionary-logo-en.svg *Source:* https://upload.wikimedia.org/wikipedia/commons/f/f8/Wiktionary-logo-en.svg *License:* Public domain *Contributors:* Vector version of Image:Wiktionary-logo-en.png. *Original artist:* Vectorized by Fvasconcellos (talk · contribs), based on original logo tossed together by Brion Vibber
- File:WilhelmyPlate.jpg *Source:* https://upload.wikimedia.org/wikipedia/commons/2/21/WilhelmyPlate.jpg *License:* Public domain *Contributors:* Own work *Original artist:* Major measure
- File:WilliamHWhyteJrGroupthinkFortuneMarch1952Page114.png *Source:* https://upload.wikimedia.org/wikipedia/en/2/2a/ WilliamHWhyteJrGroupthinkFortuneMarch1952Page114.png *License:* Fair use *Contributors:*
  Scanned from: Whyte, William H., Jr. "Groupthink." Fortune magazine, March 1952, pages 114–117, 142, and 146.
  Original artist: ?
- File:Wpm02_05.JPG *Source:* https://upload.wikimedia.org/wikipedia/commons/e/ec/Wpm02_05.JPG *License:* CC BY-SA 3.0 *Contributors:* Own work *Original artist:* Ziko-C

49.5.3 Content license

- Creative Commons Attribution-Share Alike 3.0

www.ingramcontent.com/pod-product-compliance
Lightning Source LLC
Chambersburg PA
CBHW080649190526
45169CB00006B/2045